T0212032

Lecture Notes in Computer Science   **9616**

Commenced Publication in 1973
Founding and Former Series Editors:
Gerhard Goos, Juris Hartmanis, and Jan van Leeuwen

More information about this series at http://www.springer.com/series/7409

Marc Gyssens · Guillermo Simari (Eds.)

Foundations of Information and Knowledge Systems

9th International Symposium, FoIKS 2016
Linz, Austria, March 7–11, 2016
Proceedings

 Springer

Editors
Marc Gyssens
Faculteit Wetenschappen
Universiteit Hasselt
Hasselt
Belgium

Guillermo Simari
Departamento de Ciencias e Ingeniería
de la Computación
Universidad Nacional del Sur
Bahía Blanca
Argentina

ISSN 0302-9743   ISSN 1611-3349   (electronic)
Lecture Notes in Computer Science
ISBN 978-3-319-30023-8   ISBN 978-3-319-30024-5   (eBook)
DOI 10.1007/978-3-319-30024-5

Library of Congress Control Number: 2016931282

LNCS Sublibrary: SL3 – Information Systems and Applications, incl. Internet/Web, and HCI

This Springer imprint is published by SpringerNature
The registered company is Springer International Publishing AG Switzerland

Preface

This volume contains the articles that were presented at the 9th International Symposium on Foundations of Information and Knowledge Systems (FoIKS 2016) held in Linz, Austria, during March 7–11, 2016.

The FoIKS symposia provide a biennial forum for presenting and discussing theoretical and applied research on information and knowledge systems. The goal is to bring together researchers with an interest in this subject, share research experiences, promote collaboration, and identify new issues and directions for future research. Speakers are given sufficient time to present their ideas and results within the larger context of their research. Furthermore, participants are asked in advance to prepare a first response to a contribution of another author in order to initiate discussion.

Previous FoIKS symposia were held in Bordeaux (France) in 2014, Kiel (Germany) in 2012, Sofia (Bulgaria) in 2010, Pisa (Italy) in 2008, Budapest (Hungary) in 2006, Vienna (Austria) in 2004, Schloss Salzau near Kiel (Germany) in 2002, and Burg/Spreewald near Berlin (Germany) in 2000. FoIKS took up the tradition of the conference series Mathematical Fundamentals of Database Systems (MFDBS), which initiated East–West collaboration in the field of database theory. Former MFDBS conferences were held in Rostock (Germany) in 1991, Visegrád (Hungary) in 1989, and Dresden (Germany) in 1987.

FoIKS 2016 solicited original contributions on foundational aspects of information and knowledge systems. This included submissions that apply ideas, theories, or methods from specific disciplines to information and knowledge systems. Examples of such disciplines are discrete mathematics, logic and algebra, model theory, information theory, complexity theory, algorithmics and computation, statistics, and optimization. Suggested topics included, but were not limited to the following:

- Big data: models for data in the cloud, programming languages for big data, query processing
- Database design: formal models, dependencies and independencies
- Dynamics of information: models of transactions, concurrency control, updates, consistency preservation, belief revision
- Information fusion: heterogeneity, views, schema dominance, multiple source information merging, reasoning under inconsistency
- Integrity and constraint management: verification, validation, consistent query answering, information cleaning
- Intelligent agents: multi-agent systems, autonomous agents, foundations of software agents, cooperative agents, formal models of interactions, logical models of emotions
- Knowledge discovery and information retrieval: machine learning, data mining, formal concept analysis and association rules, text mining, information extraction

- Knowledge representation, reasoning and planning: non-monotonic formalisms, probabilistic and non-probabilistic models of uncertainty, graphical models and independence, similarity-based reasoning, preference modeling and handling, argumentation systems
- Logics in databases and AI: classic and non-classic logics, logic programming, description logic, spatial and temporal logics, probability logic, fuzzy logic
- Mathematical foundations: discrete structures and algorithms, automata, abstract machines, graphs, grammars, finite model theory, information theory, coding theory, complexity theory, randomness
- Security in information and knowledge systems: identity theft, privacy, trust, intrusion detection, access control, inference control, secure Web services, secure Semantic Web, risk management
- Semi-structured data and XML: data modelling, data processing, data compression, data exchange
- Social computing: collective intelligence and self-organizing knowledge, collaborative filtering, computational social choice, Boolean games, coalition formation, reputation systems
- The Semantic Web and knowledge management: languages, agents, adaptation, intelligent algorithms, ontologies
- The WWW: models of Web databases, Web dynamics, Web services, Web transactions and negotiations

The call for papers resulted in the submission of 23 articles. Each one was carefully reviewed by at least three international experts. The 12 articles judged best by the Program Committee were accepted for long presentation. In addition, two articles were accepted for short presentation. This volume contains versions of these articles that have been revised by their authors according to the comments provided in the reviews. After the conference, authors of a few selected articles were asked to prepare extended versions of their articles for publication in a special issue of the journal *Annals of Mathematics and Artificial Intelligence*.

We wish to thank all authors who submitted papers and all conference participants for fruitful discussions. We are grateful to our keynote speakers Christoph Beierle, Joachim Biskup, Reinhard Pichler, Henry Prakken, and José Maria Turull-Torres; this volume also contains articles for four of the five invited talks. We would like to thank the Program Committee members and additional reviewers for their timely expertise in carefully reviewing the submissions. We want to thank Maria Vanina Martinez for her work as publicity chair. The support of the conference provided by the European Association for Theoretical Computer Science (EATCS) and by the Software Competence Center Hagenberg is greatfully acknowledged. Last but not least, special thanks go to the local organization chair, Flavio Ferrarotti, and his dedicated team consisting of Andreea Buga, Tania Nemeş, Loredana Tec, and Mircea Boris Vleju for their support and for being our hosts during the wonderful days in Linz.

March 2016   Marc Gyssens
Guillermo Simari

Conference Organization

FoIKS 2016 was organized by the Software Competence Center Hagenberg.

Program Committee Chairs

Marc Gyssens   Hasselt University, Belgium
Guillermo Simari   Universidad Nacional del Sur, Argentina

Program Committee

José Júlio Alferes	Universidade Nova de Lisboa, Portugal
Paolo Atzeni	Università Roma Tre, Italy
Peter Baumgartner	NICTA, Australia
Christoph Beierle	University of Hagen, Germany
Salem Benferhat	Université d'Artois, France
Leopoldo Bertossi	Carleton University, Canada
Philippe Besnard	CNRS/IRIT, France
Meghyn Bienvenu	CNRS and Université Paris-Sud, France
Joachim Biskup	University of Dortmund, Germany
François Bry	Ludwig Maximilian University, Munich, Germany
James Cheney	University of Edinburgh, UK
Marina De Vos	University of Bath, UK
Michael Dekhtyar	Tver State University, Russia
Tommaso Di Noia	Politecnico di Bari, Italy
Juergen Dix	Clausthal University of Technology, Germany
Thomas Eiter	Vienna University of Technology, Austria
Ronald Fagin	IBM Almaden Research, San Jose, CA, USA
Flavio Ferrarotti	Software Competence Center Hagenberg, Austria
Sergio Flesca	University of Calabria, Italy
Dov Gabbay	King's College, University of London, UK
Lluis Godo	Artificial Intelligence Research Institute (IIIA - CSIC), Spain
Bart Goethals	University of Antwerp, Belgium
Gianluigi Greco	University of Calabria, Italy
Claudio Gutierrez	Universidad de Chile, Chile
Stephen J. Hegner	Umeå University, Sweden
Andreas Herzig	CNRS/IRIT, France
Anthony Hunter	University College, University of London, UK
Gabriele Kern-Isberner	University of Dortmund, Germany
Sébastien Konieczny	CNRS/CRIL, France
Adila A. Krisnadhi	Wright State University, OH, USA
Jérôme Lang	LAMSADE, France

Nicola Leone	University of Calabria, Italy
Mark Levene	Birkbeck, University of London, UK
Sebastian Link	University of Auckland, New Zealand
Weiru Liu	Queen's University Belfast, UK
Thomas Lukasiewicz	University of Oxford, UK
Carsten Lutz	University of Bremen, Germany
Sofian Maabout	LaBRI, University of Bordeaux, France
Sebastian Maneth	NICTA and University of New South Wales, Australia
Pierre Marquis	Université d'Artois, France
Wolfgang May	University of Göttingen, Germany
Carlo Meghini	ISTI-CNR Pisa, Italy
Thomas Natschläger	Software Competence Center Hagenberg, Austria
Bernhard Nebel	University of Freiburg, Germany
Frank Neven	Hasselt University, Belgium
Wilfred S.H. Ng	Hong Kong University of Science and Technology, Hong Kong, SAR China
David Poole	University of British Columbia, Canada
Henri Prade	CNRS/IRIT, France
Andrea Pugliese	University of Calabria, Italy
Gavin Rens	UKZN/CSIR Meraka Center for Artificial Intelligence Research, South Africa
Miguel Romero	Universidad de Chile, Chile
Riccardo Rosati	Sapienza Università di Roma, Italy
Sebastian Rudolph	University of Dresden
Attila Sali	Alfréd Rényi Institute, Hungarian Academy of Sciences, Hungary
Luigi Sauro	University of Naples, Italy
Francesco Scarcello	University of Calabria, Italy
Klaus-Dieter Schewe	Software Competence Center Hagenberg, Austria
Nicolas Spyratos	Université Paris-Sud, France
Umberto Straccia	ISTI-CNR Pisa, Italy
Tony Tan	National Taiwan University, Taiwan
Letizia Tanca	Politecnico di Milano, Italy
Bernhard Thalheim	University of Kiel, Germany
Alex Thomo	University of Victoria, BC, Canada
Miroslaw Truszczynski	University of Kentucky, USA
José Maria Turull-Torres	Universidad Nacional de La Matanza, Argentina, and Massey University, New Zealand
Dirk Van Gucht	Indiana University, USA
Victor Vianu	University of California San Diego, USA
Peter Vojtas	Charles University Prague, Czech Republic
Qing Wang	Information Science Research Center, New Zealand
Jef Wijsen	University of Mons, Belgium
Mary-Anne Williams	University of Technology, Sydney, Australia
Stefan Woltran	Vienna University of Technology, Austria
Yuqing Wu	Pomona College, Claremont, CA, USA

Additional Reviewers

Mina Catalano
Thomas Grubinger
Tobias Kaminski
Dominique Laurent
Jean-Guy Mailly
Ryan McConville

Bernhard Moser
Viet Phan-Luong
Francesco Ricca
Loredana Tec
Stefan Wölfl

Local Organization Chair

Flavio Ferrarotti   Software Competence Center Hagenberg, Austria

Other Members of the Local Organization Team

Andreea Buga
Tania Nemeş

Loredana Tec
Mircea Boris Vleju

Publicity Chair

Maria Vanina Martinez   Universidad Nacional del Sur and CONICET, Argentina

Sponsors

Software Competence Center Hagenberg, Austria
European Association for Theoretical Computer Science (EATCS)

Keynote Speakers

Henry Prakken, *Utrecht University* and *University of Groningen, The Netherlands*

Short biography: Henry Prakken is Lecturer in AI at the Department of Information and Computing Sciences, Utrecht University, and Professor in Legal Informatics and Legal Argumentation at the Law Faculty, University of Groningen, from which he holds master's degrees in law (1985) and philosophy (1988). In 1993, he obtained his PhD (*cum laude*) at the Free University Amsterdam. His main research interests include computational models of argumentation and their application in multi-agent systems and legal reasoning. Prakken is past president of the International Association of AI & Law and current president of the JURIX Foundation for Legal Knowledge-Based Systems and of the steering committee of the COMMA Conferences on Computational Models of Argument. He is on the editorial board of journals such as *Artificial Intelligence*.

Keynote talk: *Some Recent Trends in Argumentation Research*

Summary: Argumentation is an important topic in symbolic AI research today, especially in the study of nonmonotonic reasoning and the study of inter-agent communication. Argumentation makes explicit the reasons for the conclusions that are drawn and how conflicts between these reasons are resolved. This provides a natural mechanism to handle inconsistent and uncertain information and to resolve conflicts of opinion between intelligent agents. In this talk, an overview will be given of some of the main current research issues, including the relation between abstract and structured models of argumentation and the relation between argumentation and probability theory.

Christoph Beierle, *University of Hagen, Germany*

Short biography: Christoph Beierle is professor of computer science and head of the knowledge-based systems group in the Faculty of Mathematics and Computer Science at the University of Hagen. In 1985, he received his PhD in computer science from the University of Kaiserslautern. He was senior researcher at the Scientific Center of IBM Germany, and is a recipient of an IBM Outstanding Innovation Award. He has been working on algebraic specifications and formal approaches for software development and on methods for knowledge-based systems and their applications. His current research interests include modeling and reasoning with uncertain knowledge.

Keynote talk: *Systems and Implementations for Solving Reasoning Problems in Conditional Logics*

Summary: Default rules like "If A, then usually B" or probabilistic rules like "If A, then B with probability x" are powerful constructs for knowledge representation. Such rules can be formalized as conditionals, denoted by $(B|A)$ of $(B|A)[x]$, and a conditional knowledge base consists of a set of conditionals. Different semantical models have been proposed for conditional knowledge bases, and the most important reasoning problems for conditional knowledge bases are to determine whether a knowledge base is consistent and to determine what a knowledge base entails. We present an overview on systems and implementations our group has been working on for solving reasoning problems in various semantics that have been developed for conditional knowledge bases. These semantics include quantitative, semi-quantitative, and qualitative conditional logics, based on both propositional logic and on first-order logic.

Reinhard Pichler, *Vienna University of Technology, Austria*

Short biography: Reinhard Pichler holds a master's degree in mathematics from the University of Innsbruck and a master's degree in mathematical computation from the University of London, QMW College. In 2000, he received his PhD in computer science from the Vienna University of Technology. From 1992 to 2005, he worked as software developer at the Program and Systems Engineering Department (PSE) of Siemens AG Austria. Since 2005, he has been Professor at the Faculty of Informatics of the Vienna University of Technology where he leads the Database and Artificial Intelligence Group. His main research interests in recent years have been in database theory—mainly on information integration and on foundational aspects of the Semantic Web query language SPARQL.

Keynote talk: *The Challenge of Optional Matching in SPARQL*

Summary: Conjunctive queries (or, equivalently, SELECT-FROM-WHERE queries in SQL) are arguably the most widely used querying mechanism in practice and the most intensively studied one in database theory. Answering a conjunctive query (CQ) comes down to matching all atoms of the CQ simultaneously into the database. As a consequence, a CQ fails to provide any answer if the pattern described by the query does not exactly match the data. CQs might thus be too restrictive as a querying mechanism for data on the Web, which is considered as inherently incomplete. The Semantic Web query language SPARQL therefore contains the OPTIONAL operator as a crucial feature. It allows the user to formulate queries that try to match parts of the query over the data if available, but do not destroy answers of the remaining query otherwise.

In this talk, we will have a closer look at this optional matching feature of SPARQL. More specifically, we will concentrate on an interesting fragment of SPARQL: the so-called well-designed SPARQL graph patterns. They extend CQs by optional matching while imposing certain restrictions on how variables are allowed to occur in the query. We recall recent results that even in this small fragment of SPARQL most of the fundamental computational tasks become significantly harder than for conjunctive queries. For instance, query evaluation is now on the second level of the Polynomial Hierarchy and basic static analysis tasks such as containment or equivalence testing become even undecidable for well-designed SPARQL graph patterns. Also the semantics of query answering in the presence of ontologies (referred to as entailment regimes in SPARQL) has to be reconsidered in order to give intuitive results. It turns out that the seemingly small extension of CQs by optional matching has created several interesting research opportunities.

Joachim Biskup, *University of Dortmund, Germany*

Short biography: Joachim Biskup received his master's degree in mathematics from the Technical University of Hannover and his PhD in computer science from the RWTH in Aachen, Germany. He has been Professor of Computer Science at the University of Dortmund, University of Hildesheim, and University of Dortmund again. He has performed research in areas such as recursion and complexity theory, information systems with an emphasis on schema design, query optimization and mediation, and various aspects of security, in particular access control and inference control.

Keynote talk: *Selected Results and Related Issues of Confidentiality-Preserving Controlled Interaction Execution*

Summary: Controlled interaction execution has been developed as a security server for a specific kind of inference control shielding an isolated, logic-oriented information system when interacting over time with a client by means of messages, in particular for query and transaction processing. The control aims at provably preserving confidentiality in a fully formalized sense, intuitively and simplifying rephrased as follows: even when having (assumed) a priori knowledge, recording the interaction history, being aware of the details of the control mechanism, and unrestrictedly rationally reasoning, the client should never be able to infer the validity of any sentence declared as a potential secret in the security server's confidentiality policy. To enforce this goal, for each of a rich variety of specific situations, a dedicated censor has been designed. As far as needed, a censor distorts a functionally expected reaction message such that suitably weakened or even believably incorrect information is communicated to the client.

We consider selected results of recent and ongoing work and discuss several issues for further research and development. The topics covered range from the impact of the underlying logic, whether propositional, first-order, or non-monotonic about belief or an abstraction from any specific one, over the kinds of interactions, whether only queries or also views, updates, revisions, or even procedural programs, to the dynamic representation of control states, whether by simply logging or adapting the policy.

José Maria Turull-Torres, *Universidad Nacional de La Matanza, Argentina* and *Massey University, New Zealand*

Short biography: After 20 years of professional work in informatics in Argentina and Mexico, a further 23 years followed of academic work with research in the areas of database theory, finite model theory, and complexity, in Argentina and New Zealand. Currently, José Maria Turull-Torres is Professor in the Department of Engineering at the Universidad Nacional de La Matanza, Argentina, and holds an Honorary Research Fellowship at Massey University, New Zealand. He has been a member of the Program Committee of many international conferences, co-chair of FoIKS 2004, and is often invited as keynote speaker. His main research collaboration is with the universities of Helsinki, Joensuu, and Tampere, in Finland, the University of Warsaw, the University of Toronto, the University of Cantabria in Spain, the Ecole Polytechnique de Paris, and the Software Competence Center Hagenberg in Austria.

Keynote talk: *Relational Complexity and Higher-Order Logics*

Summary: Relational machines (RM) were introduced in 1991 as abstract machines that compute queries to (finite) relational structures, or relational database instances (dbis), which are generic (i.e., that preserve isomorphisms), and hence are more appropriate than Turing machines (TM) for query computation. RMs are TMs endowed with a relational store that holds the input dbi, as well as work relations, that can be queried and updated through first-order logic (FO) formulas in their finite control. Consequently, k-ary RMs are incapable of computing the size of the input. However, they can compute its $size_k$, i.e, the number of FO^k types of k-tuples in the dbi. Consequently, a new notion of complexity suitable for RMs had to be defined. Relational complexity was also introduced in 1991 as a complexity theory where the input dbi to a query is measured as its $size_k$, and complexity classes mirroring computational complexity classes were defined. Relational complexity turned out to be a theoretical framework in which we can characterize exactly the expressive power of the well-known fixed-point quantifiers of a wide range of sorts. In 1997, several equivalences between fixed-point quantifiers (added to FO) and different relational complexity classes were proved, classifying them as either deterministic, non-deterministic, or alternating, and either inflationary or non-inflationary. These characterizations are actually very interesting and meaningful, given that it was already known that if we restrict the input to only ordered dbis, the same equivalences with computational complexity classes also hold.

Regarding the characterization of relational complexity classes with other logics, it was proved that RMs have the same computation, or expressive power, as the (effective fragment of the) well-known infinitary logic with finitely many variables. Besides, some fragments of second- and third-order logic, defined as semantic restrictions of the corresponding logic, have been proved to characterize several classes, and there is ongoing work in that direction. One interesting consequence of this is that RMs are strong enough to simulate the existence of third-order relations in their relational store. An important application of the creation of new logics to complexity theory is the search for lower bounds of problems with respect to those logics, aiming to separate computational complexity classes.

In this talk, we will give a description of RMs and NRMs, define the basic notions of relational complexity, and discuss its motivations and the tight relationship between the main classes and different fixed-point logics and fragments of second- and third-order logics.

Contents

Reasoning about Beliefs, Uncertainty, Incompleteness, and Inconsistency

A Study of Argument Acceptability Dynamics Through Core
and Remainder Sets.   3
   Martín O. Moguillansky

Anytime Algorithms for Solving Possibilistic MDPs and Hybrid MDPs.   24
   Kim Bauters, Weiru Liu, and Lluís Godo

Possibilistic Conditional Tables.   42
   Olivier Pivert and Henri Prade

Inference and Problem Solving

Skeptical Inference Based on C-Representations and Its Characterization
as a Constraint Satisfaction Problem   65
   Christoph Beierle, Christian Eichhorn, and Gabriele Kern-Isberner

Systems and Implementations for Solving Reasoning Problems
in Conditional Logics   83
   Christoph Beierle

Equivalence Between Answer-Set Programs Under (Partially) Fixed Input . . .   95
   Bernhard Bliem and Stefan Woltran

Querying and Pattern Mining

A k-Means-Like Algorithm for Clustering Categorical Data Using an
Information Theoretic-Based Dissimilarity Measure.   115
   Thu-Hien Thi Nguyen and Van-Nam Huynh

Discovering Overlapping Quantitative Associations by Density-Based
Mining of Relevant Attributes.   131
   Thomas Van Brussel, Emmanuel Müller, and Bart Goethals

Semantic Matching Strategies for Job Recruitment: A Comparison of New
and Known Approaches.   149
   Gábor Rácz, Attila Sali, and Klaus-Dieter Schewe

The Challenge of Optional Matching in SPARQL.   169
   Shqiponja Ahmetaj, Wolfgang Fischl, Markus Kröll, Reinhard Pichler,
   Mantas Šimkus, and Sebastian Skritek

Maintenance of Queries Under Database Changes:
A Unified Logic Based Approach   191
   Elena V. Ravve

Dealing with Knowledge

Selected Results and Related Issues of Confidentiality-Preserving
Controlled Interaction Execution   211
   Joachim Biskup

Integrity Constraints for General-Purpose Knowledge Bases   235
   Luís Cruz-Filipe, Isabel Nunes, and Peter Schneider-Kamp

A Knowledge Based Framework for Link Prediction in Social Networks   255
   Pooya Moradian Zadeh and Ziad Kobti

Logics and Complexity

Approximation and Dependence via Multiteam Semantics   271
   Arnaud Durand, Miika Hannula, Juha Kontinen, Arne Meier,
   and Jonni Virtema

The Complexity of Non-Iterated Probabilistic Justification Logic   292
   Ioannis Kokkinis

Relational Complexity and Higher Order Logics   311
   José Maria Turull-Torres

A Logic for Non-deterministic Parallel Abstract State Machines   334
   Flavio Ferrarotti, Klaus-Dieter Schewe, Loredana Tec, and Qing Wang

Author Index   355

Reasoning about Beliefs, Uncertainty, Incompleteness, and Inconsistency

A Study of Argument Acceptability Dynamics Through Core and Remainder Sets

Martín O. Moguillansky[1,2(✉)]

[1] CONICET, Institute for Research in Computer Science and Engineering (ICIC),
Universidad Nacional Del Sur (UNS), Bahía Blanca, Argentina
mom@cs.uns.edu.ar
[2] AI R&D Lab (LIDIA), Department of Computer Science and Engineering (DCIC),
Universidad Nacional Del Sur (UNS), Bahía Blanca, Argentina

Abstract. We analyze the acceptability dynamics of arguments through the proposal of two different kinds of minimal sets of arguments, core and remainder sets which are somehow responsible for the acceptability/rejection of a given argument. We develop a study of the consequences of breaking the construction of such sets towards the acceptance, and/or rejection, of an analyzed argument. This brings about the proposal of novel change operations for abstract argumentation first, and for logic-based argumentation, afterwards. The analysis upon logic-based argumentation shows some problems regarding the applicability of the standard semantics. In consequence, a reformulation of the notion of admissibility arises for accommodating the standard semantics upon logic-based argumentation. Finally, the proposed model is formalized in the light of the theory of belief revision by characterizing the corresponding operations through rationality postulates and representation theorems.

Keywords: Argumentation · Belief revision · Argumentation dynamics

1 Introduction

Argumentation theory [13] allows to reason over conflicting pieces of knowledge, *i.e.*, arguments. This is done by replacing the usual meaning of inference from classical logic by *acceptability* in argumentation: evaluation of arguments' interaction through conflict for deciding which arguments prevail. To that end, argumentation theory relies upon argumentation semantics and acceptance criteria. Semantics can be implemented through determination of extensions, *i.e.*, different kinds of conflict-free sets of arguments. For studying theoretic properties, like semantics, it is possible to abstract away from any particular representation of knowledge or structuring for building arguments. This is referred as abstract argumentation. On the other hand, the concretization of an argumentation framework (**AF**) to some specific logic and argument structure is called

---

Supported by UNS (PGI 24/ZN18) and CONICET (PIP 112-200801-02798).

M. Gyssens and G. Simari (Eds.): FoIKS 2016, LNCS 9616, pp. 3–23, 2016.
DOI: 10.1007/978-3-319-30024-5_1

logic-based argumentation [2,6,15]. Investigations based upon abstract argumentation usually simplify the study of some specific problem, and may bring solid fundamentals for studying afterwards its application upon logic-based argumentation. However, adapting theories, and results, from abstract to logic-based argumentation may be not straightforward.

The classic theory of belief revision [1] studies the dynamics of knowledge, coping with the problem of how to change beliefs standing for the conceptualization of a modeled world, to reflect its evolution. Revisions, as the most important change operations, concentrate on the incorporation of new beliefs in a way that the resulting base ends up consistently. When considering AFs for modeling situations which are immersed in a naturally dynamic context, it is necessary to provide models for handling acceptability dynamics of arguments. That is, models for studying change in argumentation (for instance, [4,5,10]) for providing a rationalized handling of the dynamics of a set of arguments and their implications upon the acceptability condition of arguments [19]. This led us to investigate new approaches of belief revision, which operate over paraconsistent semantics –like argumentation semantics– avoiding consistency restoration.

Argumentation provides a theoretic framework for modeling paraconsistent reasoning, a subject of utmost relevance in areas of research like medicine and law. For instance, legal reasoning can be seen as the intellectual process by which judges draw conclusions ensuring the rationality of legal doctrines, legal codes, binding prior decisions like jurisprudence, and the particularities of a deciding case. This definition can be broaden to include the act of making laws. Observe that the evolution of a normative system –for modeling promulgation of laws– would imply the removal/incorporation of norms for ensuring some specific purpose but keeping most conflicts from the original AF unaffected.

Upon such motivation, we study new forms to handle acceptability dynamics of arguments, firstly on abstract argumentation, and afterwards upon logic-based argumentation. By relying upon extension semantics, we define two different sorts of sets for recognizing acceptance or rejection of arguments: core and remainder sets, respectively. Afterwards we propose a model of change towards the proposal of an acceptance revision operation which deals with the matter of incorporating a new argument while ensuring its acceptance. This is done first, from an abstract perspective, and afterwards upon logic-based argumentation. This unveils some specific problems regarding the applicability of standard semantics to this kind of argumentation. We propose then a reformulation on the notion of admissibility to overcome from such drawback by analyzing argumentation postulates from [2]. Finally, the rationality of the proposed change operators is provided through its axiomatic characterization and corresponding representation theorem according to classic belief revision and argument-based belief revision models like Argument Theory Change (ATC) [19].

2   Fundamentals for Abstract Frameworks

An *abstract argumentation framework* (AF) will be assumed as a pair $\langle \mathbf{A}, \mathbf{R_A} \rangle$, where $\mathbf{A}$ is a finite *set of arguments*, and the set $\mathbf{R_A} \subseteq \mathbf{A} \times \mathbf{A}$ identifies the finitary *defeat relation* between pairs of arguments $a \in \mathbf{A}$ and $b \in \mathbf{A}$, such that $(a,b) \in \mathbf{R_A}$ implies that argument a *defeats* argument b, or equivalently, a is a *defeater* of b. In this part of the article, arguments are deemed as abstract since we do not specify any concrete logic, nor inner-structure, for constructing them. Thus, arguments will be considered as indivisible elements of AFs. On the other hand, we will assume the defeat relation $\mathbf{R_A}$ to be obtained through a functional construction $\mathbf{R_A} : \wp(\mathbf{A}) \longrightarrow \wp(\mathbf{A} \times \mathbf{A})^1$. This makes presumable the existence of a *defeating function* $\varepsilon : \mathbf{A} \times \mathbf{A} \longrightarrow \{\texttt{true}, \texttt{false}\}$, such that:

$$\text{for any pair of arguments } a, b \in \mathbf{A}, \ \varepsilon(a,b) = \texttt{true} \ \textit{iff} \ (a,b) \in \mathbf{R_A} \qquad (1)$$

Definition 1 (Argumentation Framework Generator). *Let $\mathbf{A}$ be a finite set of arguments, an operator $\mathbb{F_A}$ is an **argumentation framework generator** from $\mathbf{A}$ (or just, AF generator) iff $\mathbb{F_A}$ is an AF $\langle \mathbf{A}, \mathbf{R_A} \rangle$.*

Our intention is to simplify AFs at the greatest possible level in order to concentrate firstly on specific matters for dealing with the acceptability dynamics of arguments, and afterwards, from Sect. 5, we will analyze the proposed theory for argumentation dynamics in the light of logic-based frameworks, where arguments will be constructed upon a specific logic $\mathcal{L}$. Consequently, when necessary, we will abstract away the construction of an AF $\mathbb{F_A}$ from any set of arguments $\mathbf{A}$, by simply referring to an AF τ. In such a case, we will refer to the set of arguments of τ by writing $\mathbf{A}(\tau)$ and to the set of defeats of τ by writing $\mathbf{R}(\tau)$.

Next, we introduce some well known concepts from argumentation theory [13] that makes possible the acceptability analysis of arguments through the usage of argumentation semantics. Given an AF $\mathbb{F_A}$, for any $\Theta \subseteq \mathbf{A}$ we say that:

- Θ **defeats** an argument $a \in \mathbf{A}$ *iff* there is some $b \in \Theta$ such that b defeats a.
- Θ **defends** an argument $a \in \mathbf{A}$ *iff* Θ defeats every defeater of a.
- Θ is **conflict-free** *iff* $\mathbf{R}_\Theta = \emptyset$.
- Θ is **admissible** *iff* it is conflict-free and defends all its members.

Given an AF $\mathbb{F_A}$, for any set $\mathbf{E} \subseteq \mathbf{A}$ of arguments, we say that:

1. $\mathbf{E}$ is a **stable extension** if $\mathbf{E}$ is conflict-free and defeats any $a \in \mathbf{A} \setminus \mathbf{E}$
2. $\mathbf{E}$ is a **complete extension** if $\mathbf{E}$ is admissible and contains every argument it defends
3. $\mathbf{E}$ is a **preferred extension** if $\mathbf{E}$ is a maximal (wrt. set incl.) admissible set
4. $\mathbf{E}$ is the **grounded extension** if $\mathbf{E}$ is the minimal (wrt. set incl.) complete extension, *i.e.*, $\mathbf{E}$ is the least fixed point of $\mathcal{F}(X) = \{a \in \mathbf{A} | X \text{ defends } a\}$
5. $\mathbf{E}$ is a **semi-stable extension** if $\mathbf{E}$ is a complete extension and the set $\mathbf{E} \cup \{a \in \mathbf{A} | \mathbf{E} \text{ defeats } a\}$ is maximal wrt. set inclusion
6. $\mathbf{E}$ is the **ideal extension** if $\mathbf{E}$ is the maximal (wrt. set incl.) admissible set that is contained in every preferred extension

---

^{1} Observe that we use the notation $\wp(\Theta)$ for referring to the powerset of Θ.

The above six notions are known as *extension semantics*. It is possible to have no stable extensions, and also that there may be more than a single stable, complete, preferred and semi-stable extensions, but only one grounded and one ideal extension. The set $\mathbb{E}_\mathfrak{s}(\tau)$ identifies the *set of $\mathfrak{s}$-extensions* $\mathbf{E}$ from the AF $\tau = \mathbb{F}_\mathbf{A}$, where an *$\mathfrak{s}$-extension* is an extension in τ according to some extension semantics $\mathfrak{s}$, and where $\mathfrak{s}$ adopts a value from $\{\mathtt{st}, \mathtt{co}, \mathtt{pr}, \mathtt{gr}, \mathtt{ss}, \mathtt{id}\}$ corresponding to the stable ($\mathtt{st}$), complete ($\mathtt{co}$), preferred ($\mathtt{pr}$), grounded ($\mathtt{gr}$), semi-stable ($\mathtt{ss}$), and ideal ($\mathtt{id}$) semantics. For instance, the set $\mathbb{E}_{\mathtt{pr}}(\tau)$ will contain all the preferred extensions in τ. Observe that any extension $\mathbf{E} \in \mathbb{E}_\mathfrak{s}(\tau)$ is an admissible set. The relation among extension semantics is shown as $\mathbb{E}_{\mathtt{st}}(\tau) \subseteq \mathbb{E}_{\mathtt{ss}}(\tau) \subseteq \mathbb{E}_{\mathtt{pr}}(\tau) \subseteq \mathbb{E}_{\mathtt{co}}(\tau)$, and also $\mathbb{E}_{\mathtt{gr}}(\tau) \subseteq \mathbb{E}_{\mathtt{co}}(\tau)$ and $\mathbb{E}_{\mathtt{id}}(\tau) \subseteq \mathbb{E}_{\mathtt{co}}(\tau)$.

3   Preliminaries for Studying Dynamics of Arguments

We refer as *acceptance criterion* to the determination of acceptance of arguments in either a *sceptical* or *credulous* way. Several postures may appear. For instance, according to [15] a sceptical set is obtained by intersecting every $\mathfrak{s}$-extension (see Eq. 2), and a credulous set resulting from the union of every $\mathfrak{s}$-extension (Eq. 4). Since the latter posture may trigger non-conflict free sets, we suggest a different alternative for credulous acceptance, for instance, one may choose a single extension due to some specific preference, like selecting among those extensions of maximal cardinality, "the best representative" one according to some criterion upon ordering of arguments (Eq. 3). Assuming an abstract AF τ:

$$\bigcap_{\mathbf{E} \in \mathbb{E}_\mathfrak{s}(\tau)} \mathbf{E} \tag{2}$$

$$\mathbf{E} \in \mathbb{E}_\mathfrak{s}(\tau) \text{ such that for any} \tag{3}$$
$$\mathbf{E}' \in \mathbb{E}_\mathfrak{s}(\tau), |\mathbf{E}| \geq |\mathbf{E}'| \text{ holds}$$

$$\bigcup_{\mathbf{E} \in \mathbb{E}_\mathfrak{s}(\tau)} \mathbf{E} \tag{4}$$

Definition 2 (Acceptance Function). *Given an AF $\tau = \mathbb{F}_\mathbf{A}$ and an extension semantics $\mathfrak{s} \in \{\mathtt{st}, \mathtt{co}, \mathtt{pr}, \mathtt{gr}, \mathtt{ss}, \mathtt{id}\}$ determining a set $\mathbb{E}_\mathfrak{s}(\tau) \subseteq \wp(\mathbf{A})$ of $\mathfrak{s}$-extensions, a function $\delta : \wp(\wp(\mathbf{A})) \longrightarrow \wp(\mathbf{A})$ is an **acceptance function** iff $\delta(\mathbb{E}_\mathfrak{s}(\tau)) \subseteq \mathbf{A}$ determines a conflict-free set of arguments from $\mathbf{A}$.*

The acceptance criterion can be applied through an *acceptance function* as defined above. Note that Eq. 4 does not fulfill the necessary conditions for an acceptance function given that it may trigger non-conflict-free sets. We will abstract away from a specific definition for an acceptance function and will only refer to δ when necessary. We refer as *(argumentation) semantics specification* $\mathcal{S}$ to a tuple $\langle \mathfrak{s}, \delta \rangle$, where $\mathfrak{s}$ stands for identifying some extension semantics and δ for an acceptance function implementing some acceptance criterion.

Definition 3 (Acceptable Set). *Given an* AF $\tau = \mathbb{F}_{\mathbf{A}}$ *and a semantics specification* $\mathcal{S} = \langle \mathfrak{s}, \delta \rangle$, *the set* $\mathcal{A}_{\mathcal{S}}(\tau) \subseteq \mathbf{A}$ *is the* **acceptable set** *of* τ *according to* $\mathcal{S}$ *iff* $\mathcal{A}_{\mathcal{S}}(\tau) = \delta(\mathbb{E}_{\mathfrak{s}}(\tau))$.

For instance, adopting an acceptance function implementing Eq. 2, the set $\mathcal{A}_{\langle \mathrm{pr}, \delta \rangle}(\tau)$ identifies the sceptical acceptance set for a preferred semantics.

Definition 4 (Argument Acceptance/Rejection). *Given an* AF $\tau = \mathbb{F}_{\mathbf{A}}$ *and a semantics specification* $\mathcal{S} = \langle \mathfrak{s}, \delta \rangle$, *an argument* $a \in \mathbf{A}$ *is* $\mathcal{S}$-**accepted** *in* τ *iff* $a \in \mathcal{A}_{\mathcal{S}}(\tau)$. *Conversely,* $a \in \mathbf{A}$ *is* $\mathcal{S}$-**rejected** *in* τ *iff* $a \notin \mathcal{A}_{\mathcal{S}}(\tau)$.

Admissible and *core sets* of an argument as the fundamental notions for recognizing the sources for the acceptability condition of a given argument.

Definition 5 (Admissible Sets of an Argument). *Given an* AF $\tau = \mathbb{F}_{\mathbf{A}}$ *and an argument* $a \in \mathbf{A}$; *for any* $\Theta \subseteq \mathbf{A}$, *we say that:*

1. Θ *is an* a-**admissible set** *in* τ *iff* Θ *is an admissible set such that* $a \in \Theta$.
2. Θ *is a* **minimal** a-**admissible set** *in* τ *iff* Θ *is* a-admissible *and for any* $\Theta' \subset \Theta$, *it follows that* Θ' *is not* a-admissible.

Definition 6 (Core Sets). *Given an* AF $\tau = \mathbb{F}_{\mathbf{A}}$ *and an argumentation semantics specification* $\mathcal{S}$, *for any* $\mathcal{C} \subseteq \mathbf{A}$, *we say that* $\mathcal{C}$ *is an* a-**core** *in* τ, *noted as* a-core$_{\mathcal{S}}$ *iff* $\mathcal{C}$ *is a minimal* a-admissible set *and* a *is* $\mathcal{S}$-accepted in τ.

Next we define *rejecting sets* of an argument a as the fundamental notion for studying and recognizing the basics for the rejecting condition of a. Intuitively, a rejecting set $\mathcal{R}$ for a should be that which ensures that a would end up $\mathcal{S}$-accepted in the AF $\mathbb{F}_{\mathbf{A} \setminus \mathcal{R}}$. Before formalizing rejecting sets through Definition 8, we propose the intermediate notions of *partially admissible* and *defeating sets*.

Definition 7 (Partially Admissible and Defeating Sets). *Given an* AF $\tau = \mathbb{F}_{\mathbf{A}}$; *for any* $\Theta \subseteq \mathbf{A}$ *and any argument* $a \in \mathbf{A}$, *we say that:*

1. b **defeats** Θ *iff* b *defeats some* $c \in \Theta$.
2. Θ *is* a-**partially admissible** *iff* $a \in \Theta$, Θ *is conflict-free, and if* $c \in \Theta$, *with* $c \neq a$ *then there is some* $b \in \mathbf{A}$ *such that* c *defeats* b *and* b *defeats* $\Theta \setminus \{c\}$.
3. Θ *is* a-**defeating** *iff there is some* a-partially admissible set Θ' *such that* $\Theta \supseteq \Upsilon \subseteq \{b \in \mathbf{A} | b \text{ defeats } \Theta'\}$.

The *partially admissible set* for a given argument a is an effort for constructing a set which would end up turning into an a-core$_{\mathcal{S}}$ after removing an appropriate a-defeating set from the worked AF. The purpose of using a superinclusion for constructing defeating sets is to capture particular situations when working with subargumentation. This will be clear in Sect. 5. Determining a correct defeating set depends on two sequential steps: firstly, it should ensure that its removal turns a into $\mathcal{S}$-accepted (see *rejecting sets* on Definition 8), and secondly, it should be minimal for such condition (*remainder sets* on Definition 9).

Definition 8 (Rejecting Sets). *Given an* AF $\tau = \mathbb{F}_{\mathbf{A}}$, *a semantics specification* $\mathcal{S}$, *and an argument* $a \in \mathbf{A}$; *for any* $\Theta \subseteq \mathbf{A}$, *we say that* Θ *is* $\mathcal{S}$-a-**rejecting** *in* τ *iff* Θ *is* a-*defeating in* τ *and* a *is* $\mathcal{S}$-*accepted in* $\mathbb{F}_{\mathbf{A} \setminus \Theta}$.

Remainder sets state "responsibility" to arguments for the non-acceptability of an argument. Intuitively, an a-remainder is a minimal $\mathcal{S}$-a-rejecting set.

Definition 9 (Remainder Sets). *Given an* AF $\tau = \mathbb{F}_{\mathbf{A}}$ *and an argumentation semantics specification* $\mathcal{S}$, *for any* $\mathcal{R} \subseteq \mathbf{A}$, *we say that* $\mathcal{R}$ *is an* a-**remainder** *in* τ, *noted as* a-**remainder**$_{\mathcal{S}}$ *iff* $\mathcal{R}$ *is a* $\mathcal{S}$-a-*rejecting set and for any* $\Theta \subset \mathcal{R}$, *it follows that* a *is* $\mathcal{S}$-*rejected in the* AF $\mathbb{F}_{\mathbf{A} \setminus \Theta}$.

Example 10. Given the AF $\tau = \mathbb{F}_{\mathbf{A}}$, where $\mathbf{A} = \{a, b, c, d, d', e, e', f, g, h\}$ and $\mathbf{R}_{\mathbf{A}}$ renders the argumentation graph depicted below on the right. Argument b is not accepted by any semantics since there is no admissible set containing it. For instance, $\mathbb{E}_{\mathrm{pr}}(\tau) = \{\{f, e, e', c, a\}, \{h, g, e, e', c, a\}\}$, and $\mathbb{E}_{\mathrm{gr}}(\tau) = \{\{e, e', c, a\}\}$.

However, it is possible to propose different alternatives of change to move towards an epistemic state in which argument b turns to accepted in the resulting AF. For instance, let us consider a semantics specification $\mathcal{S} = \langle \mathfrak{s}, \delta \rangle$, where the acceptance function δ implements Eq. 2 and $\mathfrak{s} = \mathbf{pr}$. In this case, the acceptable set would be $\mathcal{A}_{\mathcal{S}}(\tau) = \{e, e', c, a\}$. Note that $\{e\}$, $\{e'\}$, $\{c\}$ are b-remainder$_{\mathcal{S}}$ sets. This is so, given that $\{e\}$

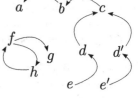

Graph of AF τ

is b-defeating for the b-partially admissible set $\{b, d\}$, in the same manner that $\{e'\}$ is for $\{b, d'\}$, and $\{c\}$ is for $\{b\}$. Note that $\{e, e', c\}$ is b-defeating for the b-partially admissible set $\{b, d, d'\}$, however while $\{e, e', c\}$ is a $\mathcal{S}$-b-rejecting set, it is not a b-remainder$_{\mathcal{S}}$ given that it is not minimal. Afterwards, considering the b-remainder$_{\mathcal{S}}$ $\{e\}$, we can build a new AF $\tau_1 = \mathbb{F}_{\mathbf{A} \setminus \{e\}}$ whose resulting acceptance set would be $\mathcal{A}_{\mathcal{S}}(\tau_1) = \{d, e', b\}$, since $\mathbb{E}_{\mathrm{pr}}(\tau_1) = \{\{f, d, e', b\}, \{h, g, d, e', b\}\}$.

Once again, considering the AF τ under the same semantic specification, note that g is not $\mathcal{S}$-accepted despite there is an extension $\{h, g, e, e', c, a\} \in \mathbb{E}_{\mathrm{pr}}(\tau)$ which contains g. The situation here arises from the acceptance function δ which requires intersecting every extension in $\mathbb{E}_{\mathrm{pr}}(\tau)$. Note also that there is a g-admissible set $\{g, h\}$. However, it is possible to propose an alternative of change to move towards an epistemic

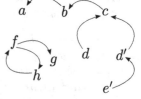

Graph of AF τ_1

state in which argument g turns to accepted in the resulting AF. To that end, we can construct two g-partially admissible sets $\{g\}$ and $\{g, h\}$. Note that, for any of them, it appears a g-defeating set $\{f\}$ which ends up being a $\mathcal{S}$-g-rejecting set and also a g-remainder$_{\mathcal{S}}$ in the resulting AF $\tau_2 = \mathbb{F}_{\mathbf{A} \setminus \{f\}}$ whose acceptance set would be $\mathcal{A}_{\mathcal{S}}(\tau_2) = \{g, h, e, e', c, a\}$, since it ends up being the unique preferred extension in $\mathbb{E}_{\mathrm{pr}}(\tau_2)$.

Observe that by considering an acceptance function δ implementing Eq. 3 under the preferred semantics, the acceptable set would be $\mathcal{A}_{\mathcal{S}}(\tau) = \{h, g, e, e', c, a\}$. Thus, it is natural to have that the unique g-remainder$_{\mathcal{S}}$ ends up being the empty set. This is so, given that although both $\emptyset$ and $\{f\}$ are g-defeating sets, and even both of them are also $\mathcal{S}$-g-rejecting sets, $\{f\}$ does not fulfill the requirements for being a g-remainder$_{\mathcal{S}}$ given that is is not minimal. This will be of utmost relevance for pursuing the verification of the well known principle of minimal change.

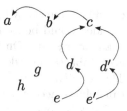

Graph of AF τ_2

Definition 11 (Set of Cores and Set of Remainders). *Given an* AF $\tau = \mathbb{F}_{\mathbf{A}}$, *a semantics specification $\mathcal{S}$, and an argument $a \in \mathbf{A}$, we say that:*

1. $\top_{\mathcal{S}}(\tau, a)$ *is the **set of cores** of a iff* $\top_{\mathcal{S}}(\tau, a)$ *contains every a-core$_{\mathcal{S}}$ $\mathcal{C} \subseteq \mathbf{A}$.*
2. $\perp_{\mathcal{S}}(\tau, a)$ *is the **set of remainders** of a iff* $\perp_{\mathcal{S}}(\tau, a)$ *contains every a-remainder$_{\mathcal{S}}$ $\mathcal{R} \subseteq \mathbf{A}$.*

Example 12 (Continues from Example 10). Considering the acceptance function implementing Eq. 2 over the preferred semantics, the set of cores for argument a ends up being $\top_{\mathcal{S}}(\tau, a) = \{\{a, c, e, e'\}\}$. Also, the corresponding set of remainders for argument b is $\perp_{\mathcal{S}}(\tau, b) = \{\{e\}, \{e'\}, \{c\}\}$. On the other hand, if we consider the b-remainder$_{\mathcal{S}}$ $\{e\}$ for analyzing the AF $\tau_1 = \mathbb{F}_{\mathbf{A}\backslash\{e\}}$, argument b turns out being $\mathcal{S}$-accepted since it is possible to identify a b-core$_{\mathcal{S}}$. In such a case, the resulting set of cores for b would be $\top_{\mathcal{S}}(\tau_1, b) = \{\{b, d\}\}$.

Proposition 13. *Given an* AF $\tau = \mathbb{F}_{\mathbf{A}}$, *a semantics specification $\mathcal{S}$, and an argument $a \in \mathbf{A}$; the following properties hold: (1) $\top_{\mathcal{S}}(\tau, a) = \emptyset$ iff $\perp_{\mathcal{S}}(\tau, a) \neq \emptyset$, (2) $a \in \mathcal{A}_{\mathcal{S}}(\tau)$ iff $\top_{\mathcal{S}}(\tau, a) \neq \emptyset$, and (3) $a \notin \mathcal{A}_{\mathcal{S}}(\tau)$ iff $\perp_{\mathcal{S}}(\tau, a) \neq \emptyset$.*

Proposition 13 states the interrelation between the sets of cores and remainders and how they relate with an argument's $\mathcal{S}$-acceptance.

4 Argumentation Dynamics Through Retractive Methods

For a rational handling of the acceptability dynamics of arguments, a change operation applied to an AF τ should provoke a controlled alteration of the acceptable set $\mathcal{A}_{\mathcal{S}}(\tau)$ towards achieving a specific purpose. For instance, a contraction operation may modify the acceptable set in order to *contract the acceptance condition* of a specific argument. The *acceptance contraction* of an argument can be achieved through the removal of arguments from the set $\mathbf{A}(\tau)$. However, observe that the acceptable set $\mathcal{A}_{\mathcal{S}}(\tau)$ has a non-monotonic construction from τ. This means that removing/incorporating arguments from/to the argumentation framework does not imply that the resulting acceptable set would be de/increased regarding the original one. Consequently, it is also possible to consider the addition of new arguments to the framework, in order to ensure an argument a to be rejected in the resulting framework. The former alternative could be achieved by breaking all a-core$_S$ sets, whereas for the latter alternative, the idea would be to incorporate new arguments towards the construction of a-remainder$_S$ sets. On the other hand, a contraction operation may modify the acceptable set in order to *contract the rejection condition* of a specific argument. The *rejection contraction* of an argument a ensures that a ends up accepted. We can achieve acceptance of an argument a either by removing arguments from **A** to break the existence of a-remainder$_S$ sets, or also by incorporating arguments to **A** to construct a-core$_S$ sets. It is possible to establish an analogy between classical belief revision, where a contraction by a formula α (resp. of, $\neg\alpha$) ensures α's truth (resp. of, falsity) is not inferred and, belief revision in argumentation, where an acceptance contraction (resp. of, rejection contraction) by an argument a ensures a is not accepted (resp. of, not rejected).

Revisions and contractions are usually defined independently with the intention to interrelate them afterwards by setting up a duality. A philosophic discussion is sustained on the matter of the nature of such independence. Some researchers assert that there is really no contraction whose existence could be justified without a revision. In fact, they state that a contraction conforms an intermediate state towards the full specification of the revision. Such an intuition fits quite well our approach. For instance, if we think the argumentation stands for a normative system, it is natural to assume that a new norm is intended to be incorporated –through a revision– for ensuring afterwards its acceptance –through some intermediate contraction. Another alternative is to assume a derogative norm, whose purpose is to enter the system –through a revision– for ensuring afterwards the rejection of an elder norm –through some intermediate contraction for ensuring the acceptance of the derogative norm. In this paper we focus on an *acceptance revision operation* obtained through the removal of arguments from the set **A**, *i.e.*, a sort of *retractive* acceptance revision. Such a revision operation retracts from the AF some a-remainder$_S$ set –for ensuring the acceptance of a new argument a– through the usage of a rejection contraction. Thus, with a retractive acceptance revision, we assume the idea of provoking change to the AF for altering the acceptable set with the intention to pursue acceptability of an argument a, which can be external to the original AF.

An operator '$\circledast$' ensures that given an AF τ and a new argument a, the acceptance revision of τ by a ends up in a new AF $\tau \circledast a$ in which a is S-accepted. We refer to an early contribution by Levi [17] to belief revision, where he related revisions to contractions. He suggested that a revision ('$*$') of a base Σ by a new information α should be achieved through two stages. Firstly, by contracting ('$-$') all possibility of deriving $\neg\alpha$ for obtaining a new base which would be consistent with α. Afterwards, it could be added ('$+$') the new information α ensuring that this stage would end up consistently. This intuition was formalized in an equivalence referred to as the *Levi identity*: $\Sigma * \alpha = (\Sigma - \neg\alpha) + \alpha$. In argumentation, it is natural to think that the new argument a should be incorporated to the AF τ through an *expansion* operator '$+$', and ensuring afterwards its acceptability through a contraction operation for breaking the rejection of a, *i.e.*, a *rejection contraction* '$\ominus_\perp$'. Note that it is mandatory to invert the two stages of the original Levi identity[2] since it is necessary for the new argument to be recognized by the framework in order to analyze its acceptability condition. This renders an equivalence between *acceptance revision* and *rejection contraction* through the generalization of the Levi identity: $\tau \circledast a = (\tau + a) \ominus_\perp a$.

We will analyze the construction of two sub-operations for achieving the acceptance revision. Firstly, we need to recognize new arguments to be incorporated to the framework. For such matter, let us assume a *domain of abstract arguments* $\mathbb{A}$, such that for any abstract AF $\mathbb{F}_\mathbf{A}$, it follows that $\mathbf{A} \subseteq \mathbb{A}$. Next we formalize the concept of *external argument*, and afterwards we define a simple *expansion operation* for incorporating an external argument to a framework.

Definition 14 (External Argument). *Given an AF $\tau = \mathbb{F}_\mathbf{A}$, an argument a is **external to** τ (or just, **external**) iff $a \in \mathbb{A}$ but $a \notin \mathbf{A}$.*

Definition 15 (Expansion). *Given an AF $\tau = \mathbb{F}_\mathbf{A}$ and an external argument $a \in \mathbb{A}$. The operator $+$ stands for an **expansion** iff $\tau + a = \mathbb{F}_{\mathbf{A} \cup \{a\}}$.*

From Proposition 13, we know that an argument a is $\mathcal{S}$-accepted *iff* there is no a-remainder$_\mathcal{S}$ set. Therefore, it is sufficient to break one single a-remainder$_\mathcal{S}$ $\mathcal{R} \in \perp_\mathcal{S}(\tau, a)$ in order to obtain a new AF in which we could construct a-core$_\mathcal{S}$ sets, implying the acceptance of a. For such purpose, we define a *remainder selection*, as a function by which it is possible to select the best option among the several a-remainder$_\mathcal{S}$ sets from $\perp_\mathcal{S}(\tau, a)$.

Definition 16 (Remainder Selection). *Given an AF $\tau = \mathbb{F}_\mathbf{A}$, a semantics specification $\mathcal{S}$, and an argument $a \in \mathbf{A}$. A **remainder selection** is obtained by a **selection function** $\gamma : \wp(\wp(\mathbf{A})) \longrightarrow \wp(\mathbf{A})$ applied over the set $\perp_\mathcal{S}(\tau, a)$ for selecting some a-remainder$_\mathcal{S}$, where $\gamma(\perp_\mathcal{S}(\tau, a)) \in \perp_\mathcal{S}(\tau, a)$ is such that for every $\mathcal{R} \in \perp_\mathcal{S}(\tau, a)$ it holds $\gamma(\perp_\mathcal{S}(\tau, a)) \preccurlyeq_\gamma \mathcal{R}$, where $\preccurlyeq_\gamma$ is a **selection criterion** by which it is possible to select the best representative a-remainder$_\mathcal{S}$ set.*

The selection criterion can be any method for ordering sets of arguments. In the sequel, we will abstract away from any specific selection criterion. Now it is easy to define the *rejection contraction* by relying upon a selection function.

Definition 17 (Rejection Contraction). *Given an AF $\tau = \mathbb{F}_\mathbf{A}$, a semantics specification $\mathcal{S}$, and an argument $a \in \mathbf{A}$. The operator $\ominus_\perp$ stands for a **rejection contraction** iff $\tau \ominus_\perp a = \mathbb{F}_{\mathbf{A} \setminus \mathcal{R}}$, where $\mathcal{R} = \gamma(\perp_\mathcal{S}(\tau, a))$.*

The *acceptance revision* may be formally given by relying upon an expansion operation and a rejection contraction determined by a selection function.

Definition 18 (Acceptance Revision). *Given an AF $\tau = \mathbb{F}_\mathbf{A}$, a semantics specification $\mathcal{S}$, and an external argument $a \in \mathbb{A}$. The operator $\circledast$ stands for an **acceptance revision** (or just, **revision**) iff $\tau \circledast a = \mathbb{F}_{\mathbf{A}'}$, where $\mathbf{A}' = \mathbf{A}(\tau + a) \setminus \gamma(\perp_\mathcal{S}(\tau + a, a))$. When necessary, we will write $\tau \circledast_\gamma a$ to identify the remainder selection function γ by which the revision $\tau \circledast a$ is obtained.*

---

[2] Inverting the Levi identity leads to an inconsistent intermediate state. This is not an issue in argumentation since we only incorporate new pairs to the defeat relation.

The axiomatization of the acceptance revision is achieved by analyzing the different characters of revisions from classical belief revision [1,16] and from ATC revision [19], for adapting the classical postulates to argumentation.

(**success**) a is S-accepted in $\tau \circledast a$
(**consistency**) $\mathcal{A}_S(\tau \circledast a)$ is conflict-free
(**inclusion**) $\tau \circledast a \subseteq \tau + a$
(**vacuity**) If a is S-accepted in $\tau + a$ then $\mathbf{A}(\tau + a) \subseteq \mathbf{A}(\tau \circledast a)$
(**core-retainment**) If $b \in \mathbf{A}(\tau) \setminus \mathbf{A}(\tau \circledast a)$ then exists an AF τ' such that $\mathbf{A}(\tau') \subseteq \mathbf{A}(\tau)$ and a is S-accepted in $\tau' + a$ but S-rejected in $(\tau' + b) + a$.

In classic belief revision, the *success* postulate states that the new information should be satisfied by the revised knowledge base. From the argumentation standpoint, this may be interpreted as the requirement of acceptability of the new argument. Through *consistency* a classic revision operation ensures that the new revised base ends up consistently always that the new belief to be incorporated is so. From the argumentation standpoint, there should be no need for ensuring a consistent (or conflict-free) set of arguments since the essence of such theory is to deal with inconsistencies. However, this requirement makes sense when thinking about the acceptable set of the framework for ensuring that the argumentation semantics allows a consistent reasoning methodology. The consistency postulate for extension semantics has been studied before in [2], among others. *Inclusion* aims at guaranteeing that the only new information to be incorporated is the object by which the base is revised. The restatement to argumentation may be seen as the sole inclusion of the external argument. *Vacuity* captures the conditions under which the revision operation has nothing to do but the sole incorporation of the new information. Its restatement to argumentation may be seen as the fact of a being S-accepted straightforwardly, with no need to remove any argument. That is, the simple expansion of the external argument would end up forming a new framework in which it is possible to construct a-core$_S$ sets. The vacuity postulate is usually referred as complementary to the inclusion postulate, thus, a change operation satisfying both postulates ends up verifying the equality $\tau \circledast a = \tau + a$ whenever the external argument is straightforwardly S-accepted in the expanded framework. Through *core-retainment* the amount of change is controlled by avoiding removals that are not related to the revision operation, *i.e.*, every belief that is lost serves to make room for the new one. In argumentation dynamics we care on the changes perpetrated to the framework in order to achieve acceptability for the external argument. Hence, any argument that is removed should be necessary for such purpose. The rational behavior of the acceptance revision operation is ensured through the following *representation theorem*.

Theorem 19. *Given an AF τ, a semantics specification S, and an external argument $a \in \mathbb{A}$; $\tau \circledast a$ is an acceptance revision iff '$\circledast$' satisfies success, consistency, inclusion, vacuity, and core-retainment.*

5   Fundamentals for Logic-Based Frameworks

We will assume a logic $\mathcal{L}$ to which the represented knowledge will correspond. In addition, we will assume an *argument domain set* referred as $\mathbb{A}_\mathcal{L}$ to which *(logic-based) arguments* containing $\mathcal{L}$ formulae will conform. Arguments will be defined upon $\mathcal{L}$ knowledge through a set of premises and a claim such that an argument $a \in \mathbb{A}_\mathcal{L}$ can be expressed through a pair, namely the *argument interface*, $\langle S, \vartheta \rangle \in \mathbb{A}_\mathcal{L}$, where $S \subseteq \mathcal{L}$ is referred as the *support* and $\vartheta \in \mathcal{L}$ as the *claim*. The logic $\mathcal{L}$ will be considered along with its corresponding inference operator $\models$, constituting a complete *deductive system* $\langle \mathcal{L}, \models \rangle$. Therefore, according to the classic notion of argument, we can assume that given an argument $\langle S, \vartheta \rangle \in \mathbb{A}_\mathcal{L}$, the basic three principles are satisfied: *(deduction)* $S \models \vartheta$, *(minimality)* there is no subset $S' \subset S$ such that $S' \models \vartheta$, and *(consistency)* S is consistent according to $\mathcal{L}$, *i.e.*, $S \not\models \bot$. Finally, we will eventually say that an argument a *supports* ϑ *from* S to informally specify that the argument claim is the formula $\vartheta \in \mathcal{L}$ and similarly that its support is given by the set $S \subseteq \mathcal{L}$, or formally that the argument $a = \langle S, \vartheta \rangle \in \mathbb{A}_\mathcal{L}$. We will rely upon two functions $\mathfrak{cl} : \mathbb{A}_\mathcal{L} \longrightarrow \mathcal{L}$ and $\mathfrak{sp} : \mathbb{A}_\mathcal{L} \longrightarrow \wp(\mathcal{L})$ to identify both the claim and support set of $\mathbb{A}_\mathcal{L}$-arguments. Hence, given an argument $a \in \mathbb{A}_\mathcal{L}$, we can refer to the claim and support set as $\mathfrak{cl}(a) \in \mathcal{L}$ and $\mathfrak{sp}(a) \subseteq \mathcal{L}$, respectively. Moreover, the function $\mathfrak{sp}$ will be overloaded as $\mathfrak{sp} : \wp(\mathbb{A}_\mathcal{L}) \longrightarrow \wp(\mathcal{L})$ in order to be applied over sets of arguments such that given a set $\Theta \subseteq \mathbb{A}_\mathcal{L}$, $\mathfrak{sp}(\Theta) = \bigcup_{a \in \Theta} \mathfrak{sp}(a)$ will identify the base determined by the set of supports of arguments contained in Θ.

A *(logic-based) argumentation framework* (AF) will be assumed as a pair $\langle \mathbf{A}, \mathbf{R_A} \rangle$, where $\mathbf{A} \subseteq \mathbb{A}_\mathcal{L}$ is a finite *set of arguments*, and the set $\mathbf{R_A} \subseteq \mathbb{A}_\mathcal{L} \times \mathbb{A}_\mathcal{L}$ identifies the finitary *defeat relation* between pairs of arguments such that:

$$\mathbf{R_A} = \{(a,b) | a, b \in \mathbf{A}, \mathfrak{sp}(a) \cup \mathfrak{sp}(b) \models \bot, \text{ and } a \succ b\} \tag{5}$$

A pair $(a, b) \in \mathbf{R_A}$ implies that $a \in \mathbf{A}$ *defeats* $b \in \mathbf{A}$, or equivalently, a is a *defeater* of b, meaning that the supports of both arguments a and b cannot be simultaneously assumed in a consistent manner, and also that a is preferred over b, according to some abstract *preference relation* $\succ$. We will keep the defeating function ε abstract, assuming that it is *valid iff* condition (1) in p. 5 is satisfied. Different instantiations of such a function has been widely studied in [15].

Since any logic-based argument is built from a set of formulae –standing for its support set– it is natural to think that any subset of the support set can be used to build another argument. This intuition describes the concepts of *sub-arguments* (and *super-arguments*). We will identify a sub-argument relation by writing $a \sqsubseteq b$ for expressing that an argument $a \in \mathbb{A}_\mathcal{L}$ is a sub-argument of argument $b \in \mathbb{A}_\mathcal{L}$ (and also that b is a super-argument of a), implying that $\mathfrak{sp}(a) \subseteq \mathfrak{sp}(b)$ holds. We will also identify the set of all sub-arguments of an argument $a \in \mathbb{A}_\mathcal{L}$ through the function $\mathfrak{subs} : \mathbb{A}_\mathcal{L} \longrightarrow \wp(\mathbb{A}_\mathcal{L})$ such that $\mathfrak{subs}(a) = \{b \in \mathbb{A}_\mathcal{L} | b \sqsubseteq a\}$, for any argument $a \in \mathbb{A}_\mathcal{L}$.

Logic-based argumentation may unveil some problems with regards to the conflict recognition between pairs of arguments. Consider the following example where arguments are constructed upon a propositional logic $\mathcal{L}$.

Example 20. Assuming $\Theta \subseteq \mathbb{A}_{\mathcal{L}}$ such that $\Theta = \{a, b, c\}$ where $a = \langle\{p\}, p\rangle$, $b = \langle\{q\}, q\rangle$, and $c = \langle\{\neg p \vee \neg q\}, \neg p \vee \neg q\rangle$. The AF generator $\mathbb{F}_\Theta$ will construct an AF with an empty set of defeats $\mathbf{R}_\Theta$. Note that Θ is admissible given that it is conflict-free and that it has no defeaters. However, $\mathfrak{sp}(\Theta) \models \perp$ holds.

The problem presented in Example 20 relies on the construction of logic-based AFs from arbitrary sets of arguments. It is necessary to build all possible arguments, including sub and super arguments, in order to ensure that the resulting AF will deliver rational responses through an argumentation semantics. We say that a set of arguments is closed whenever it contains all the sub- and super-arguments that can be constructed from its arguments. This ensures an exhaustive construction of arguments from an initial base of arguments. We provide such implementation through an *argumentation closure operator* $\mathbb{C}$.

Definition 21 (Argumentation Closure). *An operator $\mathbb{C}$ is an **argumentation closure** iff for any $\Theta \subseteq \mathbb{A}_{\mathcal{L}}$, it holds $\mathbb{C}(\Theta) = \{a \in \mathbb{A}_{\mathcal{L}} | a \sqsubseteq b, \text{ for any } b \in \Theta\} \cup \{a \in \mathbb{A}_{\mathcal{L}} | \mathfrak{subs}(a) \subseteq \Theta\}$. We say that Θ is **closed** iff it holds $\Theta = \mathbb{C}(\Theta)$.*

The following proposition shows that the closure of a set Θ of arguments triggers the complete set of arguments that can be constructed using the formulae involved in arguments contained in Θ.

Proposition 22. *Given a set of arguments $\Theta \subseteq \mathbb{A}_{\mathcal{L}}$, the underlying knowledge base $\Sigma = \mathfrak{sp}(\Theta)$, and the set $\mathbf{A}_\Sigma \subseteq \mathbb{A}_{\mathcal{L}}$ of all the possible arguments constructed from Σ. The set Θ is closed iff $\Theta = \mathbf{A}_\Sigma$.*

We refer to a structure $\langle \mathbf{A}, \mathbf{R}_\mathbf{A} \rangle$ as a *closed* AF iff it is constructed through a closed set of arguments $\mathbf{A} \subseteq \mathbb{A}_{\mathcal{L}}$, *i.e.*, $\mathbf{A} = \mathbb{C}(\mathbf{A})$. Depending on the specification of the language $\mathcal{L}$, the argumentation closure may trigger multiple different arguments with a unique support and even more, it could result infinitary, triggering an infinite set of arguments if the closure is achieved in an uncontrolled manner. Several alternatives may arise to keep a finite, and still closed, set of arguments. For instance, it is possible to restrict the claim of arguments to some specific form in order to avoid constructing several arguments with logically equivalent claims and a same support set. A nice alternative for doing this is to restrict the construction of arguments to their *canonical form* [6], in which for any argument a, its claim has the form $\mathfrak{cl}(a) = \bigwedge \mathfrak{sp}(a)$. In the sequel, and just for simplicity, we will abstract away from such specific matters involving the construction of arguments, by simply referring to a domain $\mathbb{A}_{\mathcal{L}}^* \subseteq \mathbb{A}_{\mathcal{L}}$, where $\mathbb{A}_{\mathcal{L}}^*$ is the domain of arguments of a unique representation: for any pair of arguments $a, b \in \mathbb{A}_{\mathcal{L}}^*$, it follows that if $\mathfrak{sp}(a) = \mathfrak{sp}(b)$ then $\mathfrak{cl}(a) = \mathfrak{cl}(b)$, and thus it holds $a = b$. This restriction ensures that any set $\Theta \subseteq \mathbb{A}_{\mathcal{L}}^*$ of arguments ends up in a finite closed set $\mathbb{C}^*(\Theta) = \mathbf{A}$ independently of the method used for ensuring it, where $\mathbb{C}^*(\Theta)$ is the closed set of $\mathbb{A}_{\mathcal{L}}^*$-arguments such that $\mathbb{C}^*(\Theta) = \mathbb{C}(\Theta) \cap \mathbb{A}_{\mathcal{L}}^*$. From now on, we will write $\mathbf{A}$ (or $\mathbf{A}'$) for referring only to closed sets of $\mathbb{A}_{\mathcal{L}}^*$-arguments.

In what follows, we will write $\mathbb{F}_\mathbf{A}$ for referring to the AF $\langle \mathbf{A}, \mathbf{R}_\mathbf{A} \rangle$, where $\mathbf{A} \subseteq \mathbb{A}_{\mathcal{L}}^*$ is a closed set, *i.e.*, $\mathbb{C}^*(\mathbf{A}) = \mathbf{A}$. In such a case, we say that $\mathbb{F}_\mathbf{A}$ is a *closed* AF. This will also allow us to refer to any *sub-framework* $\mathbb{F}_\Theta = \langle \Theta, \mathbf{R}_\Theta \rangle$, where $\Theta \subseteq \mathbf{A}$ is a not necessarily closed set of arguments. In such a case, we will overload the sub-argument operator '$\sqsubseteq$' by also using it for identifying sub-frameworks, writing $\mathbb{F}_\Theta \sqsubseteq \mathbb{F}_\mathbf{A}$. Observe that, if $\mathbb{C}^*(\Theta) = \mathbf{A}'$ and $\mathbf{A}' \subset \mathbf{A}$, then $\mathbb{F}_{\mathbf{A}'}$ is a closed *strict sub-framework* of $\mathbb{F}_\mathbf{A}$, i.e., $\mathbb{F}_{\mathbf{A}'} \sqsubset \mathbb{F}_\mathbf{A}$.

By relying upon closed argumentation frameworks we ensure that the acceptable set $\mathcal{A}_\mathcal{S}(\mathbb{F}_\mathbf{A})$ will trigger rational results. A closed AF $\mathbb{F}_\mathbf{A}$ will be necessary for satisfying *closure under sub-arguments* and *exhaustiveness* postulates from to [2]. On the other hand, a set $\mathbf{R}_\mathbf{A}$ as defined in Eq. 5 describes a general defeat relation which is *conflict-dependent* and *conflict-sensitive* according to [2]. This means that any minimal inconsistent set of formulae implies the construction of a pair of arguments which will necessarily be conflicting, and that any pair of conflicting arguments implies a minimal source of inconsistency. This property guarantees that the framework will satisfy the postulate referred as *closure under sub-arguments* under any of the extension semantics reviewed before. This postulate is necessary to ensure a rational framework independently of the semantics adopted given that, for any $\mathfrak{s}$-extension $\mathbf{E}$ we will ensure that if $a \in \mathbf{E}$ then for any sub-argument $b \sqsubseteq a$ it holds $b \in \mathbf{E}$. The *closure under CN* postulate [2] will not be verified given that we prevent the construction of several claims for a same argument's body through a unique representation like canonical arguments. However, it holds in a "semantic sense": closed AFs ensure drawing all such possible claims.

Example 23 (Continues from Example 20). By assuming $\mathbb{A}_\mathcal{L}^*$ as the domain of canonical arguments, the argumentation closure renders the closed set of arguments: $\mathbf{A} = \mathbb{C}^*(\Theta) = \{a, b, c, d, e, f\}$, where $d = \langle\{p, q\}, p \wedge q\rangle$, $e = \langle\{p, \neg p \vee \neg q\},$ $p \wedge (\neg p \vee \neg q)\rangle$, and $f = \langle\{q, \neg p \vee \neg q\}, q \wedge (\neg p \vee \neg q)\rangle$. For the construction of the set of defeats, we will assume that any argument in Θ is preferred over any other argument which is not in Θ, whereas when considering a pair of arguments where both are either Θ insiders or outsiders, the preference relation will be symmetric. Thus, we obtain the following pairs of defeats: $\mathbf{R}_\mathbf{A} = \{(a, f), (b, e), (c, d), (d, e), (d, f), (e, d), (f, d)\}$. Observe however that although $\mathfrak{sp}(\Theta)$ is inconsistent, Θ is still admissible.

Through the argumentation closure, we have provided a method for ensuring that a closed AF is complete given that we have all the possible arguments that can be constructed from the set of arguments and therefore all the sources of conflict will be identified through the defeat relation. However, we still have a problem: as is shown in Example 23, $\Theta \subseteq \mathbf{A}$ keeps being admissible given that it is conflict-free. Thus, it is necessary to reformulate the abstract notion for admissible sets by requiring their closure.

Definition 24 (Logic-based Admissibility). *Given an AF $\mathbb{F}_\mathbf{A}$, for any $\Theta \subseteq \mathbf{A}$ we say that Θ is **admissible** iff Θ is closed (i.e., $\Theta = \mathbb{C}^*(\Theta)$), conflict-free, and defends all its members.*

Once again, regarding postulates in [2], working with closed AFs and taking in consideration the reformulated notion of admissibility in logic-based frameworks, guarantees the *consistency* postulate which ensures that every $\mathfrak{s}$-extension contains a consistent support base, *i.e.*, for any closed AF τ, $\mathfrak{sp}(\mathbb{E}_\mathfrak{s}(\tau)) \not\models \perp$ holds.

Example 25 (Continues from Example 23). Under the new definition of admissibility, we have that Θ cannot be admissible since it is not closed. The following admissible sets appear: $\{a\}, \{b\}$, and $\{c\}$. Note that the sets $\{a, b, d\}, \{a, c, e\}$, and $\{b, c, f\}$, are not admissible given that although they are closed and conflict-free, none of them defends all its members.

Definition 24 for admissibility in logic-based frameworks makes core sets end up closed without inconvenient. However, the case of remainder sets is different. Having a set $\Theta \subseteq \mathbf{A}$, the problem is that we only can ensure that an argument a is accepted in the sub-framework $\mathbb{F}_{\mathbf{A} \setminus \Theta}$ if we can ensure that $\mathbf{A} \setminus \Theta$ is a closed set (see Example 20). This ends up conditioning Definition 9. Hence, it is necessary to provide some constructive definition for remainder sets. This allows determining which property should satisfy a set $\Theta \subseteq \mathbf{A}$ for ensuring that if $\mathbf{A}$ is a closed set then the operation $\mathbf{A} \setminus \Theta$ also determines a closed set. In Definition 27, we propose an *expansive closure* which will rely upon the identification of a set of *atomic arguments*: arguments that have no strict sub-arguments inside. That is, given an argument $a \in \mathbb{A}_\mathcal{L}^*$, a is *atomic iff* $|\mathfrak{sp}(a)| = 1$.

Definition 26 (Set of Atomic Arguments). *Given an AF $\mathbb{F}_\mathbf{A}$ and an argument $a \in \mathbf{A}$, a function* $\mathfrak{at} : \mathbb{A}_\mathcal{L}^* \longrightarrow \wp(\mathbb{A}_\mathcal{L}^*)$ *is an* **atoms function** *iff it renders the **set of atomic arguments** $\mathfrak{at}(a) \subseteq \mathbf{A}$ of a such that $\mathfrak{at}(a) = \{b \in \mathbf{A} | b \sqsubseteq a$ and there is no $c \in \mathbf{A}$ such that $c \sqsubset b\}$.*

We will overload the atoms function as $\mathfrak{at} : \wp(\mathbb{A}_\mathcal{L}^*) \longrightarrow \wp(\mathbb{A}_\mathcal{L}^*)$ to be applied over sets of arguments such that $\mathfrak{at}(\Theta) = \bigcup_{a \in \Theta} \mathfrak{at}(a)$.

Definition 27 (Expansive Closure). *Given an AF $\mathbb{F}_\mathbf{A}$ and a set $\Theta \subseteq \mathbf{A}$, an operator $\mathbb{P}$ is an* **expansive closure** *iff $\mathbb{P}(\Theta) = \{a \in \mathbf{A} | b \sqsubseteq a$, for every $b \in \mathfrak{at}(\mathbb{P}_0(\Theta))\}$, where $\mathbb{P}_0(\Theta) = \{a \in \Theta | $ there is no $b \in \Theta$ such that $b \sqsubset a\}$. We say that Θ is* **expanded** *iff it holds $\Theta = \mathbb{P}(\Theta)$.*

Note that $\mathbb{P}_0(\Theta)$ contains all the arguments from Θ having no sub-arguments in Θ, while $\mathbb{P}(\Theta)$ contains all the arguments from $\mathbf{A}$ having some atomic sub-argument of some argument in $\mathbb{P}_0(\Theta)$. The expansive closure is a sort of super-argument closure in the sense that it contains all the arguments that should disappear by removing Θ from $\mathbf{A}$. Proposition 28 verifies that if we remove from a closed set another set which is expanded then we obtain a new closed set.

Proposition 28. *Given two sets $\mathbf{A} \subseteq \mathbb{A}_\mathcal{L}^*$ and $\Theta \subseteq \mathbb{A}_\mathcal{L}^*$, where $\mathbf{A}$ is closed; if $\Theta \subseteq \mathbf{A}$ then $\mathbf{A}' = \mathbf{A} \setminus \mathbb{P}(\Theta)$ is a closed set, i.e., $\mathbf{A}' = \mathbb{C}^*(\mathbf{A}')$.*

Definition 29 (Logic-based Remainder Sets). *Given an AF $\mathbb{F}_\mathbf{A}$ and a semantics specification $\mathcal{S}$, for any $\Theta \subseteq \mathbf{A}$, we say that Θ is an a-**remainder** in $\mathbb{F}_\mathbf{A}$, noted as a-remainder$_\mathcal{S}$ iff Θ is a minimal expanded $\mathcal{S}$-a-rejecting set:*

1. Θ *is a* $\mathcal{S}$-*a-rejecting set,*
2. $\Theta = \mathbb{P}(\Theta)$, *and*
3. *for any set* $\Theta' \subset \Theta$ *such that* $\Theta' = \mathbb{P}(\Theta')$, *it holds a is* $\mathcal{S}$-*rejected in* $\mathbb{F}_{\mathbf{A}\setminus\Theta'}$.

The following example shows how a propositional logic $\mathcal{L}$ for constructing logic-based frameworks affects the notions of core and remainder sets.

Example 30. We will assume $\mathcal{L}$ as the propositional logic and $\mathbb{A}^*_{\mathcal{L}}$ as the domain of canonical arguments. Let $\Theta \subseteq \mathbb{A}^*_{\mathcal{L}}$ be a set of canonical arguments such that $\Theta = \{a, b, c, d\}$, where $a = \langle\{p \wedge q_1\}, p \wedge q_1\rangle$, $b = \langle\{p \wedge q_2\}, p \wedge q_2\rangle$, $c = \langle\{\neg p\}, \neg p\rangle$, and $d = \langle\{\neg q_2\}, \neg q_2\rangle$. The argumentation closure renders the closed set of arguments $\mathbf{A} = \mathbb{C}^*(\Theta) = \{a, b, c, d, e, f, g\}$, where:

$$e = \langle\{p \wedge q_1, p \wedge q_2\}, p \wedge q_1 \wedge q_2\rangle, \text{ where } \mathfrak{subs}(e) = \{a, b\}$$
$$f = \langle\{p \wedge q_1, \neg q_2\}, p \wedge q_1 \wedge \neg q_2\rangle, \text{ where } \mathfrak{subs}(f) = \{a, d\}$$
$$g = \langle\{\neg p, \neg q_2\}, \neg p \wedge \neg q_2\rangle, \text{ where } \mathfrak{subs}(g) = \{c, d\}$$

Then, the AF $\mathbb{F}_{\mathbf{A}}$ is closed and through a preference relation $\mathbf{R}_{\mathbf{A}} = \{(a,c),(b,c),\ (d,g),(d,b),$ $(e,c),(e,d),(b,f),\quad (f,c),(a,g),(b,g),(e,f),(e,g),$ $(f,g)\}$. Assuming $\mathcal{S} = \langle \mathsf{co}, \delta\rangle$, where δ implements Eq. 3, observe that a b-core$_{\mathcal{S}}$ $\mathcal{C}_b = \{a, b, e\}$ is constructed by the closure $\mathbb{C}^*(\{b, e\})$. Since c and d are $\mathcal{S}$-rejected, we have remainder sets for both of them: a c-remainder$_{\mathcal{S}}$ $\mathcal{R}_c = \{a, e, f\}$ and two d-remainder$_{\mathcal{S}}$ sets $\mathcal{R}_d = \{a, e, f\}$ and $\mathcal{R}'_d = \{b, e\}$. Observe that $\Upsilon = \{a, b, e, f\}$ is the result of expand-

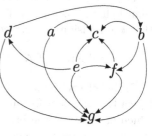

$\mathbb{F}_{\mathbf{A}}$

ing the $\mathcal{S}$-d-rejecting set $\{e\}$, *i.e.*, $\Upsilon = \mathbb{P}(\{e\})$. However Υ is not a d-remainder$_{\mathcal{S}}$ since it is not minimal: there are two d-defeating sets $\{a, e\}$ and $\{b, e\}$ whose respective expansions are $\mathbb{P}(\{a, e\}) = \mathcal{R}_d$ and $\mathbb{P}(\{b, e\}) = \mathcal{R}'_d$. Note that, although $\{e\}$ is a d-defeating set, the superinclusion in Definition 7, item 3, allows the consideration of some additional argument/s. Clearly, the only alternative for that is to incorporate some atom/s of some argument/s included in the defeating set.

6   Argumentation Dynamics in Logic-Based Frameworks

We need to consider closed logic-based frameworks which provokes a necessary reformulation of the expansion operation. This ensures a closed resulting framework after the incorporation of an external argument $a \in \mathbb{A}^*_{\mathcal{L}}$.

Definition 31 (Expansion). *Given an AF* $\mathbb{F}_{\mathbf{A}}$ *and an external argument* $a \in \mathbb{A}^*_{\mathcal{L}}$. *The operator* $+$ *stands for an* **expansion** *iff* $\mathbb{F}_{\mathbf{A}} + a = \mathbb{F}_{\mathbb{C}^*(\mathbf{A}\cup\{a\})}$.

Definitions for change operations proposed in Sect. 4 will perfectly apply for logic-based frameworks if the references to expansion operations are interpreted as logic-based expansions, according to Definition 31. Thus, a revision $\tau \circledast a$ will refer to an operation $(\tau + a) \ominus_\perp a$, where $+$ is a logic-based expansion. Change operations for logic-based frameworks incorporated the necessary consideration of $\mathcal{L}$-formulae. This brings about the necessity to discuss additional postulates for the complete rationalization of closed frameworks. In classic belief revision, the *closure* postulate states that if a base Σ is a closed set (referred as belief set) then the result of the revision should also be ensured to be closed. In this case, by closure they refer to a closure under logical consequences, obtaining, in general, infinite closed sets. This kind of closure is different from the proposed argumentation closure. As being explained before, the argumentation closure – applied over singleton construction of arguments, *i.e.*, arguments from a domain $\mathbb{A}_{\mathcal{L}}^{*}$– ensures a finite closed set of arguments. However, the purpose of the argumentation closure also differs from the closure under logical consequences in that $\mathbb{C}$ ensures the presence of all the constructible arguments (see Proposition 22) from a common knowledge respecting a specific construction $\mathbb{A}_{\mathcal{L}}^{*}$, but not the construction of all the equivalent arguments. This subject makes rationality of acceptance revision for logic-based frameworks more similar to revision of bases than belief sets. Finally, from the argumentation standpoint, we should ensure that the revision of a closed AF ends up in a new closed AF.

(**closure**) if $\mathbf{A}(\tau) = \mathbb{C}^{*}(\mathbf{A}(\tau))$ then $\mathbf{A}(\tau \circledast a) = \mathbb{C}^{*}(\mathbf{A}(\tau \circledast a))$

In belief revision, it is natural to assume that revisions applied to a base by logically equivalent formulae, have necessarily identical outcomes. The choice of which elements of the base to retain should depend on their logical relations to the new information. Therefore, if two sentences are inconsistent with the same subsets of the base, they should push out the same elements from the base. This is known as *uniformity*. Since we are considering arguments built from $\mathcal{L}$-formulae, it is natural to analyze the existing relations between two different arguments which coincide in their conflict-relations, supports and claims. For this matter it is necessary to specify an *equivalence relation* for arguments in order to ensure that the revisions $\tau \circledast a$ and $\tau \circledast b$ have equivalent outcomes (see [19]).

Definition 32 (Equivalence [19]). *For any pair of arguments $a, b \in \mathbb{A}_{\mathcal{L}}^{*}$, we say that a and b are equivalent arguments, noted as $a \equiv b$ iff $\mathfrak{cl}(a) \models \mathfrak{cl}(b)$ and $\mathfrak{cl}(b) \models \mathfrak{cl}(a)$ and for any $a' \sqsubset a$ there is $b' \sqsubset b$ such that $a' \equiv b'$.*

(**uniformity**) if $a \equiv b$ then $\mathbf{A}(\tau) \cap \mathbf{A}(\tau \circledast a) = \mathbf{A}(\tau) \cap \mathbf{A}(\tau \circledast b)$

Inspired by smooth incisions in Hansson's Kernel Contractions [16], we introduce an additional condition on remainder selection functions for guaranteeing uniformity. Under the consideration of two equivalent arguments a and b, the idea is to ensure that a remainder selection function will trigger a same remainder $\mathcal{R}$ which is common to both sets of remainders $\perp_{\mathcal{S}}(\tau + a, a)$ and $\perp_{\mathcal{S}}(\tau + b, b)$.

Definition 33 (Smooth Remainder Selection). *Given an AF τ and two external arguments $a, b \in \mathbb{A}_{\mathcal{L}}^{*}$. If $a \equiv b$ then $\gamma(\perp_{\mathcal{S}}(\tau + a, a)) = \gamma(\perp_{\mathcal{S}}(\tau + b, b))$.*

Given an AF τ and an external argument $a \in \mathbb{A}^*_{\mathcal{L}}$, we will refer to any operation $\tau \circledast_\gamma a$ as *smooth acceptance revision iff* $\tau \circledast_\gamma a$ is an acceptance revision obtained through a smooth remainder selection 'γ'. Now we are able to formalize the representation theorem for smooth acceptance revisions.

Theorem 34. *Given an AF τ, a semantics specification $\mathcal{S}$, and an external argument $a \in \mathbb{A}^*_{\mathcal{L}}$; $\tau \circledast a$ is a smooth acceptance revision iff '$\circledast$' satisfies closure, success, consistency, inclusion, vacuity, core-retainment, and uniformity.*

Example 35 [Continues from Example 30] Suppose, we have $\Theta' = \{a, b, c\}$, where $a = \langle \{p \wedge q_1\}, p \wedge q_1 \rangle$, $b = \langle \{p \wedge q_2\}, p \wedge q_2 \rangle$, and $c = \langle \{\neg p\}, \neg p \rangle$. The argumentation closure renders the set $\mathbf{A}' = \mathbb{C}^*(\Theta') = \{a, b, c, e\}$, such that $e = \langle \{p \wedge q_1, p \wedge q_2\}, p \wedge q_1 \wedge q_2 \rangle$, where $\mathfrak{subs}(e) = \{a, b\}$. Then, for the closed AF $\mathbb{F}_{\mathbf{A}'}$, the attack relation is $\mathbf{R}_{\mathbf{A}'} = \{(a, c), (b, c), (e, c)\}$. We need to revise $\tau = \mathbb{F}_{\mathbf{A}'}$ by the external argument $d = \langle \{\neg q_2\}, \neg q_2 \rangle$.

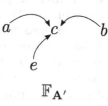

$\mathbb{F}_{\mathbf{A}'}$

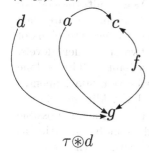

$\tau \circledast d$

Note $\tau \circledast d = \mathbb{F}_{\mathbf{A}''}$, where $\mathbf{A}'' = \mathbf{A}(\tau + d) \setminus \gamma(\perp_{\mathcal{S}}(\tau + d, d))$, is equivalent to $\tau \circledast d = (\tau + d) \ominus_{\perp} d$ through the generalization of the Levi identity (p. 11). Note that $\tau + d = \mathbb{F}_{\mathbf{A}}$ (see Example 30). We know there are two d-remainder$_{\mathcal{S}}$ sets $\mathcal{R}_d = \{a, e, f\}$ and $\mathcal{R}'_d = \{b, e\}$. Assuming a selection criterion $\mathcal{R}'_d \preccurlyeq_\gamma \mathcal{R}_d$, we have $\gamma(\perp_{\mathcal{S}}(\tau + d, d)) = \mathcal{R}'_d$ and also $\mathbf{A}'' = \mathbf{A} \setminus \mathcal{R}'_d = \{a, c, d, f, g\}$. Finally, the resulting revised framework ends up as $\tau \circledast d = \mathbb{F}_{\mathbf{A}''}$, where $\mathbf{R}_{\mathbf{A}} = \{(a, c), (d, g), (f, c), (a, g), (f, g)\}$.

Note that the acceptance revision can be seen as $\mathfrak{sp}(\mathbf{A}(\tau + d)) \setminus \mathfrak{sp}(\Theta'')$, where $\Theta'' = \mathbb{P}_0(\gamma(\perp_{\mathcal{S}}(\tau + d, d))) = \mathbb{P}_0(\mathcal{R}'_d) = \{b\}$ (see Definition 27). Hence, $\mathfrak{sp}(\mathbf{A}(\tau \circledast d)) = \mathfrak{sp}(\mathbf{A}(\tau + d)) \setminus \mathfrak{sp}(\Theta'') = \mathfrak{sp}(\{a, b, c, d, e, f, g\}) \setminus \mathfrak{sp}(\{b\}) = \{p \wedge q_1, p \wedge q_2, \neg p, \neg q_2\} \setminus \{p \wedge q_2\} = \{p \wedge q_1, \neg p, \neg q_2\}$, corresponding to the set of arguments $\{a, c, d\}$, whose closure is $\mathbb{C}^*(\{a, c, d\}) = \{a, c, d, f, g\} = \mathbf{A}''$.

The previous example shows the relation between an acceptance revision applied directly over the set of arguments, regarding a related operation upon the underlying knowledge base from which the logic-based AF is constructed.

Proposition 36. *Given an AF τ, a semantics specification $\mathcal{S}$, an external argument $a \in \mathbb{A}^*_{\mathcal{L}}$, and a smooth acceptance revision '$\circledast$'; assuming $\mathbf{A}^*_\Sigma \subseteq \mathbb{A}^*_{\mathcal{L}}$ as the set of all canonical arguments $\mathbb{A}^*_{\mathcal{L}}$ constructible from a knowledge base $\Sigma \subseteq \mathcal{L}$, it holds $\mathbf{A}(\tau \circledast a) = \mathbf{A}^*_\Sigma$, where $\Sigma = \mathfrak{sp}(\mathbf{A}(\tau + a)) \setminus \mathfrak{sp}(\mathbb{P}_0(\gamma(\perp_{\mathcal{S}}(\tau + a, a))))$.*

7 Conclusions

Related and Future Work. The expansion proposed here can be seen as a *normal expansion* [4] since we do not restrict the directionality of the new attacks which appear after a new argument is incorporated to the framework. Authors there propose some general properties for ensuring the (im)possibility of *enforcing* a set of abstract arguments, which refers to the modification of the abstract framework for achieving a specific result through some standard semantics. Logic-based argumentation is out of the scope of the article. There are two other differences with our work. Firstly, they pursue a kind of multiple expansion since they consider the addition of an entire set of arguments and the interaction through attack with the existing ones. In our work, this is possible but only from the perspective of several subarguments which are part of a single superargument. Secondly, they only consider expansions. In a subsequent work [3], authors incorporate deletion of attacks. However, only minimal change is considered which renders no possibility for a complete characterization of change through representation theorems. There, the formalization of the minimal change principle is achieved through the introduction of numerical measures for indicating how far two argumentation frameworks are. Another revision approach in an AGM spirit is presented in [10] through revision formulæ that express how the acceptability of some arguments should be changed. As a result, they derive argumentation systems which satisfy the given revision formula, and are such that the corresponding extensions are as close as possible to the extensions of the input system. The revision presented is divided in two subsequent levels: firstly, revising the extensions produced by the standard semantics. This is done without considering the attack relation. Secondly, the generation of argumentation systems fulfilling the outcome delivered by the first level. Minimal change is pursued in two different levels, firstly, by ensuring as less change as possible regarding the arguments contained in each extension, and secondly, procuring as less change as possible on the argumentation graph. The methods they provide do not provoke change upon the set of arguments, but only upon the attack relations. Similar to [3], their operator is more related to a distance based-revision which measures the differences from the actual extensions with respect to the ones obtained for verifying the revision formula. They give a basic set of rationality postulates in the very spirit of AGM, but more closed to the perspective given in [14]. They only show that the model presented satisfies the postulates without giving the complete representation theorem for which the way back of the proof, *i.e.*, from postulates to the construction, is missing. However, the very recent work [12], which is in general a refinement of [9,10], proposes a generic solution to the revision of argumentation frameworks by relying upon complete representation theorems. In addition, the revision from the perspective of argumentation frameworks is also considered. Other distance based approaches in this direction are the works by Booth *et al.* [7,8], were authors develop a general AGM-like approach for modeling the dynamics of argumentation frameworks based on the distance between conflict-free labellings for the complete semantics only. They propose the notion of *fall back beliefs* for representing the rational outcome of an AF from a constraint. A different approach, but still in an AGM spirit was presented in [5], where authors propose expansion and revision operators for Dung's abstract argumentation frameworks (AFs) based on a novel proposal called *Dung logics* with the particularity that equivalence in such logics coincides with strong equivalence for the respective argumentation semantics. The approach presents a reformulation of the AGM postulates in terms of monotonic consequence relations for AFs. They finally state that standard approaches based on measuring distance between models are not appropriate for AFs.

In general, the aforementioned works differ from ours in the perspective of dealing with the argumentation dynamics. This also renders different directions to follow for achieving rationality. To our knowledge, [19] was the first work to propose AGM postulates for rationalizing argumentation dynamics, providing also complete representation theorems for the proposed revision operations built upon logic-based argumentation. The rationalization done here is mainly inspired by such results, however change methods are pursued upon standard semantics in contrast to dialectical trees as done in [19]. Similar to the notion of remainders, in [19] and other ATC approaches like [18,20], authors recognize from the dialectical trees some sets of arguments which are identified as "responsible" for the rejection of arguments. In this paper we follow a similar intuition, however, core and remainder sets are more general notions for identifying the sources of acceptance/rejection of specific arguments upon standard semantics.

The problem of revising a framework by a set of arguments has been shown in [11] to suffer from failures regarding enforcement as originally defined in [4]. This is an interesting problem to which the theory here proposed may bring different solutions. To that end, it would be interesting to extend the acceptance revision operator for revising a framework by an unrestricted set of arguments rather than a single one, or a superargument including several subarguments. Such an operation seems to fit better as an argumentation *merge*.

Discussion. We proposed a model of change for argumentation based on the novel concepts of core and remainder sets. Core sets can be thought as minimal sets which are necessary for ensuring the acceptability of a given argument whereas remainder sets can be understood as minimal sets which are somehow responsible for the rejection of a given argument. The proposed model of change was firstly studied upon abstract argumentation and afterwards, upon logic-based argumentation. The resulting acceptance revision operation was characterized through the proposal of rationality postulates *á la* belief revision, and afterwards, the through corresponding representation theorems.

Another aspect that we wanted to demonstrate is that abstract argumentation can be counterproductive when the research is not immersed in the appropriate context of applicability. When the model, firstly proposed for abstract argumentation, was observed in the context of logic-based argumentation, several new inconveniences appeared requiring special attention, showing that abstraction can also be a path to trivialization. A conclusion that we draw is that standard semantics may not apply correctly to a logic-based argumentation system (AS). The usage of argumentation postulates [2,15] facilitates the analysis for understanding how rational a set of extensions can be. Such rationality can be achieved from two standpoints. Either from the construction of the framework, by putting special attention on how to model conflicts, or on the other hand, by tackling the problem straightforwardly from the construction of the extensions. In this sense, we proposed a new perspective for enriching the concept of admissibility for being applied over logic-based arguments through the notion of argumentation closure. We have shown that standard semantics relying on logic-based admissibility can make things easier for verifying argumentation postulates (see discussion in p. 15).

Regarding argumentation dynamics, we focus minimal change from the perspective of the knowledge base at first, and from the set of arguments, afterwards. We believe this is an appropriate manner to tackle such principle, since logic-based argumentation stands for reasoning upon inconsistencies of an underlying knowledge base. Another way to observe minimal change –which was not attended here– is from the perspective of the outcomes of the framework. A final conclusion that we draw is that although dynamics of abstract arguments can also be studied by proposing models of change affecting the set of attacks, it is not an appropriate perspective for logic-based argumentation. These sort of problems are really interesting, however they do not seem to fit well to such context of application considering that attacks are finally adjudicated in terms of logical contradictions.

References

1. Alchourrón, C., Gärdenfors, P., Makinson, D.: On the logic of theory change: partial meet contraction and revision functions. J. Symbolic Logic **50**, 510–530 (1985)
2. Amgoud, L.: Postulates for logic-based argumentation systems. Int. J. Approximate Reasoning **55**(9), 2028–2048 (2014)
3. Baumann, R.: What does it take to enforce an argument? Minimal change in abstract argumentation. In: Frontiers in Artificial Intelligence and Applications, ECAI 2012, Montpellier, France, vol. 242, pp. 127–132. IOS Press (2012)
4. Baumann, R., Brewka, G.: Expanding argumentation frameworks: enforcing and monotonicity results. In: Frontiers in Artificial Intelligence and Applications, COMMA 2010, D. del Garda, Italy, vol. 216, pp. 75–86. IOS Press (2010)
5. Baumann, R., Brewka, G.: AGM meets abstract argumentation: expansion and revision for dung frameworks. In: Yang and Wooldridge [21], pp. 2734–2740
6. Besnard, P., Hunter, A.: Elements of Argumentation. The MIT Press, Cambridge (2008)
7. Booth, R., Caminada, M., Podlaszewski, M., Rahwan, I.: Quantifying disagreement in argument-based reasoning. In: AAMAS 2012, IFAAMAS, pp. 493–500, Valencia, Spain (2012)
8. Booth, R., Kaci, S., Rienstra, T., van der Torre, L.: A logical theory about dynamics in abstract argumentation. In: Liu, W., Subrahmanian, V.S., Wijsen, J. (eds.) SUM 2013. LNCS, vol. 8078, pp. 148–161. Springer, Heidelberg (2013)
9. Coste-Marquis, S., Konieczny, S., Mailly, J.-G., Marquis, P.: A translation-based approach for revision of argumentation frameworks. In: Fermé, E., Leite, J. (eds.) JELIA 2014. LNCS, vol. 8761, pp. 397–411. Springer, Heidelberg (2014)
10. Coste-Marquis, S., Konieczny, S., Mailly, J., Marquis, P.: On the revision of argumentation systems: minimal change of arguments statuses. In: KR 2014, Vienna, Austria. AAAI Press (2014)

11. Coste-Marquis, S., Konieczny, S., Mailly, J., Marquis, P.: Extension enforcement in abstract argumentation as an optimization problem. In: Yang and Wooldridge [21], pp. 2876–2882

12. Diller, M., Haret, A., Linsbichler, T., Rümmele, S., Woltran, S.: An extension-based approach to belief revision in abstract argumentation. In: Yang and Wooldridge [21], pp. 2926–2932

13. Dung, P.M.: On the acceptability of arguments and its fundamental role in non-monotonic reasoning and logic programming and n-person games. Artif. Intell. **77**, 321–357 (1995)

14. Gärdenfors, P.: Knowledge in Flux: Modelling the Dynamics of Epistemic States. The MIT Press, Bradford Books, Cambridge (1988)

15. Gorogiannis, N., Hunter, A.: Instantiating abstract argumentation with classical logic arguments: postulates and properties. Artif. Intell. **175**(9–10), 1479–1497 (2011)

16. Hansson, S.O.: A Textbook of Belief Dynamics. Theory Change and Database Updating. Kluwer Academic, London (1999)

17. Levi, I.: Subjunctives, dispositions and chances. Synthese **34**(4), 423–455 (1977)

18. Moguillansky, M.O., Rotstein, N.D., Falappa, M.A., García, A.J., Simari, G.R.: Argument theory change applied to defeasible logic programming. In: Fox, D., Gomes, C.P. (eds.) AAAI, pp. 132–137. AAAI Press (2008)

19. Moguillansky, M.O., Wassermann, R., Falappa, M.A.: Inconsistent-tolerant base revision through argument theory change. Logic J. IGPL **20**(1), 154–186 (2012)

20. Rotstein, N.D., Moguillansky, M.O., Falappa, M.A., García, A.J., Simari, G.R.: Argument theory change: revision upon warrant. In: Proceedings of COMMA, pp. 336–347 (2008)

21. Yang, Q., Wooldridge, M. (eds.): Proceedings of the Twenty-Fourth International Joint Conference on Artificial Intelligence, IJCAI 2015. AAAI Press, Buenos Aires, Argentina, 25–31 July 2015

Anytime Algorithms for Solving Possibilistic MDPs and Hybrid MDPs

Kim Bauters[1]([⊠]), Weiru Liu[1], and Lluís Godo[2]

[1] Queen's University Belfast, Belfast, UK
{k.bauters,w.liu}@qub.ac.uk
[2] IIIA, CSIC, Bellaterra, Spain
godo@iiia.csic.es

Abstract. The ability of an agent to make quick, rational decisions in an uncertain environment is paramount for its applicability in realistic settings. Markov Decision Processes (MDP) provide such a framework, but can only model uncertainty that can be expressed as probabilities. Possibilistic counterparts of MDPs allow to model imprecise beliefs, yet they cannot accurately represent probabilistic sources of uncertainty and they lack the efficient online solvers found in the probabilistic MDP community. In this paper we advance the state of the art in three important ways. Firstly, we propose the first online planner for possibilistic MDP by adapting the Monte-Carlo Tree Search (MCTS) algorithm. A key component is the development of efficient search structures to sample possibility distributions based on the DPY transformation as introduced by Dubois, Prade, and Yager. Secondly, we introduce a hybrid MDP model that allows us to express both possibilistic and probabilistic uncertainty, where the hybrid model is a proper extension of both probabilistic and possibilistic MDPs. Thirdly, we demonstrate that MCTS algorithms can readily be applied to solve such hybrid models.

1 Introduction

A Markov Decision Process (MDP) [2] is a successful framework for dealing with sequential decision problems under uncertainty, particularly when the uncertainty is due to underlying stochastic processes. However, when dealing with uncertainty due to a lack of knowledge it is often easier to find acceptable qualitative estimates. Possibilistic counterparts of MDPs [17], referred to as π-MDP, have been introduced in recent years to tackle this problem. In some situations, optimal strategies to compute the policy of a π-(PO)MDP have even been shown to give better results than their probabilistic counterparts [3]. A limitation of π-MDP, though, is that current solvers for π-MDP rely on offline algorithms to compute the optimal policy. Conversely, state-of-the-art MDP planners are online approximate anytime planners (e.g. [12,15]). Since such planners only have to determine the next best *"enough"* action instead of coming up with a complete optimal policy, they are considerably faster. Furthermore, these online planners are often simpler to integrate with, for example, BDI systems [16] where

© Springer International Publishing Switzerland 2016
M. Gyssens and G. Simari (Eds.): FoIKS 2016, LNCS 9616, pp. 24–41, 2016.
DOI: 10.1007/978-3-319-30024-5_2

the online nature fits well with the reactive nature of such systems. MDP and π-MDP also share a common downside: both frameworks only allow for a single kind of representation for all sources of uncertainty, either all probabilistic, or all possibilistic. This is often at odds with realistic settings, where the underlying causes of uncertainty are diverse. Consider this example:

Example 1. Patrols are being used in a wildlife preserve to deter poachers. Some areas, such as the grassland, have rich statistical data available about the effects of patrolling near herds. These areas are easy to observe and various sensors are installed to monitor the herd movement. In the grassland, we know that moving a patrol to the grassland will prevent poaching in 82 % of the cases if the herd is there. Furthermore, in 11 % of the situations the herd will move to the marshes and in 7 % to the mountains when we have a patrol near them on the grassland. Otherwise, the herd stays on the grassland for the remainder of the day.

For the other areas, only the rangers' experience is available to predict the effectiveness of patrols, and their result on herd movements. For instance, in the mountains we know that it is entirely possible that the herd will stay there or that the herd moves back to the grassland. However, it is only slightly possible that the herd moves to the marches. It is only somewhat possible to deter poaching if we have a patrol around since tracking and finding the herd is hard in the mountains. The terrain with its many hiding spots also makes it entirely possible for poachers to succeed even when we have a patrol nearby.

The rewards themselves can also be quantitative or qualitative in nature. For example, we can easily express that we prefer states in which no animals are poached, but only in the grassland do we have exact numbers of how many animals would be saved by patrolling.   □

Modelling examples like these is difficult in MDP or π-MDP as it involves different types of uncertainty. In this paper we advance the state of the art in three important ways. *Firstly*, we adapt the Monte-Carlo Tree Search algorithm used in e.g. UCT [14] and PROST [12] to the π-MDP setting. To achieve this, we present in Sect 3.1 a tractable way to sample possibility distributions. The resulting algorithm is the first online planner for π-MDP, applicable to both brave and cautious reasoning. *Secondly*, we propose a hybrid MDP framework where state transitions can be described as either possibilistic or probabilistic distributions. Such a hybrid MDP, which is a proper extension of both MDP and π-MDP, allows a precise and elegant way of modelling problems such as the ones expressed in Example 1. *Thirdly*, by combining the results from our first contribution with classical MCTS we arrive at an MCTS algorithm that can be used to solve hybrid MDP. We furthermore impose rational restrictions on how qualitative and quantitative utilities of trajectories can be combined in order to guide the search. The resulting machinery provides us with a way to solve hybrid MDPs using efficient online anytime algorithms.

The remainder of the paper is organised as follows. Preliminaries are discussed in Sect. 2. We discuss how to adapt UCT to π-MDP in Sect. 3, crucially depending on our method to efficiently sample possibilistic distributions. The hybrid MDP model is presented in Sect. 4, and we also discuss an MCTS algorithm to solve such hybrid models. Related work is discussed in Sect. 6 and we draw conclusions in Sect. 7.

2 Preliminaries

MDP: A probabilistic MDP model is defined by a tuple $\langle \mathcal{S}, \mathcal{A}, T, R, \gamma \rangle$. We have that $\mathcal{S}$ is a finite set of *states* and $\mathcal{A}$ is a finite set of *actions*. The *transition function* T gives the probability distributions over states $(s_t, s_{t+1}) \in \mathcal{S}^2$ and an action $a_t \in \mathcal{A}$, such that $T(s_t, a_t, s_{t+1}) = P(s_{t+1} \mid s_t, a_t)$, i.e. the (stochastic) uncertainty of reaching the state s_{t+1} from state s_t by taking a_t. A *reward function* R associates the immediate reward value $R(s_t, a_t, s_{t+1})$ with transitioning from state s_t to s_{t+1} by using action a_t. A *discounting factor* $0 \le \gamma \le 1$ is used to discount rewards that can only potentially be obtained in the future (or, alternatively, a finite horizon is assumed). A *trajectory* τ is a sequence of states $(s_1, ..., s_h)$, where we use T_h to denote all trajectories of size h. A *policy* (δ) is a sequence of decision rules $\delta : \mathcal{S} \rightarrow \mathcal{A}$ indexed by t, i.e. $\delta(s_t) = a_t$ is the action to execute at time t. The *value*, or quantitative utility, of a policy is given by:

$$v((\delta), s_0) = E \left(\sum_{t=0}^{h-1} \gamma^t \cdot R(s_t, \delta_t(s_t), s_{t+1}) \right) \tag{1}$$

with h the *horizon*, $E(\cdot)$ the *expected reward*, and s_{t+1} the stochastic outcome of applying a_t in s_t. A policy applied in the initial state thus describes a trajectory of size h. The optimal policy is the one that maximises $v(\cdot, s_0)$, i.e. at each step t it takes the best available action to maximise the sum of future rewards.

Finding these optimal policies is an intractable problem, so approaches have been developed that can return *"good enough"* actions. One such an approach for solving MDPs is the UCT algorithm [14], which combines Monte-Carlo Tree Search (MCTS) with an upper confidence bound (UCB1) policy. MCTS is an anytime algorithm in which a search tree is built by iteratively sampling the decision space. During each iteration, the algorithm (a) selects an expandable child node; (b) expands this node using an available action; (c) performs a rollout/simulation from the expanded node to a terminal node; and (d) backpropagates the result of the simulation up to the root node to update the statistics for future searches (see Fig. 1). The MCTS algorithm closely relates to Eq. 1. The reward of a given trajectory is accumulated during the (c) rollout phase. The probability of the trajectory (implicitly assumed in Eq. 1 through the expected value) is respected during the (a) node selection/(b) expansion/(c) rollout. Finally, the total reward from traversing a node/number of times a node has been traversed, are updated during the (d) backpropagation. This allows an approximation of the expected reward of a node to be computed.

The UCB1 policy [1] can considerably speed up MCTS by addressing the exploration-exploitation trade-off during the child node selection. The action to perform in each node is the one that maximises $\overline{X}_j + B\sqrt{(\log n)/n_j}$ with $\overline{X}_j \in [0, 1]$ the average future reward by taking action a_j, B a bias parameter, n_j

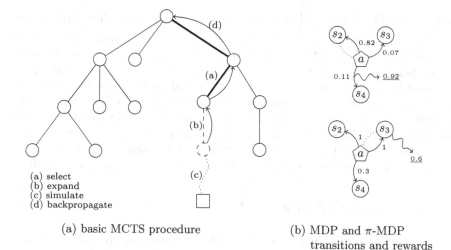

(a) basic MCTS procedure

(b) MDP and π-MDP
transitions and rewards

Fig. 1. *(left)* basic MCTS procedure with the 4 distinct phases in each iteration *(right)* MDP and π-MDP differences, where transitions are either probabilities or possibilities, rewards are over transitions, and preferences are over states

the number of times a_j has been selected in this node, and $n = \sum_{j=1}^{k} n_j$, i.e. the total number of times any of the available actions $\{a_1, ..., a_k\}$ in this node have been taken. Intuitively, the term on the left of the sum encourages exploitation, while the term on the right encourages exploring less-visited paths.

π-**MDP:** The π-MDP model [17, 18] is the possibilistic counterpart of the MDP model where the uncertainty is modelled as a qualitative possibility distribution. It is defined as a tuple $\langle \mathcal{S}, \mathcal{A}, T^\pi, M, \mathcal{L} \rangle$. Here $\mathcal{L}$ is a *possibility scale*, i.e. a finite and totally ordered set whose *greatest element* is $1_\mathcal{L}$ and whose *least element* is $0_\mathcal{L}$. Typically, it is taken as $\mathcal{L} = \{0, 1/k, 2/k, ..., k - 1/k, 1\}$ for some $k \in \mathbb{N}^+$ and it will be required to define the transition function. The possibility distribution over $\mathcal{S}$ is a function $\pi : \mathcal{S} \to \mathcal{L}$ such that $\max_s \pi(s) = 1_\mathcal{L}$, i.e. at least one state is entirely possible. Whenever $\pi(s) < \pi(s')$ it implies that s' is more plausible than s. The transition function T^π is defined over a pair of states $(s_t, s_{t+1}) \in \mathcal{S}^2$ and an action $a_t \in \mathcal{A}$ as $T^\pi(s_t, a_t, s_{t+1}) = \pi(s_{t+1} \mid s_t, a_t)$, i.e. the possibility of reaching s_{t+1} conditioned on the current state s_t and the action a_t. This reflects that the uncertainty of the effects of action a_t are due to a lack of information. Furthermore, a function $M : \mathcal{S} \to \mathcal{L}$ models the qualitative utility, or *preference*, of each state (see Fig. 1b). The *qualitative utility* of a policy in π-MDP is defined in the *cautious* setting as:

$$u_*((\delta), s_0) = \min_{\tau \in T_h} \max \{1 - \Pi(\tau \mid s_0, (\delta)), M(s_h)\} \qquad (2)$$

or, in the *brave* setting, as:

$$u^*((\delta), s_0) = \max_{\tau \in T_h} \min \{\Pi(\tau \mid s_0, (\delta)), M(s_h)\} \qquad (3)$$

with $\Pi(\tau \,|\, s_0, (\delta)) = \min_{t=0}^{h-1} \big(\pi(s_{t+1} \,|\, s_t, \delta_t(s_t)) \big)$, i.e. the possibility of the trajectory $\tau = (s_1, ..., s_h)$. The brave setting evaluates a policy based on whether there is at least one possible path that is good. The cautious setting, conversely, evaluates a policy based on how good all of the possible paths are (indeed, it selects the worst path for its utility). The utility is based on the possibility/necessity of the trajectory which starts from s_0, and the preference of the final state s_h (assuming a finite horizon). Some algorithms have been proposed to compute the solutions of π-MDP models for both decision criteria [3,17]. However, these approaches compute optimal solutions and are therefore only applicable to rather small problem spaces, contrary to the online MCTS algorithm available for MDP.

3   Adapting UCT to π-MDP

We now develop a way to apply UCT, or MCTS in general, to π-MDP. As discussed in Sect. 2, MCTS builds a search tree by iteratively sampling the decision space. The concept of sampling plays an important role in the rollout phase, but also in the selection and expansion phase due to the non-deterministic nature of the underlying model. In the probabilistic setting, the law of large numbers makes sampling straightforward. In the possibilistic setting, however, we do not have a concept similar to the law of larger numbers. Still, the idea remains similar. Through sampling, we want to select one of the effects of a given action in a π-MDP model, in accordance with the possibility associated with the various effects of that action. This idea is closely related to the idea of transforming a possibility distribution into a probability distribution, where we can sample directly from the latter. A compelling possibilistic-probabilistic transformation, the DPY transformation, was first introduced by Kaufmann in French [10], and later independently by Dubois and Prade [6] and Yager [22]. Not only has this transformation been independently developed by other authors, both in the setting of possibility theory [5,13] and Dempster-Shafer theory [20], but it also has a large number of desirable properties (see [8]). In Sect. 3.1 we focus on how we can use the DPY transformation to sample a possibility distribution, and how we can do so in a tractable way. In Sect. 3.2 we look at some of the intricacies of backpropagation and node selection in the π-MDP setting. Together, these components will allow us to use an MCTS-style algorithm to solve π-MDP.

3.1   Possibilistic Sampling

As an initial step towards sampling a possibility distribution, we transform such a distribution into an intermediate data structure with a tractable algorithm:

Definition 1. *Let π be a possibility distribution over $\mathcal{S}$. Let $\mathcal{S}_0$ be those states with a strictly positive possibility, $\mathcal{S}_0 = \{s \mid \pi(s) > 0, s \in \mathcal{S}\}$. Furthermore, we rank order elements in $\mathcal{S}_0$ as $\pi(s_0) \geq ... \geq \pi(s_k)$. Let $\mathcal{C}_\pi$ then be the list of tuples sorted in ascending order on i such that $\langle i, p_i \rangle \in \mathcal{C}_\pi$ whenever $\pi(s_i) > \pi(s_{i+1})$ with $p_i = \pi(s_i) - \pi(s_{i+1})$ and, by default, $\pi(s_{k+1}) = 0$.*

The structure C_π provides a compact representation of the DPY transformation of a possibility distribution π to a basic belief assignment m.[1] Indeed, the tuple $\langle i, p_i \rangle$ marks that $\{s_0, ..., s_i\}$ is the α-cut of π with $\alpha = \pi(s_i)$. The value p_i is the probability mass associated with this α-cut in m on the subset $A = \{s_0, ..., s_i\}$.

Example 2. Consider the following possibility distribution π:

$$\pi(s_0) = 1 \qquad \pi(s_1) = 0.7 \qquad \pi(s_2) = 0.7$$
$$\pi(s_3) = 0.3 \qquad \pi(s_4) = 0.1 \qquad \pi(s_5) = 0$$

We have $C_\pi = (\langle 0, 0.3 \rangle, \langle 2, 0.4 \rangle, \langle 3, 0.2 \rangle, \langle 4, 0.1 \rangle)$.

An algorithm to compute C_π from π is given in Algorithm 1.

Given a compact representation C_π, we can readily determine the probability distribution associated with π:

Definition 2. *Let π be a possibility distribution and C_π as in Defintion 1. For every $s_i \in S_0$ the probability of s_i is:*

$$p(s_i) = \sum_{\substack{\langle j, p_j \rangle \in C_\pi \\ j \geq i}} p_j j + 1$$

and $p(s) = 0$ for all $s \in (S \setminus S_0)$.

Of course, we only want to sample *according* to the associated probability distribution *without explicitly computing* the distribution. This can be achieved by randomly selecting a $\langle s_i, p_i \rangle \in C_\pi$ according to probability masses $\mathcal{P}$ using the principle of Probability Proportional to Size (PPS), followed by a random selection with a uniform probability $1(i+1)$ of an element $s \in \{s_0, ..., s_i\}$. This idea is reflected in Algorithm 2.

Example 3. Consider π and C_π from Example 2. Following Definition 2, we have the probability distribution p such that:

$$p(s_0) = 0.5033... \qquad p(s_1) = 0.2033... \qquad p(s_2) = 0.2033...$$
$$p(s_3) = 0.07 \qquad p(s_4) = 0.02 \qquad p(s_5) = 0$$

where e.g. $p(s_0) = 0.31 + 0.43 + 0.24 + 0.15 = 0.5033$. In other words: assume a PPS selection of $\langle s_i, p_i \rangle \in C_\pi$ has been made, followed by a random selection with a uniform probability of $s_k \in \{s_0, ..., s_i\}$. In 50.33...% of the cases, this procedure will return s_0, i.e. $k = 0$.

We now prove that using the compact representation allows for tractable sampling of a possibility distribution.

Proposition 1. *Constructing C_π for π a possibility distribution over S requires an $O(n \log n)$ algorithm with $n = |S|$. Sampling of π based on C_π can be done in constant time.*

---

[1] A basic belief assignment, or bba, is a function of the form $m : 2^S \to [0, 1]$ satisfying $m(\emptyset) = 0$ and $\sum_{A \in 2^S} m(A) = 1$.

input   : a possibility distribution π over $\mathcal{S}$
result : a compact representation $\mathcal{C}_\pi$ of the possibility distribution π
           as given in Definition 1

initialise a list $\mathcal{C}_\pi$
sort π such that $\pi(s_0) \geq \pi(s_2) \geq \ldots \geq \pi(s_{n-1})$
if $\pi(s_{n-1}) > 0$ then $\pi(s_n) \leftarrow 0$
for $i \in [0,k]$ *with* $k = |\mathcal{S}_0|$ do
  if $\pi(s_i) > \pi(s_{i+1})$ then
      append $\langle i, \pi(s_{i+1}) - \pi(s_i) \rangle$ to $\mathcal{C}_\pi$
end
return $\mathcal{C}_\pi$

Algorithm 1. tractable construction of $\mathcal{C}_\pi$

Proof. Algorithm 1 requires the sorting of the possibility distribution π, with $O(n \log n)$ time complexity, followed by an iteration over all the states in $\mathcal{S}_0$, with $O(n)$ time complexity. Hence, the algorithm is an algorithm with $O(n \log n)$ time complexity. Sampling based on Algorithm 2 relies on a PPS selection and a selection using a uniform distribution, both of which can be implemented with $O(1)$ time complexity [21].   □

input   : a possibility distribution π over $\mathcal{S}$ and its compact transformation $\mathcal{C}_\pi$
result : a state $s \in \mathcal{S}$

$\langle i, p_i \rangle \leftarrow$ PPS selection from $\mathcal{C}_\pi$ given probability masses $\mathcal{P} = \{p_j \mid \langle j, p_j \rangle \in \mathcal{C}_\pi\}$
$idx \leftarrow$ random selection from $[1,i]$ using uniform probability $1(i+1)$
return s_{idx}

Algorithm 2. constant time sampling using $\mathcal{C}_\pi$

3.2   Backpropagation and Node Selection

With an efficient way of sampling a possibility distribution, only a few other issues need to be resolved in order to apply a MCTS-style algorithm to π-MDP. The first issue has to do with the information we keep track of, and which we update during the backpropagation. Particularly, each node will need to keep track of a tuple of the form $\langle u_*, u^* \rangle$ instead of a single quantitative reward as in the MDP setting. Here, u_* denotes the cautious qualitative utility of that particular node onwards, while u^* denotes the brave qualitative utility. From Eqs. 2 and 3 we know that we also have to keep track of the possibility $\Pi(\tau \mid s_0, (\delta))$ of the trajectory τ, i.e. we have to compute $\min_{t=0}^{h-1}\left(\pi(s_{t+1} \mid s_t, \delta_t(s_t))\right)$. This can be achieved during the selection, expansion and rollout phase by keeping track of the minimum of the possibility of the transition for each chance node we encounter.[2] Further in accordance with Eqs. 2 and 3, once a terminal state is encountered, or the horizon is reached, the preference $M(s_h)$ for this state s_h is determined. Based on the values $\Pi(\tau \mid s_0, (\delta))$ and $M(s_h)$, we can readily compute a tuple $\langle u_*, u^* \rangle$ with $u_* = \max(1 - \Pi(\tau \mid s_0, (\delta)), M(s_h))$ and $u^* = \min(\Pi(\tau \mid s_0, (\delta)), M(s_h))$. This information is then backpropagated where the tuple $\langle {}_j u_*, {}_j u^* \rangle$ of each node j is updated to $\langle \min(u_*, {}_j u_*), \max(u^*, {}_j u^*) \rangle$.

The second issue is how to select the best node to expand during the selection phase. As it turns out, we can readily apply UCB1 for this, even in the possibilistic setting. However, we do have to decide for each search whether we are pursuing a cautious or brave search. For a brave search, we use ${}_j u^*$ instead of $\overline{X}_j$. For a cautious search, we instead use ${}_j u_*$. When the computational budget is reached, an action is selected for which its node is a direct child of the root and there does not exist another such node with a higher quantitative utility ${}_j u^*$ (resp. ${}_j u_*$). An MCTS algorithm to solve π-MDP is given in Algorithm 3.[3]

Note that while a possibility-probability transformation is used for the purpose of sampling a possibility distribution, the induced probability distribution is not used at any other stage. Indeed, during selection/expansion/rollout it is the *possibility* of the trajectory that is computed. This possibility is combined with the preference of the terminal state to determine the *brave and cautious qualitative utility* of the trajectory. Similarly, backpropagation only takes these qualitative utilities into account when updating the node information, and it are these qualitative utilities that guide the choice of the best node to expand during the selection phase. This ensures that no unnecessary transformation bias is introduced.

Proposition 2. *The failure probability at the root, i.e. the probability of selecting a sub-optimal action, converges to 0 as the number of samples grows to infinity for MCTS applied to π-MDP.*

Proof. (sketch for $u^*(\cdot, s_0)$) Assume $h = 1$, i.e. a search tree with only one level of alternating action and state nodes. We have that the value associated with each action a_i is given by the maximum of the qualitative utility of each outcome of a_i. This utility is in turn given by the minimum of the possibility degree of the trajectory – which corresponds with the possibility of the outcome – and the preference of the final state. Since the qualitative brave utility associated with an action never decreases through repeated sampling, and since repeated sampling will explore all branches as the number of samples grows to infinity, we know that the qualitative utility of the returned action will never decrease after additional samples and that it converges in the limit to the optimal action. Indeed, since we are assuming a finite horizon and since the number of actions/states is finite, we know that the size of the tree is bounded. For $h > 1$ the only difference is

---

[2] To deal with uncertainty in MCTS, a dual-layered approach is used in the search tree. A *decision node*, or state, allows us to choose which action to perform. A *chance node*, or action, has a number of stochastic effects which are outside our control.

[3] An implementation of the algorithm proposed in Algorithm 3 is also available online, at https://github.com/kimbauters/sparsepi.

input : a π-MDP model, C_π^a for every π describing the outcome of an action a,
and root state s_0
result : the next best action a' to execute

create root node n with state s_0
while *within computational budget* **do**
$\quad$ $\Pi(\tau) \leftarrow 1$
$\quad$ **while** n *has untried actions and* n' *has children* **do**
$\quad\quad$ $a \leftarrow \text{select}(n')$
$\quad\quad$ /* select next state by sampling C_π^a $\quad\quad\quad\quad\quad\quad\quad\quad\quad\quad\quad$ */
$\quad\quad$ $n' \leftarrow \text{sample}(n, a, C_\pi^a)$
$\quad\quad$ $\Pi(\tau) \leftarrow \min(\Pi(\tau), \pi(n' \,|\, n, a))$
$\quad\quad$ $n \leftarrow n'$
$\quad$ **end**
$\quad$ **if** n' *has untried actions* **then**
$\quad\quad$ $a \leftarrow$ select an untried action
$\quad\quad$ $n'' \leftarrow \text{sample}(n', a, C_\pi^a)$ /* expand node $\quad\quad\quad\quad\quad\quad$ */
$\quad$ **end**
$\quad$ /* rollout by sampling from C_π as needed $\quad\quad\quad\quad\quad\quad\quad$ */
$\quad$ $n_{\text{end}} \leftarrow \text{rollout}(n'')$
$\quad$ backpropagate $(n_{\text{end}}, \max(\Pi(\tau), M(n_{\text{end}})))$
end
return $best\_action(n_0)$

Algorithm 3. MCTS algorithm for brave π-MDP

the increased size of the trajectory. However, since the possibility $\Pi(\tau \,|\, s_0, (\delta))$ of the trajectory τ as calculated in the algorithm is the same as the one in Eq. 3, the results readily hold as well for $h > 1$. $\qquad\qquad\qquad\qquad\qquad\qquad$ $\square$

Initial experimental results confirm the benefit of the online planner for π-MDP. We first consider a problem space where $|\mathcal{S}| = 2^{10}$, $|\mathcal{A}| = 11$, and assume a 50 iteration budget for the online planner. Both the online planner and the offline planner are used to solve the same problem 100 times, where the offline planner first needs to compute its optimal policy but can then repeatedly use that to quickly solve the problems. The online (offline) planner took on average 34.49 ms (9.48 ms) to find a solution with an average qualitative utility of 0.602 (0.792). In small scale examples like these, the offline planner benefits from being able to apply its policy near-instantaneously once computed to navigate the problem space. When increasing the state space to $|\mathcal{S}| = 2^{12}$ the online (offline) planner took on average 37.44 ms (169.96 ms) to find a solution with an average qualitative utility of 0.632 (0.762). When $|\mathcal{S}| = 2^{15}$ the online (offline) planner took on average 37.18 ms (9961.50 ms) to find a solution with an average qualitative utility of 0.658 (0.742). Although clearly these are not conclusive experiments, they already provide indications that the online planner better qualifies to navigate the increasingly large search space, although this comes at the cost of a reduced utility due to the search space not being fully explored.

A more comprehensive experimental evaluation falls beyond the scope of this paper, yet is planned for future work.

4   Hybrid MDPs

As shown in Example 1, real-life problems can have different types of uncertainty that we cannot accurately express using only a single kind of representation. Indeed, possibility theory is not good at modelling uncertainty due to underlying stochastic processes, while probability theory is not well-suited for modelling uncertainty due to a lack of information.[4] Other theories of uncertainty, such as Dempster-Shafer theory [19], are a proper extension of both possibility and probability theory. However, due to their computational complexity, they lend themselves poorly to the use in online anytime algorithms such as MCTS. Algorithms such as MCTS rely on the repeated exploration of different trajectories, which is made feasible by the tractability of the underlying phases (see Fig. 1a). Therefore, we instead present a model in which uncertainty can be explicitly expressed as either a possibility or probability.

In the *hybrid MDP model* the actions are partitioned into those with probabilistic and a possibilistic effects, so that each action can be described using the most appropriate representation. In particular, the transition function associates with every action a either a possibilistic, or a probabilistic distribution. Furthermore, we also keep track of a total reward and preference function, which allows us to derive either a qualitative or quantitative utility as needed.

Definition 3. *A hybrid MDP is defined as a tuple $\langle S, A, T, R, M, \gamma, \mathcal{L} \rangle$ such that (i) S is a finite set of states; (ii) A is the set of actions, with $A = A_P \cup A_\pi$, where A_P is the set of actions with probabilistic effects and A_π the set of actions with possibilistic effects; (iii) T is the transition function, where $T(s_t, a_t, s_{t+1})$ is the conditional probability (resp. possibility) of reaching the state s_{t+1} by performing action a_t at state s_t if $a_t \in A_P$ (resp. if $a_t \in A_\pi$); (iv) R and M are totally specified reward and preference functions over S; (v) γ is a discounting factor such that $0 \leq \gamma \leq 1$; and (vi) $\mathcal{L}$ is a possibility scale.*

Example 4. Looking back at Example 1, the hybrid MDP model allows us to describe, on the one hand, the effect of patrolling an area on herd movement for the grassland area as a probabilistic transition and, on the other hand, for the mountain area as a possibilistic transition. We have the action *patrol_grassland* with *herd_grassland* as precondition and the three probabilistic effects ¬*poaching*, *herd_mountain*, and *herd_marsh* with resp. probability 0.82, 0.11, and 0.07. We also have the action *patrol_mountain* with *herd_mountain* as precondition and the four possibilistic effects *herd_mountain*, *herd_grassland*,

---

[4] A common approach in probability theory to try to overcome this problem is to use subjective probabilities. However, in the more general POMDP/MOMDP settings this creates difficulties in its own right as subjective probabilities from the transitions are then combined with objective probabilities from the observation function.

$\neg poaching$, and $herd\_marsh$ with resp. possibilities 1, 1, 0.4, and 0.2. Rewards and preferences are fully specified for all state transitions and states, allowing us to express e.g. that preventing a herd from being poached in the grassland is a preferred state, and it results in a reward r (e.g. based on the number of saved animals). In other words, assuming an original state s, an action a, and a state s' where animals are saved, we can have $M(s') = 1$ and $R(s, a, s') = 15$.

A policy in the hybrid MDP setting is defined in the same way as for MDP and π-MDP. However, unlike in MDP or π-MDP, the value of a policy in a hybrid MDP is not given by a single value but by a tuple consisting of both a reward and (brave/cautious) qualitative utility. We have:

Definition 4. *The utility w of a policy (δ) in a hybrid MDP model is given by:*

$$w((\delta), s_0) = \begin{cases} \langle v((\delta), s_0), u_*((\delta), s_0)\rangle & cautious\ setting \\ \langle v((\delta), s_0), u^*((\delta), s_0)\rangle & brave\ setting \end{cases}$$

with $v((\delta), s_0)$, and $u_*((\delta), s_0)$ and $u^*((\delta), s_0)$, computed over respectively the MDP and π-MDP induced by the hybrid MDP, as explained next.

An optimal policy for a hybrid MDP is defined in Sect. 5. Notice that, since we are using both $u(\cdot)$ and $v(\cdot)$ in the policy, the computation of a qualitative/quantitative reward requires an MDP/π-MDP. We thus need an efficient way of deriving the MDP and π-MDP that underlie a hybrid MDP. Obtaining the MDP $\langle \mathcal{S}, \mathcal{A}, T^*, R, \gamma \rangle$ is straightforward. Indeed, $\mathcal{S}$, $\mathcal{A}$, R, and γ are as specified in the hybrid MDP. For those actions $a \in \mathcal{A}_\pi$ that are possibilistic, we discussed in Sect. 3.1 how we can use the DPY transformation to transform a possibility distribution $\pi(\cdot) = T(s_t, a, \cdot)$, with $a \in \mathcal{A}_\pi$, into a probability distribution $p^*(\cdot) = yager(T(s_t, a, \cdot))$. As before, we do not need to explicitly compute the associated probability distribution, as we can have the MCTS algorithm indirectly rely on probabilities through the sampling process. This time around, and contrary to Sect. 3, the probabilistic sampling of a possibility distribution is taken into account (indirectly, as part of the sampling process) and used to determine the probability of the trajectory.

While deriving the MDP that underlies a hybrid MDP is straightforward, it is more complicated to derive the underlying π-MDP $\langle \mathcal{S}, \mathcal{A}, T^{**}, M, \mathcal{L} \rangle$. A simplifying factor – as in the MDP case – is that $\mathcal{S}$, $\mathcal{A}$, M, and $\mathcal{L}$ are as specified in the hybrid MDP model. Hence, the derivation of the underlying π-MDP only requires a transformation of the probability distributions used in the hybrid MDP model into possibility distributions. Many such transformations exist, applying to either objective or subjective probability distributions. Since we assume in the hybrid MDP model that subjective information is best represented using a possibility distribution, we can conversely assume that we are dealing with an objective probability distribution. As such, the transformation of a probability distribution into a possibility distribution should be based on preserving as much information as possible. One such a transformation [9] is defined as follows. Given a probability distribution p such that $p(s_0) \geq p(s_1) \geq \ldots \geq p(s_k)$ we have that the associated possibility distribution π is defined as $\pi(s_j) = \sum_{i=j,...,k} P(s_i)$. Furthermore, for equiprobable elements it is enforced that the corresponding elements are also equipossible [9]. Computing the associated possibility distribution can be done using an $O(n \log n)$ time complexity algorithm and lookups require $O(\log n)$. As a final step, all results are rounded up to the nearest element in $\mathcal{L}$. In conclusion, for those actions $a \in \mathcal{A}_P$ that are probabilistic, we transform the probability distribution $p(\cdot) = T(s_t, a, \cdot)$, with $a \in \mathcal{A}_P$, into a possibility distribution $\pi^{**}(\cdot) = dubois(T(s_t, a, \cdot))$ by using the transformation outlined in this paragraph.

Example 5. Consider the probability distribution p such that

$$p(s_0) = 0.7 \qquad p(s_1) = 0.1 \qquad p(s_2) = 0.1$$
$$p(s_3) = 0.07 \qquad p(s_4) = 0.03 \qquad p(s_5) = 0$$

Assuming $\mathcal{L} = \{0k, 1/k, ..., k/k\}$ with $k = 20$ we have the resulting π:

$$\pi(s_0) = 20/20 \qquad \pi(s_1) = 6/20 \qquad \pi(s_2) = 6/20$$
$$\pi(s_3) = 2/20 \qquad \pi(s_4) = 1/20 \qquad \pi(s_5) = 0/20$$

where π is the possibility distribution associated with p.

We can now treat a hybrid MDP as a π-MDP in a similar way as we treated a hybrid MDP as an MDP. The induced possibility distribution is used to determine the possibility of the trajectory, and the (qualitative) reward is obtained by ignoring the reward values and only relying on preference values. Importantly, it should be noted that the transformation to a possibility distribution is *only* used to compute the possibility $\Pi(\tau \mid s_0, (\delta))$ of the trajectory τ, and *not* for sampling purposes. This is because, in general, transforming a probability distribution into a possibilistic one and back to a probability distribution does not result in the same distribution, thus introducing an avoidable bias.

5   Solving Hybrid MDPs

We show in this section that an online anytime MCTS-style algorithm can be used to solve hybrid MDPs. In particular, we show that the main difficulty lies in the selection phase. Indeed, in the selection phase we need to select the next best node to expand based on the utility of a hybrid MDP, which is a tuple composed of a quantitative and qualitative utility. However, selection strategies – such as UCB1 – require a single value to assess the value of a node. We thus need methods that enforce commensurability and allow us to combine both utility values into a single value $w_\downarrow$. This ties in with Definition 4, where we so far have not defined what an optimal policy is in the hybrid MDP setting. To be able to define an optimal policy, and to define $w_\downarrow$, we propose a number of rationality postulates. These postulates are used to motivate reasonable methods to ensure commensurability. Still, as we will see, some degree of freedom remains. This is to be expected given that the meaning of the qualitative utilities, and especially their inter-relation with quantitative utilities, is domain-dependent.

To simplify the description of the postulates, we assume a straightforward transformation of the quantitative utilities to map them onto the interval $[0, 1]$. The maximum reward, or quantitative utility, is given by $\max_v v((\delta), s_0)$ with (δ) the optimal policy. In a similar way, we can define the worst reward. The transformation then consists of rescaling all rewards accordingly onto $[0, 1]$ where a value of 1 (resp. 0) denotes the most (resp. least) preferred outcome. Notice that while this transformation may appear to give the quantitative utilities a qualitative interpretation, it does not allow us to directly compare quantitative and qualitative utilities (e.g. they still have different neutral elements[5]).

We can now introduce the postulates that motivate the characteristics of $w_{\downarrow}$.

P1:   Let $w(a) = \langle 1, 1 \rangle$, then a is the most preferred action, i.e. $\nexists a' \cdot w_{(}a') > w_{\downarrow}(a)$.

   When an action has the highest possible value and is qualitatively valued higher than others, then it must be the most preferred action.

P2:   Let $w(a) = \langle 0, 0 \rangle$, then a is the least preferred action, i.e. $\nexists a' \cdot w_{\downarrow}(a') < w_{\downarrow}(a)$.

   When an action has the lowest possible value and is qualitatively valued lower than others, then it must be the least preferred action.

Postulates **P1–P2** are the base cases with a clear consensus among both utilities. The next two postulates identify the monotonicity w.r.t. a single value change:

P3:   Let $w(a) = \langle v, u \rangle$ and $w(a') = \langle v', u' \rangle$ with $v = v'$ and $u > u'$, then $w_{\downarrow}(a) > w_{\downarrow}(a')$.

P4:   Let $w(a) = \langle v, u \rangle$ and $w(a') = \langle v', u' \rangle$ with $u = u'$ and $v > v'$, then $w_{\downarrow}(a) > w_{\downarrow}(a')$.

   When the qualitative (resp. quantitative) utilities of two actions are the same, the action with the highest quantitative (resp. qualitative) utility must be the preferred one.

Commensurability of the qualitative and quantitative utility is the most difficult when there is a level of disagreement, e.g. a high reward obtained in a disliked state. Dissension between the utilities reaches its maximum when they take on their neutral elements, or the neutral elements of the other utility. In such cases, the exact interpretation of the dissension is dependent on whether we are using brave/cautious reasoning for the qualitative utility:

P5:   (brave reasoning, u^*) When $w(a) = \langle 0, 1 \rangle$, then we are ignorant (*eq.* have a weak conflict) about a. When $w(a) = \langle 1, 0 \rangle$, then we have a (strong) conflict about a.

---

[5] We use the terminology of a neutral elements loosely here to indicate that a reward of 0, and a preference of 1, are the defaults. Indeed, when rewards (resp. preferences) are omitted these are the values MDPs (resp. π-MDPs) default to.

P5′: (cautious reasoning, u_*) When $w(a) = \langle 1, 0 \rangle$, then we are ignorant (eq. have a weak conflict) about a. When $w(a) = \langle 0, 1 \rangle$, then we have a (strong) conflict about a.

Ignorance, or the situation with weak conflict, reflects that the values are their own neutral elements and thus convey no real information. Strong conflict suggests that the utilities disagree with each other by taking on the strongest (opposing) values in their scale. In the cautious setting, these notions are flipped around.

Interestingly, postulates **P1–P5** tell us nothing about how to relate two actions a, a' for which $w(a) = \langle 0, 1 \rangle$ and $w(a') = \langle 1, 0 \rangle$ (i.e. when there is a discord between the utility measures), which is the degree of freedom we have. Simple behaviour can be obtained by e.g. defining $w_\downarrow(a) = u^n \cdot v$ with $w(a) = \langle v, u \rangle$. Here, n is a parameter indicating the strength of the qualitative utility over the quantitative utility with values of $n > 1$ placing a higher importance on the qualitative utility. An implicit effect of a formula of this form is that it equates ignorance with conflict. Since we are using a product, we are also stressing the weakest component. However, simple methods like these fail to satisfy a final desirable postulate:

P6: A hybrid MDP is a proper extension of both an MDP and a π-MDP.

When the hybrid MDP only models probabilistic (resp. possibilistic) information, the result of solving the hybrid MDP should be identical to solving the identical MDP (resp. π-MDP).

Postulates **P1–P6** suggest a lexicographic ordering, based on whether we are dealing with brave or cautious qualitative reasoning, and on whether or not qualitative information takes priority over quantitative information.

Definition 5 (lexicographic ordering). *Let $w(a) = \langle v, u \rangle$ and $w(a') = \langle v', u' \rangle$. For a probabilistic lexicographic ordering, we have that $w(a) \geq_p w(a')$ iff $v > v'$, or $v = v'$ and $u \geq u'$. For a possibilistic lexicographic ordering, we instead have $w(a) \geq_\pi w(a')$ iff $u > u'$, or $u = u'$ and $v \geq v'$.*

Definition 6 (optimal policy). *Let $w_\downarrow$ be a function agreeing with Definition 5, i.e. such that $w_\downarrow(a) \geq w_\downarrow(a')$ iff $w(a) \geq w(a')$, where the last ordering is either $\geq_p$ or $\geq_\pi$. Then the optimal policy in a hybrid MDP induced by $w_\downarrow$ is the one that maximises $w_\downarrow(w(\cdot, s_0))$.*

As previously mentioned, the best choice for $w_\downarrow$ is application-specific and depends on the domain, and on the probability-to-possibility transformation used.

Proposition 3. *Let $w_\downarrow$ be a probabilistic or possibilistic lexicographic ordering. We then have that $w_\downarrow$ satisfies postulates **P1–P6**.*

Proof. It readily follows that postulates **P1** and **P2** are satisfied. Indeed, no action will be less preferred than an action a such that $w(a) = \langle 0, 0 \rangle$, or more preferred than an action a' such that $w(a') = \langle 1, 1 \rangle$ since $v, u \in [0, 1]$ for $w(\cdot) = \langle v, u \rangle$. Postulates **P3** and **P4** are satisfied directly by the choice of the lexicographic ordering. For a probabilistic lexicographic ordering we get, by definition, that $w(a) \geq_p w(a')$ if $v > v'$ which agrees with **P4**. Equivalently, a possibilistic lexicographic ordering agrees with **P3**. Postulates **P5** and **P5'** do not impose any constraints but follow directly from the underlying theories of uncertainty. Postulate **P6** holds since an MDP (resp. π-MDP) represented as a hybrid MDP can assign the neutral element for preference (resp. reward) to the action. Hence, for an MDP with a probabilistic lexicographic ordering, we get for all actions a that $w(a) = \langle 1, u \rangle$. Hence $w(a) > w(a')$ if and only if $u > u'$. The same line of reasoning applies for a π-MDP with a possibilistic lexicographic. $\square$

In the MCTS algorithm, w is used during node selection to merge the qualitative and qualitative utility of the node into a single value to be used by e.g. UCB1. The other three phases in the MCTS algorithm, expansion, rollout, and backpropagation, also need to be altered slightly. During *expansion/rollout*, both the probability and possibility of a trajectory needs to be computed. To compute the probability of the trajectory we can rely on sampling, which is available for both probabilistic and possibilistic transitions given the results in Sect. 3.1. To obtain the possibility of a trajectory, specifically when encountering a probabilistic transition during the expansion/rollout phase, the associated possibility of the trajectory is computed as discussed in the previous section. Throughout the *expansion/rollout* we also need to keep track of the rewards and preferences. Both are readily available since the reward and preference functions are total. Once a terminal node is reached, given the possibility/probability of the trajectory as well as the rewards/preferences, the qualitative/quantitative utility is calculated. The *backpropagation* phase then simply combines the backpropagation of both the MDP and π-MDP approaches to update both the qualitative/quantitative utility for each node along the trajectory.

6   Related Work and Future Work

One of the first people to discuss possibilistic MDP is Sabbadin [17,18]. In those works, the author introduces the π-MDP model in which either an optimistic or pessimistic form of reasoning can be used. Applying the optimistic approach might lead to unsatisfactory states, whereas the pessimistic version offers some guarantee that this will not happen. However, as discussed later by Drougard et al. [3], for problems where there is no risk of being blocked in an unsatisfactory state the optimistic version is generally preferred. Furthermore, optimality of an algorithm is easier to prove in the optimistic version. Our implementation of π-MDP demonstrates similar behaviour, where a brave version has a high chance of getting trapped in deadlock states, while the cautious version very often reaches its goals notwithstanding.

Another significant contribution in [17] is that the author also introduced the π-POMDP model, and shows that a finite translation of a π-POMDP to a

π-MDP exists. This is unlike the probabilistic setting where no finite translation exists between a POMDP and an MDP, and it allows algorithms used to solve a π-MDP to be used without modifications to also solve π-POMDP. In practice, however, the exponential size of the π-POMDP makes it infeasible to find solutions in a reasonable amount of time for anything but the smallest problems. These problems were addressed in [3], and later in [4], where the authors present the π-MOMDP framework. In this new framework only a subset of the states are partially observable. This significantly reduces the belief space, allowing for optimal algorithms to be applicable in practice. One such algorithm was presented in [4] based on using a factored representation of the original problem. In future work we intend to explore whether the online algorithm presented in the current paper can similarly be applied to π-MOMDP and π-POMDP.

The work in [11] is one of the earliest works to discuss how sampling techniques can be used to solve MDPs. The strength of such techniques lies in their ability to find near-optimal solutions in only a fraction of the time of other approaches. However, only after the seminal work by Kocsis and Szepesvári [14] were sampling-based planning taken seriously in the community. The main improvement proposed in [14] is to employ a multi-bandit approach, as described by the UCB1 procedure [1], to offer an effective balance between exploration and exploitation. This allows to considerably speed up the search process as the most promising part of the tree is more quickly and more profoundly explored. Not long after, the term *Monte-Carlo Tree Search* was coined to describe these sampling-based approaches to planning.

The problem of how to transform a probability distribution into a possibility distribution, and vice versa, has been addressed in a large body of papers (e.g. [5,8,9,13,20,22]). Most generally, a possibility distribution is seen as a (very large) family of probability distributions. The problem therefore boils down to choosing, and motivating, one probability distribution. Intrinsically, such a choice is based on extra information that is external to the possibility distribution. Transforming a probability distribution to a possibility distribution always implies some loss of information, leading to a range of different transformations based on different consistency principles. A concise overview of these and other issues related with both directions of the transformation is given in [7]. How the use of different transformations than the one used in this paper affects the results in a hybrid MDP setting is a topic of interest for future work.

7 Conclusions

This paper introduced a novel approximate way for solving possibilistic MDP (π-MDP) models, based on the established Monte-Carlo Tree Search (MCTS) algorithms for solving MDPs. We found that the applicability of MCTS for solving π-MDP depends on the ability to quickly sample possibility distributions. By introducing a new compact data structure that represents the DPY transformation of a possibility distribution into a probability distribution, we showed that constant time sampling is indeed feasible. Furthermore, we proposed a hybrid

MDP model in which we can encode both probabilistic, as well as possibilistic transitions. This allows us to express different facets of uncertainty in the hybrid MDP model. In addition, we showed how a modified version of MCTS can also be applied to solve hybrid MDP models. A central component of this modification is the need to relate the qualitative and quantitative utility. We showed that, while the exact procedure to combine these utilities is application-dependent, such procedures should adhere to a number of rationality postulates. In particular, the postulates enforce that any algorithm to solve a hybrid MDP can also be used to solve either a π-MDP and MDP. Finally, algorithms and computational complexity results of all the main components are presented throughout the paper to highlight the applicability of our approach.

Acknowledgements. This work is partially funded by EPSRC PACES project (Ref: EP/J012149/1). Special thanks to Steven Schockaert who read an early version of the paper and provided invaluable feedback. We also like to thank the reviewers for taking the time to read the paper in detail and provide feedback that helped to further improve the quality of the paper.

References

1. Auer, P., Cesa-Bianchi, N., Fischer, P.: Finite-time analysis of the multiarmed bandit problem. Mach. Learn. **47**(2–3), 235–256 (2002)
2. Bellman, R.: A Markovian decision process. Indiana Univ. Math. J. **6**, 679–684 (1957)
3. Drougard, N., Teichteil-Königsbuch, F., Farges, J., Dubois, D.: Qualitative possibilistic mixed-observable MDPs. In: Proceedings of the 29th Conference on Uncertainty in Artificial Intelligence (UAI 2013) (2013)
4. Drougard, N., Teichteil-Königsbuch, F., Farges, J., Dubois, D.: Structured possibilistic planning using decision diagrams. In: Proceedings of the 28th AI Conference on Artificial Intelligence (AAAI 2014), pp. 2257–2263 (2014)
5. Dubois, D., Prade, H.: On several representations of an uncertain body of evidence. In: Gupta, M.M., Sanchez, E. (eds.) Fuzzy Information and Decision Processes, pp. 167–181. North-Holland, Amsterdam (1982)
6. Dubois, D., Prade, H.: Unfair coins and necessity measures: towards a possibilistic interpretation of histograms. Fuzzy Sets Syst. **10**(1), 15–20 (1983)
7. Dubois, D., Prade, H.: Possibility theory and its application: where do we stand? Mathware Soft Comput. **18**(1), 18–31 (2011)
8. Dubois, D., Prade, H., Sandri, S.: On possibility/probability transformation. In: Proceedings of the 4th International Fuzzy Systems Association Congress (IFSA 1991), pp. 50–53 (1991)
9. Dubois, D., Prade, H., Smets, P.: New semantics for quantitative possibility theory. In: Benferhat, S., Besnard, P. (eds.) ECSQARU 2001. LNCS (LNAI), vol. 2143, pp. 410–421. Springer, Heidelberg (2001)
10. Kaufmann, A.: La simulation des sous-ensembles flous. In: Table Ronde CNRS-Quelques Applications Concrètes Utilisant les Derniers Perfectionnements de la Théorie du Flou (1980)

11. Kearns, M., Mansour, Y., Ng, A.: A sparse sampling algorithm for near-optimal planning in large Markov decision processes. In: Proceedings of the 16th International Joint Conference on Artificial Intelligence (IJCAI 1999), pp. 1324–1231 (1999)
12. Keller, T., Eyerich, P.: PROST: probabilistic planning based on UCT. In: Proceedings of the 22nd International Conference on Automated Planning and Scheduling (ICAPS 2012) (2012)
13. Klir, G.: A principle of uncertainty and information invariance. Int. J. Gen. Syst. **17**(2–3), 249–275 (1990)
14. Kocsis, L., Szepesvári, C.: Bandit based monte-carlo planning. In: Fürnkranz, J., Scheffer, T., Spiliopoulou, M. (eds.) ECML 2006. LNCS (LNAI), vol. 4212, pp. 282–293. Springer, Heidelberg (2006)
15. Kolobov, A., Mausam, Weld, D.: LRTDP versus UCT for online probabilistic planning. In: Proceedings of the 26th AAAI Conference on Artificial Intelligence (AAAI 2012) (2012)
16. Rao, A., Georgeff, M.: Modeling rational agents within a BDI-architecture. In: Proceedings of the 2nd International Conference on Principles of Knowledge Representation and Reasoning (KR 1991), pp. 473–484 (1991)
17. Sabbadin, R.: A possibilistic model for qualitative sequential decision problems under uncertainty in partially observable environments. In: Proceedings of the 15th Conference on Uncertainty in Artificial Intelligence (UAI 1999), pp. 567–574 (1999)
18. Sabbadin, R., Fargier, H., Lang, J.: Towards qualitative approaches to multi-stage decision making. Int. J. Approximate Reasoning **19**(3), 441–471 (1998)
19. Shafer, G., et al.: A Mathematical Theory of Evidence. Princeton University Press, Princeton (1976)
20. Smets, P.: Constructing the pignistic probability function in a context of uncertainty. In: Proceedings of the 5th Annual Conference on Uncertainty in Artificial Intelligence (UAI 1989), pp. 29–40 (1989)
21. Vose, M.: A linear algorithm for generating random numbers with a given distribution. IEEE Trans. Softw. Eng. **17**(9), 972–975 (1991)
22. Yager, R.: Level Sets for Membership Evaluation of Fuzzy Subset, in Fuzzy Sets and Possibility Theory - Recent Developments, pp. 90–97. Pergamon Press, NewYork (1982)

Possibilistic Conditional Tables

Olivier Pivert[1] and Henri Prade[2,3]([✉])

[1] Irisa – Enssat, University of Rennes 1, Technopole Anticipa,
22305 Lannion Cedex, France
pivert@enssat.fr
[2] IRIT, CNRS and University of Toulouse, 31062 Toulouse Cedex 9, France
[3] QCIS, University of Technology, Sydney, Australia
prade@irit.fr

Abstract. On the one hand possibility theory and possibilistic logic offer a powerful representation setting in artificial intelligence for handling uncertainty in a qualitative manner. On the other hand conditional tables (*c*-tables for short) and their probabilistic extension provide a well-known setting for representing respectively incomplete and uncertain information in relational databases. Although these two settings rely on the idea of possible worlds, they have been developed and used independently. This paper investigates the links between possibility theory, possibilistic logic and *c*-tables, before introducing possibilistic *c*-tables and discussing their relation with a recent certainty-based approach to uncertain databases and their differences with probabilistic *c*-tables.

1 Introduction

The representation and the handling of imperfect information, be it incomplete or uncertain, has led for several decades to the development of important research trends in artificial intelligence, such as nonmonotonic reasoning, belief revision, reasoning under uncertainty, or information fusion. Different formalisms have been introduced for representing uncertainty which depart from probability theory, and which are particularly of interest when information is both incomplete and uncertain, such as belief function theory or possibility theory. Moreover, possibility theory [11] may have a qualitative flavor, which may be appropriate when uncertainty is hard to assess precisely.

On its side, database research first tackled the issue of managing imperfect/incomplete information a long time ago, with the works of Codd [7] and Lipski [20] in the late seventies. Researchers first focused their attention on so-called *null* values (either totally unknown or not applicable), before considering more sophisticated cases of incompleteness. The pioneering work by Imielinski and Lipski [18] considers three models: Codd tables, *v*-tables and *c*-tables. *v*-tables are conventional instances where variables can appear in addition to constants from the considered domains. Codd tables are *v*-tables in which all the variables are distinct. They correspond roughly to the current use of nulls in SQL, while *v*-tables model "labeled" or "marked" nulls. As to *c*-tables, they are

M. Gyssens and G. Simari (Eds.): FoIKS 2016, LNCS 9616, pp. 42–61, 2016.
DOI: 10.1007/978-3-319-30024-5_3

v-tables extended with conditions on variables and constitute a very powerful model, even though they raise computational complexity issues.

The last decade witnessed a renewal of interest in the modeling and management of uncertain databases (in particular, probabilistic ones, see, e.g., [27]), which brought the c-tables model back in the spotlights. As pointed out by Antova et al. in [3], the three most prominent probabilistic database models proposed recently, namely ULDBs [4,5], U-relations [3] and the model proposed in [8] based on virtual views, are different versions of the concept of a c-table. See also the works by Green and Tannen who proposed a probabilistic c-tables model [17] and by Shimizu et al. [26].

Even though most of the literature about uncertain databases uses probability theory as the underlying uncertainty model, this type of modeling is not always so easy, as recognized in the introductory chapter of [27]: "Where do the probabilities in a probabilistic database come from? And what exactly do they mean? The answer to these questions may differ from application to application, but it is rarely satisfactory." This is one of the reasons why some authors have proposed approaches that rather rest on another model, namely possibility theory [10,28], which is appropriate for representing epistemic uncertainty.

The idea of applying possibility theory to the modeling of uncertain databases goes back to the early 1980's [23–25]. At that time, the approach was to represent ill-known attribute values by possibility distributions, and then given a query, to compute the fuzzy set of answers that are *certain* (or sure) to some extent, and the larger fuzzy set of answers that are *possible* to some extent. This was an attempt at providing a *graded* counterpart to the modal logic-based approach proposed by Lipski [20,21]. This possibility distribution-based representation of the information available about attribute values was covering the cases of null values, of classical subset restrictions on possible values, and more generally of fuzzy subset restrictions when all possible values are not equally plausible; it had nothing to do with the notion of c-tables. Note also that possibilistic logic [9], which is useful for having a possibilistic reading of c-tables as shown in the following, has been introduced later. Then the possibility distribution-based representation of ill-known attribute values has become the standard approach in possibility theory for handling databases with missing, imprecise, or uncertain information until now.

Recent advances on this approach can be found in [6]. In contrast with probability theory, one expects the following advantages when using possibility theory:

- the qualitative nature of the model makes easier the elicitation of the degrees or levels attached to candidate values, inasmuch as an ordinal scale $\mathcal{L}$ made of $k + 1$ linguistic labels may be used to assess the certainty (and possibility) levels attached to an attribute value or a tuple. For instance, with $k = 4$, one may use:

$$\alpha_0 = \text{``not at all''} < \alpha_1 = \text{``somewhat''} <$$
$$\alpha_2 = \text{``rather''} < \alpha_3 = \text{``almost''} < \alpha_4 = \text{``totally''}$$

where α_0 (resp. α_k) corresponds to 0 (resp. 1) in the unit interval when a numerical framework is used. In possibility theory [10] the necessity

(or certainty) $N(E)$ of an event E is defined as the impossibility of the opposite event, i.e., $N(E) = 1 - \Pi(\overline{E})$. Then the operation $1 - (\cdot)$ that is used when the degrees belong to the unit interval is replaced by the order reversal operation denoted by $rev(\cdot)$: $rev(\alpha_i) = \alpha_{k-i}$. In the following, however, we use the numerical scale $[0, 1]$ in order not to make the formulas too cumbersome.
 - in probability theory, the fact that the sum of the degrees from a distribution must equal 1 makes it difficult to deal with incompletely known distributions.

More recently, the authors of the present paper developed a new idea which is to use the notion of necessity from possibility theory to qualify the *certainty* that an ill-known piece of data takes a given value [22]. In contrast with both probabilistic databases *and* possibilistic ones in the sense of [6,25], — which can be referred to as the "full possibilistic" approach —, the main advantage of the certainty-based model lies in the fact that operations from relational algebra can be extended in a simple way and with a data complexity which is the same as in a classical database context (i.e., when all data are certain).

However, the model defined in [22] does not yield the same expressive power as *c*-tables inasmuch as it only aims to represent the more or less certain values (or disjunction of values) that some attributes in a tuple may take, but does not consider conditions attached to tuples as *c*-tables do. Here, our main objective is twofold (i) first to establish the link between c-tables and possibilistic representations, two settings that have remained unrelated in spite of their close concerns, and (ii) to propose a possibilistic counterpart of the probabilistic *c*-tables model defined in [17], which will make it possible to deal with more complex cases of imprecise and uncertain information. The objective of this first paper relating *c*-tables and possibility theory is to discuss the ideas underlying the two settings, illustrating them on suitable examples, and identifying the key principles for bridging them. More precisely, we intend to:

 - show how regular *c*-tables can be interpreted, then extended, in the setting of possibilistic logic;
 - establish the link between the extended relational model presented in [22], based on possibilistic certainty, and that of possibilistic *c*-tables.

The remainder of the paper is structured as follows. Section 2 is devoted to a presentation of the *c*-tables model. Section 3 provides a refresher on possibility theory. Section 4 presents the concept of a possibilistic *c*-table. It first shows how regular *c*-tables may be interpreted in terms of possibilistic logic (in the particular case where necessity degrees equal 0 or 1), then describes a gradual extension, discusses the certainty-based model as a particular case, and finally compare possibilistic *c*-tables and probabilistic ones as introduced in [17]. Finally, Sect. 5 recalls the main contributions and outlines some research perspectives.

2   Refresher on *c*-tables

As mentioned above, conditional tables (*c*-tables for short) are *v*-tables extended with conditions on variables. Let us recall the definition from [18].

Table 1. Conditional table from Example 1

Student	Course	
	$(x \neq math) \wedge (x \neq CS)$	
Sally	math	$(z = 0)$
Sally	CS	$(z \neq 0)$
Sally	x	
Alice	biology	$(z = 0)$
Alice	math	$(x = physics) \wedge (t = 0)$
Alice	physics	$(x = physics) \wedge (t \neq 0)$

Let $\mathcal{U}$ be a fixed, finite set of *attributes*. Attributes are denoted by A, B, C, and sets of attributes by X, Y, Z. Associated with every $A \in \mathcal{U}$ is an *attribute domain* $D(A)$. We denote $D = \bigcup_{A \in \mathcal{U}} D(A)$. Elements of D are called *constants*. For every $A \in \mathcal{U}$, let $V(A)$ be a countably infinite set of symbols called *variables*. It is assumed that $V(A) \cap D = \emptyset$, $V(A) \cap V(B) = \emptyset$ if $D(A) \neq D(B)$ and $V(A) = V(B)$ if $D(A) = D(B)$. Let us denote by $\mathcal{S}$ the set of all expressions built up from atomic conditions of the form $(x = a)$, $(x = y)$, *false* and *true*, where for some $A \in \mathcal{U}$, $a \in D(A)$, x, $y \in V(A)$, by means of $\neg$, $\vee$, and $\wedge$. In the following, we use the notation $x \neq a$ (resp. $x \neq y$) for $\neg(x = a)$ (resp. $\neg(x = y)$). For every condition $\Phi \in \mathcal{S}$, we say that a valuation v satisfies Φ if its assignment of constants to variables makes the formula true.

Definition 1. *By a c-tuple on a set of attributes X, we mean any mapping t defined on $X \cup \{con\}$ such that $t(X)$ is a V-tuple (i.e., any mapping t' that associates an element $t'(A) \in D(A) \cup V(A)$ with every $A \in X$) and $t(con) \in \mathcal{S}$ is the condition associated with $t(X)$. A conditional table (or c-table) on X is any finite set T of c-tuples on X.*

In this original definition, conditions may only be attached to individual tuples. An extension is to consider that a condition may also be attached to the c-table globally, see [1]. The definition then becomes:

Definition 2. *A conditional table (or c-table) on X is a pair (T, Φ_T) where T is any finite set of c-tuples on X and $\Phi_T \in \mathcal{S}$ is a global condition on T.*

In the following, we denote by φ_t the condition associated with tuple t.

Example 1. Suppose we know that Sally is taking math or computer science (CS) (but not both) and another course; Alice takes biology if Sally takes math, and math or physics (but not both) if Sally takes physics. This can be represented by the c-table depicted in Table 1.

Remark 1. Notice that it is possible to use disjunctions that are not mutually exclusive. For instance, if Sally could take math or physics or both, the first two lines of Table 1 would have to be replaced by the four lines:

Sally	math	$x = 0$
Sally	CS	$x = 0$
Sally	math	$x \neq 0 \wedge y = 0$
Sally	CS	$x \neq 0 \wedge y \neq 0$

□

Adopting the closed world assumption, a given c-table T represents a set of instances (possible worlds) as follows:

$rep(T) = \{I \mid$ there is a valuation ν satisfying Φ_T such that relation I
$\qquad$ consists exactly of those facts $\nu(t)$ for which ν satisfies $\varphi_t\}$.

In the context of uncertain databases, the notion of a *strong representation system* plays an important part. Let us recall its definition [1]. Consider some particular representation system (e.g., tables). Such a system involves a language for describing representations and a mapping *rep* that associates a set of instances with each representation. Suppose that we are interested in a particular query language $\mathcal{L}$ (e.g., relational algebra). We would like to be capable of representing the result of a query in the same system. More precisely, for each representation T and query q, there should exist a computable representation $\bar{q}(T)$ such that

$$rep(\bar{q}(T)) = q(rep(T)). \tag{1}$$

In other words, $\bar{q}(T)$ represents the possible results of q, i.e., $\{q(I) \mid I \in rep(T)\}$. In such a context, a *possible answer* (resp. a *certain answer*) to a query q is a tuple that belongs to at least one world (resp. to all of the worlds) of $\bar{q}(T)$:

$$t \in poss(q) \Leftrightarrow t \in \bigcup_{I \in rep(\bar{q}(T))} I \tag{2}$$

$$t \in cert(q) \Leftrightarrow t \in \bigcap_{I \in rep(\bar{q}(T))} I \tag{3}$$

Example 2. If we consider Table 1, the possible (resp. certain) answers to the query "find the students who take math or computer science" are {Sally, Alice} (resp. {Sally}).   ◇

If some representation system τ possesses property (1) for a query language $\mathcal{L}$, then τ is said to be a *strong representation system* for $\mathcal{L}$. It has been proven that c-tables are a strong representation system for relational algebra [18]. For each operation u of relational algebra, [18] defines an operation $\bar{u}$ on c-tables. Hereafter, we recall these definitions using the easier-to-read presentation from [17]. For projection, we have

$$\bar{\pi}_\ell(T) := \{(t' : \varphi_{t'}) \mid t \in T \text{ s.t. } \pi_l(t) = t', \ \varphi_{t'} = \bigvee \varphi_t\}$$

Table 2. Conditional table T (left) and result of $\sigma_{B \neq b}(T)$ (right)

B	C	
y	c	$(y \neq b)$
z	w	
b	x	

B	C	
y	c	$(y \neq b)$
z	w	$(z \neq b)$

Table 3. Conditional table from Example 3

Course	Room	
math	Fermat	$(u = 0)$
math	Gödel	$(u \neq 0)$
CS	Turing	$(v \neq 0)$
CS	Gödel	$(v = 0) \wedge (u = 0)$
CS	Fermat	$(v = 0) \wedge (u \neq 0)$
biology	Turing	$(v = 0)$
biology	Monod	$(v \neq 0)$
physics	Einstein	

where ℓ is a list of indexes and the disjunction is over all t in T such that $\pi_\ell(t) = t'$. For selection, we have

$$\bar{\sigma}_c(T) := \{(t : \varphi_t) \wedge c(t)) \mid (t, \varphi_t) \in T\}$$

where $c(t)$ denotes the result of evaluating the selection predicate c on the values in t (this is in general a Boolean formula on constants and variables). See Table 2 for an example.

For cross product and union, we have

$$T_1 \bar{\times} T_2 := \{(t_1 \times t_2 : \varphi_{t_1} \wedge \varphi_{t_2}) \mid t_1 \in T_1, \ t_2 \in T_2\}$$
$$T_1 \bar{\cup} T_2 := T_1 \cup T_2$$

Difference and intersection are handled similarly.

Example 3. Here is an example involving a join between two conditional tables. Let us assume that math courses take place either in Fermat room or in Gauss room. The CS course takes place in Turing room if the biology course does not use it, otherwise it takes place in Gödel room or in Fermat room. The course either takes place in Monod room or in Turing room. Physics is in Einstein room.

We are interested in knowing in which room each student may attend a course, which implies joining Tables 1 and 3. The result is represented in Table 4. For instance, the first line expresses that Sally will be in Fermat room if she takes math and math is in Fermat.   ◇

It is quite clear that c-tables *per se* are quite difficult to interpret by end-users, but they can be used as a means to answer yes/no queries. Complexity results about different types of yes/no queries on c-tables are given in [2,15]. The authors consider five types of problems:

Table 4. Resulting conditional table

Course	Room	
Sally	Fermat	$(z = 0) \wedge (u = 0)$
Sally	Gödel	$(z = 0) \wedge (u \neq 0)$
Sally	Turing	$(z \neq 0) \wedge (v \neq 0)$
Sally	Monod	$x = biology \wedge (v \neq 0)$
Sally	Einstein	$x = physics$
Alice	Fermat	$(x = physics) \wedge (t \neq 0) \wedge (u = 0)$
Alice	Gödel	$(x = physics) \wedge (t \neq 0) \wedge (u \neq 0)$
Alice	Turing	$(z = 0) \wedge (v = 0)$
Alice	Monod	$(z = 0) \wedge (v \neq 0)$
Alice	Einstein	$(x = physics) \wedge (t \neq 0)$

- containment problem $cont(q_0, q)$: is the result of a given query q_0 over a given set of c-tables S_0 included in the result of a given query q over a given set of c-tables S ?
- membership problem $memb(q)$: is a given instance I_0 a possible world of the result of a given query q over a given set of c-tables S ?
- uniqueness problem $uniq(q_0)$: is the result of a given query q_0 equal to the single given possible world $\{I\}$?
- possibility problem $poss(*, q)$: do all the facts of a given set P belong to a same possible world of the result of a given query q over a given set of c-tables S ?
- certainty problem $cert(*, q)$: do all the facts of a given set P belong to every possible world of the result of a given query q over a given set of c-tables S ?

In [2,15], the authors show that for any polynomial time computable queries q_0, q, one has:

- $cont(q_0, q)$ is in Π_2^p;
- $memb(q)$ is in NP;
- $uniq(q_0)$ is in coNP;
- $poss(*, q)$ is in NP; and
- $cert(*, q)$ is in coNP.

3   Refresher on Possibility Theory

In possibility theory [10,28], each event E — defined as a subset of a universe Ω — is associated with two measures, its possibility $\Pi(E)$ and its necessity $N(E)$. Π and N are two dual measures, in the sense that

$$N(E) = 1 - \Pi(\overline{E})$$

(where the overbar denotes complementation). This clearly departs from the probabilistic situation where $Prob(E) = 1 - Prob(\overline{E})$. So in the probabilistic case, as soon as you are not certain about E ($Prob(E)$ is small), you become rather certain about $\overline{E}$ ($Prob(\overline{E})$ is large). This is not at all the situation in possibility theory, where complete ignorance about E ($E \neq \emptyset$, $E \neq \Omega$) is allowed: This is represented by $\Pi(E) = \Pi(\overline{E}) = 1$, and thus $N(E) = N(\overline{E}) = 0$. In possibility theory, being somewhat certain about E ($N(E)$ has a high value) forces you to have $\overline{E}$ rather impossible ($1 - \Pi$ is impossibility), but it is allowed to have no certainty neither about E nor about $\overline{E}$. Generally speaking, possibility theory is oriented towards the representation of epistemic states of information, while probabilities are deeply linked to the ideas of randomness, and of betting in case of subjective probability, which both lead to an additive model such that $Prob(E) = 1 - Prob(\overline{E})$.

A possibility measure Π (as well as its dual necessity measure N) is based on a possibility distribution π, which is a mapping from a referential U to an ordered scale, say $[0, 1]$. Namely,

$$\Pi(E) = \sup_{u \in E} \pi(u) \text{ and } N(E) = \inf_{u \notin E} (1 - \pi(u)).$$

$\pi(u) = 0$ means that u is (fully) impossible, while $\pi(u) = 1$ just means u is (fully) possible, since it is important to notice that nothing prevents to have $u \neq u'$ and $\pi(u) = \pi(u') = 1$. Thus, $E = \{u\}$ is (fully) certain only if $\pi(u) = 1$ and $\forall u' \neq u$, $\pi(u') = 0$. A possibility distribution π is normalized as soon as $\exists u$, $\pi(u) = 1$; it expresses a form of consistency, since it is natural to have at least one alternative fully possible as soon as the referential is exhaustive (this is the counterpart in possibility theory of having the sum of the probabilities in a probability distribution equal to 1).

Conversely, if we know that $N(E) \geq \alpha$, which means that we are certain (at least) at level α that E is true, there are several possibility distributions that can be compatible with this constraint, but it can be shown that the largest one (the one that allocates the greatest possible possibility to each $u \in U$) is unique, and is such that $\pi(u) = 1$ if $u \in E$ and $\pi(u) = 1 - \alpha$ if $u \notin E$. So, if we are α-certain that Bob lives in Paris or Lyon, this is represented by the distribution $\pi(Paris) = 1 = \pi(Lyon)$, and $\pi(u) = 1 - \alpha$ for any other city u.

Representing a possibility distribution with more than two levels in terms of constraints of the form $N(E) \geq \alpha$ requires several constraints. For instance, if it is possible that Peter lives in Brest, Paris, Lyon, or another city with respective possibility levels $1 > \alpha > \alpha' > \alpha''$ (i.e., $\pi(Brest) = 1$, $\pi(Paris) = \alpha$, $\pi(Lyon) = \alpha'$, $\pi(u) = \alpha''$ for any other city u), then, it corresponds to the constraints $N(\{Brest, Paris, Lyon\}) \geq 1 - \alpha''$, $N(\{Brest, Paris\}) \geq 1 - \alpha'$ and $N(\{Brest\}) \geq 1 - \alpha$. More generally, any possibility distribution with a finite number of levels $1 = \alpha_1 > \cdots > \alpha_n > 0 = \alpha_{n+1}$ can be represented by a collection of n constraints of the form $N(E_i) > 1 - \alpha_{i+1}$ with $E_i = \{u \mid \pi(u) \geq \alpha_i\}$.

Constraints such as $N(E) \geq \alpha$ where E stands for the set of models of a proposition p can be handled inside possibilistic logic under the form of a pair (p, α) made of the classical logic proposition p and a level α belonging to a linearly ordered scale [9,12]. The semantics of a possibilistic logic base K, i.e., a set of such a pairs, $K = \{(p_i \; \alpha_i) | i = 1, \cdots, n\}$ is expressed by means of the possibility distribution π_K defined by

$$\forall \omega \in \Omega, \pi_K(\omega) = \min_{i=1,\cdots,n} \pi_{\{(p_i \; \alpha_i)\}}(\omega) \text{ with } \pi_{\{(p_i \; \alpha_i)\}}(\omega) = \max([p_i](\omega), 1 - \alpha_i)$$

where Ω is the set of interpretations of the language induced by the literals of the formulas in K, and $[p_i](\omega) = 1$ if $\omega \models p_i$ (i.e., ω is a model of K) and $[p_i](\omega) = 0$ otherwise. The semantic entailment is then defined by

$$K \models (p, \alpha) \text{ if and only if } N_K([p]) \geq \alpha \Leftrightarrow \forall \omega \; \pi_K(\omega) \leq \max([p](\omega), 1 - \alpha)$$

where N_K is the necessity measure defined from π_K.

The syntactic inference machinery of possibilistic logic, using the resolution rule

$$(\neg p \vee q, \; \alpha), (p \vee \neg r, \; \beta) \vdash (q \vee \neg r, \; \min(\alpha, \beta))$$

and refutation (it amounts to adding $(\neg \varphi, 1)$, put in clausal form, to K, and using this rule repeatedly to show that $K \cup (\neg \varphi, 1) \vdash (\bot, a))$, has been proved to be sound and complete with respect to the semantics. Algorithms and complexity evaluation (similar to the one of classical logic) can be found in [19]. It is worth mentioning that the repeated use of the probabilistic resolution rule $P(p \vee q) \geq \alpha, P(\neg p \vee r) \geq \beta \vdash P(q \vee r) \geq \max(0, \alpha + \beta - 1)$ does not always provide the tightest bounds that can be obtained by probability computation and thus does not lead to a complete calculus. Moreover, a mixed resolution rule that involves necessity and possibility bounds holds in possibilistic logic [9], here written in terms of semantic constraints:

$$N(\neg p \vee q) \geq \alpha, \Pi(p \vee \neg r) \geq \beta \models \Pi(q \vee \neg r) \geq \beta \text{ provided that } \alpha > 1 - \beta.$$

Lastly, it is worth mentioning that a formula such as $(\neg p \vee q, \; \alpha)$ is semantically equivalent to $(q, \min(\alpha, [p]))$ (with $[p] = 1$ if p is true and $[p] = 0$ if p is false), where $(q, \min(\alpha, [p]))$ can be read q is α-certain provided that p is true. Then it can be checked that the following resolution rule holds $(q, \min(\alpha, [p]))$, $(p, \min(\beta, [r])) \vdash (q, \min(\alpha, \beta, [r]))$. The interested reader may find more details about possibilistic logic and its various applications in [12].

4   Possibilistic c-tables

In his pioneering work [20, 21] on databases with incomplete information Lipski distinguishes between answers that are certain and answers that are only possible for a given query in a modal logic setting. Inspired by this work, the use of possibility theory for modeling incomplete and uncertain information in databases was first proposed in [23–25], and later revisited in [6]. In these approaches, each attribute value in a tuple is represented by means of a possibility distribution *defined on the domain of the attribute*. This possibility distribution restricts the more or less possible values of the attribute for the considered tuple according to the available information (as, e.g., in the previous section example of Peter

Table 5. Conditional table from Example 4

Student	Course	
Sally	math	$(z = 0)$
Sally	CS	$(z \neq 0)$

living in Brest, Paris, Lyon, or another city). However these works were making no reference to c-tables. In the following, we show how any c-table can be directly understood as a possibility distribution over a set of possible database worlds, and thus expressed by a possibilistic logic database. We first do it in the case where possibility is not graded, and then consider the general case of graded possibility that enables us to accommodate uncertainty. We then relate possibilistic c-tables to the particular case of the certainty-based approach to uncertain databases that we recently developed, and finally compare possibilistic c-tables to probabilistic c-tables.

4.1   Regular c-tables Interpreted in Terms of Possibilistic Logic

As already mentioned, the conditions associated with tuples in a c-table are specifying possible worlds. Thus, they implicitly define possibility distributions over mutually exclusive situations.

Example 4. If we take the first part of Example 1, namely representing the information $(take(Sally, math) \vee (take(Sally, CS))$, this corresponds to the possibility distribution $\pi_{take(Sally,\cdot)}(math) = 1 = \pi_{take(Sally,\cdot)}(CS)$, or if we prefer $\pi(z = 0) = \pi(z \neq 0) = 1$. Note that the situation in Remark 1 where $math$ and CS are no longer mutually exclusive would require a possibility distribution defined on the power set of $\{math, CS\}$. Still both cases can be easily captured in possibilistic logic. Indeed "Sally is taking math or computer science" is expressed by

$$(take(Sally, math) \vee take(Sally, CS), 1)$$

and the additional constraint "but not both" by

$$(\neg take(Sally, math) \vee \neg take(Sally, CS), 1).$$

Let us now examine the rest of Example 1. We can take for the domain of attribute $Course$ the set $D_{Course} = \{math, CS, biology, physics, others\}$ that involves all the topics mentioned in the example and leave room for others. Then the information "Sally takes another course" (apart from "math" or "CS") writes in possibilistic logic

$$(take(Sally, physics) \vee take(Sally, biology) \vee take(Sally, others), 1)$$

while "Alice takes biology if Sally takes math and math or physics (but not both) if Sally takes physics" writes

$$(take(Alice, biology), [take(Sally, math)])$$
$$(take(Alice, math) \vee take(Alice, physics), [take(Sally, physics)])$$
$$(\neg take(Alice, math) \vee \neg take(Alice, physics), 1).$$

We could equivalently write $(take(Alice, biology) \lor \neg take(Sally, math), 1)$ in place of $(take(Alice, biology), [take(Sally, math)])$, and this applies as well to possibilistic formula after. Having $[take(Sally, math)]$ (or $[take(Sally, physics)]$) in the certainty level slot of the formulas puts the main focus on Alice.   ◇

Thus, the conditional table represented in Table 5 translates easily in a possibilistic logic base. This base is semantically associated to a possibility distribution, which can be obtained by applying the general result recalled in Sect. 3. This would enable us to explicit the possibility distribution underlying Table 5. This is a $\{0, 1\}$-valued possibility distribution; however, if binary-valued functions are used in the certainty slots (as $[take(Sally, math)]$ in the above example), some possibility degrees will receive a conditional value (such as $1 - [take(Sally, math)] \in \{0, 1\}$).

A query such as "find the x's such that condition Q is true" is processed by refutation, adding the formulas corresponding to $\neg Q(x) \lor answer(x)$ to the base, using a small trick due to [16] (see [9]). Let us take a first example.

Example 5. In the previous Example 4, let us consider the query "find the students who take math or computer science", which translates into
$$\{(\neg take(x, math) \lor answer(x), 1), (\neg take(x, CS) \lor answer(x), 1)\}.$$

It can be checked that it yields $(answer(Sally), 1)$
and $(answer(Alice) \lor take(Alice, physics), [take(Sally, physics))])$ (or equivalently $(answer(Alice), \min([\neg take(Alice, physics)], [take(Sally, physics)]))$
$\Leftrightarrow (answer(Alice), [\neg take(Alice, physics)] \land [take(Sally, physics)]))$.   ◇

As can be seen, we may obtain two types of answers: (i) the answers x_0 of the form $(answer(x_0), 1)$, and (ii) the answers x_0 of the form $(answer(x_0), [\varphi])$ where $[\varphi]$ is a nontautological condition which may *take* values 0 and 1. The first answers are exactly those that are *certain*, while the second ones are exactly those that are *possible* without being *certain*. Let us consider an example with a join.

Example 6. Let us come back to Example 3 involving a join query. The translation in possibilistic logic is straightforward:

$(place(physics, Einstein), 1)$
$(place(math, Fermat) \lor place(math, Gödel), 1)$,
$(\neg place(math, Fermat) \lor \neg place(math, Gödel), 1)$
$(place(biology, Turing) \lor place(biology, Monod), 1)$,
$(\neg place(biology, Turing) \lor \neg place(biology, Monod), 1)$
$(\neg place(CS, Turing) \lor \neg place(biology, Turing), 1)$
$(\neg place(biology, Turing) \lor \neg place(math, Fermat) \lor (place(CS, Gödel), 1)$
$(\neg place(biology, Turing) \lor \neg place(math, Gödel) \lor (place(CS, Fermat), 1)$
$(place(CS, Turing) \lor place(CS, Gödel) \lor (place(CS, Fermat), 1)$

Let us now consider the question "who is in Monod room?", which translates into $(\neg take(x, y) \lor \neg place(y, Monod) \lor answer(x), 1)$. Then from the possibilistic logic counterparts of Tables 1 and 3, it can be checked that we can infer

Table 6. Possibilistic conditional tables from Example 7 (r left, s right)

Student	Course	
$x = 0$ (1) $\oplus$ $x = 1$ (0.6)		
Sally	CS	$x = 0$
Sally	math	$x = 1$

Course	Room
$y = Fermat$ (1) $\oplus$ $y = Turing$ (0.3)	
math	y
CS	Turing

$\{Sally, Alice\}$, as the set of possible answers. Indeed we get

$$(answer(Sally), \min([\neg take(Sally, physics)], [\neg take(Sally, others)],$$
$$[\neg place(biology, Turing)]]))$$

$$(answer(Alice), \min([take(Sally, math)], [\neg place(biology, Turing)]]))$$

which is in agreement with Table 4.

As expected, the conjunctive structure of the combined certainty levels reflects the conjunctions performed when the join of the two c-tables are computed.   ◇

4.2 Gradual Possibilistic c-tables

Obviously, we are not obliged to use binary-valued possibility and certainty levels taking values '0' or '1' only. We can thus express for instance that "Sally is taking math or computer science (but not both)", but we are somewhat certain that it is computer science. This corresponds to the possibility distribution $\pi_{take(Sally, \cdot)}(math) = 1 - \alpha$; $\pi_{take(Sally, \cdot)}(CS) = 1$ if we are certain at level α that it is computer science. Similarly, we may want to express that it is α-certain that "Alice takes math or physics (but in any case not, both) if Sally takes physics". This latter information translates in the possibilistic formulas

$$(take(Alice, math) \vee take(Alice, physics), \min(\alpha, [take(Sally, physics)]))$$

and $(\neg take(Alice, math) \vee \neg take(Alice, physics), 1)$.

Let us now formally define the concept of a possibilistic c-table.

Definition 3. *A possibilistic c-table on X is a triple $(T, \Phi_T, \mathcal{P})$ where T is any finite set of c-tuples on X, $\Phi_T \in \mathcal{S}$ is a global condition on T, and $\mathcal{P}$ is a set of possibility distributions on some variables of $V = \bigcup_{A \in X} V(A)$.*

In this definition, $\mathcal{P}$ corresponds to the set of "soft constraints" (possibilistic restrictions) bearing on some variables of the c-table. First observe that the following property trivially holds.

Property 1. In the special case where possibility degrees take their values in $\{0, 1\}$, possibilistic c-tables reduce to regular c-tables.

Example 7. Let us consider the two following possibilistic c-tables (Table 6).

Table 7 represents the result of $q = \pi_{\{Student, Room\}}(r \bowtie_{Course=Course} s)$. The tuple $t = \langle Sally, Turing \rangle$ is a possible (resp. certain) answer to the degree 1 (resp. 0.4). Indeed, the most possible world such that the result of q contains t (resp. does not contain t) corresponds to the valuation $\{x = 0, y = Fermat\}$ (resp. $\{x = 1, y = Fermat\}$) whose possibility degree equals 1 (resp. 0.6).   ◇

Table 7. Result of the query from Example 7

Student	Room	
	$x = 0$ (1) $\oplus$ $x = 1$ (0.6)	
	$y = Fermat$ (1) $\oplus$ $y = Turing$ (0.3)	
Sally	Turing	$x = 0 \vee (x = 1 \wedge y = Turing)$
Sally	Fermat	$x = 1 \wedge y = Fermat$

Remark 2. It is important to emphasize that as soon as candidate values may be attached a degree of possibility, the worlds of $rep(T)$ become more or less possible (they would be more or less probable if a probabilistic database model were used). This has an impact on the notion of possible and certain answers (cf. Eqs. 2 and 3). In the graded possibilistic c-tables model we introduce, the degree of possibility (resp. certainty) associated with an answer t to a query q corresponds to the maximum of the possibility degrees attached to the worlds of $rep(\bar{q}(T))$ that contain t (resp. 1 minus the maximum of the possibility degrees attached to the worlds of $rep(\bar{q}(T))$ that do not contain t).   □

When a consequence of interest is of the form (p, α), where α is a certainty level that depends on nothing, we have an α-certain answer. Besides, if we get only $(p, [q])$, and more generally $(p, \min(\alpha, [q])$, and if $[q]$ is not known as being equal to 1, the answer p can be regarded as being *possible* at level 1. Indeed the possibility distribution associated with $(p, \min(\alpha, [q])$ is such that $\pi_{\{(p,\min(\alpha,[q]))\}}(pq) = 1; \pi_{\{(p,\min(\alpha,[q]))\}}(p\neg q) = 1; \pi_{\{(p,\min(\alpha,[q]))\}}(\neg p \neg q) = 1$ and $\pi_{\{(p,\min(\alpha,[q]))\}}(\neg pq) = 1 - \alpha$, and thus $\Pi(p) = \max_{\omega \models p} \pi_{\{(p,\min(\alpha,[q]))\}}(\omega) = 1$. This indicates that the certainty level-based possibilistic logic cannot alone account for intermediary possibility levels, as further discussed now. What is computed here is only an upper bound of the possibility level, exploiting a part of the information only.

Example 8. Consider again the situation where Sally is taking math with possibility $1 - \alpha$ or computer science with possibility 1 (but not both, with full certain-ty), which writes in possibilistic logic $(take(Sally, CS), \alpha)$ and $(take(Sally, CS) \vee take(Sally, math), 1)$ (this latter formula acknowledges that Sally studies either CS or math). Let us evaluate the query "find the students who take math", which translates into $(\neg take(x, math) \vee answer(x), 1)$; we obtain

$$(answer(Sally), [\neg take(Sally, CS)]),$$

but we do not retrieve $\pi_{take(Sally,\cdot)}(math) = 1 - \alpha$. This can be only done by applying the mixed resolution pattern recalled in Sect. 3, namely here

$$N(\neg take(x, math) \vee answer(x)) = 1, \Pi(take(Sally, math)) = 1 - \alpha$$
$$\models \Pi(answer(Sally)) \geq 1 - \alpha. \diamond$$

However, although it does not appear on this very simple example, the evaluation of the possibility levels associated with the possibility distribution underlying the uncertain and imprecise database may lead to complicated expressions and heavy computation, as in the probabilistic case.

Theorem 1. *Possibilistic c-tables are a strong representation system for relational algebra.*

Sketch of the Proof. We need to prove (cf. Formula 1) that: $rep(\bar{q}(T)) = q(rep(T))$. In the possibilistic c-tables model, an element of $rep(T)$ is a pair $(W_i, \Pi(W_i))$, where W_i is a possible world of T and $\Pi(W_i)$ is its associated possibility degree. Then, $q(rep(T))$ corresponds to the weighted set of worlds $\{\Pi(W_i)/q(W_i) \mid W_i \in rep(T)\}$ (remark: we keep the max of the possibility degrees in case of duplicate worlds). Let us denote by $\mathcal{W}(T')$ the possible worlds (without the associated possibility degrees) represented by the possibilistic c-table T'. Since regular c-tables are a strong representation system for relational algebra (see [18]), and since the definitions of the algebraic operators remain unchanged, we have $\mathcal{W}(rep(\bar{q}(T))) = \mathcal{W}(q(rep(T)))$ where $\mathcal{W}(q(rep(T))) = \{q(W_i) \mid W_i \in rep(T)\}$. Now, all one needs is a sound way to compute the possibility degree of a world generated by a possibilistic c-table (i.e. of a valuation that satisfies the conditions in the c-table). This way is provided by the axioms of possibility theory regarding conjunction and disjunction, and the computation is based on the possibility distributions attached to the possibilistic c-table on the one hand, and the conditions attached to the tuples on the other hand.   □

Let us now consider the counterparts of the yes-no queries discussed at the end of Sect. 2 and their associated complexity. In the possibilistic c-tables framework these queries are not of type yes-no anymore but are of the form "to which extent is it possible (resp. certain) that ...". For instance, the containment problem $cont(q_0, q)$ now corresponds to the question: to which extent is it possible (resp. certain) that the result of a given query q_0 over a given set of c-tables S_0 is included in the result of a given query q over a given set of c-tables S? Just as the complexity of possibilistic logic is the one of classical logic multiplied by the logarithm of the number of levels used in the scale [19], the complexity here remains in the same class as the one for regular c-tables.

4.3 The Particular Case of the Certainty-Based Model

In [22], we defined a model that we called "certainty-based" for representing relational databases containing uncertain attribute values, when some knowledge is available about the more or less certain value (or disjunction of values) that a given attribute in a tuple may *take*.

As the possibilistic model described in [6], the certainty-based model [22] relies on possibility theory [28]. However, it only keeps pieces of information that are more or less certain and leaves aside what is just possible. This corresponds to the most important part of information (a possibility distribution is "summarized" by keeping its most plausible elements, associated with a certainty level). For instance, $\langle 037, John, (40, \alpha) \rangle$ denotes the existence of a person named John, whose age is 40 with certainty α. Then the possibility that his age differs from 40 is upper bounded by $1 - \alpha$ without further information.

The model can also deal with disjunctive uncertain values. For instance, $\langle 3, Peter, (Newton \vee Quincy, 0.8)\rangle$ represents the fact that it is 0.8-certain that the person number 3 named Peter lives in Newton or in Quincy. Then, the underlying possibility distributions π are of the form $\pi(u) = \max(A(u), 1 - \alpha)$ where A is an α-certain subset of the attribute domain and $A(u)$ equals 1 if $u \in A$, 0 otherwise.

Moreover, since some operations (e.g., the selection) may create "maybe tuples", each tuple t from an uncertain relation r has to be associated with a degree N expressing the certainty that t exists in r. It will be denoted by N/t.

Example 9. Let us consider a relation r of schema ($\#id$, $Name$, $City$) containing tuple $t_1 = \langle 1, John, (Quincy, 0.8)\rangle$, and the query "find the people who live in Quincy". Let the domain of attribute $City$ be $\{Boston, Newton, Quincy\}$. The answer contains $0.8/t_1$ since it is 0.8 certain that t_1 satisfies the requirement, while the result of the query "find the people who live in Boston, Newton or Quincy" contains $1/t_1$ since it is totally certain that t_1 satisfies the condition. ◊

To sum up, a tuple $\alpha/\langle 037, John, (Quincy, \beta)\rangle$ from relation r means that it is α certain that person 037 exists in the relation, and that it is β certain that 037 lives in Quincy (independently from the fact that it is or not in relation r).

Obviously, this model is a particular case of a gradual possibilistic c-table where the only conditions present in a relation are used to represent the more or less certain value (or disjunction of values) that an attribute in a tuple may *take*. Of course, when using a c-table to represent such a relation, there is no need for the extra attribute N since the certainty level attached to a tuple is computed by evaluating the condition associated with this tuple (interpreting the conjunction as the minimum according to the axioms of possibility theory).

We have extended relational algebra in this context and shown that the model constitutes a representation system for this set of operators. The only constraints concern (i) the join that has to be based on an equality condition, (ii) the Cartesian product and join operations that must *take* independent relations as arguments. An important result is that the data complexity of these operations is the same as in the classical database case. This is also the case of general possibilistic c-tables when it comes to computing the "compact" result of a query, i.e., its resulting possibilistic c-table, but of course not when it comes to answering generalized yes-no queries (cf. the end of Subsect. 4.2). Since the certainty-based model does not include intertuple dependencies, a table that represents the result of a query in this model is easily interpretable. On the other hand, possibilistic c-tables are more expressive but also more complex and can only be exploited by an end-user through generalized yes-no queries.

Table 8. Probabilistic conditional table from Example 10

Student	Course
	$x = math$ (0.3) $\oplus$ $x = phys$ (0.3) $\oplus$ $x = chem$ (0.4)
	$t = 0$ (0.15) $\oplus$ $t = 1$ (0.85)
Alice	x
Bob	x $\quad\quad\quad (x = physics) \oplus (x = chem)$
Theo	$math \quad\quad t = 1$

4.4 Comparison with Probabilistic c-tables

We introduce probabilistic c-tables by means of the following example drawn from [17].

Example 10. Suppose Alice takes a course that is *math* with probability 0.3, or *physics* with probability 0.3, or *chem* with probability 0.4; Bob takes the same course as Alice provided that the course is *physics* or *chem*; Theo takes *math* with probability 0.85. This can be represented by the probabilistic c-table depicted in Table 8.

We may easily imagine a possibilistic version of the example. Suppose that Alice takes a course that is *math* with possibility α, or *physics* with possibility α, or *chem* with possibility 1. Bob takes the same course as Alice provided that the course is *physics* or *chem*. Theo takes *math* with possibility 1 and does not take *math* with possibility β. Now $D_{course} = \{math, chem, physics\}$. The pieces of information above can be expressed in possibilistic logic as:

$(take(Alice, chem), 1 - \alpha)$
$(take(Bob, chem), [take(Alice, chem)])$
$(take(Bob, physics), [take(Alice, physics)])$
$(take(Theo, math), 1 - \beta)$

From which one can deduce, e.g., $(take(Bob, chem), 1 - \alpha)$.   ◇

Note that in case we would have
$x = math$ (0.25) $\oplus$ $x = phys$ (0.35) $\oplus$ $x = chem$ (0.4)
we would have to add

$(take(Alice, chem) \vee take(Alice, physics), 1 - \alpha')$ with $\alpha' < \alpha$
to $(take(Alice, chem), 1 - \alpha)$.
Moreover, in case we would have
$x = math$ (0.4) $\oplus$ $x = phys$ (0.4) $\oplus$ $x = chem$ (0.2)
we would have to replace $(take(Alice, chem), 1 - \alpha)$ by
$(take(Alice, math) \vee take(Alice, physics), 1 - \alpha)$.

The above examples suggest that probabilistic c-tables and possibilistic c-tables are quite close as a representation tool, although obeying to different inference principles. Compared to probabilistic c-tables, an argument in favor of the possibilistic c-tables model lies in its robustness (i.e., in the fact that the order of

Table 9. Relations r (top) and s (bottom) — Possibilistic c-tables

#id	name	city
		$x = Newton$ (1) $\oplus$ $x = Quincy$ (0.2)
		$y = Quincy$ (1) $\oplus$ $y = Gardner$ (0.6)
		$z = Newton$ (1) $\oplus$ $z = Gardner$ (0.8) $\oplus$ $z = Quincy$ (0.7)
		$u = Quincy$ (1) $\oplus$ $u = Gardner$ (1)
37	John	x
53	Mary	y
72	Paul	z
81	Lisa	u

city	flea market
$v = yes$ (1) $\oplus$ $v = no$ (0.3)	
$w = yes$ (0.1) $\oplus$ $w = no$ (1)	
$r = yes$ (1) $\oplus$ $r = no$ (0.4)	
Newton	v
Quincy	w
Gardner	r

the answers obtained is less sensitive to the values of the degrees in the distributions attached to the variables), which is illustrated by the toy example hereafter.

Example 11. Let us consider the possibilistic c-tables r of schema (#id, name, city), and s of schema (city, flea market) describing respectively a set of people whose city of residence is ill-know, and a set of city for which we are not sure if they have a flea market or not. Let us consider the query q asking for the people who live in a city with a flea market: $\pi_{Name}(r \bowtie_{city=city} (\sigma_{flea\,market=yes}(s)))$.

Let us first consider the possibilistic c-tables model (cf. Table 9). The resulting possibilistic c-table is represented in Table 10.

Table 10. Result of query q

name	
	$x = Newton$ (1) $\oplus$ $x = Quincy$ (0.2)
	$y = Quincy$ (1) $\oplus$ $y = Gardner$ (0.6)
	$z = Newton$ (1) $\oplus$ $z = Gardner$ (0.8) $\oplus$ $z = Quincy$ (0.7)
	$u = Quincy$ (1) $\oplus$ $u = Gardner$ (1)
	$v = yes$ (1) $\oplus$ $v = no$ (0.3)
	$w = yes$ (0.1) $\oplus$ $w = no$ (1)
	$r = yes$ (1) $\oplus$ $r = no$ (0.4)
John	$(x = Newton \wedge v = yes) \vee (x = Quincy \wedge w = yes)$
Mary	$(y = Quincy \wedge w = yes) \vee (y = Gardner \wedge r = yes)$
Paul	$(z = Newton \wedge v = yes) \vee (z = Gardner \wedge r = yes) \vee (z = Quincy \wedge w = yes)$
Lisa	$(u = Quincy \wedge w = yes) \vee (u = Gardner \wedge r = yes)$

Table 11. Relations r (top) and s (bottom) — Probabilistic c-tables

#id	name	city
		$x = Newton\ (0.8)\ \oplus\ x = Quincy\ (0.2)$
		$y = Quincy\ (0.6)\ \oplus\ y = Gardner\ (0.4)$
		$z = Newton\ (0.5)\ \oplus\ z = Gardner\ (0.3)\ \oplus\ z = Quincy\ (0.2)$
		$u = Quincy\ (0.5)\ \oplus\ u = Gardner\ (0.5)$
37	John	x
53	Mary	y
72	Paul	z
81	Lisa	u

city	flea market
$v = yes\ (0.7)\ \oplus\ v = no\ (0.3)$	
$w = yes\ (0.1)\ \oplus\ w = no\ (0.9)$	
$r = yes\ (0.6)\ \oplus\ r = no\ (0.4)$	
Newton	v
Quincy	w
Gardner	r

The answers obtained are thus:

John $\Pi = \max(\min(1, 1), \min(0.2, 0.1)) = 1,$
$N = \max(\min(1 - 0.2, 1 - 0.3), \min(1 - 1, 1 - 1)) = 0.7$

Mary $\Pi = \max(\min(1, 0.1), \min(0.6, 1)) = 0.6,$
$N = \max(\min(1 - 0.6, 1 - 1), \min(1 - 1, 1 - 0.4)) = 0$

Paul $\Pi = \max(\min(1, 1), \min(0.8, 1), \min(0.7, 0.1)) = 1,$
$N = \max(\min(1 - 0.8, 1 - 0.3), \min(1 - 1, 1 - 0.4), \min(1 - 1, 1 - 1))$
$= 0.2$

Lisa $\Pi = \max(\min(1, 0.1), \min(1, 1)) = 1,$
$N = \max(\min(1 - 1, 1 - 1), \min(1 - 1, 1 - 0.4)) = 0.$

John is a completely possible answer ($\Pi = 1$) since it is completely possible that (i) he lives in Newton, and (ii) Newton has a flea market. On the other hand, it is only 0.7 certain that he is an answer since it is 0.3 possible that Newton does not have a flea market. Since one has $N > 0 \Rightarrow \Pi = 1$, one may rank the answers in decreasing order of N first, then, for those such that $N = 0$, in decreasing order of Π. We get the following ranking: *John $\succ$ Paul $\succ$ Lisa $\succ$ Mary*.

Finally, let us use a probabilistic model. In Table 11, the probability values are roughly specified, in agreement with the uncertainty ordering specified in the previous possibilistic tables. The result of q is the same as in the possibilistic case (cf. Table 10) except for the global conditions in which the degrees are different. Finally, we get the answers:

John $pr = 0.8 \times 0.7 + 0.2 \times 0.1 = 0.58$
Mary $pr = 0.6 \times 0.1 + 0.4 \times 0.6 = 0.3$
Paul $pr = 0.5 \times 0.7 + 0.3 \times 0.6 + 0.2 \times 0.1 = 0.55$
Lisa $pr = 0.5 \times 0.1 + 0.5 \times 0.6 = 0.35$

As can be seen, we again obtain the same ranking as in the possibilistic
c-tables model. However, a rather slight modification of the probability values
may lead to a modification of the ranking. For instance, if the probability distribution associated with Paul's city were changed into $\{0.55/Newton, 0.35/Gardner,$
$0.1/Quincy\}$, Paul would get the degree 0.605 and would be ranked first (before
John). This contrasts with the possibilistic situation where the result remains
stable as long as the ordering of the possibilistic values is not changed.   ◇

5   Conclusion

Possibility theory and c-tables have appeared at about the same time, at the
end of the 1970's, in two different areas of information processing. Curiously,
they had remained unrelated until now. This paper provides a first study in
order to bridge them. Indeed, c-tables, as a convenient way of describing possible worlds, can be easily extended to a possibilistic modeling of uncertain and
imprecise information. This provides a general setting that appears quite appropriate for handling uncertainty in a qualitative way. The qualitative nature of
possibility theory makes simpler the elicitation of the possibility and certainty
degrees, and leads to a modeling less sensitive to modifications of the values of
the degrees than in a probabilistic framework. Moreover, the particular case of
the certainty-based approach is especially tractable. Besides, the existing relation
between answer-set programming and generalized possibilistic logic [13] and the
underlying role of possibilistic logic with respect to possibilistic c-tables suggests
to study a possibilistic version of Datalog [2,14] in the future.

References

1. Abiteboul, S., Hull, R., Vianu, V.: Foundations of Databases. Addison-Wesley,
   Reading (1995)
2. Abiteboul, S., Kanellakis, P.C., Grahne, G.: On the representation and querying
   of sets of possible worlds. Theor. Comput. Sci. **78**(1), 158–187 (1991)
3. Antova, L., Jansen, T., Koch, C., Olteanu, D.: Fast and simple processing of uncertain data. In: Proceedings of the 24th International Conference on Data Engineering (ICDE 2008), pp. 983–992 (2008)
4. Benjelloun, O., Das Sarma, A., Halevy, A., Theobald, M., Widom, J.: Databases
   with uncertainty and lineage. VLDB J. **17**(2), 243–264 (2008)
5. Benjelloun, O., Das Sarma, A., Halevy, A., Widom, J.: ULDBs: databases with
   uncertainty and lineage. In: Proceedings of VLDB 2006, pp. 953–964 (2006)
6. Bosc, P., Pivert, O.: About projection-selection-join queries addressed to possibilistic relational databases. IEEE T. Fuzzy Syst. **13**(1), 124–139 (2005)
7. Codd, E.F.: Extending the relational database model to capture more meaning.
   ACM Trans. Database Syst. **4**(4), 397–434 (1979)
8. Dalvi, N., Suciu, D.: Management of probabilistic data: foundations and challenges.
   In: Proceedings of PODS 2007, pp. 1–12 (2007)
9. Dubois, D., Lang, J., Prade, H.: Possibilistic logic. In: Gabbay, D.M., Hogger, C.J.,
   Robinson, J.A., Nute, D. (eds.) Handbook of Logic in Artificial Intelligence and
   Logic Programming, vol. 3, pp. 439–513. Oxford University Press, Oxford (1994)

10. Dubois, D., Prade, H.: Possibility Theory: An Approach to Computerized Processing of Uncertainty. Plenum Press, New York (1988). (with the collaboration of H. Farreny, R. Martin-Clouaire, and C. Testemale)
11. Dubois, D., Prade, H.: Possibility theory: qualitative and quantitative aspects. In: Gabbay, D.M., Smets, P. (eds.) Quantified Representation of Uncertainty and Imprecision. Handbook of Defeasible Reasoning and Uncertainty Management Systems, vol. 1, pp. 169–226. Kluwer Academic Publishers, Dordrecht (1998)
12. Dubois, D., Prade, H.: Possibilistic logic – an overview. In: Siekmann, J.H. (ed.) Computational Logic. Handbook of the History of Logic, vol. 9, pp. 283–342. Elsevier, Amsterdam (2014)
13. Dubois, D., Prade, H., Schockaert, S.: Stable models in generalized possibilistic logic. In: Brewka, G., Eiter, T., McIlraith, S.A. (eds.) Proceedings of the 13th International Conference on Principles of Knowledge Representation and Reasoning (KR 2012), pp. 519–529. AAAI Press, Menlo Park (2012)
14. Gallaire, H., Minker, J.J. (eds.): Advances in Data Base Theory. In: Proceedings of the Symposium on Logic and Data Bases, Centre d' Etudes et de Recherches de Toulouse, 1977. Plenum Press (1978)
15. Grahne, G.: The Problem of Incomplete Information in Relational Databases. LNCS, vol. 554. Springer, Heidelberg (1991)
16. Green, C.: Theorem-proving by resolution as a basis for question-answering systems. In: Michie, D., Meltzer, B. (eds.) Machine Intelligence, vol. 4, pp. 183–205. Edinburgh University Press, Edinburgh (1969)
17. Green, T.J., Tannen, V.: Models for incomplete and probabilistic information. In: Proceedings of the IIDB 2006 Workshop, pp. 278–296 (2006)
18. Imielinski, T., Lipski, W.: Incomplete information in relational databases. J. ACM 31(4), 761–791 (1984)
19. Lang, J.: Possibilistic logic: complexity and algorithms. In: Kohlas, J., Moral, S. (eds.) Algorithms for Uncertainty and Defeasible Reasoning. Handbook of Defeasible Reasoning and Uncertainty Management Systems, vol. 5, pp. 179–220. Kluwer Academic Publisher, Dordrecht (2001). (Gabbay, D.M. and Smets, Ph., eds.)
20. Lipski, W.: On semantic issues connected with incomplete information databases. ACM Trans. Database Syst. 4(3), 262–296 (1979)
21. Lipski, W.: On databases with incomplete information. J. ACM 28, 41–70 (1981)
22. Pivert, O., Prade, H.: A certainty-based model for uncertain databases. IEEE Trans. Fuzzy Syst. 23(4), 1181–1196 (2015)
23. Prade, H.: The connection between Lipski's approach to incomplete information data bases and Zadeh's possibility theory. In: Proceedings of the International Conference on Systems Methodology, Washington, D.C., 5–9 January, pp. 402–408 (1982)
24. Prade, H.: Lipski's approach to incomplete information data bases restated and generalized in the setting of Zadeh's possibility theory. Inf. Syst. 9(1), 27–42 (1984)
25. Prade, H., Testemale, C.: Generalizing database relational algebra for the treatment of incomplete/uncertain information and vague queries. Inf. Sci. 34(2), 115–143 (1984)
26. Shimizu, S., Ishihara, Y., Takarabe, T., Ito, M.: A probabilistic database model with representability of dependency among tuples. In: Proceedings of the 4th World Multiconference on Systemics, Cybernetics and Informatics (SCI 2000), pp. 221–225 (2000)
27. Suciu, D., Olteanu, D., Ré, C., Koch, C.: Probabilistic Databases. Synthesis Lectures on Data Management. Morgan & Claypool Publishers (2011)
28. Zadeh, L.: Fuzzy sets as a basis for a theory of possibility. Fuzzy Sets Syst. 1(1), 3–28 (1978)

Inference and Problem Solving

Skeptical Inference Based on C-Representations and Its Characterization as a Constraint Satisfaction Problem

Christoph Beierle[1]([✉]), Christian Eichhorn[2], and Gabriele Kern-Isberner[2]

[1] Department of Computer Science, University of Hagen, 58084 Hagen, Germany
beierle@fernuni-hagen.de
[2] Department of Computer Science, TU Dortmund, 44221 Dortmund, Germany

Abstract. The axiomatic system P is an important standard for plausible, nonmonotonic inferences that is, however, known to be too weak to solve benchmark problems like irrelevance, or subclass inheritance (so-called *Drowning Problem*). Spohn's ranking functions which provide a semantic base for system P have often been used to design stronger inference relations, like Pearl's system Z, or c-representations. While each c-representation shows excellent inference properties and handles particularly irrelevance and subclass inheritance properly, it is still an open problem which c-representation is the best. In this paper, we focus on the generic properties of c-representations and consider the skeptical inference relation (*c-inference*) that is obtained by taking all c-representations of a given knowledge base into account. In particular, we show that c-inference preserves the properties of solving irrelevance and subclass inheritance which are met by every single c-representation. Moreover, we characterize skeptical c-inference as a constraint satisfaction problem so that constraint solvers can be used for its implementation.

1 Introduction

Calculating inductive inferences based on knowledge bases of conditional rules is an important task in nonmonotonic reasoning. Here, calculi like Adams' system P [1], probabilistic approaches like p-entailment [7], or possibilistic inference methods [5] have been developed, as well as the inductive methods of Pearl's system Z [13] or c-representations [10,11]. The latter two rely on Spohn's ordinal conditional functions [14,15] for calculating inferences which means that the underlying preferential model [12] always is a total ordering of the set of possible worlds. In this paper, we define a novel inductive inference relation, called *c-inference,* as a skeptical inference over the (infinitely many) c-representations of a knowledge base. We show that this inference relation, even if set up upon a partial ordering of the worlds, exceeds system P and handles important benchmarks of plausible reasoning, like the Drowning Problem or irrelevance, properly. We model c-representations correctly and completely as a constraint satisfaction problem (CSP, cf. [2]) and, on top of this model, also characterize c-inference as

© Springer International Publishing Switzerland 2016
M. Gyssens and G. Simari (Eds.): FoIKS 2016, LNCS 9616, pp. 65–82, 2016.
DOI: 10.1007/978-3-319-30024-5_4

a CSP. This makes the skeptical inference over infinitely many c-representations not only being calculable, but also implementable.

This paper is organized as follows: After this introduction, we recall the basics of conditionals, ordinal conditional functions (OCF), plausible inference, system P, system Z and c-representations which we need as formal background of this paper. In Sect. 3, we prove that the CSP for c-representations of a knowledge base proposed in [2] is a correct and complete modeling of c-representations. Section 4 defines *c-inference* as skeptical inference relation over all c-representations of a knowledge base. Here, we prove that this relation not only satisfies but exceeds system P and show that the relation handles selected benchmarks properly. Then, we characterize c-inference as a CSP in Sect. 5. We conclude in Sect. 6.

2   Conditionals, OCF, and Plausible Inference

Let $\Sigma = \{V_1, ..., V_m\}$ be a propositional alphabet. A *literal* is the positive (v_i) or negated $(\overline{v}_i)$ form of a propositional variable V_i. From these we obtain the propositional language $\mathcal{L}$ as the set of formulas of Σ closed under negation $\neg$, conjunction $\wedge$, and disjunction $\vee$, as usual; for formulas $A, B \in \mathcal{L}$, $A \Rightarrow B$ denotes the material implication and stands for $\neg A \vee B$. For shorter formulas, we abbreviate conjunction by juxtaposition (i.e., AB stands for $A \wedge B$), and negation by overlining (i.e., $\overline{A}$ is equivalent to $\neg A$). Let Ω denote the set of possible worlds over $\mathcal{L}$; Ω will be taken here simply as the set of all propositional interpretations over $\mathcal{L}$ and can be identified with the set of all complete conjunctions over Σ. For $\omega \in \Omega$, $\omega \models A$ means that the propositional formula $A \in \mathcal{L}$ holds in the possible world ω.

A *conditional* $(B|A)$ with $A, B \in \mathcal{L}$ encodes the defeasible rule "if A then normally B" and is a trivalent logical entity with the evaluation [6,10]

$$[\![(B|A)]\!]_\omega = \begin{cases} true & \text{iff} \quad \omega \models AB & \text{(verification)} \\ false & \text{iff} \quad \omega \models A\overline{B} & \text{(falsification)} \\ undefined & \text{iff} \quad \omega \models \overline{A} & \text{(not applicable)} \end{cases}$$

A *knowledge base* $\mathcal{R} = \{(B_1|A_1), ..., (B_n|A_n)\}$ is a finite set of conditionals. A conditional $(B|A)$ is *tolerated* by a set of conditionals $\mathcal{R}$ if and only if there is a world $\omega \in \Omega$ such that $\omega \models AB$ and $\omega \models \bigwedge_{i=1}^{n}(A_i \Rightarrow B_i)$, i.e., iff ω verifies $(B|A)$ and does not falsify any conditional in $\mathcal{R}$.

An *Ordinal Conditional Function* (OCF, ranking function) [14,15] is a function $\kappa : \Omega \to \mathbb{N}_0 \cup \{\infty\}$ that assigns to each world $\omega \in \Omega$ an implausibility rank $\kappa(\omega)$, that is, the higher $\kappa(\omega)$, the more surprising ω is. OCFs have to satisfy the normalization condition that there has to be a world that is maximally plausible, i.e., the preimage of 0 cannot be empty, formally $\kappa^{-1}(0) \neq \emptyset$. The rank of a formula A is defined to be the rank of the least surprising world that satisfies A, formally

$$\kappa(A) = \min\{\kappa(\omega) \mid \omega \models A\}. \tag{1}$$

The set of models of tautologies is the complete set of possible worlds, therefore the normalization condition directly gives us $\kappa(\top) = 0$. In accordance with general order-theoretical conventions, we set $\kappa(\bot) = \infty$.

An OCF κ *accepts* a conditional $(B|A)$ (denoted by $\kappa \models (B|A)$) iff the verification of the conditional is less surprising than its falsification, i.e., iff $\kappa(AB) < \kappa(A\overline{B})$. This can also be understood as a nonmonotonic inference relation between the premise A and the conclusion B: We say that A κ-*entails* B (written $A \mathrel{\mid\!\sim}^{\kappa} B$) if and only if κ accepts the conditional $(B|A)$, formally

$$\kappa \models (B|A) \quad \text{iff} \quad \kappa(AB) < \kappa(A\overline{B}) \quad \text{iff} \quad A \mathrel{\mid\!\sim}^{\kappa} B. \tag{2}$$

The acceptance relation in (2) is extended as usual to a set $\mathcal{R}$ of conditionals by defining $\kappa \models \mathcal{R}$ iff $\kappa \models (B|A)$ for all $(B|A) \in \mathcal{R}$. This is synonymous to saying that $\mathcal{R}$ is *admissible* with respect to $\mathcal{R}$ [8].

A knowledge base $\mathcal{R}$ is *consistent* iff there exists an OCF κ such that $\kappa \models \mathcal{R}$. Such an OCF can be found if and only if there is an ordered partitioning $(\mathcal{R}_0, ..., \mathcal{R}_m)$ of $\mathcal{R}$ with the property that for every $0 \leqslant i \leqslant m$ every conditional $(B|A) \in \mathcal{R}_i$ is tolerated by $\bigcup_{j=i}^{m} \mathcal{R}_j$ [8,13].

Example 1 ($\mathcal{R}_{bird}$). We illustrate the definitions and propositions in this article with the well-known penguin example. Here, the variables in the alphabet $\Sigma = \{P, B, F\}$ indicate whether something is a bird (b) or not ($\overline{b}$), can fly (f) or not ($\overline{f}$) and whether something is a penguin (p) or not ($\overline{p}$) which results in the possible worlds $\Omega = \{pbf, pb\overline{f}, p\overline{b}f, p\overline{b}\,\overline{f}, \overline{p}bf, \overline{p}b\overline{f}, \overline{p}\,\overline{b}f, \overline{p}\,\overline{b}\,\overline{f}\}$. The knowledge base $\mathcal{R}_{bird} = \{\delta_1, \delta_2, \delta_3, \delta_4\}$ consists of the four conditionals:

δ_1 : $(f|b)$ "If something is a bird, it usually can fly."
δ_2 : $(\overline{f}|p)$ "If something is a penguin, it usually cannot fly."
δ_3 : $(\overline{f}|pb)$ "If something is a penguin bird, it usually cannot fly."
δ_4 : $(b|p)$ "If something is a penguin, it usually is a bird."

This knowledge base is consistent: For $\mathcal{R}_0 = \{(f|b)\}$ and $\mathcal{R}_1 = \mathcal{R}_{bird} \setminus \mathcal{R}_0$ we have the ordered partitioning $(\mathcal{R}_0, \mathcal{R}_1)$ such that every conditional in $\mathcal{R}_0$ is tolerated by $R_0 \cup \mathcal{R}_1 = \mathcal{R}_{bird}$ and every conditional in $\mathcal{R}_1$ is tolerated by $\mathcal{R}_1$. For instance, $(f|b)$ is tolerated by $\mathcal{R}_{bird}$ since there is, for example, the world $\overline{p}bf$ with $\overline{p}bf \models bf$ as well as $\overline{p}bf \models (p \Rightarrow \overline{f}) \wedge (pb \Rightarrow \overline{f}) \wedge (p \Rightarrow b)$.

The following *p-entailment* is an established inference in the area of ranking functions.

Definition 1 (p-entailment [8]). *Let $\mathcal{R}$ be a conditional knowledge base and let A, B be formulas. A p-entails B in the context of $\mathcal{R}$, written $A \mathrel{\mid\!\sim}^{p}_{\mathcal{R}} B$, if and only if $A \mathrel{\mid\!\sim}^{\kappa} B$ for all $\kappa \models \mathcal{R}$.*

P-entailment can be easily characterized:

Proposition 1 [8]. *Let $\mathcal{R}$ be a conditional knowledge base and let A, B be formulas. A p-entails B in the context of a knowledge base $\mathcal{R}$, if and only if $\mathcal{R} \cup \{(\overline{B}|A)\}$ is inconsistent.*

Example 2. We illustrate p-entailment with the running example. Here knowledge base $\mathcal{R}_{bird}$ p-entails, for instance, that not-flying penguins are birds, formally, $p\overline{f} \hspace{1pt}\vdash^p_{\mathcal{R}} b$: Using Proposition 1, we observe that $\mathcal{R}_{bird} \cup \{(\overline{b}|p\overline{f})\}$ is inconsistent because every world ω that verifies the conditional $(\overline{b}|p\overline{f})$, i.e. $\omega \models p\overline{b}\,\overline{f}$, violates $(b|p)$, and every world ω that verifies $(b|p)$, i.e. $\omega \models pb$, violates $(\overline{b}|p\overline{f})$. Therefore, the conditional $(\overline{b}|p\overline{f})$ is neither tolerated by $\mathcal{R}_{bird}$ nor does it tolerate $\mathcal{R}_{bird}$ and hence $\mathcal{R}_{bird} \cup \{(\overline{b}|p\overline{f})\}$ is inconsistent. Hence by Definition 1 we obtain $p\overline{f} \hspace{1pt}\vdash^p_{\mathcal{R}} b$.

Nonmonotonic inference relations are usually evaluated by means of properties. In particular, the axiom system P [1] provides an important standard for plausible, nonmonotonic inferences. With $\vdash$ being a generic nonmonotonic inference operator and A, B, C being formulas in $\mathfrak{L}$, the six properties of system P are defined as follows:

(REF) Reflexivity		for all $A \in \mathfrak{L}$ it holds that $A \vdash A$
(LLE) Left Logical Equivalence		$A \equiv B$ and $B \vdash C$ imply $A \vdash C$
(RW) Right weakening		$B \models C$ and $A \vdash B$ imply $A \vdash C$
(CM) Cautious Monotony		$A \vdash B$ and $A \vdash C$ imply $AB \vdash C$
(CUT)		$A \vdash B$ and $AB \vdash C$ imply $A \vdash C$
(OR)		$A \vdash C$ and $B \vdash C$ imply $(A \vee B) \vdash C$

We refer to Dubois and Prade [4] for the relation between p-entailment and system P:

Proposition 2 [4]. *Let A, B be formulas and let $\mathcal{R}$ be a conditional knowledge base. B follows from A in the context of $\mathcal{R}$ with the rules of system P if and only if A p-entails B in the context of $\mathcal{R}$.*

So, given a knowledge base $\mathcal{R}$, system P inference is the same as p-entailment.

Two inference relations which are defined by specific OCFs obtained inductively from a knowledge base $\mathcal{R}$ have received some attention: system Z and c-representations, or induced inference relations, respectively, both show excellent inference properties. We recall both approaches briefly.

System Z [13] is based upon the ranking function κ^Z, which is the unique Pareto-minimal OCF that accepts $\mathcal{R}$. The system is set up by forming an ordered partition $(\mathcal{R}_0, ..., \mathcal{R}_m)$ of $\mathcal{R}$, where each $\mathcal{R}_i$ is the (with respect to set inclusion) maximal subset of $\bigcup_{j=i}^{m} \mathcal{R}_j$ that is tolerated by $\bigcup_{j=i}^{m} \mathcal{R}_j$. This partitioning is unique due to the maximality. The resulting OCF κ^Z is defined by assigning to each world ω a rank of 1 plus the maximal index $1 \leqslant i \leqslant m$ of the partition that contains conditionals falsified by ω or 0 if ω does not falsify any conditional in $\mathcal{R}$. Formally, for all $(B|A) \in \mathcal{R}$ and for $Z(B|A) = i$ if and only if $(B|A) \in \mathcal{R}_i$, the OCF κ^Z is given by:

$$\kappa^Z(\omega) = \begin{cases} 0 & \text{iff } \omega \text{ does not falsify any conditional in } \mathcal{R} \\ \max\{Z(B|A)|(B|A) \in \mathcal{R}, \omega \models A\overline{B}\} + 1 & \text{otherwise.} \end{cases}$$

Other than system Z, the approach of c-representations does not use the most severe falsification of a conditional, but assigns an individual impact to each conditional and generates the world ranks as a sum of impacts of falsified conditionals.

Definition 2 (c-representation [10,11]). *Let $\mathcal{R}$ be a knowledge base. A c-representation of $\mathcal{R}$ is a ranking function κ constructed from integer impacts $\eta_i \in \mathbb{N}_0$ assigned to each conditional $(B_i|A_i)$ such that κ accepts $\mathcal{R}$ and is given by:*

$$\kappa(\omega) = \sum_{\substack{1 \leqslant i \leqslant n \\ \omega \models A_i \overline{B}_i}} \eta_i \tag{3}$$

Examples of system Z and c-representations are given in the following sections.

3   Correctness and Completeness of a CSP Modeling of C-Representations

In [2], a modeling of c-representations as solutions of a constraint satisfaction problem is proposed and employed for computing c-representations using constraint logic programming. In this section, we first recall this modeling, and then prove its correctness and completeness.

Definition 3 ($CR(\mathcal{R})$ [2]). *Let $\mathcal{R} = \{(B_1|A_1), \ldots, (B_n|A_n)\}$. The constraint satisfaction problem for c-representations of $\mathcal{R}$, denoted by $CR(\mathcal{R})$, is given by the conjunction of the constraints, for all $i \in \{1, \ldots, n\}$:*

$$\eta_i \geqslant 0 \tag{4}$$

$$\eta_i > \min_{\omega \models A_i B_i} \sum_{\substack{j \neq i \\ \omega \models A_j \overline{B}_j}} \eta_j - \min_{\omega \models A_i \overline{B}_i} \sum_{\substack{j \neq i \\ \omega \models A_j \overline{B}_j}} \eta_j \tag{5}$$

A solution of $CR(\mathcal{R})$ is an n-tuple $(\eta_1, \ldots, \eta_n)$ of natural numbers. For a constraint satisfaction problem CSP, the set of solutions is denoted by $Sol(CSP)$. Thus, with $Sol(CR(\mathcal{R}))$ we denote the set of all solutions of $CR(\mathcal{R})$.

Example 3. The verification/falsification behaviour of the conditionals in $\mathcal{R}_{bird}$ from Example 1 is given in Table 1. Based on these evaluations, the constraints in $CR(\mathcal{R}_{bird})$ according to (5) are

$$\eta_1 > \min\{\eta_2 + \eta_3, 0\} - \min\{0, 0\} = 0 \tag{6}$$

$$\eta_2 > \min\{\eta_1, \eta_4\} - \min\{\eta_3, \eta_4\} \tag{7}$$

$$\eta_3 > \eta_1 - \eta_2 \tag{8}$$

$$\eta_4 > \min\{\eta_2 + \eta_3, \eta_1\} - \min\{\eta_2, 0\} = \min\{\eta_2 + \eta_3, \eta_1\} \tag{9}$$

because $\eta_i \geqslant 0$ for all $1 \leqslant i \leqslant n$. The inequality (8) is equivalent to $\eta_2 + \eta_3 > \eta_1$, which together with (9) gives us $\eta_4 > \eta_1$, and we finally obtain

$$
\begin{aligned}
&\eta_1 > 0 && \eta_3 > \eta_1 - \eta_2 \\
&\eta_2 > \eta_1 - \min\{\eta_3, \eta_4\} && \eta_4 > \eta_1.
\end{aligned}
\tag{10}
$$

Possible solutions for this system that also satisfy the constraint (4) are $\vec{\eta}^{(1)} = (1,1,1,2)$, $\vec{\eta}^{(2)} = (1,2,0,2)$ and $\vec{\eta}^{(3)} = (1,0,3,2)$.

Table 1. Verification/falsification behavior of $\mathcal{R}_{bird}$: $(+)$ indicates verification, $(-)$ falsification and an empty cell non-applicability.

	pbf	$pb\overline{f}$	$p\overline{b}f$	$p\overline{b}\,\overline{f}$	$\overline{p}bf$	$\overline{p}b\overline{f}$	$\overline{p}\,\overline{b}f$	$\overline{p}\,\overline{b}\,\overline{f}$
$\delta_1 = (f\|b)$	+	−			+	−		
$\delta_2 = (\overline{f}\|p)$	−	+	−	+				
$\delta_3 = (\overline{f}\|pb)$	−	+						
$\delta_4 = (b\|p)$	+	+	−	−				

Proposition 3 (Correctness of $CR(\mathcal{R})$). *For $\mathcal{R} = \{(B_1|A_1), \ldots, (B_n|A_n)\}$ let $\vec{\eta} = (\eta_1, \ldots, \eta_n) \in Sol(CR(\mathcal{R}))$. Then the function κ defined by the equation system given by (3) is a c-representation that accepts $\mathcal{R}$.*

Proof. To show that the modeling given by the constraint satisfaction problem $CR(\mathcal{R})$ is correct we have to show that every solution of the constraints given by (4) and (5) is a c-representation of $\mathcal{R}$. We will use the techniques for showing c-representation properties given in [10] to show that κ accepts $\mathcal{R}$, i.e., for all $(B_i|A_i) \in \mathcal{R}$

$$
\kappa(A_i B_i) < \kappa(A_i \overline{B}_i).
\tag{11}
$$

Using the definition of ranks of formulas for each $1 \leqslant i \leqslant n$, (11) gives us

$$
\min_{\omega \models A_i B_i} \{\kappa(\omega)\} < \min_{\omega \models A_i \overline{B}_i} \{\kappa(\omega)\}.
\tag{12}
$$

We now use (3) and get:

$$
\min_{\omega \models A_i B_i} \left\{ \sum_{\substack{1 \leqslant j \leqslant n \\ \omega \models A_j \overline{B}_j}} \eta_i \right\} < \min_{\omega \models A_i \overline{B}_i} \left\{ \underbrace{\sum_{\substack{1 \leqslant j \leqslant n \\ \omega \models A_j \overline{B}_j}} \eta_i}_{(a)} \right\}
\tag{13}
$$

In each line i, η_i is a summand of (a) and hence this summand can be extracted from the minimum, yielding:

$$\min_{\omega \models A_i B_i} \left\{ \sum_{\substack{1 \leqslant j \leqslant n \\ \omega \models A_j \overline{B_j}}} \eta_i \right\} < \min_{\omega \models A_i \overline{B_i}} \left\{ \sum_{\substack{1 \leqslant j \leqslant n; i \neq j \\ \omega \models A_j \overline{B_j}}} \eta_i \right\} + \eta_i \tag{14}$$

We rearrange the inequality in (14) and get

$$\eta_i > \underbrace{\min_{\omega \models A_i B_i} \left\{ \sum_{\substack{1 \leqslant j \leqslant n \\ \omega \models A_j \overline{B_j}}} \eta_i \right\} - \min_{\omega \models A_i \overline{B_i}} \left\{ \sum_{\substack{1 \leqslant j \leqslant n; i \neq j \\ \omega \models A_j \overline{B_j}}} \eta_i \right\}}_{(b)}. \tag{15}$$

In the first minimum in (15), $(B_i | A_i)$ is never falsified, so i can be removed from the range of (b), which gives us

$$\eta_i > \min_{\omega \models A_i B_i} \left\{ \sum_{\substack{1 \leqslant j \leqslant n; i \neq j \\ \omega \models A_j \overline{B_j}}} \eta_i \right\} - \min_{\omega \models A_i \overline{B_i}} \left\{ \sum_{\substack{1 \leqslant j \leqslant n; i \neq j \\ \omega \models A_j \overline{B_j}}} \eta_i \right\} \tag{16}$$

which is (5). Therefore, every solution of $CR(\mathcal{R})$ ensures that the resulting κ defined by (3) in Definition 2 is a c-representation that accepts $\mathcal{R}$. □

Definition 4 ($\kappa_{\vec{\eta}}$). *For $\vec{\eta} \in Sol(CR(\mathcal{R}))$ and κ as in Eq. (3), κ is the OCF induced by $\vec{\eta}$ and is denoted by $\kappa_{\vec{\eta}}$.*

Example 4. Using (3) with the solutions for the CSP calculated in Example 3 gives us the OCFs shown in Table 2. All of these OCFs accept the knowledge base $\mathcal{R}_{bird}$; for $\vec{\eta}^{(1)} = (1, 1, 1, 2)$ we have, for instance,

$$\kappa_{\vec{\eta}^{(1)}}(bf) = \min\{\kappa_{\vec{\eta}^{(1)}}(pbf), \kappa_{\vec{\eta}^{(1)}}(\overline{p}bf)\} = 0$$
$$\kappa_{\vec{\eta}^{(1)}}(b) = \min\{\kappa_{\vec{\eta}^{(1)}}(pbf), \kappa_{\vec{\eta}^{(1)}}(pb\overline{f}), \kappa_{\vec{\eta}^{(1)}}(\overline{p}bf), \kappa_{\vec{\eta}^{(1)}}(\overline{p}b\overline{f})\} = 0$$
$$\kappa_{\vec{\eta}^{(1)}}(b\overline{f}) = \min\{\kappa_{\vec{\eta}^{(1)}}(pb\overline{f}), \kappa_{\vec{\eta}^{(1)}}(\overline{p}b\overline{f})\} = 1$$

and hence:

$$\kappa_{\vec{\eta}^{(1)}}(f|b) = \kappa_{\vec{\eta}^{(1)}}(bf) - \kappa_{\vec{\eta}^{(1)}}(b) = 0 < 1 = \kappa_{\vec{\eta}^{(1)}}(b\overline{f}) - \kappa_{\vec{\eta}^{(1)}}(b) = \kappa_{\vec{\eta}^{(1)}}(\overline{f}|b)$$

The other ranks for the verification resp. falsification of the conditionals given these ranking functions are given in Table 3 from which we can see that each of the three induced OCFs accepts $\mathcal{R}_{bird}$.

Proposition 4 (Completeness of $CR(\mathcal{R})$). *Let κ be a c-representation for a knowledge base $\mathcal{R} = \{(B_1|A_1), \ldots, (B_n|A_n)\}$, i.e., $\kappa \models \mathcal{R}$. Then there is a vector $\vec{\eta} \in Sol(CR(\mathcal{R}))$ such that $\kappa = \kappa_{\vec{\eta}}$.*

Table 2. Ranking functions for the penguin example $\mathcal{R}_{bird}$.

ω	pbf	$pb\overline{f}$	$p\overline{b}f$	$p\overline{b}\,\overline{f}$	$\overline{p}bf$	$\overline{p}b\overline{f}$	$\overline{p}\,\overline{b}f$	$\overline{p}\,\overline{b}\,\overline{f}$
$\kappa_{\vec{\eta}^{(1)}}(\omega)$	2	1	3	2	0	1	0	0
$\kappa_{\vec{\eta}^{(2)}}(\omega)$	2	1	4	2	0	1	0	0
$\kappa_{\vec{\eta}^{(3)}}(\omega)$	3	1	2	2	0	1	0	0

Table 3. Acceptance of conditionals in the penguin example $\mathcal{R}_{bird}$ by the OCFs in Table 2. Note that $\kappa(B|A) = \kappa(AB) - \kappa(A)$ for all $A, B \in \mathfrak{L}$.

	$\kappa_{\vec{\eta}^{(1)}}$			$\kappa_{\vec{\eta}^{(2)}}$			$\kappa_{\vec{\eta}^{(3)}}$		
	verif.		falsif. accpt?	verif.		falsif. accpt?	verif.		falsif. accpt?
$(f|b)$	0	<	1 ✓	0	<	1 ✓	0	<	1 ✓
$(\overline{f}|p)$	0	<	1 ✓	0	<	1 ✓	0	<	1 ✓
$(\overline{f}|pb)$	0	<	1 ✓	0	<	1 ✓	0	<	2 ✓
$(b|p)$	2	<	1 ✓	0	<	1 ✓	0	<	1 ✓

Proof. A ranking function is a c-representation for $\mathcal{R}$ if and only if it is composed by (3) and accepts $\mathcal{R}$. For the proof of Proposition 3 we have shown that these two conditions are equivalent to the impacts being chosen to satisfy (4) and (5). Therefore for every c-representation κ for $\mathcal{R}$ there is a vector $\vec{\eta} \in Sol(CR(\mathcal{R}))$ such that $\kappa = \kappa_{\vec{\eta}}$, as proposed.   □

It has been shown that there is a c-representation for a knowledge base $\mathcal{R}$ if and only if $\mathcal{R}$ is consistent [10,11]. The completeness and correctness results in Propositions 3 and 4 give us that $CR(\mathcal{R})$ is solvable if and only if there is a c-representation for $\mathcal{R}$. This gives us an additional criterion for the consistency of a knowledge base which we formalize as follows.

Corollary 1 (Consistency). *A knowledge base $\mathcal{R}$ is consistent iff the constraint satisfaction problem $CR(\mathcal{R})$ is solvable.*

Applying a constraint satisfaction solver, Corollary 1 gives us an implementable alternative to the tolerance test algorithm in [13].

4   Skeptical Inference Based on C-Representations

Equation (2) defines an inference relation $\vdash^{\kappa}$ based on a single OCF κ. For a given knowledge base $\mathcal{R}$ and two formulas A, B we will now introduce a novel skeptical inference relation based on all c-representations.

Definition 5 (c-inference, $\vdash^{c}_{\mathcal{R}}$). *Let $\mathcal{R}$ be a knowledge base and let A, B be formulas. B is a (skeptical) c-inference from A in the context of $\mathcal{R}$, denoted by $A \vdash^{c}_{\mathcal{R}} B$, iff $A \vdash^{\kappa} B$ holds for all c-representations κ for $\mathcal{R}$.*

We will show that skeptical c-inference is different from p-entailment which is equivalent to the skeptical inference relation obtained by considering all OCFs that accept $\mathcal{R}$, and that it is able to preserve high-quality inference properties that inference based on single c-representations has.

Example 5 ($\mathrel{|\!\sim}_{cr}$). Consider the three OCFs $\kappa_{\vec{\eta}^{(1)}}$, $\kappa_{\vec{\eta}^{(2)}}$, and $\kappa_{\vec{\eta}^{(3)}}$ from Table 2 calculated in Example 4 that are induced by the solutions $\vec{\eta}^{(1)}$, $\vec{\eta}^{(2)}$, and $\vec{\eta}^{(3)}$ of $CR(\mathcal{R}_{bird})$ given in Example 3. In Table 4, their acceptance properties with respect to some conditionals that are not contained in $\mathcal{R}_{bird}$ are given. From the acceptance properties in Table 4, we conclude that b is not a c-inference of pf in the context of $\mathcal{R}_{bird}$, denoted by $pf \mathrel{|\!\not\sim}^{c}_{\mathcal{R}_{bird}} b$ since e.g. $\kappa_{\vec{\eta}^{(3)}} \not\models (b|pf)$. We also have $pf \mathrel{|\!\not\sim}^{c}_{\mathcal{R}_{bird}} \overline{b}$ since e.g. $\kappa_{\vec{\eta}^{(2)}} \not\models (\overline{b}|pf)$. Thus, by c-inference we can neither infer that flying penguins are birds nor can we infer that flying penguins are not birds.

Table 4. Acceptance properties for different ranking functions that accept $\mathcal{R}_{bird}$ with respect to some conditionals not contained in $\mathcal{R}_{bird}$

	$\kappa_{\vec{\eta}^{(1)}}$			$\kappa_{\vec{\eta}^{(2)}}$			$\kappa_{\vec{\eta}^{(3)}}$		
	verif.	falsif.	accpt.	verif.	falsif.	accpt.	verif.	falsif.	accpt.
$(b|pf)$	0	< 1	✓	0	< 2	✓	1	> 0	↯
$(\overline{b}|pf)$	1	> 0	↯	2	> 0	↯	0	< 1	✓
$(b\overline{f}|p)$	0	< 1	✓	0	< 1	✓	0	< 1	✓

On the other hand, c-inference allows the plausible conclusion that flying birds are no penguins, i.e., we have $bf \mathrel{|\!\sim}^{c}_{\mathcal{R}_{bird}} \overline{p}$. For a first illustration, to see that the conditional $(\overline{p}|pb)$ is accepted by the three OCFs given in Example 4, observe that $\kappa_{\vec{\eta}^{(1)}} \models (\overline{p}|bf)$ since $\kappa_{\vec{\eta}^{(1)}}(\overline{p}bf) = 0 < 2 = \kappa_{\vec{\eta}^{(1)}}(pbf)$, $\kappa_{\vec{\eta}^{(2)}} \models (\overline{p}|bf)$ since $\kappa_{\vec{\eta}^{(2)}}(\overline{p}bf) = 0 < 2 = \kappa_{\vec{\eta}^{(2)}}(pbf)$ and $\kappa_{\vec{\eta}^{(3)}} \models (\overline{p}|bf)$ since $\kappa_{\vec{\eta}^{(3)}}(\overline{p}bf) = 0 < 3 = \kappa_{\vec{\eta}^{(3)}}(pbf)$. More generally, the conditional $(\overline{p}|bf)$ is accepted by all c-representations of $\mathcal{R}_{bird}$ since we have $\kappa_{\vec{\eta}}(\overline{p}bf) = 0$ and $\kappa_{\vec{\eta}}(pbf) = \eta_2 + \eta_3$, for every solution $\vec{\eta}$ of the CSP because of (3) and Table 1. From the system of inequalities (10) in Example 3 we obtain that $\eta_2 + \eta_3 > \eta_1 > 0$. Therefore $\kappa_{\vec{\eta}}(\overline{p}bf) < \kappa_{\vec{\eta}}(pbf)$ which implies $\kappa_{\vec{\eta}} \models (\overline{p}|bf)$ for all solutions $\vec{\eta}$, by which we obtain $bf \mathrel{|\!\sim}^{c}_{\mathcal{R}_{bird}} \overline{p}$ using Definition 5. Another c-inference of $\mathcal{R}_{bird}$ is that non-flying penguins are birds, i.e., we have $p\overline{f} \mathrel{|\!\sim}^{c}_{\mathcal{R}_{bird}} b$: According to (3) and Table 1 we have $\kappa(pb\overline{f}) = \eta_1$ and $\kappa(p\overline{b}\,\overline{f}) = \eta_4$. Equation (10) in Example 3 gives us that $\eta_4 > \eta_1$ and therefore we have $p\overline{f} \mathrel{|\!\sim}^{\kappa} b$ by (2) for every c-representation κ of $\mathcal{R}_{bird}$ and hence $p\overline{f} \mathrel{|\!\sim}^{c}_{\mathcal{R}_{bird}} b$.

Thus, overall from this example we obtain that for $\mathcal{R}_{bird}$ we have, for instance, $p\overline{f} \mathrel{|\!\sim}^{c}_{\mathcal{R}_{bird}} b$ but neither $pf \mathrel{|\!\not\sim}^{c}_{\mathcal{R}_{bird}} b$ nor $pf \mathrel{|\!\not\sim}^{c}_{\mathcal{R}_{bird}} \overline{b}$.

Comparing Definitions 1 to 5, we find that c-inference is defined in full analogy to p-entailment but with the set of OCF that accept $\mathcal{R}$ being restricted to c-representations of $\mathcal{R}$. An obvious question is what the exact relationship between c-inference and p-entailment is, and which features of c-representations still hold for c-inference. First, we show that c-inference satisfies system P but allows for additional inferences.

Proposition 5. *Let $\mathcal{R}$ be a knowledge base and let A, B be formulas. If B can be inferred from A in the context of $\mathcal{R}$ using system P, then it can also be c-inferred from A in the context of $\mathcal{R}$.*

Proof. By Definition 1, B can be p-entailed from A in the context of $\mathcal{R}$ if and only if the conditional $(B|A)$ is accepted by every OCF that accepts $\mathcal{R}$. Naturally, if $(B|A)$ is accepted by all OCFs that accept $\mathcal{R}$, then $(B|A)$ is also accepted by every subset of all OCFs that accept $\mathcal{R}$. Since every c-representation accepts $\mathcal{R}$, we obtain that if B is p-entailed from A given $\mathcal{R}$, then it is also c-inferred. Since p-entailment is equivalent to system P inference (cf. Proposition 2) we conclude that every system P inference from $\mathcal{R}$ can also be drawn using c-inference on $\mathcal{R}$. □

To show that c-inference allows for inferences beyond system P we consider the following example.

Example 6 ($\mathcal{R}'_{bird}$). We use the knowledge base

$$\mathcal{R}'_{bird} = \{\delta_1 : (f|b), \ \delta_4 : (b|p)\}$$

which is a proper subset of $\mathcal{R}_{bird}$ from Example 1. For each impact vector $\vec{\eta} = (\eta_1, \eta_4)$ for $\mathcal{R}'_{bird}$, we obtain the inequalities $\eta_1 > 0$ and $\eta_4 > 0$ by the verification/falsification behavior from Table 5, implying

$$\kappa_{\vec{\eta}}(pf) = \min\{\kappa_{\vec{\eta}}(pbf), \kappa_{\vec{\eta}}(p\overline{b}f)\} = \min\{0, \eta_4\} = 0$$
$$\kappa_{\vec{\eta}}(p\overline{f}) = \min\{\kappa_{\vec{\eta}}(pb\overline{f}), \kappa_{\vec{\eta}}(p\overline{b}\,\overline{f})\} = \min\{\eta_1, \eta_4\} > 0$$

and hence $\kappa_{\vec{\eta}}(pf) < \kappa_{\vec{\eta}}(p\overline{f})$ for the OCF induced by $\vec{\eta}$. Thus, $\kappa(pf) < \kappa(p\overline{f})$ for every c-representation κ of $\mathcal{R}'_{bird}$, giving us $p \vdash^c_{\mathcal{R}'_{bird}} f$. Note that this inference is reasonable with respect to $\mathcal{R}'_{bird}$, since $\mathcal{R}'_{bird}$ does not contain any information that can inhibit this chaining of rules.

Table 5. Verification/falsification behaviour and abstract weights for $\mathcal{R}'_{bird}$ from Example 6 and the OCF $\kappa_{(\eta_1,\eta_4)}$ induced by an impact vector (η_1, η_4).

	pbf	$pb\overline{f}$	$p\overline{b}f$	$p\overline{b}\overline{f}$	$\overline{p}bf$	$\overline{p}b\overline{f}$	$\overline{p}\overline{b}f$	$\overline{p}\overline{b}\overline{f}$
$\delta_1 = (f|b)$	+	−			+	−		
$\delta_4 = (b|p)$	+	+	−	−				
$\kappa_{(\eta_1,\eta_4)}(\omega)$	0	η_1	η_4	η_4	0	η_1	0	0

Proposition 6. *There are knowledge bases $\mathcal{R}$ and propositions A, B such that B is c-entailed, but not p-entailed, from A in the context of $\mathcal{R}$.*

Proof. $\mathcal{R}'_{bird}$ from Example 6 is an example for such a knowledge base. Here, we have $p \mathrel{\vmid^c_{\mathcal{R}'_{bird}}} f$. From Proposition 1 we obtain that if we had $p \mathrel{\vmid^p_{\mathcal{R}'_{bird}}} b$, then $\mathcal{R}'_{bird} \cup \{(\overline{f}|p)\}$ would be inconsistent. This is not the case since $\mathcal{R}'_{bird} \cup \{(\overline{f}|p)\}$ is consistent (e.g. with the tolerance partitioning $(\{(f|b)\}, \{(\overline{f}|p), (b|p)\})$), which gives us that $(f|p)$ is c-entailed, but not p-entailed from $\mathcal{R}'_{bird}$.   □

From these two propositions we conclude:

Corollary 2. *Every system P entailment of a knowledge base $\mathcal{R}$ is also a c-inference of $\mathcal{R}$; the reverse is not true in general.*

We have seen that c-inference exceeds system P. In the following we examine benchmarks for plausible inference relations, namely subclass inheritance, irrelevance, and rule chaining, which we illustrate using the following modification of the running example.

Example 7 ($\mathcal{R}^*_{bird}$). We extend the alphabet $\Sigma = \{P, B, F\}$ of our running example knowledge base $\mathcal{R}_{bird}$ from Example 1 with the variables W for *having wings* (w) or not ($\overline{w}$), A for *being airborne* (a) or not ($\overline{a}$), and R for *being red* (r) or not ($\overline{r}$) to obtain the alphabet $\Sigma^* = \{P, B, F, W, A, R\}$. We use the knowledge base

$$\mathcal{R}^*_{bird} = \{\delta_1 : (f|b), \ \delta_2 : (\overline{f}|p), \ \delta_4 : (b|p), \ \delta_5 : (w|b), \ \delta_6 : (a|f)\}$$

where the conditional $\delta_5 = (w|b)$ encodes the rule that birds usually have wings and the conditional $\delta_6 = (a|f)$ encodes the rule that flying things are usually airborne; the other three conditionals $\delta_1, \delta_2, \delta_4$ are the same as in $\mathcal{R}_{bird}$. The verification/falsification behavior of the worlds for the knowledge base $\mathcal{R}^*_{bird}$ is given in Table 6. For each impact vector $\vec{\eta} = (\eta_1, \eta_2, \eta_4, \eta_5, \eta_6)$ for $\mathcal{R}^*_{bird}$, the constraints defined by (4) and (5) give us the following system of inequations:

$$\eta_1 > 0 \tag{17}$$
$$\eta_2 > \min\{\eta_1, \eta_4\} \tag{18}$$
$$\eta_4 > \min\{\eta_1, \eta_2\} \tag{19}$$
$$\eta_5 > 0 \tag{20}$$
$$\eta_6 > 0. \tag{21}$$

If we assume $\eta_1 \geqslant \eta_2$ then (19) would give us $\eta_4 > \eta_2$ which would imply that $\eta_1 < \eta_4$ by (18). But then, (18) would also require $\eta_1 < \eta_2$ in contradiction to the assumption. Therefore, we conclude $\eta_1 \not\geqslant \eta_2$ and hence $\eta_1 < \eta_2$, which gives us the inequalities

$$\begin{array}{lll} \eta_1 > 0 & \eta_2 > \eta_1 & \eta_4 > \eta_1 \\ \eta_5 > 0 & \eta_6 > 0. & \end{array} \tag{22}$$

Table 6. Verification/falsification behavior of the worlds for the knowledge base $\mathcal{R}^*_{bird}$ (Example 7) and ranking function κ^Z obtained from $\mathcal{R}^*_{bird}$ using System Z.

ω	verifies	falsifies	κ^Z	ω	verifies	falsifies	κ^Z
$p\,b\,f\,w\,a\,r$	$\delta_1,\delta_4,\delta_5,\delta_6$	δ_2	2	$\bar{p}\,b\,f\,w\,a\,r$	$\delta_1,\delta_5,\delta_6$	—	0
$p\,b\,f\,w\,a\,\bar{r}$	$\delta_1,\delta_4,\delta_5,\delta_6$	δ_2	2	$\bar{p}\,b\,f\,w\,a\,\bar{r}$	$\delta_1,\delta_5,\delta_6$	—	0
$p\,b\,f\,w\,\bar{a}\,r$	$\delta_1,\delta_4,\delta_5$	δ_2,δ_6	2	$\bar{p}\,b\,f\,w\,\bar{a}\,r$	δ_1,δ_5	δ_6	1
$p\,b\,f\,w\,\bar{a}\,\bar{r}$	$\delta_1,\delta_4,\delta_5$	δ_2,δ_6	2	$\bar{p}\,b\,f\,w\,\bar{a}\,\bar{r}$	δ_1,δ_5	δ_6	1
$p\,b\,f\,\bar{w}\,a\,r$	$\delta_1,\delta_4,\delta_6$	δ_2,δ_5	2	$\bar{p}\,b\,f\,\bar{w}\,a\,r$	δ_1,δ_6	δ_5	1
$p\,b\,f\,\bar{w}\,a\,\bar{r}$	$\delta_1,\delta_4,\delta_6$	δ_2,δ_5	2	$\bar{p}\,b\,f\,\bar{w}\,a\,\bar{r}$	δ_1,δ_6	δ_5	1
$p\,b\,f\,\bar{w}\,\bar{a}\,r$	δ_1,δ_4	$\delta_2,\delta_5,\delta_6$	2	$\bar{p}\,b\,f\,\bar{w}\,\bar{a}\,r$	δ_1	δ_5,δ_6	1
$p\,b\,f\,\bar{w}\,\bar{a}\,\bar{r}$	δ_1,δ_4	$\delta_2,\delta_5,\delta_6$	2	$\bar{p}\,b\,f\,\bar{w}\,\bar{a}\,\bar{r}$	δ_1	δ_5,δ_6	1
$p\,b\,\bar{f}\,w\,a\,r$	$\delta_2,\delta_4,\delta_5$	δ_1	1	$\bar{p}\,b\,\bar{f}\,w\,a\,r$	δ_5	δ_1	1
$p\,b\,\bar{f}\,w\,a\,\bar{r}$	$\delta_2,\delta_4,\delta_5$	δ_1	1	$\bar{p}\,b\,\bar{f}\,w\,a\,\bar{r}$	δ_5	δ_1	1
$p\,b\,\bar{f}\,w\,\bar{a}\,r$	$\delta_2,\delta_4,\delta_5$	δ_1	1	$\bar{p}\,b\,\bar{f}\,w\,\bar{a}\,r$	δ_5	δ_1	1
$p\,b\,\bar{f}\,w\,\bar{a}\,\bar{r}$	$\delta_2,\delta_4,\delta_5$	δ_1	1	$\bar{p}\,b\,\bar{f}\,w\,\bar{a}\,\bar{r}$	δ_5	δ_1	1
$p\,b\,\bar{f}\,\bar{w}\,a\,r$	δ_2,δ_4	δ_1,δ_5	1	$\bar{p}\,b\,\bar{f}\,\bar{w}\,a\,r$	—	δ_1,δ_5	1
$p\,b\,\bar{f}\,\bar{w}\,a\,\bar{r}$	δ_2,δ_4	δ_1,δ_5	1	$\bar{p}\,b\,\bar{f}\,\bar{w}\,a\,\bar{r}$	—	δ_1,δ_5	1
$p\,b\,\bar{f}\,\bar{w}\,\bar{a}\,r$	δ_2,δ_4	δ_1,δ_5	1	$\bar{p}\,b\,\bar{f}\,\bar{w}\,\bar{a}\,r$	—	δ_1,δ_5	1
$p\,b\,\bar{f}\,\bar{w}\,\bar{a}\,\bar{r}$	δ_2,δ_4	δ_1,δ_5	1	$\bar{p}\,b\,\bar{f}\,\bar{w}\,\bar{a}\,\bar{r}$	—	δ_1,δ_5	1
$p\,\bar{b}\,f\,w\,a\,r$	δ_6	δ_2,δ_4	2	$\bar{p}\,\bar{b}\,f\,w\,a\,r$	δ_6	—	0
$p\,\bar{b}\,f\,w\,a\,\bar{r}$	δ_6	δ_2,δ_4	2	$\bar{p}\,\bar{b}\,f\,w\,a\,\bar{r}$	δ_6	—	0
$p\,\bar{b}\,f\,w\,\bar{a}\,r$	—	$\delta_2,\delta_4,\delta_6$	2	$\bar{p}\,\bar{b}\,f\,w\,\bar{a}\,r$	—	δ_6	1
$p\,\bar{b}\,f\,w\,\bar{a}\,\bar{r}$	—	$\delta_2,\delta_4,\delta_6$	2	$\bar{p}\,\bar{b}\,f\,w\,\bar{a}\,\bar{r}$	—	δ_6	1
$p\,\bar{b}\,f\,\bar{w}\,a\,r$	δ_6	δ_2,δ_4	2	$\bar{p}\,\bar{b}\,f\,\bar{w}\,a\,r$	δ_6	—	0
$p\,\bar{b}\,f\,\bar{w}\,a\,\bar{r}$	δ_6	δ_2,δ_4	2	$\bar{p}\,\bar{b}\,f\,\bar{w}\,a\,\bar{r}$	δ_6	—	0
$p\,\bar{b}\,f\,\bar{w}\,\bar{a}\,r$	—	$\delta_2,\delta_4,\delta_6$	2	$\bar{p}\,\bar{b}\,f\,\bar{w}\,\bar{a}\,r$	—	δ_6	1
$p\,\bar{b}\,f\,\bar{w}\,\bar{a}\,\bar{r}$	—	$\delta_2,\delta_4,\delta_6$	2	$\bar{p}\,\bar{b}\,f\,\bar{w}\,\bar{a}\,\bar{r}$	—	δ_6	1
$p\,\bar{b}\,\bar{f}\,w\,a\,r$	δ_2	δ_4	2	$\bar{p}\,\bar{b}\,\bar{f}\,w\,a\,r$	—	—	0
$p\,\bar{b}\,\bar{f}\,w\,a\,\bar{r}$	δ_2	δ_4	2	$\bar{p}\,\bar{b}\,\bar{f}\,w\,a\,\bar{r}$	—	—	0
$p\,\bar{b}\,\bar{f}\,w\,\bar{a}\,r$	δ_2	δ_4	2	$\bar{p}\,\bar{b}\,\bar{f}\,w\,\bar{a}\,r$	—	—	0
$p\,\bar{b}\,\bar{f}\,w\,\bar{a}\,\bar{r}$	δ_2	δ_4	2	$\bar{p}\,\bar{b}\,\bar{f}\,w\,\bar{a}\,\bar{r}$	—	—	0
$p\,\bar{b}\,\bar{f}\,\bar{w}\,a\,r$	δ_2	δ_4	2	$\bar{p}\,\bar{b}\,\bar{f}\,\bar{w}\,a\,r$	—	—	0
$p\,\bar{b}\,\bar{f}\,\bar{w}\,a\,\bar{r}$	δ_2	δ_4	2	$\bar{p}\,\bar{b}\,\bar{f}\,\bar{w}\,a\,\bar{r}$	—	—	0
$p\,\bar{b}\,\bar{f}\,\bar{w}\,\bar{a}\,r$	δ_2	δ_4	2	$\bar{p}\,\bar{b}\,\bar{f}\,\bar{w}\,\bar{a}\,r$	—	—	0
$p\,\bar{b}\,\bar{f}\,\bar{w}\,\bar{a}\,\bar{r}$	δ_2	δ_4	2	$\bar{p}\,\bar{b}\,\bar{f}\,\bar{w}\,\bar{a}\,\bar{r}$	—	—	0

An inference relation suffers from the *Drowning Problem* [3,13] if it does not allow to infer properties of a superclass for a subclass that is exceptional with respect to another property because the respective conditional is "drowned" by others. E.g., penguins are exceptional birds with respect to flying but not with respect to having wings. So we would reasonably expect that penguins have wings. However, system Z is known to suffer from the Drowning Problem, as the following example shows.

Example 8. System Z partitions the knowledge base $\mathcal{R}^*_{bird}$ of Example 7 into $(\mathcal{R}_0 = \{\delta_1, \delta_5, \delta_6\}, \mathcal{R}_1 = \{\delta_2, \delta_4\})$ which results in the ranking function κ^Z given in Table 6 (rightmost columns). Here we have $\kappa^Z(pw) = 1 = \kappa^Z(p\overline{w})$ and therefore we cannot infer whether penguins have wings: Every world $\omega \models p$ falsifies a conditional (cf. Table 6), and the minimal rank of every p satisfying world is 1. Since the conditional $(w|b)$ is in $\mathcal{R}_0$ and system Z always takes this maximal index of the partitions containing conditionals falsified by ω, this conditional will never contribute to the rank of any such world ω; its effect is "drowned" in the effects of the other conditionals.

The Drowning Problem distinguishes between inference relations that allow for subclass inheritance only for non-exceptional subclasses (like System Z inference) and inference relations that allow for subclass inheritance for exceptional subclasses (like inference with minimal c-representations, cf. [9,16]). Here we show that this property is preserved by c-inference, the skeptical inference over all c-representations.

Observation 1. *Skeptical c-inference does not suffer from the Drowning Problem in Example 7.*

Proof. From the observations for c-representations of $\mathcal{R}^*_{bird}$ in Example 7 together with Definition 2 we obtain, for each impact vector $\vec{\eta} = (\eta_1, \eta_2, \eta_4, \eta_5, \eta_6)$ for $\mathcal{R}^*_{bird}$ and the correspondingly induced OCF $\kappa_{\vec{\eta}}$,

$$\kappa_{\vec{\eta}}(pw) = \min\{\eta_2, \eta_2 + \eta_6, \eta_1, \eta_2 + \eta_4, \eta_2 + \eta_4 + \eta_6, \eta_4\}$$
$$= \eta_1$$

according to (22) and

$$\kappa_{\vec{\eta}}(p\overline{w}) = \min\{\underbrace{\eta_2 + \eta_5}_{>\eta_1}, \underbrace{\eta_2 + \eta_5 + \eta_6}_{>\eta_1}, \underbrace{\eta_1 + \eta_5}_{>\eta_1}, \underbrace{\eta_2 + \eta_4}_{>\eta_1}, \underbrace{\eta_2 + \eta_4 + \eta_6}_{>\eta_1}, \underbrace{\eta_4}_{>\eta_1}\}.$$

This implies $\kappa_{\vec{\eta}}(pw) < \kappa_{\vec{\eta}}(p\overline{w})$ for all impact vectors $\vec{\eta}$ of $\mathcal{R}^*_{bird}$, and therefore $\kappa(pw) < \kappa(p\overline{w})$ for all c-representations of $\mathcal{R}^*_{bird}$ and hence $p \mathrel{\vdash^c_{\mathcal{R}^*_{bird}}} w$. That is, in difference to system Z (see Example 8), using c-inference we can infer that penguins have wings in the context of $\mathcal{R}^*_{bird}$, even if they are exceptional birds with respect to flying.   □

It is straightforward to explain more generally why c-inference does not suffer from a Drowning Problem. c-inference is the skeptical inference of all c-representations of a knowledge base $\mathcal{R}$. These OCFs are set up such that every rank of every world takes the impact of every single conditional into account independently, i.e., a world that falsifies a conditional is usually less plausible than a world that, ceteris paribus, does not falsify this conditional. If we presuppose that all η_i are strictly positive, then we can definitely exclude the Drowning Problem; this means that in c-representations with strictly positive impacts, no conditional can simply "drown" in a set of others.

Another benchmark for plausible reasoning is *irrelevance*. It is safe to assume that a variable is not relevant for an inference based on a knowledge base if the variable does not appear in any conditional of the knowledge base.

Proposition 7 (c-inference and irrelevance). *Variables that do not appear in the knowledge base do not change the outcome of the inferences drawn with c-inference.*

Proof. Let Σ be a propositional alphabet and $D \in \Sigma$, and let $\mathcal{R}$ be a conditional knowledge base where there is no conditional $(B_i|A_i) \in \mathcal{R}$ such that either d or $\overline{d}$ appears in the conjunction $A_i B_i$. Let ω, ω' be a pair of worlds such that $\omega = \mathbf{o} \wedge d$ and $\omega' = \mathbf{o} \wedge \overline{d}$. Since neither d nor $\overline{d}$ is a member of any conjunction $A_i B_i$ of the conditionals $(B_i|A_i) \in \mathcal{R}$, the sets of conditionals falsified by ω and by ω', respectively, are identical. By Definition 2 this means that $\kappa(\omega) = \kappa(\omega')$. This implies that for every two formulas A, B, which are composed from the language of the alphabet $\Sigma \setminus \{D\}$, and for every configuration $\dot{d}$ of D, the conjunction AB (respectively $A\overline{B}$) falsifies a conditional $(B_i|A_i)$ if and only if $AB\dot{d}$ (respectively $A\overline{B}\dot{d}$) falsifies the conditional, and therefore $\kappa(AB) = \min\{\kappa(ABd), \kappa(AB\overline{d})\} = \kappa(AB\dot{d})$ and also $\kappa(A\overline{B}) = \min\{\kappa(A\overline{B}d), \kappa(A\overline{B}\overline{d})\} = \kappa(A\overline{B}\dot{d}) = \kappa(A\overline{B}\dot{d})$. This means for all c-representations κ of $\mathcal{R}$, if $\kappa(AB) < \kappa(A\overline{B})$, then also $\kappa(AB\dot{d}) < \kappa(A\overline{B}\dot{d})$. Thus, if $A \mathrel{\vdash^c_{\mathcal{R}}} B$, then also $A\dot{d} \mathrel{\vdash^c_{\mathcal{R}}} B$. □

We illustrate the behavior of $\mathrel{\vdash^c_{\mathcal{R}}}$ regarding variables that are not relevant using Example 7:

Example 9 (c-inference and irrelevance). Table 6 gives us that the behavior of all worlds ω for $\mathcal{R}^*_{bird}$ such that $\omega \models r$ is, ceteris paribus, identical to the behavior of all worlds ω with $\omega \models \overline{r}$. Thus, we conclude directly that for all fixed configurations $\dot{p}, \dot{b}, \dot{f}, \dot{w}, \dot{a}$ of $\{P, B, F, W, A\}$, we have $\kappa(\dot{p}\dot{b}\dot{f}\dot{w}\dot{a}r) = \kappa(\dot{p}\dot{b}\dot{f}\dot{w}\dot{a}\overline{r})$. This means that, for instance, since in the context of $\mathcal{R}^*_{bird}$ we can infer that birds can fly $(b \mathrel{\vdash^c_{\mathcal{R}^*_{bird}}} f)$, we can also infer that red birds can fly $(br \mathrel{\vdash^c_{\mathcal{R}^*_{bird}}} f)$.

Combining the conditionals in a knowledge base by rule chaining is a natural element of plausible reasoning and is, e.g., the base of syllogisms. However, we know that transitivity is not a general inference rule in nonmonotonic logics. But we would expect that chaining rules yields plausible inferences as long as there is no reason to believe the opposite.

Example 10 (c-inference and chaining rules). We use again the knowledge base $\mathcal{R}^*_{bird}$ from Example 7. Given that we have $(f|b)$ and $(a|f)$ in the knowledge base, and no interference between b and a, we would expect that chaining these rules is reasonable and that we can infer that birds are usually airborne. With Table 6 and (22) in Example 7 we obtain that $\kappa(b\overline{a}) = \min\{\eta_1, \eta_6\} > 0 = \kappa(ba)$ and hence $b \mathrel{\vert\kern-0.3em\sim}^c_{\mathcal{R}^*_{bird}} a$, as supposed.

Note that Example 10 illustrates that c-inference does not rely on a total, but a partial ordering of the impacts imposed by (4) and (5) and therefore, via Definition 2, a partial ordering of the worlds for drawing inferences. In the example, nothing can be derived about the concrete values of η_1 and η_6 except that they are positive (see (22)). This is sufficient to guarantee the considered skeptical inference.

5   Characterizing C-Inference by a CSP

For a given OCF κ, the relation $\kappa \models (B|A)$ can be checked by determining $\kappa(AB)$ and $\kappa(A\overline{B})$ according to (2). For checking the relation $A \mathrel{\vert\kern-0.3em\sim}^c_{\mathcal{R}} B$ the countable infinitely large set of all c-representations for $\mathcal{R}$ has to be taken into account. In the previous section, we showed that such a c-inference can not be reduced to the inconsistency of $\mathcal{R} \cup \{(\overline{B}|A)\}$. In the following, we will show that the relation $\mathrel{\vert\kern-0.3em\sim}^c_{\mathcal{R}}$ can be characterized by a constraint satisfaction problem, implying that $\mathrel{\vert\kern-0.3em\sim}^c_{\mathcal{R}}$ can be computed using a constraint-based approach.

Definition 6 ($CR_{\mathcal{R}}(B|A), \neg CR_{\mathcal{R}}(B|A)$). *Let $\mathcal{R} = \{(B_1|A_1), \ldots, (B_n|A_n)\}$ and $(B|A)$ be a conditional. The acceptance constraint for $(B|A)$ with respect to $\mathcal{R}$, denoted by $CR_{\mathcal{R}}(B|A)$, is:*

$$\min_{\substack{\omega \models AB \\ 1 \leqslant i \leqslant n \\ \omega \models A_i \overline{B_i}}} \sum \eta_i \;<\; \min_{\substack{\omega \models A\overline{B} \\ 1 \leqslant i \leqslant n \\ \omega \models A_i \overline{B_i}}} \sum \eta_i \tag{23}$$

Likewise, $\neg CR_{\mathcal{R}}(B|A)$ denotes the negation of (23), i.e., it denotes the constraint:

$$\min_{\substack{\omega \models AB \\ 1 \leqslant i \leqslant n \\ \omega \models A_i \overline{B_i}}} \sum \eta_i \;\geqslant\; \min_{\substack{\omega \models A\overline{B} \\ 1 \leqslant i \leqslant n \\ \omega \models A_i \overline{B_i}}} \sum \eta_i \tag{24}$$

Note that $CR_{\mathcal{R}}(B|A)$ is a constraint on the constraint variables $\eta_1, \ldots, \eta_n$ which are used in the CSP $CR(\mathcal{R})$, but it does not introduce any new variables not already contained in $CR(\mathcal{R})$; this observation also holds for the constraint $\neg CR_{\mathcal{R}}(B|A)$.

The following proposition shows that the skeptical c-inference relation $\mathrel{\vert\kern-0.3em\sim}_{cr}$ can be modeled by a CSP.

Proposition 8 (c-inference as a CSP). *Let $\mathcal{R} = \{(B_1|A_1), \ldots, (B_n|A_n)\}$ be a consistent knowledge base and $(B|A)$ be a conditional. Then the following holds:*

$$A \mathrel{\vert\kern-0.3em\sim}^c_{\mathcal{R}} B \text{ iff } CR(\mathcal{R}) \cup \{\neg CR_{\mathcal{R}}(B|A)\} \text{ is not solvable.} \tag{25}$$

Proof. Assume that $A \mathrel{|\!\sim}^{c}_{\mathcal{R}} B$ holds, i.e., $\kappa \models (B|A)$ holds for all c-representations κ for $\mathcal{R}$. If $CR(\mathcal{R}) \cup \{\neg CR_{\mathcal{R}}(B|A)\}$ were solvable with a solution $\vec{\eta} = (\eta_1, \dots, \eta_n)$ then $\kappa_{\vec{\eta}} \models \mathcal{R}$ according to Proposition 3 where

$$\kappa_{\vec{\eta}}(\omega) = \sum_{\substack{1 \leqslant i \leqslant n \\ \omega \models A_i \overline{B}_i}} \eta_i \tag{26}$$

due to (3). Furthermore, since $\vec{\eta}$ also solves $\neg CR_{\mathcal{R}}(B|A)$, Eq. (24) holds. Applying (26) to (24) yields

$$\min_{\omega \models AB} \kappa_{\vec{\eta}}(\omega) \geqslant \min_{\omega \models A\overline{B}} \kappa_{\vec{\eta}}(\omega) \tag{27}$$

and further applying Eqs. (3) to (27) yields

$$\kappa_{\vec{\eta}}(AB) \geqslant \kappa_{\vec{\eta}}(A\overline{B}). \tag{28}$$

Using Eq. (2), this implies

$$\kappa_{\vec{\eta}} \not\models (B|A), \tag{29}$$

contradicting the assumption $A \mathrel{|\!\sim}^{c}_{\mathcal{R}} B$. Thus, $CR(\mathcal{R}) \cup \{\neg CR_{\mathcal{R}}(B|A)\}$ is unsolvable.

For the other direction, we use contraposition and assume that $A \mathrel{|\!\not\sim}^{c}_{\mathcal{R}} B$ holds. Therefore, since $\mathcal{R}$ is consistent, there is a c-representation κ with $\kappa \models \mathcal{R}$ and $\kappa \not\models (B|A)$. According to Proposition 4, there is a solution $\vec{\eta} = (\eta_1, \dots, \eta_n) \in Sol(CR(\mathcal{R}))$ such that $\kappa = \kappa_{\vec{\eta}}$. From $\kappa_{\vec{\eta}} \not\models (B|A)$ we get:

$$\kappa_{\vec{\eta}}(AB) \geqslant \kappa_{\vec{\eta}}(A\overline{B}) \tag{30}$$

Applying Eqs. (1) to (30) yields

$$\min\{\kappa_{\vec{\eta}}(\omega) \mid \omega \models AB\} \geqslant \min\{\kappa_{\vec{\eta}}(\omega) \mid \omega \models A\overline{B}\} \tag{31}$$

and further applying Eqs. (3) to (31) yields

$$\min\{\sum_{\substack{1 \leqslant i \leqslant n \\ \omega \models A_i \overline{B}_i}} \eta_i \mid \omega \models AB\} \geqslant \min\{\sum_{\substack{1 \leqslant i \leqslant n \\ \omega \models A_i \overline{B}_i}} \eta_i \mid \omega \models A\overline{B}\} \tag{32}$$

which is equivalent to (24). Thus, $CR(\mathcal{R}) \cup \{\neg CR_{\mathcal{R}}(B|A)\}$ is solvable, completing the proof. $\qquad \square$

In Sect. 4 we already discussed that c-inference and p-entailment are defined in analogy, but using the set of all c-representations of a knowledge base $\mathcal{R}$ rather than the set of all OCFs that accept $\mathcal{R}$ when defining inference leads to the differences shown above. While our CSP modeling of the inference closely resembles the characterization of p-entailment given in Proposition 1, there is a major difference: While the characterization in Proposition 1 tests whether an augmented knowledge base is consistent, the characterization in Proposition 8 tests for the solvability of an augmentation of the CSP specifying the c-representations of the knowledge base. If we compare both approaches, Corollary 1 gives us that $\mathcal{R} \cup (\overline{B}|A)$ is consistent iff $CR(\mathcal{R} \cup (\overline{B}|A))$ is solvable, hence the not-solvability of $CR(\mathcal{R} \cup (\overline{B}|A))$ is, by Proposition 1, equivalent to the question whether the entailment $A \hspace{1pt}\vmid\hspace{-7pt}\sim^p_{\mathcal{R}} B$ holds.

Since we have shown that c-inference $A \hspace{1pt}\vmid\hspace{-7pt}\sim^c_{\mathcal{R}} B$ is characterized by $CR(\mathcal{R}) \cup \{\neg CR_{\mathcal{R}}(B|A)\}$ not being solvable in Proposition 8, and c-inference to exceed System P in Corollary 2, we conclude:

Corollary 3. *Let $\mathcal{R}$ be a conditional knowledge base and let A, B be formulas. If $CR(\mathcal{R} \cup (\overline{B}|A))$ is not solvable, then $CR(\mathcal{R}) \cup \{\neg CR_{\mathcal{R}}(B|A)\}$ is not solvable; the reverse is not true in general.*

6   Conclusions and Future Work

In this paper we defined the novel inference relation *c-inference* as the skeptical inference over all c-representations of a given conditional knowledge base $\mathcal{R}$. We proved that c-inference exceeds the skeptical inference of all OCFs that accept $\mathcal{R}$, the latter being equivalent to Adams' system P. In particular, we showed that c-inference shares important benchmark properties with inference based on single c-representations namely subclass inheritance for exceptional subclasses (the "Drowning Problem") and irrelevance, and also allows for rule chaining in a rational way. This is all the more remarkable because, even if by inferring skeptically over all c-representations, we abandon the total ordering of the worlds induced by single c-representations for a partial ordering.

By characterizing c-inference as a constraint satisfaction problem we could guarantee that the novel skeptical inference over infinitely many c-representations is not only calculable, but also implementable by using a constraint solver. Since there is a c-representation for a knowledge base if and only if the knowledge base is consistent, this CSP modeling additionally allows for an additional implementable consistency test apart from the tolerance test [13]. Implementing the calculation of c-representations using constraint solvers has been demonstrated successfully in [2]; extending this implementation to c-inference is part of our current work, as well as evaluating it empirically and investigating the complexity of this approach.

Acknowledgment. This work was supported by DFG-Grant KI1413/5-1 of Prof. Dr. Gabriele Kern-Isberner as part of the priority program "New Frameworks of Rationality" (SPP 1516). Christian Eichhorn is supported by this Grant. This work benefitted very much from discussions led during Dagstuhl Seminar 15221 "Multi-disciplinary approaches to reasoning with imperfect information and knowledge - a synthesis and a roadmap of challenges".

References

1. Adams, E.W.: The Logic of Conditionals: An Application of Probability to Deductive Logic. Synthese Library. Springer, Dordrecht (1975)
2. Beierle, C., Kern-Isberner, G.: A declarative approach for computing ordinal conditional functions using constraint logic programming. In: Tompits, H., Abreu, S., Oetsch, J., Pührer, J., Seipel, D., Umeda, M., Wolf, A. (eds.) INAP/WLP 2011. LNCS (LNAI), vol. 7773, pp. 175–192. Springer, Heidelberg (2013)
3. Benferhat, S., Cayrol, C., Dubois, D., Lang, J., Prade, H.: Inconsistency management and prioritized syntax-based entailment. In: Proceedings of the Thirteenth International Joint Conference on Artificial Intelligence (IJCAI 1993), vol. 1, pp. 640–647. Morgan Kaufmann Publishers, San Francisco (1993)
4. Dubois, D., Prade, H.: Conditional objects as nonmonotonic consequence relations. In: Principles of Knowledge Representation and Reasoning: Proceedings of the Fourth International Conference (KR 1994), pp. 170–177. Morgan Kaufmann Publishers, San Francisco (1996)
5. Dubois, D., Prade, H.: Possibility theory and its applications: where do we stand? In: Kacprzyk, J., Pedrycz, W. (eds.) Springer Handbook of Computational Intelligence, pp. 31–60. Springer, Heidelberg (2015)
6. Finetti, B.D.: Theory of Probability, vol. 1,2. Wiley, New York (1974)
7. Goldszmidt, M., Pearl, J.: On the consistency of defeasible databases. Artif. Intell. **52**(2), 121–149 (1991)
8. Goldszmidt, M., Pearl, J.: Qualitative probabilities for default reasoning, belief revision, and causal modeling. Artif. Intell. **84**(1–2), 57–112 (1996)
9. Kern-Isberner, G., Eichhorn, C.: Structural inference from conditional knowledge bases. In: Unterhuber, M., Schurz, G. (eds.) Logic and Probability: Reasoning in Uncertain Environments, pp. 751–769 (2014). No. 102(4) in Studia Logica. Springer, Dordrecht (2014)
10. Kern-Isberner, G.: Conditionals in Nonmonotonic Reasoning and Belief Revision. LNCS (LNAI), vol. 2087. Springer, Heidelberg (2001)
11. Kern-Isberner, G.: A thorough axiomatization of a principle of conditional preservation in belief revision. Ann. Math. Artif. Intell. **40**, 127–164 (2004)
12. Makinson, D.: General patterns in nonmonotonic reasoning. In: Gabbay, D.M., Hogger, C.J., Robinson, J.A. (eds.) Handbook of Logic in Artificial Intelligence and Logic Programming, vol. 3, pp. 35–110. Oxford University Press, New York (1994)
13. Pearl, J.: System Z: a natural ordering of defaults with tractable applications to nonmonotonic reasoning. In: Proceedings of the 3rd Conference on Theoretical Aspects of Reasoning About Knowledge (TARK1990), pp. 121–135. Morgan Kaufmann Publishers Inc., San Francisco (1990)
14. Spohn, W.: Ordinal conditional functions: a dynamic theory of epistemic states. In: Harper, W.L., Skyrms, B. (eds.) Causation in Decision, Belief Change and Statistics: Proceedings of the Irvine Conference on Probability and Causation. The Western Ontario Series in Philosophy of Science, vol. 42, pp. 105–134. Springer, Dordrecht (1988)
15. Spohn, W.: The Laws of Belief: Ranking Theory and Its Philosophical Applications. Oxford University Press, Oxford (2012)
16. Thorn, P.D., Eichhorn, C., Kern-Isberner, G., Schurz, G.: Qualitative probabilistic inference with default inheritance for exceptional subclasses. In: PROGIC 2015: The Seventh Workshop on Combining Probability and Logic (2015)

Systems and Implementations for Solving Reasoning Problems in Conditional Logics

Christoph Beierle[(✉)]

Faculty of Mathematics and Computer Science,
University of Hagen, 58084 Hagen, Germany
beierle@fernuni-hagen.de

Abstract. Default rules like *"If A, then normally B"* or probabilistic rules like *"If A, then B with probability x"* are powerful constructs for knowledge representation. Such rules can be formalized as conditionals, denoted by $(B|A)$ or $(B|A)[x]$, and a conditional knowledge base consists of a set of conditionals. Different semantical models have been proposed for conditional knowledge bases, and the most important reasoning problems are to determine whether a knowledge base is consistent and to determine what a knowledge base entails. We present an overview on systems and implementations our group has been working on for solving reasoning problems in various semantics that have been developed for conditional knowledge bases. These semantics include quantitative, semi-quantitative, and qualitative conditional logics, based on both propositional logic and on first-order logic.

1 Introduction

When studying concepts and methods for nonmonotonic reasoning, actually implemented and operational systems realizing the developed approaches can be very helpful. Besides providing a proof-of-concept, such systems may also yield the basis for practical applications. In recent years, our group at the University of Hagen has been involved in the development of several software systems implementing reasoning tasks for conditional logics. The types of conditional logics covered by these systems comprise pure qualitative logics providing default rules like *"If A, then normally B"* and also quantitative probabilistic logics with rules like *"If A, then B with probability x"*, based either on an underlying propositional language or on a first-order language. The purpose of this paper is to provide a brief overview of some of these systems and to illustrate the reasoning tasks they address.

In Sect. 2, after sketching syntax and models of several propositional conditional logics, systems dealing with these logics are presented, both for qualitative logics and for probabilistic logics. Along the same dimensions, Sect. 3 deals with first-order conditionals. In Sect. 4, we conclude and point out future work.

© Springer International Publishing Switzerland 2016
M. Gyssens and G. Simari (Eds.): FoIKS 2016, LNCS 9616, pp. 83–94, 2016.
DOI: 10.1007/978-3-319-30024-5_5

2   Propositional Conditional Logics

2.1   Unquantified and Quantified Conditionals

We start with a propositional language $\mathcal{L}$, generated by a finite set Σ of atoms $a, b, c, \ldots$. The formulas of $\mathcal{L}$ will be denoted by uppercase Roman letters $A, B, C, \ldots$. For conciseness of notation, we may omit the logical *and*-connective, writing AB instead of $A \wedge B$, and overlining formulas will indicate negation, i.e. $\overline{A}$ means $\neg A$. Let Ω denote the set of possible worlds over $\mathcal{L}$; Ω will be taken here simply as the set of all propositional interpretations over $\mathcal{L}$ and can be identified with the set of all complete conjunctions over Σ. For $\omega \in \Omega$, $\omega \models A$ means that the propositional formula $A \in \mathcal{L}$ holds in the possible world ω.

By introducing a new binary operator $|$, we obtain the set

$$(\mathcal{L} \mid \mathcal{L}) = \{(B|A) \mid A, B \in \mathcal{L}\}$$

of unquantified *conditionals* over $\mathcal{L}$. A conditional $(B|A)$ formalizes "*if A then (normally) B*" and establishes a plausible, probable, possible etc. connection between the *antecedent* A and the *consequence* B. By attaching a probability value to an unquantified conditional, we obtain the set

$$(\mathcal{L} \mid \mathcal{L})^{prob} = \{(B|A)[x] \mid A, B \in \mathcal{L}, x \in [0,1]\}$$

of all *probabilistic conditionals* (or *probabilistic rules*) over $\mathcal{L}$. A *knowledge base* $\mathcal{R}$ is a set of conditionals from $(\mathcal{L} \mid \mathcal{L})$ or from $(\mathcal{L} \mid \mathcal{L})^{prob}$, respectively.

Example 1 (Qualitative conditional knowledge base).   Suppose we have the propositional atoms f - *flying*, b - *birds*, p - *penguins*, w - *winged* animals, k - *kiwis*. Let the set

$$\mathcal{R}_{bird} = \{(f|b), (b|p), (\overline{f}|p), (w|b), (b|k)\}$$

consist of the following five conditionals:

$$
\begin{array}{lll}
r_1 : & (f|b) & \textit{birds fly} \\
r_2 : & (b|p) & \textit{penguins are birds} \\
r_3 : & (\overline{f}|p) & \textit{penguins do not fly} \\
r_4 : & (w|b) & \textit{birds have wings} \\
r_5 : & (b|k) & \textit{kiwis are birds}
\end{array}
$$

Example 2 (Probabilistic conditional knowledge base).   We use the well-known Léa Sombé example (see e.g. [47]) and consider the three propositional variables s - *being a student*, y - *being young*, and u - *being unmarried*. *Students* and *unmarried people* are mostly *young*. This commonsense knowledge an agent may have can be expressed by the probabilistic knowledge base

$$\mathcal{R}_{syu} = \{(y|s)[0.8], (y|u)[0.7]\}$$

containing the two conditionals:

$$
\begin{array}{lll}
r_1 : & (y|s)[0.8] & \textit{students are young with probability 0.8} \\
r_2 : & (y|u)[0.7] & \textit{unmarried people are young with probability 0.7}
\end{array}
$$

2.2   Models of Propositional Conditional Knowledge Bases

In order to give appropriate semantics to conditionals, they are usually considered within richer structures such as *epistemic states*. Besides certain (logical) knowledge, epistemic states also allow the representation of preferences, beliefs, assumptions of an intelligent agent. Basically, an epistemic state allows one to compare formulas or worlds with respect to plausibility, possibility, necessity, probability, etc.

In a quantitative framework with probabilistic conditionals, obvious representations of epistemic states are provided by *probability distributions* $P : \Omega \to [0,1]$ with $\sum_{\omega \in \Omega} P(\omega) = 1$. The probability of a formula $A \in \mathcal{L}$ is given by $P(A) = \sum_{\omega \models A} P(\omega)$, and the probability of a conditional $(B|A) \in (\mathcal{L} \mid \mathcal{L})$ with $P(A) > 0$ is defined as $P(B|A) = \dfrac{P(AB)}{P(A)}$, the corresponding conditional probability. Thus, the satisfaction relation $\models^{prob}$ between probability distributions over Ω and conditionals from $(\mathcal{L} \mid \mathcal{L})^{prob}$ is defined by:

$$P \models^{prob} (B|A)[x] \text{ iff } P(A) > 0 \text{ and } P(B|A) = \frac{P(AB)}{P(A)} = x \qquad (1)$$

As usual, this relation is extended to a set $\mathcal{R}$ of conditionals by defining $P \models^{prob} \mathcal{R}$ iff $P \models^{prob} (B|A)[x]$ for all $(B|A)[x] \in \mathcal{R}$; for all satisfaction relations considered in the rest of this paper, we will tacitly assume the corresponding extension to sets of conditionals.

Example 3. For the propositional language used in Example 2, let P^* be the probability distribution given by:

ω	$P^*(\omega)$	ω	$P^*(\omega)$	ω	$P^*(\omega)$	ω	$P^*(\omega)$
syu	0.1950	$sy\overline{u}$	0.1758	$s\overline{y}u$	0.0408	$s\overline{y}\,\overline{u}$	0.0519
$\overline{s}yu$	0.1528	$\overline{s}y\overline{u}$	0.1378	$\overline{s}\,\overline{y}u$	0.1081	$\overline{s}\,\overline{y}\,\overline{u}$	0.1378

It is easy to check that $P^* \models^{prob} \mathcal{R}_{syu}$; for instance since $P^*(ys) = 0.3708$ and $P^*(s) = 0.4635$, we have $P^*(y|s) = {}^{0.3708}/{}_{0.4635} = 0.8$ and thus $P^* \models^{prob} (y|s)[0.8]$.

Various types of models have been proposed to interpret qualitative conditionals $(B|A)$ adequately within a logical system (cf. e.g. [39]). One of the most prominent approaches is the *system-of-spheres* model of Lewis [38] which makes use of a notion of similarity between possible worlds. Other, more fine-grained semantics for conditionals use numbers to compare different degrees of "plausibility" between the verification and the falsification of a conditional. In these qualitative frameworks, a conditional $(B|A)$ is *accepted* (or *verified*), if its confirmation, AB, is more plausible, possible etc. than its refutation, $A\overline{B}$; a suitable degree of acceptance is calculated from the degrees associated with AB and $A\overline{B}$. Here, two of the most popular approaches to represent epistemic states are *ordinal conditional functions, OCFs,* (also called *ranking functions*)

ω	$\kappa(\omega)$	ω	$\kappa(\omega)$	ω	$\kappa(\omega)$	ω	$\kappa(\omega)$
$pbfwk$	2	$p\overline{b}fwk$	5	$\overline{p}bfwk$	0	$\overline{p}\,\overline{b}fwk$	1
$pbfw\overline{k}$	2	$p\overline{b}fw\overline{k}$	4	$\overline{p}bfw\overline{k}$	0	$\overline{p}\,\overline{b}fw\overline{k}$	0
$pbf\overline{w}k$	3	$p\overline{b}f\overline{w}k$	5	$\overline{p}bf\overline{w}k$	1	$\overline{p}\,\overline{b}f\overline{w}k$	1
$pbf\overline{w}\,\overline{k}$	3	$p\overline{b}f\overline{w}\,\overline{k}$	4	$\overline{p}bf\overline{w}\,\overline{k}$	1	$\overline{p}\,\overline{b}f\overline{w}\,\overline{k}$	0
$pb\overline{f}wk$	1	$p\overline{b}\,\overline{f}wk$	3	$\overline{p}b\overline{f}wk$	1	$\overline{p}\,\overline{b}\,\overline{f}wk$	1
$pb\overline{f}w\overline{k}$	1	$p\overline{b}\,\overline{f}w\overline{k}$	2	$\overline{p}b\overline{f}w\overline{k}$	1	$\overline{p}\,\overline{b}\,\overline{f}w\overline{k}$	0
$pb\overline{f}\,\overline{w}k$	2	$p\overline{b}\,\overline{f}\,\overline{w}k$	3	$\overline{p}b\overline{f}\,\overline{w}k$	2	$\overline{p}\,\overline{b}\,\overline{f}\,\overline{w}k$	1
$pb\overline{f}\,\overline{w}\,\overline{k}$	2	$p\overline{b}\,\overline{f}\,\overline{w}\,\overline{k}$	2	$\overline{p}b\overline{f}\,\overline{w}\,\overline{k}$	2	$\overline{p}\,\overline{b}\,\overline{f}\,\overline{w}\,\overline{k}$	0

Fig. 1. OCF κ accepting $\mathcal{R}_{bird}$ from Example 1

[49,50], and *possibility distributions* [11,14], assigning degrees of plausibility, or of possibility, respectively, to formulas and possible worlds.

In the following, we will focus on OCFs [49]. An OCF κ is a function $\kappa : \Omega \to \mathbb{N} \cup \{\infty\}$ with $\kappa^{-1}(0) \neq \emptyset$. The smaller $\kappa(\omega)$, the less suprising or the more plausible the world ω. For formulas $A \in \mathcal{L}$, $\kappa(A)$ is given by:

$$\kappa(A) = \begin{cases} \min\{\kappa(\omega) \mid \omega \models A\} & \text{if } A \text{ is satisfiable} \\ \infty & \text{otherwise} \end{cases}$$

The satisfaction relation between OCFs and qualitative conditionals from $(\mathcal{L} \mid \mathcal{L})$, denoted by $\models^{ocf}$, is defined by:

$$\kappa \models^{ocf} (B|A) \text{ iff } \kappa(AB) < \kappa(A\overline{B})$$

Thus, a conditional $(B|A)$ is *accepted* by the ordinal conditional function κ iff its confirmation AB is less surprising than its refutation $A\overline{B}$.

Example 4. For the propositional language used in Example 1, let κ be the OCF given in Fig. 1. For the conditional $(\overline{f}|p) \in \mathcal{R}_{bird}$, we have $\kappa(p\overline{f}) = 1 < 2 = \kappa(pf)$ and thus $\kappa \models^{ocf} (\overline{f}|p)$. Similarly, it is easy to check that κ also accepts the other conditionals in $\mathcal{R}_{bird}$, implying $\kappa \models^{ocf} \mathcal{R}_{bird}$.

2.3 Systems for Reasoning with Propositional Conditional Knowledge Bases

Reasoning with respect to a conditional knowledge base $\mathcal{R}$ means to determine what $\mathcal{R}$ entails. While in classical logic, entailment is defined with respect to all models, for probabilistic conditional knowledge bases this approach is very restrictive since it may yield only uninformative answers. Therefore, entailment may be defined with respect to a set of some best or preferred models.

In probabilistic conditional logic, the *principle of maximum entropy* (*ME principle*) has been advocated [28,30,40,41]. While in general, each model of a probabilistic conditional knowledge base $\mathcal{R}$ determines a particular way of extending and completing the probabilistic knowledge expressed in $\mathcal{R}$ to a full probability distribution, the ME principle selects the distribution that accepts $\mathcal{R}$ and that is as unbiased as possible. Formally, given a knowledge base $\mathcal{R} = \{(B_1|A_1)[x_1], \ldots, (B_n|A_n)[x_n]\}$, $ME(\mathcal{R})$ is the unique probability distribution that satisfies all constraints specified by $\mathcal{R}$ and has the highest entropy $\mathcal{H}(P) = -\sum_{\omega \in \Omega} P(\omega) \log P(\omega)$ among all models P of $\mathcal{R}$:

$$ME(\mathcal{R}) = \arg \max_{P \models \mathcal{R}} \mathcal{H}(P) \tag{2}$$

Reasoning in probabilistic conditional logic by employing the principle of maximum entropy [28,40] requires solving the numerical optimization problem given in Eq. (2). MECoRe [19] is a software system implementing maximum entropy reasoning. While MECoRe does not employ a junction-tree modelling as in the expert system shell SPIRIT [48], but a straightforward representation of the complete probability distribution, its focus is on flexibly supporting different basic knowledge and belief management functions like revising or updating probabilistic beliefs, or hypothetical reasoning in what-if mode. In addition, there is a component checking the consistency of a knowledge base $\mathcal{R}$, i.e., checking whether the set of models of $\mathcal{R}$ is non-empty. A query asking for the probability of $(B|A)$ in the context of $\mathcal{R}$ is answered with respect to the uniquely defined maximum entropy model $ME(\mathcal{R})$, i.e., $(B|A)[x]$ is ME-entailed from $\mathcal{R}$ iff $ME(\mathcal{R})(B|A) = x$. The distribution P^* given in Example 3 is in fact the ME distribution computed by MECoRe for $\mathcal{R}_{syu}$, i.e., we have $P^* = ME(\mathcal{R}_{syu})$. MECoRe can be controlled by a text command interface or by script files containing command sequences. It features an expressive command language which allows, e.g., to manipulate knowledge bases, and to automate sequences of updates and revisions. Besides this, a Java software interface allows to integrate MECoRe in other programs. In [3,33], the functionalities of MECoRe are illustrated in applications of ME modelling and reasoning in the medical domain.

The methodological theory of conditionals developed by Kern-Isberner [29,30] allows to describe the aim of *knowledge discovery* in a very general sense: to reveal structures of knowledge which can be seen as structural relationships being represented by conditionals. In this setting, knowledge discovery is understood as a process which is inverse to inductive knowledge representation. By applying this theory, an algorithm that computes sets of propositional probabilistic conditionals from distributions was developed and implemented in the system CONDORCKD [22,23,34] using the functional programming language Haskell.

For propositional qualitative conditional logic using OCFs, *p-entailment* [25] is an inference relation defined with respect to all OCF models of a knowledge base $\mathcal{R}$: If A, B are formulas, then A *p-entails* B in the context of $\mathcal{R}$ iff $\kappa \models (B|A)$ for all κ such that $\kappa \models \mathcal{R}$. System P [1] provides a kind of gold standard for plausible, nonmonotonic inferences, and in [13] it is shown that, given a knowledge base $\mathcal{R}$, system P inference is the same as p-entailment.

There are also inference relations which are defined with respect to specific OCFs obtained inductively from a knowledge base $\mathcal{R}$. System Z [42] is based upon the ranking function which is the unique minimal OCF that accepts $\mathcal{R}$; this ranking function is obtained from an ordered partition $(\mathcal{R}_0, ..., \mathcal{R}_m)$ of $\mathcal{R}$ defined by the notion of *tolerance* [42]. Other OCFs accepting $\mathcal{R}$ that have favourable inference properties are c-representations [30, 31]. A *c-representation* of $\mathcal{R}$ is a ranking function κ constructed from integer impacts $\eta_i \in \mathbb{N}_0$ assigned to each conditional $(B_i|A_i) \in \mathcal{R}$ such that κ accepts $\mathcal{R}$ and is given by [31]:

$$\kappa(\omega) = \sum_{\substack{1 \leqslant i \leqslant n \\ \omega \models A_i \overline{B_i}}} \eta_i \tag{3}$$

CONDOR@AsmL [6] is a software system that implements automated reasoning with qualitative default rules employing c-representations. Based on a characterization theorem for c-representations and c-revisions and an approach to compute c-representations and c-revisions using the tolerance-induced partition of $\mathcal{R}$ [31], inference is done with respect to the OCF thus obtained from $\mathcal{R}$. CONDOR@AsmL provides functionalities for advanced knowledge management tasks like belief revision and update or diagnosis and hypothetical what-if-analysis for qualitative conditionals. CONDOR@AsmL implements the abstract CONDOR specification given in [4] and was developed in AsmL [26], allowing for a high-level implementation that minimizes the gap between the mathematical specification of the underlying concepts and the executable code and supports the formal verification of the implemented system [5].

While CONDOR@AsmL computes a c-representation for any $\mathcal{R}$ that is consistent, this c-representation may not be minimal. Unlike in system Z where there is a unique minimal OCF, there may be more than one minimal c-representation. In [7], the set of all c-representations for $\mathcal{R}$ is specified as the set of all solutions of a constraint satisfaction problem $CR(\mathcal{R})$, and a high-level declarative approach using constraint logic programming (CLP) techniques for solving the constraint satisfaction problem $CR(\mathcal{R})$ is presented. In particular, the approach developed in [7] supports the generation of all minimal solutions; these minimal solutions are of special interest as they provide a preferred basis for model-based inference from $\mathcal{R}$. Moreover, different notions of minimality are investigated and the flexibility of the approach is demonstrated by showing how alternative minimality concepts can be taken into account by slight modifications of the CLP implementation. In [2], a skeptical inference relation taking all c-representations of $\mathcal{R}$ into account is introduced, and it is demonstrated that it can be implemented as a constraint satisfaction problem that extends $CR(\mathcal{R})$.

3   First-Order Conditional Logics

As an illustration for first-order probabilistic conditionals, consider the following example, adapted from [12], modelling the relationships among elephants in a zoo and their keepers. Elephants usually like their keepers, except for keeper

Fred. But elephant Clyde gets along with everyone, and therefore he also likes Fred. The knowledge base $\mathcal{R}_{ek}$ consists of the following conditionals:

$$ek_1 : \; (likes(E, K) \mid elephant(E), keeper(K))[0.9]$$
$$ek_2 : \; (likes(E, fred) \mid elephant(E), keeper(fred))[0.05]$$
$$ek_3 : \; (likes(clyde, fred) \mid elephant(clyde), keeper(fred)[0.85]$$

Conditional ek_1 models statistical knowledge about the general relationship between elephants and their keepers ("*elephants like their keeper with probability 0.9*"), whereas conditional ek_2 represents knowledge about the exceptional keeper Fred and his relationship to elephants in general ("*elephants like keeper Fred only with probability 0.05*"). Conditional ek_3 models subjective belief about the relationship between the elephant Clyde and keeper Fred ("*elephant Clyde likes keeper Fred with probability 0.85*"). From a common-sense point of view, the knowledge base $\mathcal{R}_{ek}$ makes perfect sense: conditional ek_2 is an exception of ek_1, and ek_3 is an exception of ek_2.

However, assigning a formal semantics to $\mathcal{R}_{ek}$ is not straightforward. For instance, for transforming the propositional approach employed in Eq. (1) to the relational case with free variables as in $\mathcal{R}_{ek}$, the exact role of the variables has to be specified. While there are various approaches dealing with a combination of probabilities with a first-order language (e.g. [24,27,35,36]) here we focus on two semantics for probabilistic relational conditionals, the *aggregating semantics* [36] proposed by Kern-Isberner and the grounding semantics employed in the logic FO-PCL [21].

While the two approaches are related in the sense that they refer to a (finite) set of constants when interpreting the variables in the conditionals, there is also a major difference. FO-PCL requires all groundings of a conditional to have the same probability x given in the conditional, and in general, FO-PCL needs to restrict the possible instantiations for the variables occurring in a conditional by providing constraint formulas like $U \neq V$ or $U \neq a$ in order to avoid inconsistencies. Thus, while the aggregating semantics uses probabilistic conditionals $(B|A)[x]$ with relational formulas A, B, these conditionals are extended by a constraint formula C to $\langle (B|A)[x], C \rangle$ in FO-PCL. The models of a knowledge base $\mathcal{R}$ consisting of such first-order probabilistic conditionals are again probability distributions over the possible worlds, where a possible world is a subset of the Herbrand base induced by the predicates and constants used for $\mathcal{R}$.

The satisfaction relation $\models_{\otimes}$ for FO-PCL is defined by

$$P \models_{\otimes} \langle (B|A)[x], C \rangle \quad \text{iff} \quad \frac{P(\theta(AB))}{P(\theta(A))} = x \tag{4}$$

$$\text{for all } \theta \in \Theta^{adm}(\langle (B|A)[x], C \rangle)$$

where $\Theta^{adm}(\langle (B|A)[x], C \rangle)$ is the set of all *admissible* ground substitutions θ for the given conditional, i.e. where $\theta(C)$ evaluates to *true*. Thus, a probability distribution P $\otimes$-satisfies a conditional $\langle (B|A)[x], C \rangle$ if it satisfies each admissible individual instantiation of it. In contrast, the satisfaction relation $\models_{\odot}$ for aggregating semantics [36] is less strict with respect to probabilities of ground instances, since it is capable of balancing the probabilities of ground instances in order to ensure the probability x given by a conditional; $\models_\odot$ is defined by

$$P \models_\odot (B|A)[x] \quad \text{iff} \quad \frac{\sum\limits_{\theta \in \Theta((B|A)[x])} P(\theta(AB))}{\sum\limits_{\theta \in \Theta((B|A)[x])} P(\theta(A))} = x \qquad (5)$$

where $\Theta((B|A)[x])$ is the set of all ground substitutions of $(B|A)[x]$.

The principle of maximum entropy used in the propositional setting (Equation (2)) has been extended to first-order knowledge bases for aggregating semantics and for FO-PCL [21,36] by defining

$$ME_\bullet(\mathcal{R}) = \arg \max_{P \models_\circ \mathcal{R}} \mathcal{H}(P) \qquad (6)$$

where $\bullet \in \{\otimes, \odot\}$. Since for FO-PCL grounding and for aggregating semantics the set of models is convex, the optimization problem in (6) yields a unique solution for every consistent $\mathcal{R}$. Thus, analogously to the propositional case, reasoning can be done with respect to the maximum entropy model $ME_\bullet(\mathcal{R})$.

Software components for these inference tasks have been implemented in KREATOR[1] [20], an integrated development environment for representing, reasoning, and learning with relational probabilistic knowledge. In particular, KREATOR provides specific plugins for an optimized computation of the ME model under aggregating semantics (cf. [16–18]) that exploits the conditional structure of $\mathcal{R}$ and its induced equivalence classes [30,37]. The KREATOR plugin for FO-PCL semantics employs a simplification of the ME model computation by transforming $\mathcal{R}$ into an equivalent knowledge base $\mathcal{R}'$ that is parametrically uniform [8–10,21]. Furthermore, algorithms for solving various reasoning problems for probabilistic conditional logics that also take inconsistent information into account have been implemented in the Log4KR library[2] [43–46].

In [37], ranking functions for qualitative first-order conditionals are introduced, and in [32], a system Z-like approach for first-order default reasoning is developed. Unlike propositional system Z, the first-order approach of [32] may yield more than one minimal solution; an implementation of the approach in [32] using Log4KR is given in [15].

4   Conclusions and Future Work

Conditionals play a major role in logic-based knowledge representation and reasoning. In this paper, we gave a brief survey on different versions of conditional logics and illustrated corresponding reasoning tasks addressed by software systems that have been implemented within our research projects in recent years. Our current work includes the further exploitation of conditional structures for

---

[1] KREATOR can be found at http://kreator-ide.sourceforge.net/.

[2] https://www.fernuni-hagen.de/wbs/research/log4kr/index.html.

relational probabilistic inference under maximum entropy, and the investigation of the precise properties of inference with c-representations using OCFs in the propositional case and with the system Z-like approach in the relational case.

Acknowledgments. A large part of the work reported here was done in cooperation and in joint projects with Gabriele Kern-Isberner and her research group at TU Dortmund University, Germany. I am also very grateful to all members of the project teams involved, in particular to Marc Finthammer, Jens Fisseler, Nico Potyka, Matthias Thimm, and numerous students for their contributions.

References

1. Adams, E.W.: The Logic of Conditionals: An Application of Probability to Deductive Logic. Synthese Library. Springer Science+Business Media, Dordrecht (1975)
2. Beierle, C., Eichhorn, C., Kern-Isberner, G.: Skeptical inference based on c-representations and its characterization as a constraint satisfaction problem. In: Gyssens, M., Simari, G. (eds.) FoIKS 2016. LNCS, vol. 9161, pp. 65–82. Springer, Switzerland (2016)
3. Beierle, C., Finthammer, M., Potyka, N., Varghese, J., Kern-Isberner, G.: A case study on the application of probabilistic conditional modelling and reasoning to clinical patient data in neurosurgery. In: van der Gaag, L.C. (ed.) ECSQARU 2013. LNCS, vol. 7958, pp. 49–60. Springer, Heidelberg (2013)
4. Beierle, C., Kern-Isberner, G.: Modelling conditional knowledge discovery and belief revision by abstract state machines. In: Börger, E., Gargantini, A., Riccobene, E. (eds.) ASM 2003. LNCS, vol. 2589, pp. 186–203. Springer, Heidelberg (2003)
5. Beierle, C., Kern-Isberner, G.: A verified AsmL implementation of belief revision. In: Börger, E., Butler, M., Bowen, J.P., Boca, P. (eds.) ABZ 2008. LNCS, vol. 5238, pp. 98–111. Springer, Heidelberg (2008)
6. Beierle, C., Kern-Isberner, G., Koch, N.: A high-level implementation of a system for automated reasoning with default rules (system description). In: Armando, A., Baumgartner, P., Dowek, G. (eds.) IJCAR 2008. LNCS (LNAI), vol. 5195, pp. 147–153. Springer, Heidelberg (2008)
7. Beierle, C., Kern-Isberner, G., Södler, K.: A declarative approach for computing ordinal conditional functions using constraint logic programming. In: Tompits, H., Abreu, S., Oetsch, J., Pührer, J., Seipel, D., Umeda, M., Wolf, A. (eds.) INAP/WLP 2011. LNCS, vol. 7773, pp. 168–185. Springer, Heidelberg (2013)
8. Beierle, C., Krämer, A.: Achieving parametric uniformity for knowledge bases in a relational probabilistic conditional logic with maximum entropy semantics. Ann. Math. Artif. Intell. **73**(1–2), 5–45 (2015)
9. Beierle, C., Kuche, S., Finthammer, M., Kern-Isberner, G.: A software system for the computation, visualization, and comparison of conditional structures for relational probabilistic knowledge bases. In: Proceeding of the Twenty-Eigth International Florida Artificial Intelligence Research Society Conference (FLAIRS 2015), pp. 558–563. AAAI Press, Menlo Park (2015)
10. Beierle, C., Finthammer, M., Kern-Isberner, G.: Relational probabilistic conditionals and their instantiations under maximum entropy semantics for first-order knowledge bases. Entropy **17**(2), 852–865 (2015)

11. Benferhat, S., Dubois, D., Prade, H.: Representing default rules in possibilistic logic. In: Proceedings 3th International Conference on Principles of Knowledge Representation and Reasoning KR 1992, pp. 673–684 (1992)
12. Delgrande, J.: On first-order conditional logics. Artif. Intell. **105**, 105–137 (1998)
13. Dubois, D., Prade, H.: Conditional objects as nonmonotonic consequence relations: main results. In: Principles of Knowledge Representation and Reasoning: Proceedings of the Fourth International Conference (KR 1994), pp. 170–177. Morgan Kaufmann Publishers, San Francisco (1994)
14. Dubois, D., Prade, H.: Possibility Theory and Its Applications: Where Do We Stand? In: Kacprzyk, J., Pedrycz, W. (eds.) Springer Handbook of Computational Intelligence, pp. 31–60. Springer, Heidelberg (2015)
15. Falke, T.: Computation of ranking functions for knowledge bases with relational conditionals. M.Sc. Thesis, Dept. of Computer Science, University of Hagen, Germany (2015)
16. Finthammer, M., Beierle, C.: Using equivalences of worlds for aggregation semantics of relational conditionals. In: Glimm, B., Krüger, A. (eds.) KI 2012. LNCS, vol. 7526, pp. 49–60. Springer, Heidelberg (2012)
17. Finthammer, M., Beierle, C.: A two-level approach to maximum entropy model computation for relational probabilistic logic based on weighted conditional impacts. In: Straccia, U., Calì, A. (eds.) SUM 2014. LNCS, vol. 8720, pp. 162–175. Springer, Heidelberg (2014)
18. Finthammer, M., Beierle, C.: Towards a more efficient computation of weighted conditional impacts for relational probabilistic knowledge bases under maximum entropy semantics. In: Hölldobler, S., Krötzsch, M., Peñaloza, R., Rudolph, S. (eds.) KI 2015. LNCS, vol. 9324, pp. 72–86. Springer, Heidelberg (2015). doi:10.1007/978-3-319-24489-1_6
19. Finthammer, M., Beierle, C., Berger, B., Kern-Isberner, G.: Probabilistic reasoning at optimum entropy with the MEcore system. In: Lane, H.C., Guesgen, H.W. (eds.) Proceedings 22nd International FLAIRS Conference, FLAIRS 2009, pp. 535–540. AAAI Press, Menlo Park (2009)
20. Finthammer, M., Thimm, M.: An integrated development environment for probabilistic relational reasoning. Logic J. IGPL **20**(5), 831–871 (2012)
21. Fisseler, J.: First-order probabilistic conditional logic and maximum entropy. Logic J. IGPL **20**(5), 796–830 (2012)
22. Fisseler, J., Kern-Isberner, G., Beierle, C.: Learning uncertain rules with CondorCKD. In: Proceedings 20th International FLAIRS Conference, FLAIRS 2007. AAAI Press, Menlo Park (2007)
23. Fisseler, J., Kern-Isberner, G., Beierle, C., Koch, A., Müller, C.: Algebraic knowledge discovery using haskell. In: Hanus, M. (ed.) PADL 2007. LNCS, vol. 4354, pp. 80–93. Springer, Heidelberg (2007). http://dx.doi.org/10.1007/978-3-540-69611-7_5
24. Getoor, L., Taskar, B. (eds.): Introduction to Statistical Relational Learning. MIT Press, Cambridge (2007)
25. Goldszmidt, M., Pearl, J.: Qualitative probabilities for default reasoning, belief revision, and causal modeling. Artif. Intell. **84**(1–2), 57–112 (1996)
26. Gurevich, Y., Rossman, B., Schulte, W.: Semantic essence of AsmL. Theoret. Comput. Sci. **343**(3), 370–412 (2005)
27. Halpern, J.Y.: Reasoning About Uncertainty. MIT Press, Cambridge (2005)
28. Kern-Isberner, G.: Characterizing the principle of minimum cross-entropy within a conditional-logical framework. Artif. Intell. **98**, 169–208 (1998)

29. Kern-Isberner, G.: Solving the inverse representation problem. In: Proceedings 14th European Conference on Artificial Intelligence. ECAI 2000, pp. 581–585. IOS Press, Berlin (2000)
30. Kern-Isberner, G.: Conditionals in Nonmonotonic Reasoning and Belief Revision. LNCS (LNAI), vol. 2087. Springer, Heidelberg (2001)
31. Kern-Isberner, G.: A thorough axiomatization of a principle of conditional preservation in belief revision. Annals Math. Artif. Intell. **40**(1–2), 127–164 (2004)
32. Kern-Isberner, G., Beierle, C.: A system Z-like approach for first-order default reasoning. In: Eiter, T., Strass, H., Truszczyński, M., Woltran, S. (eds.) Advances in Knowledge Representation. LNCS, vol. 9060, pp. 81–95. Springer, Heidelberg (2015)
33. Kern-Isberner, G., Beierle, C., Finthammer, M., Thimm, M.: Comparing and evaluating approaches to probabilistic reasoning: theory, implementation, and applications. Trans. Large-Scale Data Knowl.-Centered Syst. **6**, 31–75 (2012)
34. Kern-Isberner, G., Fisseler, J.: Knowledge discovery by reversing inductive knowledge representation. In: Proceedings of the Ninth International Conference on the Principles of Knowledge Representation and Reasoning, KR-2004, pp. 34–44. AAAI Press (2004)
35. Kern-Isberner, G., Lukasiewicz, T.: Combining probabilistic logic programming with the power of maximum entropy. Artif. Intell. **157**(1–2), 139–202 (2004). Special Issue on Nonmonotonic Reasoning
36. Kern-Isberner, G., Thimm, M.: Novel semantical approaches to relational probabilistic conditionals. In: Lin, F., Sattler, U., Truszczynski, M. (eds.) Proceedings Twelfth International Conference on the Principles of Knowledge Representation and Reasoning, KR 2010, pp. 382–391. AAAI Press (2010)
37. Kern-Isberner, G., Thimm, M.: A ranking semantics for first-order conditionals. In: De Raedt, L., Bessiere, C., Dubois, D., Doherty, P., Frasconi, P., Heintz, F., Lucas, P. (eds.) Proceedings 20th European Conference on Artificial Intelligence, ECAI-2012, pp. 456–461. No. 242 in Frontiers in Artificial Intelligence and Applications. IOS Press (2012)
38. Lewis, D.: Counterfactuals. Harvard University Press, Cambridge (1973)
39. Nute, D.: Topics in Conditional Logic. D. Reidel Publishing Company, Dordrecht (1980)
40. Paris, J.: The Uncertain Reasoner's Companion - A Mathematical Perspective. Cambridge University Press, Cambridge (1994)
41. Paris, J., Vencovska, A.: In defence of the maximum entropy inference process. Int. J. Approximate Reasoning **17**(1), 77–103 (1997)
42. Pearl, J.: System Z: A natural ordering of defaults with tractable applications to nonmonotonic reasoning. In: Proceeding of the 3rd Conference on Theoretical Aspects of Reasoning About Knowledge (TARK 1990), pp. 121–135. Morgan Kaufmann Publ. Inc., San Francisco (1990)
43. Potyka, N.: Linear programs for measuring inconsistency in probabilistic logics. In: Proceedings KR 2014, pp. 568–578. AAAI Press (2014)
44. Potyka, N.: Solving Reasoning Problems for Probabilistic Conditional Logics with Consistent and Inconsistent Information. Ph.D. thesis, Fernuniversität Hagen, Germany (2015)
45. Potyka, N., Thimm, M.: Consolidation of probabilistic knowledge bases by inconsistency minimization. In: Proceedings ECAI 2014, pp. 729–734. IOS Press (2014)
46. Potyka, N., Thimm, M.: Probabilistic reasoning with inconsistent beliefs using inconsistency measures. In: Proceeding of the International Joint Conference on Artificial Intelligence 2015 (IJCAI 2015), pp. 3156–3163 (2015)

47. Rödder, W., Kern-Isberner, G.: Léa sombé und entropie-optimale informationsver-
    arbeitung mit der expertensystem-shell SPIRIT. OR Spektrum **19**(3), 41–46 (1997)
48. Rödder, W., Reucher, E., Kulmann, F.: Features of the expert-system-shell
    SPIRIT. Logic J. IGPL **14**(3), 483–500 (2006)
49. Spohn, W.: Ordinal conditional functions: a dynamic theory of epistemic states. In:
    Harper, W., Skyrms, B. (eds.) Causation in Decision, Belief Change, and Statistics,
    II, pp. 105–134. Kluwer Academic Publishers (1988)
50. Spohn, W.: The Laws of Belief: Ranking Theory and Its Philosophical Applications.
    Oxford University Press, Oxford (2012)

Equivalence Between Answer-Set Programs Under (Partially) Fixed Input

Bernhard Bliem[1]([✉]) and Stefan Woltran[1]

Institute of Information Systems, TU Wien, Vienna, Austria
{bliem,woltran}@dbai.tuwien.ac.at

Abstract. Answer Set Programming (ASP) has become an increasingly popular formalism for declarative problem solving. Among the huge body of theoretical results, investigations of different equivalence notions between logic programs play a fundamental role for understanding modularity and optimization. While strong equivalence between two programs holds if they can be faithfully replaced by each other in any context (facts and rules), uniform equivalence amounts to equivalent behavior of programs under any set of facts. Both notions (as well as several variants thereof) have been extensively studied. However, the somewhat reverse notion of equivalence which holds if two programs are equivalent under the addition of any set of proper rules (i.e., all rules except facts) has not been investigated yet. In this paper, we close this gap and give a thorough study of this notion, which we call rule equivalence (RE), and its parameterized version where we allow facts over a given restricted alphabet to appear in the context. RE is thus a relationship between two programs whose input is (partially) fixed but where additional proper rules might still be added. Such a notion might be helpful in debugging of programs. We give full characterization results and a complexity analysis for the propositional case of RE. Moreover, we show that RE is decidable in the non-ground case.

1 Motivation

In the area of Answer Set Programming [1] investigations of different equivalence notions have been a major research topic within the last 15 years. This is due to the fact that the straightforward notion of equivalence, which holds if two programs possess the same answer sets, is too weak to guarantee faithful replacements. In other words, replacing within a program R a subprogram P by program P' equivalent to that subprogram might change the answer sets of the thus modified program $R[P/P']$. Despite this effect being implicit to any nonmonotonic formalism, little was known how to characterize a sufficiently strong notion of equivalence in order to guarantee such faithful replacements.

In their seminal paper on strong equivalence between logic programs, Lifschitz et al. [12], have found a strikingly simple and elegant solution. First, they explicitly defined the required notion: P and Q are strongly equivalent iff, for each further so-called context program R, $P \cup R$ and $Q \cup R$ possess the same

© Springer International Publishing Switzerland 2016
M. Gyssens and G. Simari (Eds.): FoIKS 2016, LNCS 9616, pp. 95–111, 2016.
DOI: 10.1007/978-3-319-30024-5_6

answer sets. Secondly, they gave a characterization in terms of a monotonic non-classical logic: P and Q are strongly equivalent iff P and Q are equivalent in the logic of here-and-there; this characterization was later reformulated to be applied more directly to programs and their reducts by Turner [17] who introduced the notion of SE-models.

The difference between strong and ordinary equivalence motivated investigations of equivalence notions in between. Uniform equivalence, originally introduced by Sagiv [15] as an approximation for datalog equivalence, is such an example. Uniform equivalence tests whether, for each set F of facts, $P \cup F$ and $Q \cup F$ possess the same answer sets. This notion has been adapted to answer set programming by Eiter and Fink [3], who also provided a characterization that is based on a subset of SE-models (called UE-models); alternative characterizations can be found in [7]. A further direction is known as relativized equivalence or hyperequivalence. Here the atoms allowed to occur in the context programs are restricted to stem from a given alphabet [6,8,16,18,19].

Concerning further notions of equivalence, it was claimed that any "reasonable" attempt to syntactically restrict the context programs (i.e., where the restriction is defined rule-wise, for instance only allowing context programs with Horn rules) coincides with either ordinary, strong, or uniform equivalence (see, e.g., [14]). The reason behind this claim is the fact that strong equivalence coincides with a much simpler notion of equivalence. Define the class of unary program as consisting of facts and rules of the form $a \leftarrow b$ only. It can be shown that programs P and Q are strongly equivalent, if and only if, for any unary program R, the answer sets of $P \cup R$ and $Q \cup R$ coincide.

However, there is still room for finding another equivalence notion. What if we allow in the context programs R of an equivalence only proper rules, but no facts? Surprisingly, this form of equivalence has not been investigated yet. Indeed such a notion can prove useful as a test for replacement of program modules under fixed known input and appears to be more natural than other notions of modular equivalence from the literature [9,13,16].

As a second contribution we go towards relativized equivalence and relax our notion by allowing potential addition of some facts (similar to relativized uniform equivalence). It turns out that this equivalence notion boils down to strong equivalence if both programs under consideration already contain facts, but is a different concept otherwise.

Understanding different notions of equivalence in ASP not only provided fundamental insights into the nature of nonmonotonic formalisms, it also gave pointers for static program optimization (see, e.g., [2,4]). We see our contribution mainly as a missing link between established equivalence notions. As our results suggest, rule equivalence is somewhat closer to strong equivalence than to uniform equivalence. This becomes in particular apparent by the observation that for programs containing at least one fact, strong and rule equivalence coincide. Moreover, in the non-ground case it was proven that strong equivalence is decidable while uniform equivalence is not [5]. We complement the picture and prove decidability of rule equivalence in the non-ground case.

To summarize, the main contribution of the paper is to provide a general and uniform semantic characterization for the newly introduced notions of rule equivalence. Our characterization will follow the tradition in ASP and is based on SE-models. We show some basic properties of rule equivalence and also include a complexity analysis. While our paper is focused on the ground case of ASP, we also prove that rule equivalence remains decidable for the non-ground case.

2 Background

Throughout the paper we assume an arbitrary but fixed universe $\mathcal{U}$ of atoms. A propositional disjunctive logic program (or simply, a program) is a finite set of rules of the form

$$a_1 \vee \cdots \vee a_l \leftarrow b_1, \ldots, b_m, \textit{not } b_{m+1}, \ldots, \textit{not } b_n, \tag{1}$$

($l \geq 0$, $n \geq m \geq 0$), and where all a_i and b_j are propositional atoms from $\mathcal{U}$ and *not* denotes default negation; for $l = 1$ and $n = 0$, we usually identify the rule (1) with the atom a_1, and call it a *fact*. A rule of the form (1) is called a *constraint* if $l = 0$; *proper* if $n \geq 1$; *positive* if $m = n$; *normal* if $l \leq 1$; and *unary* if $l = 1$ and $m = n \leq 1$. A program is proper (resp., positive, normal, unary) iff all its rules are proper (resp., positive, normal, unary). If all atoms occurring in a program P are from a given alphabet $\mathcal{A} \subseteq \mathcal{U}$ of atoms, we say that P is a program *over* (alphabet) $\mathcal{A}$.

For a rule r of the form (1), we identify its head by $H(r) = \{a_1, \ldots, a_l\}$ and its body by $B^+(r) = \{b_1, \ldots, b_m\}$ and $B^-(r) = \{b_{m+1}, \ldots, b_n\}$. We shall write rules of the form (1) also as $H(r) \leftarrow B^+(r), \textit{not } B^-(r)$. Moreover, we use $B(r) = B^+(r) \cup B^-(r)$. Finally, for a program P and $\alpha \in \{H, B, B^+, B^-\}$, let $\alpha(P) = \bigcup_{r \in P} \alpha(r)$.

The relation $Y \models r$ between an interpretation Y and a program r is defined as usual, i.e., $Y \models r$ iff $H(r) \cap Y \neq \emptyset$ whenever jointly $B^+(r) \subseteq Y$ and $B^-(r) \cap Y = \emptyset$ hold; for a program P, $Y \models P$ holds iff for each $r \in P$, $Y \models r$. If $Y \models P$ holds, Y is called a *model* of P. An interpretation Y is an *answer set* of a program P iff it is a minimal (w.r.t. set inclusion) model of the *reduct* $P^Y = \{H(r) \leftarrow B^+(r) \mid Y \cap B^-(r) = \emptyset\}$ of P w.r.t. Y. The set of all answer sets of a program P is denoted by $\mathrm{AS}(P)$.

Next, we review some prominent notions of equivalence, which have been studied under the answer-set semantics: Programs P, Q are

- *strongly equivalent* [12], iff, for any program R, $\mathrm{AS}(P \cup R) = \mathrm{AS}(Q \cup R)$;
- P and Q are *uniformly equivalent* [3], iff, for any set F of facts, $\mathrm{AS}(P \cup F) = \mathrm{AS}(Q \cup F)$.

Relativizations of these notions are as follows [6,18]: For a given alphabet $\mathcal{A} \subseteq \mathcal{U}$, we call programs P, Q *strongly equivalent relative to* $\mathcal{A}$, iff, for any program R over $\mathcal{A}$, it holds that $\mathrm{AS}(P \cup R) = \mathrm{AS}(Q \cup R)$; P, Q are *uniformly equivalent relative to* $\mathcal{A}$, iff, for any set $F \subseteq \mathcal{A}$ of facts, $\mathrm{AS}(P \cup F) = \mathrm{AS}(Q \cup F)$. In case of strong equivalence (also in the relativized case), it was shown [12,18] that the syntactic class of *counterexamples* (i.e., programs R such that $AS(P \cup R) \neq AS(Q \cup R)$) can always be restricted to the class of unary programs. All equivalence notions mentioned have been generalized to a uniform setting in [19]: Given alphabets $\mathcal{A}, \mathcal{B} \subseteq \mathcal{U}$, programs P, Q are called $(\mathcal{A}, \mathcal{B})$ equivalent iff, for any program R with $H(R) \subseteq \mathcal{A}$ and $B(R) \subseteq \mathcal{B}$, $AS(P \cup R) = AS(Q \cup R)$.

Given a program P, we call a pair (X, Y) SE-model of P (in symbols $(X, Y) \in SE(P)$) if $X \subseteq Y$, $X \models P^Y$ and $Y \models P$. We call SE-models (X, Y) total if $X = Y$, and non-total otherwise. SE-models characterize strong equivalence in the following sense [17]. Two program P and Q are strongly equivalent iff $SE(P) = SE(Q)$. Based on SE-models similar (however, more involved) characterizations for the aforementioned equivalence notions have been introduced [6,19]. We review here only the result for uniform equivalence [3]. Call an SE-model (X, Y) of P UE-model of P (in symbols $(X, Y) \in UE(P)$) if for each $(X', Y) \in SE(P)$ with $X \subset X' \subseteq Y$, $X' = Y$. Then, programs P and Q are uniformly equivalent iff $UE(P) = UE(Q)$.

3   Definition and Characterization

We now define the novel equivalence notion we are interested in here.

Definition 1. *Given an alphabet $\mathcal{A} \subseteq \mathcal{U}$, we say that two programs P and Q are* rule-equivalent relative to $\mathcal{A}$ *and write $P \equiv_{\mathcal{A}} Q$ if for any set of rules R composed of*

1. *arbitrary proper rules and*
2. *facts $S \subseteq \mathcal{A}$,*

$AS(P \cup R) = AS(Q \cup R)$ holds. If $P \equiv_{\mathcal{A}} Q$ holds for $\mathcal{A} = \emptyset$, we occasionally say that P and Q are proper-rule-equivalent. *Whenever we leave $\mathcal{A}$ unspecified, we simply talk about rule equivalence* between programs.

The main idea is that we assume knowledge about facts from $\mathcal{U} \setminus \mathcal{A}$ as already fixed (i.e., they will not be altered in the context R) while no further restriction is imposed in the equivalence test.

By definition, strong equivalence implies rule equivalence relative to any $\mathcal{A}$. If we set $\mathcal{A} = \mathcal{U}$ the notions coincide. In the other extreme case, $\mathcal{A} = \emptyset$, no facts will be added to the two programs under consideration. This particular notion of rule equivalence thus can be seen as an antipode to uniform equivalence. Hence, it is a natural question whether rule and uniform equivalence are incomparable notions.

Example 1. Let us for the moment fix $\mathcal{U} = \{a, b\}$ and consider the following programs:

$$P = \{a \leftarrow b, \quad \leftarrow not\ a, \quad \leftarrow not\ b\}$$
$$Q = \{a \leftarrow, \quad \leftarrow not\ b\}$$
$$R = \{a \leftarrow b, \quad b \leftarrow a, \quad \leftarrow not\ a, \quad \leftarrow not\ b\}$$

Their SE-models[1] are as follows:

$$SE(P) = \{(ab, ab), (a, ab), (\emptyset, ab)\}$$
$$SE(Q) = \{(ab, ab), (a, ab)\}$$
$$SE(R) = \{(ab, ab), (\emptyset, ab)\}$$

By definition of UE-models $UE(P) = UE(Q)$, and thus P and Q are uniformly equivalent. However, they are not rule-equivalent relative to $\mathcal{A}$, even for $\mathcal{A} = \emptyset$. Consider $S = \{b \leftarrow a\}$. As is easily checked, $AS(P \cup S) = \emptyset$ while $AS(Q \cup S) = \{\{a, b\}\}$. On the other hand, P and R are not uniformly equivalent ($UE(P) = \{(ab, ab), (a, ab)\} \neq UE(R) = \{(ab, ab), (\emptyset, ab)\}$), while they are rule-equivalent relative to $\mathcal{A}$ with $\mathcal{A} \subseteq \{b\}$. We will see next how the latter result can be verified due to a suitable characterization. ◇

Also, observe that the notion of rule equivalence is *not* captured by the general $(\mathcal{A}, \mathcal{B})$ equivalence framework of [19], since our notion allows for any atoms in heads and bodies of proper rules, but restricts the atoms in heads only for facts, i.e., rules with empty bodies.

We now continue with the characterization for our main result.

Definition 2. *An SE-model (X, Y) of a program P is called an $\mathcal{A}$-RE-model of P if:*

(a) $(X = \emptyset$ *and* $(Y \cap \mathcal{A}) \neq \emptyset)$ *or*
(b) $(\emptyset, Y) \notin SE(P)$ *or*
(c) $(X \cap \mathcal{A}) \neq \emptyset$ *or*
(d) $Y = \emptyset$.

We write $(X, Y) \in RE^{\mathcal{A}}(P)$ to indicate that (X, Y) is an $\mathcal{A}$-RE-model of P.

For the case $\mathcal{A} = \emptyset$ we can give a simpler characterization.

Definition 3. *An SE-model (X, Y) of a program P is called an RE-model if $(\emptyset, Y) \notin SE(P)$ holds whenever $Y \neq \emptyset$. We write $(X, Y) \in RE(P)$ to indicate that (X, Y) is an RE-model of P.*

Lemma 1. *For any program P the following relations hold:*

1. $RE(P) = RE^{\emptyset}(P)$
2. $SE(P) = RE^{\mathcal{U}}(P)$.

Proof. 1 is straightforward from Definition 2. For 2, we observe the following implication from Definition 2: $(X, Y) \in RE^{\mathcal{U}}(P)$ iff $(X, Y) \in SE(P)$ and $((a)$ $(X = \emptyset$ and $Y \neq \emptyset)$ or (b) $(\emptyset, Y) \notin SE(P)$ or (c) $X \neq \emptyset$ or (d) $Y = \emptyset)$ iff $(X, Y) \in SE(P)$. □

---

[1] Within SE-models, we denote interpretations $\{a_1, \ldots, a_n\}$ by juxtaposition $a_1 \cdots a_n$ of their elements.

As a final simple observation, we have the following result that generalizes the well-known property of SE-models that $(X, Y) \in SE(P)$ implies $(Y, Y) \in SE(P)$ for any program P.

Lemma 2. *For any* $\mathcal{A} \subseteq \mathcal{U}$ *and program* P: $(X, Y) \in RE^{\mathcal{A}}(P)$ *implies* $(Y, Y) \in RE^{\mathcal{A}}(P)$.

Proof. Towards a contradiction, assume $(X, Y) \in RE^{\mathcal{A}}(P)$ and $(Y, Y) \notin RE^{\mathcal{A}}(P)$. From $(X, Y) \in RE^{\mathcal{A}}(P)$, $(X, Y) \in SE(P)$ and thus also $(Y, Y) \in SE(P)$. Hence, in order to have $(Y, Y) \notin RE^{\mathcal{A}}(P)$, we need, in particular, $Y \neq \emptyset$, $(Y \cap \mathcal{A}) = \emptyset$, and $(\emptyset, Y) \in SE(P)$. We conclude that $(X \cap \mathcal{A}) = \emptyset$, which together with the above observation contradicts $(X, Y) \in RE^{\mathcal{A}}(P)$ by definition of $\mathcal{A}$-RE-models. $\square$

We now present our main result. In the spirit of the seminal results for strong equivalence [12,17] we not only show that our characterization decides the equivalence notion but also that the equivalence notion boils down to unary rules.

Theorem 1. *For any programs* P, Q, *and alphabet* $\mathcal{A} \subseteq \mathcal{U}$, *the following statements are equivalent:*

(i) P *and* Q *are rule-equivalent relative to* $\mathcal{A}$.
(ii) *For any set* R *of facts from* $\mathcal{A}$ *and proper unary rules,* $AS(P \cup R) = AS(Q \cup R)$.
(iii) $RE^{\mathcal{A}}(P) = RE^{\mathcal{A}}(Q)$.

Proof. $(i) \Rightarrow (ii)$ is clear.

$(ii) \Rightarrow (iii)$. W.l.o.g. suppose $(X, Y) \in RE^{\mathcal{A}}(P) \setminus RE^{\mathcal{A}}(Q)$. First consider the case $(Y, Y) \notin RE^{\mathcal{A}}(Q)$. Then we define $R = (Y \cap \mathcal{A}) \cup \{a \leftarrow b \mid a, b \in Y\}$. We show $Y \in AS(P \cup R)$ but $Y \notin AS(Q \cup R)$. For the former observe that $Y \models P \cup R$ (since $(X, Y) \in SE(P)$). If $Y = \emptyset$ or $Y \cap \mathcal{A} \neq \emptyset$, then $Y \in AS(P \cup R)$ as no proper subset of Y would satisfy R^Y. On the other hand, if $Y \neq \emptyset$ and $Y \cap \mathcal{A} = \emptyset$ hold, $\emptyset$ is the only proper subset of Y satisfying R^Y. By Lemma 2, $(Y, Y) \in RE^{\mathcal{A}}(P)$, and by definition of RE-models (b) applies and we obtain $(\emptyset, Y) \notin SE(P)$. Thus, $\emptyset$ is not model of $P^Y \cup R = (P \cup R)^Y$. This shows $Y \in AS(P \cup R)$. To see that $Y \notin AS(Q \cup R)$, we distinguish the two possible reasons for $(Y, Y) \notin RE^{\mathcal{A}}(Q)$: First, $(Y, Y) \notin SE(Q)$. Then, $Y \not\models Q$. Otherwise, we have

(a) $(Y \neq \emptyset$ or $Y \cap \mathcal{A} = \emptyset)$ and
(b) $(\emptyset, Y) \in SE(Q)$ and
(c) $(Y \cap \mathcal{A}) = \emptyset$ and
(d) $Y \neq \emptyset$.

Thus, in particular (b), (c) and (d). We observe that $\emptyset \models Q^Y \cup R = (Q \cup R)^Y$. For both cases, thus $Y \notin AS(Q \cup R)$.

It remains to consider the case $(Y,Y) \in \mathrm{RE}^{\mathcal{A}}(Q)$. Recall that we have $(X,Y) \notin \mathrm{RE}^{\mathcal{A}}(Q)$ and $(X,Y) \in \mathrm{RE}^{\mathcal{A}}(P)$, which implies $(Y,Y) \in \mathrm{RE}^{\mathcal{A}}(P)$ and $X \subset Y$.

Suppose $X \cap \mathcal{A} = \emptyset$. Since $(X,Y) \in \mathrm{RE}^{\mathcal{A}}(P)$, $(X,Y) \in \mathrm{SE}(P)$ holds and

(a1) $(X = \emptyset$ and $(Y \cap \mathcal{A}) \neq \emptyset)$ or
(b1) $(\emptyset,Y) \notin \mathrm{SE}(P)$.

Since $(X,Y) \notin \mathrm{RE}^{\mathcal{A}}(Q)$ either $(X,Y) \notin \mathrm{SE}(Q)$ or jointly:

(a2) $(X \neq \emptyset$ or $(Y \cap \mathcal{A}) = \emptyset)$ and
(b2) $(\emptyset,Y) \in \mathrm{SE}(Q)$.

Suppose (a1) holds. Then (a2) is false and thus $(X,Y) \notin \mathrm{SE}(Q)$ with $X = \emptyset$. We can take $R = \{a \leftarrow b \mid a,b \in Y\}$ and obtain $Y \in \mathrm{AS}(Q \cup R) \setminus \mathrm{AS}(P \cup R)$. Otherwise, we have $(\emptyset,Y) \notin \mathrm{SE}(P)$, and at least one of $(\emptyset,Y) \in \mathrm{SE}(Q)$ and $(X,Y) \notin \mathrm{SE}(Q)$. For $(\emptyset,Y) \in \mathrm{SE}(Q)$, take $R = \{a \leftarrow b \mid a,b \in Y\}$ as before. Now however, $Y \in \mathrm{AS}(P \cup R) \setminus \mathrm{AS}(Q \cup R)$. For the case $(\emptyset,Y) \notin \mathrm{SE}(Q)$ and $(X,Y) \notin \mathrm{SE}(Q)$, take

$$R = \{a \leftarrow b \mid a,b \in X\} \cup \{c \leftarrow d \mid c \in Y, d \in Y \setminus X\}.$$

Now, $Y \in \mathrm{AS}(Q \cup R) \setminus \mathrm{AS}(P \cup R)$.

It remains to consider the case $X \cap \mathcal{A} \neq \emptyset$. Here we take

$$R = (X \cap \mathcal{A}) \cup \{a \leftarrow b \mid a,b \in X\} \cup \{c \leftarrow d \mid c \in Y, d \in Y \setminus X\}$$

and observe that the only models of R being a subset of Y are Y and X. We already know $(X,Y) \in \mathrm{SE}(P)$. From $(X,Y) \notin \mathrm{RE}^{\mathcal{A}}(Q)$ and $(X \cap \mathcal{A}) \neq \emptyset$, we get $(X,Y) \notin \mathrm{SE}(Q)$. This suffices to see that $Y \in \mathrm{AS}(Q \cup R) \setminus \mathrm{AS}(P \cup R)$.

$(iii) \Rightarrow (i)$. Assume existence of R being a set of proper rules and facts $F \subseteq \mathcal{A}$, such that w.l.o.g. $Y \in \mathrm{AS}(P \cup R)$ and $Y \notin \mathrm{AS}(Q \cup R)$. From the former we get $Y \models P$ (hence, $(Y,Y) \in \mathrm{SE}(P)$) and $Y \models R$.

We first show $(Y,Y) \in \mathrm{RE}^{\mathcal{A}}(P)$. Towards a contradiction, suppose this is not the case. By definition of $\mathcal{A}$-RE-models, we then have $(Y \cap \mathcal{A}) = \emptyset$, $(\emptyset,Y) \in \mathrm{SE}(P)$, $Y \neq \emptyset$. From the former we observe that R must not contain facts (otherwise $Y \not\models R$), hence $\emptyset$ is a model of R^Y. Together with $(\emptyset,Y) \in \mathrm{SE}(P)$, $\emptyset \models (P \cup R)^Y$. Contradiction to $Y \in \mathrm{AS}(P \cup R)$.

Hence $(Y,Y) \in \mathrm{RE}^{\mathcal{A}}(P)$, and we proceed by distinguishing two cases.

1. $Y \not\models Q \cup R$. Since we already have seen $Y \models R$, this amounts to $Y \not\models Q$. It follows that $(Y,Y) \notin \mathrm{SE}(Q)$, and consequently, $(Y,Y) \notin \mathrm{RE}^{\mathcal{A}}(Q)$.
2. $Y \models Q$ and some $X \subset Y$ is model of $(Q \cup R)^Y = Q^Y \cup R^Y$; hence $(X,Y) \in \mathrm{SE}(Q)$; and moreover, $(X,Y) \notin \mathrm{SE}(P)$ – otherwise $Y \in \mathrm{AS}(P \cup R)$ cannot hold. Since $(X,Y) \notin \mathrm{SE}(P)$, $(X,Y) \notin \mathrm{RE}^{\mathcal{A}}(P)$. Also recall that $(Y,Y) \in \mathrm{RE}^{\mathcal{A}}(P)$.

In case $(X,Y) \in \mathrm{RE}^{\mathcal{A}}(Q)$ or $(Y,Y) \notin \mathrm{RE}^{\mathcal{A}}(Q)$, we immediately obtain $\mathrm{RE}^{\mathcal{A}}(P) \neq \mathrm{RE}^{\mathcal{A}}(Q)$, so suppose $(X,Y) \notin \mathrm{RE}^{\mathcal{A}}(Q)$ and $(Y,Y) \in \mathrm{RE}^{\mathcal{A}}(Q)$. Since $(X,Y) \in \mathrm{SE}(Q)$ and $(X,Y) \notin \mathrm{RE}^{\mathcal{A}}(Q)$ the following jointly hold:

(a) $X \neq \emptyset$ or $(Y \cap \mathcal{A}) = \emptyset$

(b) $(\emptyset, Y) \in \mathrm{SE}(Q)$

(c) $(X \cap \mathcal{A}) = \emptyset$

Then, $(Y,Y) \in \mathrm{RE}^{\mathcal{A}}(Q)$ must be due to $(Y \cap \mathcal{A}) \neq \emptyset$, which implies $(\emptyset, Y) \in \mathrm{RE}^{\mathcal{A}}(Q)$ by Definition 2 together with (b). We already have seen that $X \models R^Y$. Together with (c) $X \cap \mathcal{A} = \emptyset$, we can conclude that R must not contain facts (no facts from X are allowed due to (c), other facts violate $X \models R^Y$). It follows that $\emptyset$ is a model of R. We conclude $(\emptyset, Y) \notin \mathrm{SE}(P)$ – otherwise $Y \in \mathrm{AS}(P \cup R)$ cannot hold. Consequently, $(\emptyset, Y) \notin \mathrm{RE}^{\mathcal{A}}(P)$. Hence, for the case that $(Y,Y) \in \mathrm{RE}^{\mathcal{A}}(P)$, $(X,Y) \notin \mathrm{RE}^{\mathcal{A}}(P)$, $(Y,Y) \in \mathrm{RE}^{\mathcal{A}}(Q)$, $(X,Y) \notin \mathrm{RE}^{\mathcal{A}}(Q)$, we have shown that $(\emptyset, Y) \notin \mathrm{RE}^{\mathcal{A}}(P)$ and $(\emptyset, Y) \in \mathrm{RE}^{\mathcal{A}}(Q)$. $\square$

4 Properties of Rule Equivalence

In this section, we compare the notions of strong, uniform, and proper-rule equivalence. We also provide some complexity analysis.

Let us first proceed with our example from above.

Example 2. Recall programs P, Q, R from Example 1. We already have observed that P and Q are not proper-rule-equivalent. In fact, the RE-models according to Definition 3 are given by $\mathrm{RE}(P) = \emptyset$ and $\mathrm{RE}(Q) = \mathrm{SE}(Q) = \{(ab, ab), (a, ab)\}$. On the other hand, we have claimed that P and R are rule-equivalent relative to $\mathcal{A}$ with $\mathcal{A} \subseteq \{b\}$. We can now verify this by observing $\mathrm{RE}(R) = \emptyset$. If we now set $\mathcal{A} = \{b\}$, we observe that $\mathrm{RE}^{\mathcal{A}}(P) = \{(ab, ab), (\emptyset, ab)\} = \mathrm{RE}^{\mathcal{A}}(R)$. For $\mathcal{A} \supseteq \{a\}$ however, we observe $\mathrm{RE}^{\mathcal{A}}(P) = \mathrm{SE}(P) \neq \mathrm{SE}(R) = \mathrm{RE}^{\mathcal{A}}(R)$, since in the case of P, condition (c) of Definition 2 applies for the SE-model (a, ab). ◇

Let us say that two notions of equivalence $\equiv$ and $\equiv'$ are incomparable if there exist programs P, P', Q, Q', such that $P \equiv Q$ and $P \not\equiv' Q$, and moreover $P' \not\equiv Q'$ and $P' \equiv' Q'$. Our examples thus imply the following result.

Theorem 2. *Proper-rule equivalence and uniform equivalence are incomparable notions; this holds already for normal programs.*

Interestingly, if we disallow negation, uniform equivalence turns out to be a strictly stronger notion than proper-rule equivalence.

Theorem 3. *For positive programs, uniform equivalence implies proper-rule equivalence, but proper-rule equivalence does not imply uniform equivalence.*

Proof. Recall that for positive programs, strong and uniform equivalence coincide [6], thus the first direction is obvious. On the other hand, program $\{a \leftarrow b; \ b \leftarrow a\}$ can be seen to be proper-rule-equivalent to $\{a \leftarrow c; \ c \leftarrow a\}$, but these two programs are not uniformly equivalent. □

As a final result in this section, we define one simple class of programs for which rule equivalence coincides with strong equivalence, but where uniform equivalence remains a weaker concept.

Definition 4. *Call a program P factual if it contains at least one fact, i.e., at least one rule with empty body.*

It is rather obvious that programs with at least one fact a, are able to "simulate" the presence of further facts b via proper rules $b \leftarrow a$. The forthcoming result links this observation to equivalence notions.

Theorem 4. *For factual programs, proper-rule equivalence implies strong equivalence (and thus uniform equivalence), but uniform equivalence does not imply proper-rule equivalence.*

Proof. The first direction is due to the easy observation that factual programs do not have SE-models of the form $(\emptyset, Y)$ and thus by Definition 3, SE-models and RE-models coincide. For the second direction consider programs $\{a \leftarrow; \ b \vee c \leftarrow\}$ and $\{a \leftarrow; \ b \leftarrow not \ c; \ c \leftarrow not \ b\}$. The programs are uniformly equivalent but neither strongly nor rule-equivalent; in fact, just add $R = \{b \leftarrow c, c \leftarrow b\}$. □

We observe that the above counter-example also shows that shifting atoms from the head into the body of a rule is not a faithful manipulation w.r.t. rule equivalence, while it is known that shifting is faithful w.r.t. uniform equivalence [6].

Finally, we study here the following decision problem (RE-equivalence): Given disjunctive programs P, Q and set of propositional atoms $\mathcal{A}$, are P and Q rule-equivalent relative to $\mathcal{A}$?

Theorem 5. *RE-equivalence is coNP-complete. Hardness holds even for fixed $\mathcal{A}$, P being positive or normal and Q Horn.*

Proof. Hardness follows from the same idea as the proof of Theorem 6.17 in [6]. Membership can be seen via the following algorithm for the complementary problem. Thanks to Theorem 1 we know that P and Q are not rule-equivalent relative to $\mathcal{A}$ iff they possess different $\mathcal{A}$-RE-models. Thus, we guess a pair (X, Y) over the atoms from $P \cup Q$ and check whether it is $\mathcal{A}$-RE-model of exactly one program. Verifying whether (X, Y) is SE-model of a program and likewise checking conditions (a)–(d) from Definition 2 are all easy tests. □

Thus the complexity for checking rule equivalence matches the one of strong equivalence and is easier than uniform equivalence, which is known to Π_2^P-complete. However, it is notable that membership in coNP also holds in the relativized case of rule equivalence for arbitrary $\mathcal{A}$, while strong equivalence relative to $\mathcal{A}$ yields an increase in complexity up to Π_2^P. For an overview of the mentioned known complexity results, we refer to [6].

5   Rule Equivalence in the Non-Ground Case

Our final result is concerned with the question whether rule equivalence remains decidable in the non-ground case of ASP. We emphasize that the answer to this question is not obvious. While strong equivalence is known to be decidable, uniform equivalence is in fact undecidable [5].

We will restrict ourselves here to the case of proper-rule equivalence, i.e., rule equivalence relative to $\mathcal{A} = \emptyset$. Thus, with some little abuse of notation, we will from now on refer to rule equivalence when we mean rule equivalence relative to $\mathcal{A} = \emptyset$. Our result is based on a transformation τ such that programs P and Q are rule-equivalent iff $\tau(P)$ and $\tau(Q)$ are strongly equivalent. We shall first introduce τ for the propositional case and then adapt the idea to the non-ground case.

5.1   Reducing Rule Equivalence to Strong Equivalence

For the sake of presentation, we first present τ as a translation from disjunctive programs (the class we have focused on in this paper) to the more general class of programs with nested expressions [11]. In this class, rule bodies can be an arbitrary Boolean combination of atoms and default-negated atoms.[2] Concerning semantics, it is only the definition of the reduct that changes. Formally, given a nested logic program P and an interpretation I, define P^I as obtained from P by replacing every occurrence of a literal *not a* by $\bot$ if $a \in I$, and by $\top$ otherwise. As before, an interpretation I is an answer set of a program P iff it is a minimal model (w.r.t. set inclusion) of P^I. With the modified reduct at hand, SE-models are likewise easily generalized to nested logic programs.

We now turn to the expected behavior of the translation τ. In what follows, we let V be the set of atoms occurring in a program P and $\overline{V}$, V', $\overline{V}'$ be fresh copies of V.

Definition 5. *A translation τ mapping disjunctive programs to programs with nested expressions satisfies the* RE-to-SE *property if the following holds for any program P:*

1. for any $I \subseteq V$,

$$I \cup (\overline{V \setminus I}) \cup V' \cup \overline{V}' \models \tau(P)$$

---

[2] In fact, the programs we use here form a proper subclass of programs compared to [11], where, e.g., also double negation is allowed. However, for our purpose it is sufficient to consider this weaker class.

2. *for any* $J \subseteq I \subseteq V$,

$$I \cup (\overline{V \setminus I}) \cup J' \cup (\overline{V \setminus J})' \models \tau(P)^{I \cup (\overline{V \setminus I}) \cup V' \cup \overline{V}'}$$

$$\Longleftrightarrow$$

$$(J, I) \in SE(P) \text{ and either } (\emptyset, I) \notin SE(P) \text{ or } I = \emptyset$$

3. *all models of* $\tau(P)$ *and reducts of* $\tau(P)$ *are implicitly given above; formally for any* X, Y *such that* $Y \models \tau(P)$ *and* $X \models \tau(P)^Y$ *it holds that (i)* Y *is of the form* $I_Y \cup V' \cup \overline{V}'$ *for some* $I \subseteq V$ *with* $I_Y = I \cup (\overline{V \setminus I})$; *(ii) if* $X \neq Y$ *then* X *is of the form* $I_Y \cup J' \cup (\overline{V \setminus J})'$ *for some* J *such that* $(J, I) \in SE(P)$ *and either* $(\emptyset, I) \notin SE(P)$ *or* $I = \emptyset$.

Intuitively, the idea is to embed all RE-models (J, I) of P as models of reducts of $\tau(P)$ using the unprimed atoms to express I and the primed ones to express J. More formally, for each RE-model (J, I) of P we will have that $I \cup (\overline{V \setminus I}) \cup J' \cup (\overline{V \setminus J})' \models \tau(P)^{I \cup (\overline{V \setminus I}) \cup V' \cup \overline{V}'}$. The reason why we need to express RE-models in this rather cumbersome way is that we need to a express a negative test $(\emptyset, I) \notin SE(P)$ with $\tau(P)$; although the test is conceptually easy, in order to express it via ASP, we have to make use of a certain saturation[3] encoding to realize it.

For the forthcoming theorem, observe that the fact that P and Q contain exactly the same set of atoms is not a severe restriction, since we can always add a "tautological rule" $a \leftarrow a$ for a missing atom a.

Theorem 6. *Given programs* P, Q *both containing the same set of atoms* V *and function* τ *satisfying the RE-to-SE property, we have that* P *and* Q *are rule-equivalent iff* $\tau(P)$ *and* $\tau(Q)$ *are strongly equivalent.*

Proof (sketch). The proof follows from the following observations: (1) the total SE-models (Y, Y) of $\tau(P)$ and $\tau(Q)$ coincide; (2) for any $J, I \subseteq V$ we have by definition that $(J, I) \in RE(R)$ iff $(X_J, Y_I) \in SE(\tau(R))$ with $Y_I = I \cup (\overline{V \setminus I}) \cup V' \cup \overline{V}'$ and $X_J = I \cup (\overline{V \setminus I}) \cup J' \cup (\overline{V \setminus J})'$; (3) there are no other non-total SE-models of $\tau(P)$ or $\tau(Q)$ than those corresponding to an RE-model. $\qquad \square$

We now actually give a translation τ satisfying the RE-to-SE property. Below we intuitively explain the functioning of the three parts.

Definition 6. *Given* P *over* V, *we define*

$$\tau(P) = \tau_G(P) \cup \tau_M(P) \cup \tau_R(P)$$

---

[3] The concept of saturation refers to a programming technique, where reasons for a candidate answer set I to be ruled out are not explicitly stated via constraints, but in terms of rules which ensure that a certain model $J \subset I$ of the program's reduct with respect to I exists, see, e.g., [10].

where

$$\tau_G(P) = \{v \vee \overline{v} \leftarrow; \ \leftarrow v, \overline{v}; \ \leftarrow not \ v'; \ \leftarrow not \ \overline{v}'; \tag{2}$$

$$v' \vee \overline{v}' \leftarrow; \ v \leftarrow v' \mid v \in V\} \tag{3}$$

$$\tau_M(P) = \{v' \leftarrow B^+(r), \overline{B^-(r)}, \overline{H(r)}; \tag{4}$$

$$\overline{v}' \leftarrow B^+(r), \overline{B^-(r)}, \overline{H(r)}; \tag{5}$$

$$v' \leftarrow (B^+(r))', \overline{B^-(r)}, (\overline{H(r)})'; \tag{6}$$

$$\overline{v}' \leftarrow (B^+(r))', \overline{B^-(r)}, (\overline{H(r)})' \mid r \in P, v \in V\} \tag{7}$$

$$\tau_R(P) = \{v' \leftarrow w, \bigwedge_{r \in P, B^+(r)=\emptyset} \bigvee_{y \in B^-(r)} y; \tag{8}$$

$$\overline{v}' \leftarrow w, \bigwedge_{r \in P, B^+(r)=\emptyset} \bigvee_{y \in B^-(r)} y \mid v, w \in V\} \tag{9}$$

$\tau_G(P)$ is responsible for the guess of interpretations; (2) forces the classical models of $\tau(P)$ to be of the required form $I_Y \cup V' \cup \overline{V}'$ for $I \subseteq V$ with $I_Y = I \cup (\overline{V \setminus I})$. (3) restricts the proper submodels of reducts $\tau(P)^{I_Y \cup V' \cup \overline{V}'}$ to be of the form $J \cup (\overline{V \setminus J}) \cup I_Y$ for $J \subseteq I$. $\tau_M(P)$ is responsible for eliminating models of the latter kind in case (J, I) is not SE-model of P, i.e., either $I \not\models P$ (rules (4)+(5)) or $J \not\models P^I$ (rules (6)+(7)). The final rules from $\tau_R(P)$ now do the following. They eliminate reduct models $J \cup (\overline{V \setminus J}) \cup I_Y$ in case $J \neq \emptyset$ and $(\emptyset, I) \in SE(P)$. To this end, in rules (8)+(9) we use atom w to make the rules applicable only if $J \neq \emptyset$. Moreover, in order to have $(\emptyset, I) \in SE(P)$ we need that each rule r of P with an empty positive body, i.e., $B^+(r) = \emptyset$, is removed in the construction of P^I, i.e., for each such r, $B^-(r) \cap I \neq \emptyset$ has to hold. We model this via the disjunction in the rule bodies of rules (8)+(9).

Lemma 3. *Translation τ from Definition 6 satisfies the RE-to-SE property.*

In fact, it is only $\tau_R(P)$ which uses non-standard rules, i.e., rules with nested Boolean expressions. As known from [11], these rule can be transformed to standard rules without changing the SE-models, by applying standard laws of distributivity within rule bodies and a so-called splitting rule

$$\{A \leftarrow B \vee C\} \ \Rightarrow \ \{A \leftarrow B; \ A \leftarrow C\}.$$

Applying this transformation rules to $\tau_R(P)$ sufficiently often, we end up with a standard disjunctive logic program. This can be given as follows: Let $P_0 = \{r_1, \ldots, r_k\}$ be the set of all rules of P with empty positive body and $P_0^- = \{\langle y_1, \ldots y_k \rangle \mid y_1 \in B^-(r_1), \ldots, y_k \in B^-(r_k)\}$ any selection of negative body atoms of rules in P_0. We then can rewrite $\tau_R(P)$ as:

$$\tau_R'(P) = \{ \ v' \leftarrow w, y_1, \ldots, y_k;$$
$$\overline{v}' \leftarrow w, y_1, \ldots, y_k \mid v, w \in V, \langle y_1, \ldots y_k \rangle \in P_0^- \}$$

Let $\tau'(\cdot)$ be the result of replacing in $\tau(\cdot)$ the program $\tau_R(\cdot)$ by $\tau_R'(\cdot)$.

Corollary 1. *For any finite programs* P, Q *over the same atoms* V, *it holds that* P *and* Q *are rule-equivalent iff* $\tau'(P)$ *and* $\tau'(Q)$ *are strongly equivalent.*

We observe that $\tau'(P)$ is exponential in the size of the original program P (whereas $\tau(P)$ is efficiently constructible). Although our complexity results suggest that better translations from disjunctive programs to disjunctive programs might exist, we leave this question of an efficient reduction for future work. Indeed, since we are mainly interested in a reduction for the non-ground case in order to show decidability, the only issue that counts is that we have a computable translation.

5.2 Decidability in the Non-Ground Case

We briefly recall syntax and semantics of non-ground ASP. Here, programs are formulated in a language containing a set $\mathcal{Q}$ of *predicate symbols*, a set $\mathcal{V}$ of *variables*, and a countably infinite set $\mathcal{C}$ of *constants*. Each predicate symbol p has an associated arity $n(p)$. Rules are of the form (1), but now an atom is an expression of form $p(t_1,\ldots,t_n)$, where $p \in \mathcal{Q}$ is a predicate of arity n and $t_i \in \mathcal{C} \cup \mathcal{V}$, for $1 \leq i \leq n$. An atom is *ground* if no variable occurs in it. A rule r is *safe* if each variable occurring in $H(r) \cup B^{\langle}r)$ also occurs in $B^+(r)$; r is *ground*, if all atoms occurring in r are ground. Given a rule r and $C \subseteq \mathcal{C}$, we define $\mathsf{G}(r, C)$ as the set of all rules obtained from r by all possible substitutions of elements of C for the variables in r. Moreover, we define $\mathsf{G}(P, C) = \bigcup_{r \in P} \mathsf{G}(r, C)$. Interpretations are now sets of ground atoms, and the concept of satisfaction ($\models$) is analogous to the ground case; likewise the reduct on ground programs is defined as expected. Finally, a set I of ground atoms is an *answer set* of a non-ground program P iff I is a subset-minimal set satisfying $\mathsf{G}(P, C_P)^I$, where C_P denotes the so-called active domain of P, i.e., the set of all constant symbols in P (if no such constant symbol exists, $C_P = \{c\}$ with an arbitrary constant symbol c).

Equivalence notions for non-ground programs are defined as expected as well. Note that for strong equivalence it is thus possible that the context programs R extend the active domain from C_P to $C_{P \cup R}$. This also holds for rule equivalence where a context program R can contain any rule of the form $p(c) \leftarrow q(X)$ where $c \in \mathcal{C}$ is any constant not occurring in the programs that are compared to each other. Since $\mathcal{C}$ is not bounded, decidability is thus not trivial.

We now define the translation for the non-ground case. We use some abbreviations: $dom(X, n)$ is a shorthand for the sequence $dom(X_1),\ldots dom(X_n)$, and $p(X)$ denotes $p(X_1,\ldots X_{n(p)})$. As before we use new predicate symbols $\bar{p}$, p', $\bar{p}'$. The main observation is that we have a new set of rules τ_D^* to collect all constant symbols occurring in the program plus in the possibly added rules of the equivalence test. The remaining parts basically just generalize the program for the propositional case from Definition 6 to the non-ground case, using the *dom* predicate to ensure safety of rules.

Definition 7. *Given a non-ground program P over predicates $\mathcal{Q}$, we define*

$$\tau^*(P) = \tau_D^*(P) \cup \tau_G^*(P) \cup \tau_M^*(P) \cup \tau_R^*(P)$$

where

$$\tau_D^*(P) = \{dom(X_i) \leftarrow p(X_1, \ldots, X_{n(p)}) \mid p \in \mathcal{Q}, 1 \leq i \leq n(p)\}$$
$$\tau_G^*(P) = \{p(X) \vee \overline{p}(X) \leftarrow dom(X, n(p));$$
$$\leftarrow p(X), \overline{p}(X);$$
$$\leftarrow not\ p'(X), dom(X, n(p));$$
$$\leftarrow not\ \overline{p}'(X), dom(X, n(p));$$
$$p'(X) \vee \overline{p}'(X) \leftarrow dom(X, n(p));$$
$$p(X) \leftarrow p'(X) \mid p \in \mathcal{Q}\}$$
$$\tau_M^*(P) = \{p'(X) \leftarrow B^+(r), \overline{B^-(r)}, \overline{H(r)}, dom(X, n(p));$$
$$\overline{p}'(X) \leftarrow B^+(r), \overline{B^-(r)}, \overline{H(r)}, dom(X, n(p));$$
$$p'(X) \leftarrow (B^+(r))', \overline{B^-(r)}, (\overline{H(r)})', dom(X, n(p));$$
$$\overline{p}'(X) \leftarrow (B^+(r))', \overline{B^-(r)}, (\overline{H(r)})', dom(X, n(p)) \mid r \in P, p \in \mathcal{Q}\}$$
$$\tau_R^*(P) = \{p'(X) \leftarrow w(Y), y_1, \ldots, y_k, dom(X, n(p));$$
$$\overline{p}'(X) \leftarrow w(Y), y_1, \ldots, y_k, dom(X, n(p)) \mid p, w \in \mathcal{Q}, \langle y_1, \ldots y_k \rangle \in P_0^-\}$$

and $P_0^- = \{\langle y_1, \ldots y_k \rangle \mid y_1 \in B^-(r_1), \ldots, y_k \in B^-(r_k)\}$ is any selection of negative body atoms of rules in P with empty positive body.

As before, we mention that the forthcoming theorem is not restricted by the fact that the compared programs need to contain the same predicate and constant symbols. Again, adding tautological rules of the form $p(a) \leftarrow p(a)$ is possible without changing any equivalence notion.

Theorem 7. *Let P and Q be non-ground programs over the same set of predicate symbols and constants. Then, P and Q are rule-equivalent iff $\tau^*(P)$ and $\tau^*(Q)$ are strongly equivalent.*

Proof (sketch). The proof idea relies on the fact the we can reduce the non-ground setting to the ground one which will allow us to employ Corollary 1. However, some care is needed to restrict ourselves to finite ground programs. Recall that we assumed an infinite set $\mathcal{C}$ of constant symbols (otherwise the rule equivalence problem would be decidable by trivial means).

For the if-direction, suppose $\tau^*(P)$ and $\tau^*(Q)$ are not strongly equivalent. From the results in [5], it follows that then there is a finite $C \subseteq \mathcal{C}$ such that $G(\tau^*(P), C)$ is not strongly equivalent to $G(\tau^*(Q), C)$, and moreover that for the witnessing SE-model (X, Y) in the symmetric difference $SE(G(\tau^*(P), C)) \triangle SE(G(\tau^*(Q), C))$ we can assume w.l.o.g. that all constants from C occur the set of ground atoms Y. Let us now denote by P^* the program obtained from $G(\tau^*(P) \setminus \tau_D^*(P), C)$ by removing all atoms of the form $dom(a)$. Analogously, define Q^*. Inspecting the usage of the *dom* atoms in $\tau^*(\cdot)$

and using the fact that all constants of C occur in atoms of Y, one can show that (X, Y) remains a witnessing SE-model, i.e., $(X, Y) \in \mathrm{SE}(P^*) \triangle \mathrm{SE}(Q^*)$. Moreover, observe that $P^* = \tau(\mathsf{G}(P, C))$ and $Q^* = \tau(\mathsf{G}(Q, C))$. We thus can now apply Corollary 1 and obtain that $\mathsf{G}(P, C)$ is not rule-equivalent to $\mathsf{G}(Q, C)$. From this observation, the result that P is not rule-equivalent to Q follows quite easily.

For the only-if direction, suppose P and Q are not rule-equivalent, hence there exists a set of proper non-ground rules R such that $\mathrm{AS}(P \cup R) \neq \mathrm{AS}(Q \cup R)$. Let C_0 be the set of constants occurring in P (and thus in Q) and C be the set of constants occurring in $P \cup R$ (and thus in $Q \cup R$). By definition of answer sets, we have $\mathrm{AS}(\mathsf{G}(P \cup R, C)) \neq \mathrm{AS}(\mathsf{G}(Q \cup R, C))$. Note that $\mathsf{G}(P \cup R, C) = \mathsf{G}(P, C) \cup \mathsf{G}(R, C)$ and likewise, $\mathsf{G}(Q \cup R, C) = \mathsf{G}(Q, C) \cup \mathsf{G}(R, C)$. Hence, $\mathsf{G}(P, C)$ and $\mathsf{G}(Q, C)$ are not rule-equivalent. In case C is not finite, we need some additional argument. Since $\mathsf{G}(P, C)$ and $\mathsf{G}(Q, C)$ are ground, we can employ our characterization from Theorem 1 and observe that there exists a pair (X, Y) which is RE-model of exactly one of the programs. Inspecting the definition of RE-models, one can then show that we can safely restrict C to C' containing C_0 plus k further constants where k is the maximal number of variables in the rules of P and Q. Thus C' is finite. Moreover, $(X|_{C'}, Y|_{C'})$ is then RE-model of exactly one of the programs $\mathsf{G}(P, C')$ and $\mathsf{G}(Q, C')$, where $Z|_{C'}$ denotes the set of all ground atoms in Z which are built from constants in C' only. We conclude (again by Theorem 1) that $\mathsf{G}(P, C')$ and $\mathsf{G}(Q, C')$ are not rule-equivalent. In case C was finite, we continue with $C' = C$ and in both cases apply Corollary 1, which yields that $\tau(\mathsf{G}(P, C'))$ and $\tau(\mathsf{G}(Q, C'))$ are not strongly equivalent. Hence, there exists a pair (U, Z) that is SE-model of exactly one of the two programs. Let $D = \{dom(a) \mid a \in C'\}$. It can be shown that then $(U \cup D, Z \cup D)$ is SE-models of exactly one of the programs $\tau^*(\mathsf{G}(P, C'))$ and $\tau^*(\mathsf{G}(Q, C'))$. Moreover, $\tau^*(\mathsf{G}(P, C')) = \mathsf{G}(\tau^*(P, C'))$ and likewise, $\tau^*(\mathsf{G}(Q, C')) = \mathsf{G}(\tau^*(Q, C'))$. Thus $\mathsf{G}(\tau^*(P, C'))$ is not strongly equivalent to $\mathsf{G}(\tau^*(Q, C'))$. From this, it easily follows that $\tau^*(P)$ is not strongly equivalent to $\tau^*(Q)$. $\square$

From the above result we can immediately conclude the following.

Theorem 8. *Deciding rule equivalence between non-ground programs is decidable.*

6   Conclusion

In this paper, we have studied an equivalence notion in ASP that has been overlooked in previous work. This notion weakens strong equivalence in the sense that programs are compared under scenarios where arbitrary proper rules are allowed to be added but the addition of facts is restricted, or even forbidden. Our results show that an SE-model-based characterization for this problem is quite involved but remains on the coNP complexity level, thus showing the same complexity as strong equivalence. This correspondence also carries over to non-ground ASP, where we have shown decidability for (the unrelativized version of) rule equivalence.

Future work includes the following research questions. First, we aim for a mix of relativized strong equivalence and rule equivalence where one alphabet restricts the proper rules and a second alphabet restricts the facts in the potential context programs. Second, we are interested in finding better translations from rule equivalence to strong equivalence, avoiding the exponential blow-up we have witnessed in Sect. 5.2. Finally, we claim that decidability also holds for the relativized variants of rule equivalence, but a formal proof is subject of ongoing work.

Acknowledgments. This work was supported by the Austrian Science Fund (FWF) projects P25607 and Y698.

References

1. Brewka, G., Eiter, T., Truszczyński, M.: Answer set programming at a glance. Communications of the ACM **54**(12), 92–103 (2011)
2. Eiter, T., Fink, M., Tompits, H., Woltran, S.: Simplifying Logic Programs Under Uniform and Strong Equivalence. In: Lifschitz, V., Niemelä, I. (eds.) LPNMR 2004. LNCS (LNAI), vol. 2923, pp. 87–99. Springer, Heidelberg (2003)
3. Eiter, T., Fink, M.: Uniform Equivalence of Logic Programs under the Stable Model Semantics. In: Palamidessi, C. (ed.) ICLP 2003. LNCS, vol. 2916, pp. 224–238. Springer, Heidelberg (2003)
4. Eiter, T., Fink, M., Pührer, J., Tompits, H., Woltran, S.: Model-based recasting in answer-set programming. J. Appl. Non-Classical Logics **23**(1–2), 75–104 (2013). http://dx.org/10.1080/11663081.2013.799318
5. Eiter, T., Fink, M., Tompits, H., Woltran, S.: Strong and uniform equivalence in answer-set programming: characterizations and complexity results for the non-ground case. In: Proceedings of the 20th National Conference on Artificial Intelligence (AAAI 2005), pp. 695–700. AAAI Press (2005)
6. Eiter, T., Fink, M., Woltran, S.: Semantical characterizations and complexity of equivalences in answer set programming. ACM Trans. Comput. Log. **8**(3), 1–53 (2007). http://doi.acm.org/10.1145/1243996.1244000
7. Fink, M.: A general framework for equivalences in answer-set programming by countermodels in the logic of here-and-there. Theory Pract. Logic Programm. **11**(2–3), 171–202 (2011)
8. Inoue, K., Sakama, C.: Equivalence of logic programs under updates. In: Alferes, J.J., Leite, J. (eds.) JELIA 2004. LNCS (LNAI), vol. 3229, pp. 174–186. Springer, Heidelberg (2004)
9. Janhunen, T., Oikarinen, E., Tompits, H., Woltran, S.: Modularity aspects of disjunctive stable models. J. Artif. Intell. Res. (JAIR) **35**, 813–857 (2009). http://dx.org/10.1613/jair.2810
10. Leone, N., Pfeifer, G., Faber, W., Eiter, T., Gottlob, G., Perri, S., Scarcello, F.: The DLV system for knowledge representation and reasoning. ACM Trans. Comput. Log. **7**(3), 499–562 (2006)
11. Lifschitz, V., Tang, L., Turner, H.: Nested expressions in logic programs. Ann. Math. Artif. Intell. **25**(3–4), 369–389 (1999)
12. Lifschitz, V., Pearce, D., Valverde, A.: Strongly equivalent logic programs. ACM Trans. Comput. Logic **2**(4), 526–541 (2001)

13. Oikarinen, E., Janhunen, T.: Modular equivalence for normal logic programs. In: Proceedings of the 17th European Conference on Artificial Intelligence (ECAI 2006), pp. 412–416. IOS Press (2006)
14. Pearce, D.J., Valverde, A.: Uniform equivalence for equilibrium logic and logic programs. In: Lifschitz, V., Niemelä, I. (eds.) LPNMR 2004. LNCS (LNAI), vol. 2923, pp. 194–206. Springer, Heidelberg (2003)
15. Sagiv, Y.: Optimizing datalog programs. In: Minker, J. (ed.) Foundations of Deductive Databases and Logic Programming, pp. 659–698. Morgan Kaufmann, USA (1988)
16. Truszczynski, M., Woltran, S.: Relativized hyperequivalence of logic programs for modular programming. TPLP **9**(6), 781–819 (2009). http://dx.org/10.1017/S1471068409990159
17. Turner, H.: Strong equivalence made easy: nested expressions and weight constraints. Theor. Pract. Logic Program. **3**(4–5), 602–622 (2003)
18. Woltran, S.: Characterizations for relativized notions of equivalence in answer set programming. In: Alferes, J.J., Leite, J. (eds.) JELIA 2004. LNCS (LNAI), vol. 3229, pp. 161–173. Springer, Heidelberg (2004)
19. Woltran, S.: A common view on strong, uniform, and other notions of equivalence in answer-set programming. TPLP **8**(2), 217–234 (2008). http://dx.org/10.1017/S1471068407003250

Querying and Pattern Mining

A k-Means-Like Algorithm for Clustering Categorical Data Using an Information Theoretic-Based Dissimilarity Measure

Thu-Hien Thi Nguyen and Van-Nam Huynh[✉]

School of Knowledge Science, Japan Advanced Institute of Science and Technology,
Nomi, Japan
{nguyen.hien,huynh}@jaist.ac.jp

Abstract. Clustering large datasets is one of the important research problems for many machine learning applications. The k-means is very popular and widely used due to its ease of implementation, linear time complexity in size of the data, and almost surely convergence to local optima. However, working only on numerical data prohibits it from being used for clustering categorical data. In this paper, we aim to introduce an extension of k-means algorithm for clustering categorical data. Basically, we propose a new dissimilarity measure based on an information theoretic definition of similarity that considers the amount of information of two values in the domain set. The definition of cluster centers is generalized using kernel density estimation approach. Then, the new algorithm is proposed by incorporating a feature weighting scheme that automatically measures the contribution of individual attributes for the clusters. In order to demonstrate the performance of the new algorithm, we conduct a series of experiments on real datasets from UCI Machine Learning Repository and compare the obtained results with several previously developed algorithms for clustering categorical data.

Keywords: Cluster analysis · Categorical data clustering · K-means · Dissimilarity measures

1 Introduction

During the last decades, data mining has emerged as a rapidly growing interdisciplinary field, which merges together databases, statistics, machine learning and other related areas in order to extract useful knowledge from data [11]. Cluster analysis or simply clustering is one of fundamental tasks in data mining that aims at grouping a set of data objects into multiple clusters, such that objects within a cluster are similar one another, yet dissimilar to objects in other clusters. Dissimilarities and similarities between objects are assessed based on those attribute values describing the objects and often involve distance measures.

Typically, objects can be considered as vectors in n-dimensional space, where n is the number of features. When objects are described by numerical features,

© Springer International Publishing Switzerland 2016
M. Gyssens and G. Simari (Eds.): FoIKS 2016, LNCS 9616, pp. 115–130, 2016.
DOI: 10.1007/978-3-319-30024-5_7

the distance measure based on geometric concept such as Euclid distance or Manhattan distance can be used to define similarity between objects. However, these geometric distance measures are not applicable for categorical data which contains values, for instance, from gender, locations, etc. Recently, clustering data with categorical attributes have increasingly gained considerable attention [7–10,13,14]. As for categorical data, the comparison measure is most naturally used [13]. However, this metric does not distinguish between the different values taken by the attribute, since we only measure the equality between pair of values, as argued in [18].

In this paper we propose a new extension of the k-means algorithm for clustering categorical data. In particular, as for measuring dissimilarity between categorical objects, we make use of the information theoretic definition of similarity proposed in [20], which is intuitively defined based on the amount of information contained in the statement of commonality between values in the domain set of a categorical attribute. On the other hand, the definition of cluster centers is generalized using the kernel-based density estimates for categorical clusters as similarly considered in [6], instead of using the frequency estimates as originally in [24]. We then develop a new clustering algorithm by incorporating a feature weighting scheme that automatically measures the contribution of individual attributes to formation of the clusters.

The rest of this paper is organized as follows. Section 2 briefly describes the related work. Section 3 first introduces the k-means algorithm, and then presents its existing extensions for clustering categorical data. The proposed method is discussed in Sect. 4, and the experimental results are presented in Sect. 5. Finally, Sect. 6 concludes the paper.

2   Related Work

Probably, the k-means clustering [21] is the most well-known approach for clustering data sets with numerical attributes. It is a traditional partitioning based approach which starts with k random centroids and the centroids are updated iteratively by computing the average of the numerical features in each cluster. Each observation or object is assigned to clusters based upon the nearest distance to the means of the clusters. The iteration continues until the assignment is stable, that is, the clusters formed in the current stage are the same as those formed in the previous stage. The k-means is very popular due to its ease of implementation, linear time complexity in size of the data, and almost surely convergence to local optima [25]. However, in real life many data sets are categorical, of which k-means algorithm cannot be directly applied.

In recent years several attempts have been made in order to overcome the numerical-only limitation of k-means algorithm so as to make it applicable to clustering for categorical data, such as k-modes algorithm [14] and k-representative algorithm [24]. Particularly, in the k-modes algorithm [14], the simple matching similarity measure is used to compute distance between categorical objects, and "modes" are used instead of means for cluster centers. The mode of a cluster is a data point, in which the value of each attribute is assigned the most frequent value of the attribute's domain set appearing in the cluster. Furthermore, Huang also combined the k-modes algorithm with k-means algorithm in order to deal with mixed numerical and categorical databases. These extensions allow us to efficiently cluster very large data sets from real world applications. It is worth, however, noting that a cluster can have more than one mode and the performance of k-mode algorithm depends strongly on the selection of modes during the clustering process. In an attempt to overcome this drawback, San et al. [24] introduced a new notion of "cluster centers" called representatives for categorical objects. In particular, the representative of a cluster is defined making use of the distributions of categorical values appearing in clusters. Then, the dissimilarity between a categorical object and the representative of a cluster is easily defined based on relative frequencies of categorical values within the cluster and the simple matching measure between categorical values. In such a way, the resulting algorithm called k-representative algorithm is then formulated in a similar fashion to the k-means algorithm. In fact, it has been shown that the k-representative algorithm is very effective in clustering categorical data [22].

More recently, Chen and Wang [6] have proposed a new kernel density based method for defining cluster centers in central clustering of categorical data. Then the so-called k-centers algorithm that incorporates the new formulation of cluster centers and the weight attributes calculation scheme has been also developed. The experimental results have shown that the k-centers algorithm has good performance especially for the task of recognizing biological concepts in DNA sequences.

3 k-Means Algorithm and Its Extensions for Categorical Data

Assume that DB is a data set consisting of N objects, each of which is characterized by a set of D attributes with finite domains $O_1, \ldots, O_D$, respectively. That is, each object in DB is represented by a tuple $t \in O_1 \times \ldots \times O_D$, and the dth attribute takes $|O_d|(> 1)$ discrete values. In addition, the categories in O_d will be denoted by o_{dl}, for $l = 1, \ldots, |O_d|$, and each data object in DB will be denoted by X, with subscript if necessary, which is represented as a tuple $X = (x_1, ..., x_D) \in O_1 \times ... \times O_D$. Let $\mathcal{C} = \{C_1, \ldots, C_k\}$ be the set of k clusters of DB, i.e. we have

$$C_j \cap C_{j'} = \emptyset \text{ if } j \neq j' \text{ and } DB = \bigcup_{j=1}^{k} C_j$$

Regarding the clustering problem discussed in this paper, we consider two types of data: *numeric* and *categorical*. The domain of numerical attributes consists of continuous real values. Thus, the distance measure based on geometric concept such as the Euclid distance or Manhattan distance can be used.

A domain O_d is defined as categorical if it is finite and unordered, so that only a comparison operation is allowed in O_d. It means, for any $x, y \in O_d$, we have either $x = y$ or $x \neq y$.

3.1 k-Means Algorithm

The k-means algorithm [21] is one of the most popular algorithm in *partitional* or *non-hierarchical* clustering methods. Given a set DB of N numerical data objects, a natural number $k \leq N$, and a distance measure dis$(\cdot, \cdot)$, the k-means algorithm searches for a partition of DB into k non-empty disjoint clusters that minimizes the overall sum of the squared distances between data objects and their cluster centers. Mathematically, the problem can be formulated in terms of an optimization problem as follows:

Minimize

$$P(U, V) = \sum_{j=1}^{k} \sum_{i=1}^{N} u_{i,j} \text{dis}(X_i, V_j) \tag{1}$$

subject to

$$\sum_{j=1}^{k} u_{i,j} = 1, \quad 1 \leq i \leq N,$$

$$u_{i,j} \in \{0,1\}, \quad 1 \leq i \leq N, 1 \leq j \leq k, \tag{2}$$

where $U = [u_{i,j}]_{N \times k}$ is a partition matrix ($u_{i,j}$ take value 1 if object X_i is in cluster C_j, and 0 otherwise), $V = \{V_1, \ldots, V_k\}$ is a set of cluster centers, and dis$(\cdot, \cdot)$ is the squared Euclidean distance between two objects.

The problem P can be solved by iteratively solving two problems:

– Fix $V = \hat{V}$ then solve the reduced problem $P(U, \hat{V})$ to find $\hat{U}$.
– Fix $U = \hat{U}$ then solve the reduced problem $P(\hat{U}, V)$.

Basically, the k-means algorithm iterates through a three-step process until $P(U, V)$ converges to some local minimum:

1. Select an initial $V^{(0)} = V_1^{(0)}, \ldots, V_k^{(0)}$, and set $t = 0$.
2. Keep $V^{(t)}$ fixed and solve $P(U, V^{(t)})$ to obtain $U^{(t)}$. That is, having the cluster centers, we then assign each object to the cluster of its nearest cluster center.
3. Keep $U^{(t)}$ fixed and generate $V^{(t+1)}$ such that $P(U^{(t)}, V^{(t+1)})$ is minimized. That is, construct new cluster centers according to the current partition.
4. In the case of convergence or if a given stopping criterion is fulfilled, output the result and stop. Otherwise, set $t = t + 1$ and go to step 2.

In numerical clustering problem, the Euclidean norm is often chosen as a natural distance measure in the k-means algorithm. With this distance measure, we calculate the partition matrix in step 2 as below, and the cluster center is computed by the mean of cluster's objects.

$$\text{if dis}(X_i, V_j) \leq \text{dis}(X_i, V_p) \text{ then}$$

$$u_{i,j} = 1, \text{and } u_{i,p} = 0, \quad \text{for } 1 \leq p \leq k, p \neq j \tag{3}$$

3.2   Extensions of k-Means for Categorical Data

k-Modes Algorithm. It was also shown in [13] that the k-means method can be extended to categorical data by using a simple matching distance measure for categorical objects and the most frequent values to define the "cluster centers" called modes. Let X_1, X_2 are two categorical objects in DB, with $X_1 = (x_{11}, \ldots, x_{1D})$ and $X_2 = (X_{21}, \ldots, X_{2D})$. The dissimilarity between X_1 and X_2 can be computed by the total matching of the corresponding attribute values of the two objects. Formally,

$$\mathrm{dis}(X_1, X_2) = \sum_{d=1}^{D} \delta(x_{1d}, x_{2d}) \tag{4}$$

where

$$\delta(x_{1d}, x_{2d}) = \begin{cases} 0 & \text{if } x_{1d} = x_{2d}, \\ 1 & \text{if } x_{1d} \neq x_{2d}. \end{cases}$$

Given a cluster $\{X_1, \ldots, X_p\}$ of categorical objects, with $X_i = (x_{i1}, \ldots, x_{iD})$, $1 \leq i \leq p$, its mode $V = (o_1, \ldots, o_D)$ is defined by assigning o_d, $1 \leq d \leq D$, the value most frequently appeared in $\{x_{1d}, \ldots, x_{pd}\}$. With these modifications, Huang [14] developed the k-modes algorithm that mimics the k-means method to cluster categorical data. However, as mentioned previously, by definition the mode of a cluster is not in general unique. This makes the algorithm unstable depending on the selection of modes during the clustering process.

k-Representative Algorithm. In stead of using modes for cluster centers as in [13], San et al. [24] proposed the notion of *representatives* for clusters defined as follows.

Again, let $C = \{X_1, \ldots, X_p\}$ be a cluster of categorical objects and

$$X_i = (x_{i1}, \ldots, x_{iD}), 1 \leq i \leq p.$$

For each $d = 1, \ldots, D$, let us denote O_d^C the set forming from categorical values $x_{1d}, \ldots, x_{pd}$. Then the representative of C is defined by $V_C = (v_1^C, \ldots, v_D^C)$, with

$$v_d^C = \{(o_{dl}, f_C(o_{dl})) \mid o_{dl} \in O_d^C\}, \tag{5}$$

where $f_C(o_{dl})$ is the relative frequency of category o_{dl} within C, i.e.

$$f_C(o_{dl}) = \frac{\#_C(o_{dl})}{p} \tag{6}$$

where $\#_C(o_{dl})$ is the number of objects in C having the category o_{dl} at d^{th} attribute. More formally, each v_d^C is a distribution on O_d^C defined by relative frequencies of categorical values appearing within the cluster.

Then, the dissimilarity between object $X = (x_1, \ldots, x_D)$ and representative V_C is defined based on the simple matching measure δ by

$$\mathrm{dis}(X, V_C) = \sum_{d=1}^{D} \sum_{o_{dl} \in O_d^C} f_C(o_{dl}) \cdot \delta(x_d, o_{dl}) \tag{7}$$

As such, the dissimilarity $\mathrm{dis}(X, V_C)$ is mainly dependent on the relative frequencies of categorical values within the cluster and simple matching between categorical values.

k-Centers Algorithm. More generally, Chen and Wang [6] have recently proposed a generalized definition for centers of categorical clusters as follows. The center of a cluster C_j is defined as

$$V_j = [\boldsymbol{\nu}_{j1}, \ldots, \boldsymbol{\nu}_{jD}] \tag{8}$$

in which the dth element $\boldsymbol{\nu}_{jd}$ is a probability distribution on O_d estimated by a kernel density estimation method [1]. More particularly, let denote X_d a random variable associated with observations x_{id}, for $i = 1, \ldots, |C_j|$, appearing in C_j at d^{th} attribute, and $p(X_d)$ its probability density. Let O_{jd} be the set forming from categorical values $\{x_{id}\}_{i=1}^{|C_j|}$. Then the kernel density based estimate of $p(X_d)$, denoted by $\hat{p}(X_d, \lambda_j | C_j)$, is of the following form (see, e.g., [27]):

$$\hat{p}(X_d, \lambda_j | C_j) = \sum_{o_{dl} \in O_{jd}} f_j(o_{dl}) K(X_d, o_{dl} | \lambda_j) \tag{9}$$

where $K(\cdot, o_{dl} | \lambda_j)$ is a so-called kernel function, $\lambda_j \in [0, 1]$ is a smoothing parameter called the bandwidth, and f_j is the frequency estimator for C_j, i.e.

$$f_j(o_{dl}) = \frac{\#_j(o_{dl})}{|C_j|} \tag{10}$$

with $\#_j(o_{dl})$ being the number of o_{dl} appearing in C_j. Note that another equivalent form of (9) was used in [6] for defining a kernel density estimate of $p(X_d)$.

Also, Chen and Wang [6] used a variation of Aitchison and Aitken's kernel function [1] defined by

$$K(X_d, o_{dl} | \lambda_j) = \begin{cases} 1 - \frac{|O_d| - 1}{|O_d|} \lambda_j & \text{if } X_d = o_{dl} \\ \frac{1}{|O_d|} \lambda_j & \text{if } X_d \neq o_{dl} \end{cases} \tag{11}$$

to derive the estimate $\hat{p}(X_d, \lambda_j | C_j)$, which is then used to define $\boldsymbol{\nu}_{jd}$.

It is worth noting here that the kernel function $K(X_d, o_{dl} | \lambda_j)$ is defined in terms of the cardinality of the whole domain O_d but not in terms of the cardinality of the subdomain O_{jd} of the given cluster C_j.

From (9)–(11), it easily follows that $\boldsymbol{\nu}_{jd}$ can be represented as

$$\boldsymbol{\nu}_{jd} = \left[P_{jd}(o_{d1}), \ldots, P_{jd}(o_{dl}), \ldots, P_{jd}(o_{d|O_d|}) \right]$$

where

$$P_{jd}(o_{dl}) = \lambda_j \frac{1}{|O_d|} + (1 - \lambda_j)f_j(o_{dl}) \tag{12}$$

and $\lambda_j \in [0, 1]$ is the bandwidth for C_j.

When $\lambda_j = 0$, the center degenerates to the pure frequency estimator, which is originally used in the k-representative algorithm to define the center of a categorical cluster.

To measure the dissimilarity between a data object and its center, each data object X_i is represented by a set of vectors $\{y_{id}\}_{d=1}^D$, with

$$y_{id} = \left[I(x_{id} = o_{d1}), \ldots, I(x_{id} = o_{dl}), \ldots, I(x_{id} = o_{d|O_d|})\right]$$

Here $I(\cdot)$ is an indicator function whose value is either 1 or 0, indicating whether x_{id} is the same as $o_{dl} \in O_d$ or not. The dissimilarity on the dth dimension is then measured by

$$\text{dis}_d(X_i, V_j) = ||y_{id} - \nu_{jd}||_2 \tag{13}$$

We can see that, k-centers uses the different way to calculate the dissimilarities between objects and cluster centers, but the idea of comparing two categorical values is still based on the simple matching method (represented by indicator function $I(\cdot)$). The remains of the k-center mimics the idea of k-means algorithm.

4 The Proposed Algorithm

In this section we will introduce a new extension of the k-means clustering algorithm for categorical data by combining a slightly modified concept of cluster centers based on Chen and Wang's kernel-based estimation method and an information theoretic based dissimilarity measure.

4.1 Representation of Cluster Centers

Similar as in k-centers algorithm [6], for each cluster C_j, let us define the center of C_j as

$$V_j = [\nu_{j1}, \ldots, \nu_{jD}]$$

where ν_{jd} is a probability distribution on O_d estimated by a kernel density estimation method.

As our aim is to derive a kernel density based estimate $\hat{p}(X_d, \lambda_j | C_j)$ for the dth attribute of cluster C_j, instead of directly using Chen and Wang's kernel function defined in terms of the cardinality of the domain O_d as above, we use a slightly modified version as follows.

For any $o_{dl} \in O_d$, if $o_{dl} \in O_{jd}$ then we define

$$K(X_d, o_{dl}|\lambda_j) = \begin{cases} 1 - \frac{|O_{jd}|-1}{|O_{jd}|}\lambda_j & \text{if } X_d = o_{dl} \\ \frac{1}{|O_{jd}|}\lambda_j & \text{if } X_d \neq o_{dl} \end{cases} \qquad (14)$$

otherwise, i.e. $o_{dl} \notin O_{jd}$, we let $K(X_d, o_{dl}|\lambda_j) = 0$. Then, from (9), (10) and (14) it easily follows that $\boldsymbol{\nu}_{jd}$ can be obtained as

$$\boldsymbol{\nu}_{jd} = [P_{jd}(o_{d1}), \ldots, P_{jd}(o_{dl}), \ldots, P_{jd}(o_{d|O_d|})] \qquad (15)$$

where

$$P_{jd}(o_{dl}) = \begin{cases} \lambda_j \frac{1}{|O_{jd}|} + (1 - \lambda_j)f_j(o_{dl}) & \text{if } o_{dl} \in O_{jd} \\ 0 & \text{otherwise} \end{cases} \qquad (16)$$

and $\lambda_j \in [0, 1]$ is the smoothing parameter for C_j.

The parameter λ_j is selected using the least squares cross validation (LSCV) as done in [6], which is based on the principle of selecting a bandwidth that minimizes the total error of the resulting estimation over all the data objects. Specifically, the optimal λ_j^* is determined by the following equation:

$$\lambda_j^* = \frac{1}{|C_j| - 1} \frac{\sum_{d=1}^{D}(1 - \sum_{o_{dl} \in O_{jd}}[f_j(o_{dl})]^2)}{\sum_{d=1}^{D}(\sum_{o_{dl} \in O_{jd}}[f_j(o_{dl})]^2 - \frac{1}{|O_{jd}|})} \qquad (17)$$

4.2   Dissimilarity Measure

Instead of using the simple matching measure as in [13,24] or the Euclidean norm as in [6], we first introduce a dissimilarity measure for categorical values of each attribute domain based on an information-theoretic definition of similarity proposed by Lin [20], and then propose a new method for computing the distance between categorical objects and cluster centers, making use of the kernel density based definition of centers and the information-theoretic based dissimilarity measure for categorical data.

In [20], Lin developed an information-theoretic framework for similarity within which a formal definition of similarity can be derived from a set of underlying assumptions. Basically, Lin's definition of similarity is stated in information theoretic terms, as quoted "the similarity between A and B is measured by the ratio between the amount of information needed to state the commonality of A and B and the information needed to fully describe what A and B are." Formally, the similarity between A and B is generally defined as

$$\text{sim}(A, B) = \frac{\log P(\text{common}(A, B))}{\log P(\text{description}(A, B))} \qquad (18)$$

where $P(s)$ is the probability of a statement s. To show the universality of the information-theoretic definition of similarity, Lin [20] also discussed it in different settings, including ordinal domain, string similarity, word similarity and semantic similarity.

In 2008, Boriah et al. [5] applied Lin's framework to the categorical setting and proposed a similarity measure for categorical data as follows. Let DB be a data set consisting of objects defined over a set of D categorical attributes with finite domains denoted by $O_1, \ldots, O_D$, respectively. For each $d = 1, \ldots, D$, the similarity between two categorical values $o_{dl}, o_{dl'} \in O_d$ is defined by

$$\text{sim}_d(o_{dl}, o_{dl'}) = \begin{cases} 2 \log f_d(o_{dl}) & \text{if } o_{dl} = o_{dl'} \\ 2 \log(f_d(o_{dl}) + f_d(o_{dl'})) & \text{otherwise} \end{cases} \quad (19)$$

where

$$f_d(x) = \frac{\#(x)}{|DB|}$$

with $\#(x)$ being the number of objects in DB having the category x at d^{th} attribute. In fact, Boriah et al. [5] also proposed another similarity measure derived from Lin's framework and conducted an experimental evaluation of many different similarity measures for categorical data in the context of outlier detection.

It should be emphasized here that the similarity measure $\text{sim}_d(\cdot, \cdot)$ does not satisfy the Assumption 4 assumed in Lin's framework [20], which states that the similarity between a pair of identical object is 1. Particularly, the range of $\text{sim}_d(o_{dl}, o_{dl'})$ for $o_{dl} = o_{dl'}$ is $[-2 \log |DB|, 0]$, with the minimum being attained when o_{dl} occurs only once and the maximum being attained when $O_d = \{o_{dl}\}$. Similarly, the range of $\text{sim}_d(o_{dl}, o_{dl'})$ for $o_{dl} \neq o_{dl'}$ is $\left[-2 \log \frac{|DB|}{2}, 0\right]$, with the minimum being attained when o_{dl} and $o_{dl'}$ each occur only once, and the maximum value is attained when o_{dl} and $o_{dl'}$ each occur $\frac{|DB|}{2}$ times, as pointed out in [5].

Based on the general definition of similarity given in (18) and its application to similarity between ordinal values briefly discussed in [20], we introduce another similarity measure for categorical values as follows.

For any two categorical values $o_{dl}, o_{dl'} \in O_d$, their similarity, denoted by $\text{sim}_d^*(o_{dl}, o_{dl'})$, is defined by

$$\text{sim}_d^*(o_{dl}, o_{dl'}) = \frac{2 \log f_d(\{o_{dl}, o_{dl'}\})}{\log f_d(o_{dl}) + \log f_d(o_{dl'})} \quad (20)$$

where

$$f_d(\{o_{dl}, o_{dl'}\}) = \frac{\#(\{o_{dl}, o_{dl'}\})}{|DB|}$$

with $\#(\{o_{dl}, o_{dl'}\})$ being the number of categorical objects in DB that receive the value belonging to $\{o_{dl}, o_{dl'}\}$ at the dth attribute. Clearly, we have $\text{sim}_d^*(o_{dl}, o_{dl'}) = 1$ if o_{dl} and $o_{dl'}$ are identical, which satisfies the Assumption 4 stated as above.

Then, the dissimilarity measure between two categorical values $o_{dl}, o_{dl'} \in O_d$ can be defined by

$$\text{dis}_d^*(o_{dl}, o_{dl'}) = 1 - \text{sim}_d^*(o_{dl}, o_{dl'}) = 1 - \frac{2 \log f_d(\{o_{dl}, o_{dl'}\})}{\log f_d(o_{dl}) + \log f_d(o_{dl'})} \qquad (21)$$

Let $X_i = [x_{i1}, x_{i2}, \ldots, x_{iD}] \in DB$ and $V_j = [\nu_{j1}, \ldots, \nu_{jD}]$ be the center of cluster C_j. We are now able to extend the dissimilarity between categorical values of O_d to the dissimilarity on the dth attribute between X_i and V_j, i.e. the dissimilarity between the dth component $x_{id} \in O_d$ of X_i and the dth component ν_{jd} of the center V_j, as follows. Without danger of confusion, we shall also use dis_d^* to denote this dissimilarity and

$$\text{dis}_d^*(X_i, V_j) = \sum_{o_{dl} \in O_{jd}} P_{jd}(o_{dl}) \text{dis}_d^*(x_{id}, o_{dl}) \qquad (22)$$

4.3   Algorithm

With the modifications just made above, we are now ready to formulate the problem of clustering categorical data in a similar way as k-means clustering. Adapted from Huang's W-k-means algorithm [16], we also use a weighting vector $W = [w_1, w_2, \ldots, w_D]$ for D attributes and β being a parameter for attribute weight, where $0 \le w_d \le 1$ and $\sum_d w_d = 1$. The principal for attribute weighting is to assign a larger weight to an attribute that has a smaller sum of the within cluster distances and a smaller one to an attribute that has a larger sum of the within cluster distances. More details of this weighting scheme can be found in [16]. Then, the weighted dissimilarity between data object X_i and cluster center V_j, denoted by $\text{dis}^*(X_i, V_j)$, is defined by

$$\text{dis}^*(X_i, V_j) = \sum_{d=1}^{D} w_d^\beta \text{dis}_d^*(X_i, V_j) = \sum_{d=1}^{D} w_d^\beta \sum_{o_{dl} \in O_{jd}} P_{jd}(o_{dl}) \text{dis}_d^*(x_{id}, o_{dl}) \qquad (23)$$

Based on these definitions, the clustering algorithm now aims to minimize the following objective function:

$$J(U, \mathcal{V}, W) = \sum_{j=1}^{k} \sum_{i=1}^{N} \sum_{d=1}^{D} u_{i,j} w_d^\beta \text{dis}_d^*(X_i, V_j) \qquad (24)$$

subject to

$$\begin{array}{ll} \sum_{j=1}^{k} u_{i,j} = 1, & 1 \le i \le N \\ u_{i,j} \in \{0,1\}, & 1 \le i \le N, 1 \le j \le k \\ \sum_{d=1}^{D} w_d = 1, & 0 \le w_d \le 1 \end{array}$$

where $U = [u_{i,j}]_{N \times k}$ is a partition matrix.

The proposed algorithm is formulated as below.

Algorithm 1.1. The Proposed Algorithm

Select an initial $\mathcal{V}^{(0)} = \{V_1^{(0)}, \ldots, V_k^{(0)}\}$, and set $t = 0$, $\lambda_j = 0$ for $j = 0, \ldots, k$, set $W^{(0)} = [1/D, \ldots, 1/D]$.

repeat

  Keep $\mathcal{V}^{(t)}$ and $W^{(t)}$ fixed, generate $U^{(t)}$ to minimize the distances between objects and cluster mode (using Eq. (23)).

  Keep $U^{(t)}$ fixed, update $\mathcal{V}^{(t+1)}$ using Eq. (16) and Eq. (17).

  Generate $W^{(t+1)}$ using formulas from [16].

  $t = t + 1$.

until The partition does not changed.

5 Experiments Results

In this section, we will provide experiments conducted to compare the clustering performances of *k*-modes, *k*-representatives and three modified versions of *k*-representatives briefly described as below.

- In the first modified version of *k*-representatives (namely, Modified 1), we replace the simple matching dissimilarity measure with the information theoretic-based dissimilarity measure defined by Eq. (21).
- In the second modified version of *k*-representatives (namely, Modified 2), we combine the new dissimilarity measure with the concept of cluster centers proposed by Chen and Wang [6], i.e. the Algorithm 1.1 uses Eq. (12) instead of Eq. (16) to update the cluster centers).
- The third modified version of *k*-representatives (namely, Modified 3) is exactly Algorithm 1.1, which incorporates the new dissimilarity measure with our modified representation of cluster centers.

5.1 Datasets

For the evaluation, we used real world data sets downloaded from the UCI Machine Learning Repository [4]. The main characteristics of the datasets are summarized in Table 1. These datasets are chosen to test our algorithm because of their public availability and since all attributes can be treated as categorical ones.

5.2 Clustering Quality Evaluation

Evaluating the clustering quality is often a hard and subjective task [18]. Generally, objective functions in clustering are purposely designed so as to achieve high intra-cluster similarity and low inter-cluster similarity. This can be viewed as an internal criterion for the quality of a clustering. However, as observed in the literature, good scores on an internal criterion do not necessarily translate into good effectiveness in an application. Here, by the same way as in [19], we use three external criteria to evaluate the results: Purity, Normalized Mutual

Table 1. Categorical datasets

Datasets	Number of instances	Number of attributes	Number of classes
Car	1728	6	4
Soybean (Small)	47	35	4
Soybean (Large)	683	35	19
Breast cancer	286	9	2
Nursery	12960	8	5
Mushroom	8124	22	2

Information (NMI) and Adjusted Rand Index (ARI). These methods make use of the original class information of each object and the cluster to which the same objects have been assigned to evaluate how well the clustering result matches the original classes.

We denote by $C = \{C_1, \ldots, C_J\}$ the partition of the dataset built by the clustering algorithm, and by $P = \{P_1, \ldots, P_I\}$ the partition inferred by the original classification. J and I are respectively the number of clusters $|C|$ and the number of classes $|P|$. We denote by N the total number of objects.

Purity Metric. Purity is a simple and transparent evaluation measure. To compute purity, each cluster is assigned to the class which is most frequent in the cluster, and then the accuracy of this assignment is measured by counting the number of correctly assigned objects and dividing by the number of objects in the dataset. High purity is easy to achieve when the number of clusters is large. Thus, we cannot use purity to trade off the quality of the clustering against the number of clusters.

$$Purity(C, P) = \frac{1}{N} \sum_j \max_i |C_j \cap P_i| \tag{25}$$

NMI Metric. The second metric (NMI) provides an information that is independent from the number of clusters [26]. This measure takes its maximum value when the clustering partition matches completely the original partition. NMI is computed as the average mutual information between any pair of clusters and classes

$$NMI(C, P) = \frac{\sum_{i=1}^{I} \sum_{j=1}^{J} |C_j \cap P_i| \log \frac{N|C_j \cap P_i|}{|C_j||P_i|}}{\sqrt{\sum_{j=1}^{J} |C_j| \log \frac{|C_j|}{N} \sum_{i=1}^{I} |P_i| \log \frac{|P_i|}{N}}} \tag{26}$$

ARI Metric. The third metric is the adjusted Rand index [17]. Let a be the number of object pairs belonging to the same cluster in C and to the same class in P. This metric captures the deviation of a from its expected value corresponding to the hypothetical value of a obtained when C and P are two random, independent partitions.

The expected value of a denoted by $E[a]$ is computed as follows:

$$E[a] = \frac{\pi(C)\pi(P)}{N(N-1)/2} \qquad (27)$$

where $\pi(C)$ and $\pi(P)$ denote respectively the number of object pairs from the same clusters in C and from the same class in P. The maximum value for a is defined as:

$$\max(a) = \frac{1}{2}(\pi(C) + \pi(P)) \qquad (28)$$

The agreement between C and P can be estimated by the adjusted rand index as follows:

$$ARI(C,P) = \frac{a - E[a]}{\max(a) - E[a]} \qquad (29)$$

when $ARI(C,P) = 1$, we have identical partitions.

In many previous studies, only purity metric has been used to analyze the performance of clustering algorithm. However, purity is easy to achieve whens the number of cluster is large. In particular, purity is 1 if each object data gets its own cluster. Beside, many partitions have the same purity but they are different from each other e.g., the number of object data in each clusters, and which objects constitute the clusters. Therefore, we need the other two metrics to have the overall of how our clustering results matches the original classes.

5.3 Results

The experiments were run on a Mac with a 3.66 GHz Intel QuadCore processor, 8 GB of RAM running Mac OSX 10.10. For each categorical dataset, we run 300 times per algorithm. We provide the parameter k equals to the number of classes in each dataset. The performance of three evaluation metrics are calculated by the average after 300 times of running. The weighting exponent β was set to 8 as experimentally recommended in [16].

As we can see from Tables 2, 3 and 4, the modified versions 2 and 3 produce the best results in five out of six datasets. The results are remarkably good in the soybean (small) dataset, mushroom dataset, car dataset, soybean (large) dataset (when modified version 3 outperformed in all three metric) and breast cancer dataset (when the purity is slightly lower than the best one but the other two criteria are significantly higher). Comparing the performance of modified versions 2 and 3, we can see that the proposed approach yields better results in many cases, especially in NMI values and ARI values. In conclusion, the new approach has been proved to enhance the performance of previously developed k-means like algorithms for clustering categorical data.

Table 2. Purity results of categorical dataset algorithms

Datasets	K-mode	K-representative	Modified 1	Modified 2	Modified 3
Car	0.7	0.7	0.703	**0.705**	**0.705**
Soybean (Small)	0.873	0.961	0.967	0.981	**0.986**
Soybean (Large)	0.538	0.595	0.6	0.69	**0.71**
Breast-cancer	0.702	0.707	**0.713**	0.71	0.709
Nursery	0.409	0.425	0.435	0.451	**0.468**
Mushroom	0.518	0.83	0.864	**0.87**	**0.87**

Table 3. NMI results categorical dataset algorithms

Datasets	K-mode	K-representative	Modified 1	Modified 2	Modified 3
Car	0.051	0.077	0.101	0.119	**0.125**
Soybean (Small)	0.823	0.96	0.965	0.98	**0.981**
Soybean (Large)	0.533	0.708	0.71	0.72	**0.732**
Breast-cancer	0.0015	0.036	0.039	0.051	**0.057**
Nursery	0.044	0.047	0.055	**0.071**	**0.071**
Mushroom	9.26E-05	0.448	0.493	0.51	**0.521**

Table 4. Ajusted rand index results categorical dataset algorithms

Datasets	K-mode	K-representative	Modified 1	Modified 2	Modified 3
Car	0.028	0.024	0.043	0.049	**0.052**
Soybean (Small)	0.729	0.929	0.938	0.963	**0.975**
Soybean (Large)	0.306	0.379	0.38	0.44	**0.46**
Breast-cancer	-0.002	0.082	0.085	0.128	**0.135**
Nursery	0.034	0.028	0.034	0.046	**0.049**
Mushroom	4.62E-06	0.5	0.554	**0.589**	0.586

6   Conclusions

In this paper, we have proposed a new k-means like algorithm for clustering categorical data based on an information theoretic based dissimilarity measure and a kernel density estimate-based concept of cluster centers for categorical objects. Several variations of the proposed algorithm have been also discussed. The experimental results on real datasets from UCI Machine Learning Repository have shown that the proposed algorithm outperformed the k-means like algorithms previously developed for clustering categorical data. For the future work, we are planning to extend the proposed approach to the problem of clustering mixed numeric and categorical datasets as well as fuzzy clustering.

References

1. Aitchison, J., Aitken, C.G.G.: Multivariate binary discrimination by the kernel method. Biometrika **63**(3), 413–420 (1976)
2. Andritsos, P., Tsaparas, P., Miller, R.J., Sevcik, K.C.: LIMBO: A Scalable Algorithm to Cluster Categorical Data (2003)
3. Barbara, D., Couto, J., Li, Y.: COOLCAT: an entropy-based algorithm for categorical clustering. In: Proceedings of the Eleventh International Conference on Information and Knowledge Management, pp. 582–589 (2002)
4. Blake, C.L., Merz, C.J.: UCI Repository of Machine Learning Databases. Dept. of Information and Computer Science, University of California at Irvine (1998). http://www.ics.uci.edu/mlearn/MLRepository.html
5. Boriah, S., Chandola, V., Kumar V.: Similarity measures for categorical data: a comparative evaluation. In: Proceedings of the SIAM International Conference on Data Mining, SDM 2008, pp. 243–254 (2008)
6. Chen, L., Wang, S.: Central clustering of categorical data with automated feature weighting. In: Proceedings of the Twenty-Third International Joint Conference on Artificial Intelligence, pp. 1260–1266 (2013)
7. Ganti, V., Gehrke, J., Ramakrishnan, R.: CATUS–clustering categorical data using summaries. In: Proceedings of the International Conference on Knowledge Discovery and Data Mining, San Diego, CA, pp. 73–83 (1999)
8. Gibson, D., Kleinberg, J., Raghavan, P.: Clustering categorical data: an approach based on dynamic systems. In: Proceedings of the 24th International Conference on Very Large Databases, New York, pp. 311–323 (1998)
9. Guha, S., Rastogi, R., Shim, K.: CURE: an efficient clustering algorithm for large databases. In: Proceedings of ACM SIGMOD International Conference on Management of Data, New York, pp. 73–84 (1998)
10. Guha, S., Rastogi, R., Shim, K.: ROCK: a robust clustering algorithm for categorical attributes. Inf. Syst. **25**(5), 345–366 (2000)
11. Han, J., Kamber, M.: Data Mining: Concepts and Techniques. Morgan Kaufmann Publishers, San Francisco (2001)
12. Hathaway, R.J., Bezdek, J.C.: Local convergence of the *c*-means algorithms. Pattern Recogn. **19**, 477–480 (1986)
13. Huang, Z.: Clustering large data sets with mixed numeric and categorical values. In: Lu, H., Motoda, H., Luu, H. (eds.) KDD: Techniques and Applications, pp. 21–34. World Scientific, Singapore (1997)
14. Huang, Z.: Extensions to the k-means algorithm for clustering large data sets with categorical values. Data Min. Knowl. Discov. **2**, 283–304 (1998)
15. Huang, Z., Ng, M.K.: A fuzzy *k*-modes algorithm for clustering categorical data. IEEE Trans. Fuzzy Syst. **7**, 446–452 (1999)
16. Huang, Z., Ng, M.K., Rong, H., Li, Z.: Automated variable weighting in *k*-means type clustering. IEEE Trans. Pattern Anal. Mach. Intell. **27**(5), 657–668 (2005)
17. Hubert, L., Arabie, P.: Comparing partitions. J. Classif. **2**(1), 193–218 (1995)
18. Ienco, D., Pensa, R.G., Meo, R.: Context-based distance learning for categorical data clustering. In: Adams, N.M., Robardet, C., Siebes, A., Boulicaut, J.-F. (eds.) IDA 2009. LNCS, vol. 5772, pp. 83–94. Springer, Heidelberg (2009)
19. Ienco, D., Pensa, R.G., Meo, R.: From context to distance: learning dissimilarity for categorical data clustering. ACM Trans. Knowl. Discov. Data **6**(1), 1–25 (2012)
20. Lin, D.: An information-theoretic definition of similarity. In: Proceedings of the 15th International Conference on Machine Learning, pp. 296–304 (1998)

21. MacQueen, J.B.: Some methods for classification, analysis of multivariate obser-
    vations. In: Proceedings of the Fifth Symposium on Mathematical Statistics and
    Probability, Berkelely, CA, vol. 1(AD 669871), pp. 281–297 (1967)
22. Ng, M.K., Li, M.J., Huang, J.Z., He, Z.: On the impact of dissimilarity measure in
    k-modes clustering algorithm. IEEE Trans. Pattern Anal. Mach. Intell. **29**, 503–507
    (2007)
23. Ralambondrainy, H.: A conceptual version of the k-means algorithm. Pattern
    Recog. Lett. **16**, 1147–1157 (1995)
24. San, O.M., Huynh, V.N., Nakamori, Y.: An alternative extension of the k-means
    algorithm for clustering categorical data. Int. J. Appl. Math. Comput. Sci. **14**,
    241–247 (2004)
25. Selim, S.Z., Ismail, M.A.: k-Means-type algorithms: a generalized convergence the-
    orem and characterization of local optimality. IEEE Trans. Pattern Anal. Mach.
    Intell. **6**(1), 81–87 (1984)
26. Strehl, A., Ghosh, J.: Cluster ensembles–a knowledge reuse framework for combin-
    ing multiple partitions. J. Mach. Learn. Res. **3**, 583–617 (2003)
27. Titterington, D.M.: A comparative study of kernel-based density estimates for
    categorical data. Technometrics **22**(2), 259–268 (1980)

Discovering Overlapping Quantitative Associations by Density-Based Mining of Relevant Attributes

Thomas Van Brussel[1]([⊠]), Emmanuel Müller[1,2], and Bart Goethals[1]

[1] University of Antwerp, Antwerp, Belgium
thomas.vanbrussel@uantwerpen.be
[2] Hasso-Plattner-Institute, Potsdam, Germany

Abstract. Association rule mining is an often used method to find relationships in the data and has been extensively studied in the literature. Unfortunately, most of these methods do not work well for numerical attributes. State-of-the-art quantitative association rule mining algorithms follow a common routine: (1) discretize the data and (2) mine for association rules. Unfortunately, this two-step approach can be rather inaccurate as discretization partitions the data space. This misses rules that are present in overlapping intervals.

In this paper, we explore the data for quantitative association rules hidden in overlapping regions of numeric data. Our method works without the need for a discretization step, and thus, prevents information loss in partitioning numeric attributes prior to the mining step. It exploits a statistical test for selecting relevant attributes, detects relationships of dense intervals in these attributes, and finally combines them into quantitative association rules. We evaluate our method on synthetic and real data to show its efficiency and quality improvement compared to state-of-the-art methods.

1 Introduction

Ever since its introduction [1], association rule mining has been a popular method for discovering interesting patterns in databases. However, the methods used to find these rules were originally only intended for boolean or categorical data, that is, rules of the form BUY[TOOTHBRUSH] → BUY[TOOTHPASTE]. However, since more and more of the data we gather is numerical, an interesting extension to classical association rule mining is quantitative association rule mining [3, 19, 21, 25]. In their purest form, quantitative association rules (QAR) denote relations in numeric data but can be extended to cover categorical items as well. Two examples of such association rules including numeric data are the following:

Example 1.

$$\text{AGE}[16, 90] \wedge \text{US-CITIZEN}[\text{YES}] \rightarrow \text{DRIVERS-LICENSE}[\text{YES}]$$

© Springer International Publishing Switzerland 2016
M. Gyssens and G. Simari (Eds.): FoIKS 2016, LNCS 9616, pp. 131–148, 2016.
DOI: 10.1007/978-3-319-30024-5_8

and

$$\text{AGE}[0, 21] \wedge \text{US-CITIZEN}[\text{YES}] \rightarrow \text{ALCOHOL-ALLOWED}[\text{NO}]$$

In a real-life database containing such information, such rules would have high support and confidence [1]: (1) US citizens are allowed to get a drivers license from the age of 16 and above, and (2) US citizens are not allowed to drink alcohol under the age of 21. The overlap in the numeric age attribute is a very natural and desired one. In contrast to these natural rules, discretization of the data in traditional methods would miss one of these two rules due to pre-selected intervals (e.g. by equal-width discretization) and their disjoint partitioning before mining the quantitative association rules.

The first attempts to mine data for quantitative association rules fall prey to this limitation. The first publication by Srikant et al. [21] transforms the original (numeric) data such that it could be handled by traditional association rule mining algorithms. This is done by a pre-processing step, in which each attribute is partitioned into an equal number of disjoint bins. This is still the most common procedure for quantitative association rules [11,23]. However, the disjoint partitioning remains a major unresolved challenge for all of these techniques. Two regions that overlap each other in a certain attribute are not detectable after discretizing the data (cf. Example 1). As shown in our example, two rules might be naturally present in overlapping intervals. The association rule mining algorithm should select these relevant attributes, and further, detect the most appropriate interval for each rule individually.

Many techniques [3,7,19] try to bypass this problem and can generally be seen as optimization strategies. These strategies attempt to find optimal rules with respect to specific rule templates. Hence, they are limited to certain criteria, such as limited dimensionality, specific shapes, etc. Furthermore, choosing the optimal boundaries for this partitioning has been shown to be an NP-complete and intractable problem [26].

In our work, we focus on overlapping quantitative association rules. We propose a method that works without the need for a discretization step, and thus, prevents information loss in partitioning numeric attributes prior to the mining step. It exploits a statistical test for selecting relevant attributes for association rule mining. We select attributes that show a dependency with an interval from a different dimension. We then detect relationships of dense intervals in these attributes, and finally combine them into quantitative association rules.

In summary, our approach called DRule has the following contributions, (i) it does not require a partitioning of the data as pre-processing step; and detects overlapping rules. (ii) It does not impose any restriction on the type of rule that can be found; and hence provides a more flexible framework for quantitative association rules. (iii) It selects statistically relevant attributes and dense regions in each of these attributes while mining association rules. (iv) It provides an efficient computation of overlapping quantitative association rules for large databases.

2 Formal Problem Statement

Database. For a database $\mathcal{DB}$ of mixed attribute types (binary, categorical, and numerical) we consider each object $o \in \mathcal{DB}$ to be represented by a vector

$$o \in \{\mathbb{R} \cup \{c_1, \ldots, c_k\} \cup \{\mathbf{false}, \mathbf{true}\}\}^m .$$

We denote $o(A_i)$ the value of object o for attribute A_i and call the set of all attributes $\mathcal{A}$. We use the notation $n = |\mathcal{DB}|$ as the number of objects in the database and $m = |\mathcal{A}|$ the dimensionality of the database.

Quantitative Predicate. Let $\mathcal{A} = \{A_1, \ldots, A_m\}$ be the set of all attributes in the database. A *quantitative predicate* is then defined as follows:

Definition 1 (Quantitative Predicate). *Given one attribute $A_i \in \mathcal{A}$ and lower and upper bounds $(l, u) \in \mathrm{dom}^2(A_i)$,*

$$A[l, u] \text{ defines a quantitative predicate}$$

with $l \le u$ in case of numeric attribute A_i and with a constant c equal to both bounds $l = u = c$ in the case of categorical or binary attributes.

This general definition covers all attribute types. We distinguish between two notations: $A[l, u]$ for numeric attributes and $A[c]$ for binary and categorical ones. Further, we say that an object $o \in \mathcal{DB}$ is covered by a quantitative predicate $A_i[l, u]$ iff $l \le o(A_i) \le u$ and $A_i[c]$ iff $c = o(A_i)$.

Set of Predicates and Quantitative Association Rules. Based on the above definition, a *set of predicates* is simply defined as a conjunction of predicates. An object (or transaction) is covered by a set of predicates *iff* it is covered by each predicate in the set. For a predicate set $\mathcal{X}$ we denote its attributes by $attr(\mathcal{X}) \subseteq \mathcal{A}$.

Furthermore, the objects that are covered by such a set of predicates are defined as $I(\mathcal{X})$, the function that returns all $o \in \mathcal{DB}$ that are covered by all predicates given in $\mathcal{X}$:

$$I(\mathcal{X}) = |\{o \in \mathcal{DB} \mid o \text{ is covered by } \mathcal{X}\}|$$

We define the *support* and frequency of $\mathcal{X}$ as

$$\mathrm{supp}(\mathcal{X}) = |I(\mathcal{X})| \qquad \mathrm{freq}(\mathcal{X}) = \frac{|I(\mathcal{X})|}{|\mathcal{DB}|}$$

Definition 2 (Quantitative Association Rule). *A quantitative association rule (QAR) R is defined as*

$$\mathcal{P} \to \mathcal{Q}$$

with $\mathcal{P}$ and $\mathcal{Q}$ predicate sets, $\mathcal{Q} \ne \emptyset$, and $\mathcal{P} \cap \mathcal{Q} = \emptyset$.

The support of a rule $R : \mathcal{P} \to \mathcal{Q}$ is defined as $\text{supp}(R) = \text{supp}(\mathcal{P} \cup \mathcal{Q})$, the number of objects that satisfy the predicates in both $\mathcal{P}$ and $\mathcal{Q}$. The confidence is defined as $\text{conf}(R) = \frac{\text{supp}(R)}{\text{supp}(\mathcal{P})}$. A rule has to fulfill $\text{supp}(R) \geq minSupport$ and $\text{conf}(R) \geq minConfidence$ parameters as in traditional association rule mining [1].

Note that support and confidence are traditional measures for assessing association rules, but they are not sufficient for detecting interesting intervals in numeric data. Both support and confidence can be naively optimized by selecting the entire domain of a numeric attribute. For a more reasonable detection of interesting intervals, we introduce *density* in numeric data. We will further introduce this in Sect. 3. We have chosen for density as a measure as it is well understood and easy to interpret and inspiration can be drawn from existing efficient algorithms. This also meshes well with our approach to search space reduction introduced in the next section.

Generalized Problem Statement. Mining association rules according to Definition 2 is a general problem statement for quantitative association rule mining: (1) it is a flexible definition in terms of choice of attributes, (2) It allows different intervals for each individual rule, (3) and it allows overlapping intervals for the numeric attributes. Let us discuss these three properties formally and contrast them to the more restricted definitions found in the literature:

Property 1 (Free choice of attributes). The attribute set $\mathcal{P} \cup \mathcal{Q}$ of a rule can be any subset of all given attributes $\mathcal{A}$. We do not impose any restrictions on the number of attributes in a rule nor the configuration in which they appear.

In contrast to this free choice, other methods are not as flexible: *pre-selection of the attributes* by the user is a widely used restriction [19]. It is limited in both LHS and RHS of the rule, which are selected by the user beforehand. Such restriction does not allow for rule discovery outside this template of pre-selected LHS and RHS. Other methods perform restrictions on the size of a rule. For example, the restriction to mine 2D rules only [3, 25]. Such methods are designed for a specific goal and can not be easily generalised.

Property 2 (Individual intervals for each rule). For a numeric attribute $A_i \in \mathcal{P} \cup \mathcal{Q}$ of a rule, its corresponding interval can be any $[l, u] \in dom^2(A_i)$, i.e. we do not have any restriction on the intervals and they can be determined for each rule individually.

While we allow for arbitrary intervals, related algorithms already mentioned before choose their interval bounds from a set of predetermined disjoint intervals [15, 21]. Formally, $[l, u] \in \{[l_1, u_1), [l_2, u_2), \dots, [l_k, u_k]\}$ is fixed for each attribute A_i. These interval bounds create a true partioning of the data space. That is, the possible intervals are fixed for each attribute A_i for all rules. In contrast to such a limited choice of intervals, we aim at an on-the-fly selection of intervals within the rule mining process. That is, for each rule, we aim at density-based intervals directly on the numeric data, which allows for individual and better intervals per rule. In addition, a third key property automatically follows from the previous one. In addition to finding individual intervals for each rule, it also allows overlap between these intervals as well. Techniques that rely on a pre-discretization of numeric attributes [15, 21] do not allow for such overlap.

3   DRule

A straightforward solution for the problem described in the previous section would be exhaustive search. Such an approach would detect interesting intervals by checking each combination of dimensions and even checking every possible pair of interval bounds. This is obviously an *intractable* solution. However, due to its exhaustive nature it provides us all the rules contained in the data.

With DRule we propose an efficient method that allows us to find the same results as exhaustive search would, i.e. without missing any rules, while avoiding the tremendous runtime complexity. In order to find all rules, we start from each dimension, just like exhaustive search does. However, we quickly prune parts of the search space to prevent uninteresting rules or regions from slowing down our algorithm. To this end, while going through the search space, we demand that dimensions that are considered together are not uniformly related to each other. As such, uninteresting rules can be pruned.

In the following sections we will describe DRule in more detail. Our algorithm achieves the detection of overlapping QARs by using an approach that can be split into three parts, namely detecting interesting regions through *search space reduction*, *mining for individual intervals*, and *generating quantitative association rules*. In each of these three parts we ensure flexibility in both the selection of relevant attributes and the detection of individual (possibly overlapping) intervals for each rule. These three parts will be further explored in the following sections.

3.1   Mining for Intervals

To successfully create rules, we need meaningful interval bounds. As mentioned before, fixing these interval bounds beforehand severely limits the rules we can find, while considering all possible intervals is infeasible. To solve this problem, we use a method that allows us to efficiently generate interval bounds on-the-fly for each rule.

Algorithm Overview. The general idea of DRule is to mine all possible predicate sets in a depth-first search. For efficient mining of intervals $A_i[l, u]$, we first detect interesting (i.e. dense) regions. These regions are hyper-rectangles in the multi-dimensional space and approximate the individual intervals of one rule. These hyper-rectangles, which we shall refer to as *interesting regions*, describe areas in the data that show promise for rule generation from a density-based viewpoint. When generating interesting regions, we perform an analysis to determine which combinations of attributes can provide us with interesting results. That is, we already filter out those combinations of attributes that leave us with predictable and uninteresting results (cf. Sect. 3.2). Given one of these interesting regions (i.e. one hyper-rectangle), DRule refines this hyper-rectangle by identifying the dense regions in each dimension and providing a list of dense intervals. These can be used as a quantitative predicate; however, we still need to visit the other dimensions in order to produce meaningful rules. We proceed the search in a depth-first manner for each of the dense regions detected using the dense region as a dataset for the next step. After each recursive step, we generate rules for the intervals detected so far. An overview of DRule is given in Algorithm 1. Note that Algorithm 1 is called from each dimension in a region R.

Algorithm 1. DRULE (Interesting region R, dimension d)

1: Find all dense regions in R in dimension d
2: **for all** Dense regions r **do**
3:    Generate_Rules(r, minimum confidence)
4:    **for all** Dimensions d' interesting to dimension d in R **do**
5:       **if** d' has not yet been visited **then**
6:          DRULE (r, d')
7:       **end if**
8:    **end for**
9: **end for**

Comparison to Existing Processing Schemes. There are three important sides to the algorithm that have not been explicitly stated. One, the algorithm does not guarantee that the hyper-rectangle forms a dense region in all dimensions. That is, it only guarantees that each interval discovered by the algorithm is dense. This requirement is less restrictive as it allows us to find a larger set of possibly interesting rules. Secondly, the order in which we visit dimensions is important. This is not considered by other algorithms (e.g. for quantitative frequent itemsets [24]): first visiting dimension 1, followed by dimension 2 can lead to different results compared to visiting them in reverse order. Failure to do so can result in many missed patterns.

The previous remark is one of the key aspects in allowing us to find overlapping regions and allows us to distinguish ourselves from our competitors, such as QFIMiner [24]. Working on a per dimension basis and not taking the ordering of dimensions into account, can result in regions not being split properly. Taking this into account, our strategy allows us to find overlapping regions (conform Property 3).

3.2   Search Space Reduction

In order to reduce the infinite search space for intervals in a numeric data space, we propose to mine for interesting regions. These interesting regions should have high density (similar to the notion of high frequency for traditional itemset mining). Please note that this step might be considered as a sort of discretization as we have to fix intervals at some point. However, it is by far more flexible than pre-discretization as it allows for on-the-fly generation of density-based regions within the rule mining. Unfortunately, such a density computation in arbitrary attribute combinations is known to be computationally very expensive [18]. Thus, we will restrict to one-dimensional density computation using DBSCAN [6,12], similar to colossal pattern mining for frequency assessment [27]. This heuristic is well-known for approximate subspace clustering [12,16,17] and we will exploit its efficient search for finding overlapping QARs. Our extension to this reduction scheme can handle both numeric and categorical attributes, and reduces the search even further by assessing statistical dependence between pairs of attributes. Our detection method is flexible in using any subset of the given attributes (conform Property 1) and any intervals in the respective attribute domain (conform Property 2). Overall, this allows us to reduce the potential interval boundaries based on the data distribution observed in any attribute combination. For more information about this approach and more detailed descriptions of the choices, we refer to the original paper on this kind of search space reduction [12].

One-Dimensional Clustering. Our search space reduction works as follows. Start by clustering each dimension separately according to some clustering algorithm. In our case we have chosen to use the well-known DBSCAN algorithm [6] for density-based clustering. Since DRule only needs dense regions, we can substitute DBSCAN for any algorithm that finds these kinds of regions or algorithms such as spectral clustering.

We support categorical items by performing a frequency analysis on the categorical dimensions and treat them as clusters if they are sufficiently frequent. We will also refer to these 1D clusters as *base-clusters* and they will be denoted as C^1. These base-clusters form the foundation from which we will decide what points and dimensions are relevant.

Merge of Candidates. Since we are not interested in clusters that exist in only one dimension, we search for approximations of interesting regions by merging the base-clusters found in the previous step. This way, we can quickly determine (i) which points form a cluster and (ii) in which dimensions these points form a cluster. To find these approximations, we cannot simply merge base-clusters and hope for the best as this would be exponential in nature. Therefore, for our merge of cluster candidates, we employ a greedy merge of the most similar base-clusters. Let us first provide the formal definitions of this step before we continue.

Definition 3 (The Most Similar Cluster). *Given a 1D cluster $c \in C^1$, the most similar cluster (MSC) $\hat{c} \neq c$ is defined by:*

$$\forall c_i \in C^1 \setminus \{c\} : sim(c, \hat{c}) \geq sim(c, c_i)$$

with similarity instantiated by $sim(c_i, c_j) = |c_i \cap c_j|$.

If the most similar cluster $\hat{c}$ shares a large number of objects with the base cluster c, this is usually indicative of the presence of a multi-dimensional cluster, i.e. a combination of these two 1D clusters.

In order to gain a set of merge candidates (i.e. 1D clusters that can be merged), we need a merge procedure to identify multiple similar clusters. We consider merging the base cluster with its most similar 1D clusters if they share many objects.

Definition 4 (k-Most-Similar Clusters). *Given a 1D cluster $c \in C^1$, we call $MSC_k(c) \subseteq C^1$ the k-most-similar clusters iff it is the smallest subset of C^1 that contains at least k base-clusters, where the following condition holds:*
$$\forall c_i \in MSC_k(c), \forall c_j \in C^1 \setminus MSC_k(c) : sim(c, c_i) > sim(c, c_j).$$

In order to select which of these clusters should be merged, we compute the best-merge candidates (BMC) as follows.

Definition 5 (Best-Merge Candidates). *Given a 1D cluster $c \in C^1$ and $k, \mu \in \mathbb{N}^+ (\mu \leq k)$. The best-merge candidates are defined as:*
$$BMC(c) := \{x \in C^1 \mid MSC_k(c) \cap MSC_k(x)| \geq \mu\}$$

Merging all best-merge candidates filters out merge candidates that do not add a lot of value. However, due to the parameter k, this method might be too restrictive, causing a loss of cluster information. Since we want maximal-dimensional regions, we bypass this problem by merging best-merge candidate sets. That is, we decide to merge two base clusters if they share at least one base-cluster which fulfils the properties of a best-merge cluster.

Definition 6 (Best-Merge Cluster). *Given a 1D cluster $c \in C^1$ and minClu $\in \mathbb{N}^+$. Then c is called a best-merge cluster if $|BMC(c)| \geq minClu$.*

Generation of Interesting Regions. Now that we have found which clusters can be merged, we can start our search space reduction. We start by generating our approximative regions. These are the regions that will ultimately be refined into rules if possible. First, generate all best-merge clusters. This will give us a list of base clusters that are most suited to be merged. Then, find all pairs of best-merge clusters c_A and c_B where $c_A \in BMC(c_b)$ and $c_B \in BMC(c_A)$. As explained above, simply merging c_A and c_B is not enough, we therefore add all best-merge candidates of c_A and c_B to this group. This ensures that we find maximal-dimensional approximations of the regions we are interested in. The resulting set of base-clusters form our region. Formally, a region is then defined as follows.

Definition 7 (Region). *A region R is a set of 1D clusters (base-clusters),*
$$R = \{c \mid c \in (BMC(c_A) \cup BMC(c_B))\} \cup \{c_A, c_B\},$$
where $c_A, c_B \in R$, $c_A \in BMC(c_B) \wedge c_B \in BMC(c_A)$.

The points of a region are found by taking the union of all points in the base-clusters. We take the union as to not be too restrictive on the points that are part of the region.

Intuitively, a region defines a hyper-rectangle in the original data that can be seen as a smaller version of the original database. This hyper-rectangle approximates a quantitative predicate set (cf. Definition 1). That is, given a detected region, we can consider the intervals of this region to define a quantitative predicate. Of special interest is that these regions do not have to be disjoint. That is, we allow two regions R_1 and R_2 with $R_1 \cap R_2 \neq \emptyset$ (conform Property 3).

Selecting Relevant Attributes. Now that we have found our *possibly* interesting regions, we can continue with generating our rules. Before we do this, however, we first introduce an intermediary step that will help us filter out uninteresting rules. That is, given a region, we will verify that each dimension of this region is interesting w.r.t. all of its other dimensions. For example, suppose that we have found a dense region denoting all people between a certain age. It would be uninteresting to report to the user that approximately half of these people are female. To avoid considering combinations of mutually uninteresting dimensions, we propose to use an additional measure.

We extend the previous approach by introducing a pairwise independence test between all attributes using the χ^2-test for uniformity. That is, given a region R, we decide, for each pair of dimensions of this region, whether the points in this region are uniformly spread in the second dimension when viewed from the first, and vice versa.

Optimizations. Recall that to find overlapping rules, we consider the order in which we visit dimensions. However, looking at the data from every possible angle will not always result in different rules. To stop the algorithm from performing a lot of unnecessary computations, we can opt for a *short-circuiting* strategy. That is, if we find that visiting attribute A_1 from attribute A_2 yields the same result as visiting them in reverse order, we can choose to terminate our computation as we will not be able to find new results.

Another optimization that can be done is when visiting a certain set of attributes $\mathcal{X}$ yields largely the same results as visiting a set of attributes $\mathcal{Y}$. All attributes we visit starting from $\mathcal{X}$ will yield the same result as when visiting them from $\mathcal{Y}$. Therefore, we can stop our computation for one of these sets and create a link between the results, appending the results to that set in the end. Please note that this optimization, while reducing the output size of the algorithm, increases the memory consumption of the algorithm as we have to keep track of intermediary results. This also does not remove any results from the rules we would have discovered had we not run this step. We would simply halt the computation of one branch when the results would be same from that point on. The end results can then be added to the branch for which computation was halted.

3.3 Generating QARs

The final step of the algorithm is to take these interval bounds and transform them into rules. This step is the simplest of the three and boils down to a simple counting strategy. That is, given a set of intervals, for each interval compute the points of the dataset that are contained within that interval. Then, try placing each dimension of the region on the right hand side of the rule and compute the remaining left hand side. Intersecting the points of the intervals on the left hand side and comparing them with the points on the right hand side allows us to quickly compute the confidence of the rule.

Note that this step does not have to separate from the previous step as rules can be generated at each level to provide more immediate feedback.

4 Experimental Results

We ran our experiments on a machine running Mac OSX 10.9.1 with a 2.3 GHz Intel Core i7 and 8 GB of memory. The algorithm was written in Python, making heavy use of NumPy which provide bindings to efficient linear algebra libraries. We compare our method to state of the art methods such as QFIMiner [24] and QuantMiner [19] and also to an equi-width discretization technique. The source code for QuantMiner can be found online, the implementations for the other algorithms were created by us. We test the algorithms on synthetic data to provide an objective comparison, and on real data to show that our results can be used in real world situations. The synthetic data does not contain any categorical attributes as QuantMiner is not capable of handling them.

Synthetic Data

The synthetic data we use in this section was generated by creating dense regions in subsets of all attributes and using overlapping intervals in the individual attribute domains. Unless otherwise specified, we set the minimum confidence level to 60 %. That is, only rules that are true at least 60 % of the time are reported.

Efficiency. To test the efficiency and scalability of our algorithm, we compared our runtime against that of our three competitors. We generated a dataset of 50000 objects and increased its dimensionality.

Note that there are several caveats with respect to this comparison. First, to compare to QFIMiner, we have included our own routine to generate quantitative association rules. Since QFIMiner only mines the data for quantitative frequent itemsets (*QFIs*), we still need to transform this data to *QARs*. Second, while QuantMiner is included in the comparison, it is very hard to directly compare it to QFIMiner or DRule as it works in a different way. QuantMiner takes as input a dataset and a rule template. A rule template tells the algorithm which attributes to use in the LHS of the rule and which to use in the RHS of the

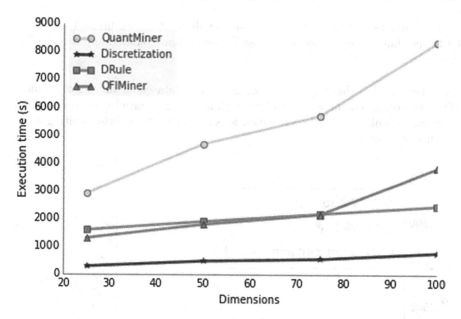

Fig. 1. Comparison of execution time between DRule, QFIMiner, QuantMiner, and discretization. The dataset in all instances contains 50000 transactions.

rule. QuantMiner then looks for optimal rules with respect to this template. To generate all rules of a certain size, the user has to loop through all possible combinations of dimensions manually. To provide a comparison, we only use QuantMiner to mine for rules in the dimensions in which the other algorithms found rules.

From Fig. 1 we can conclude that DRule performs well when compared to its competitors. The only competitor consistently outperforming our method is the method using discretization. For reasons mentioned before, this method does, however, not provide us with satisfactory results. When comparing to QFIMiner, we can see that our method performs equally well. However, QFIMiner misses some of the hidden rules, which in contrast are correctly detected by our method. Specifically, rules that can only be found when visiting dimensions in a certain order are not found by QFIMiner. An easy example is when looking at a dataset in the shape of a face with the mouth overlapping with both eyes in the x dimensions. If an algorithm first splits on the x dimension and then on the y dimension, the eyes are not separated, while this does happen when first visiting the y dimension and then x dimension. DRule is capable of generating more rules than QFIMiner while keeping up in terms of runtime. We can see that our method consistently outperforms QuantMiner, but it should be noted that QuantMiner was designed for rules using user-templates and not for an unsupervised detection of rules with a free choice of attributes.

It should be noted that it can easily be shown that we find all rules that QFIMiner finds as both algorithms use a similar density-based approach to find intervals.

Rules Output. For this experiment, we generated datasets of varying dimensionality and compared the amount of rules that were generated with and without checkout for uniformity between dimensions and using the short-circuiting strategy discussed in Sect. 3.

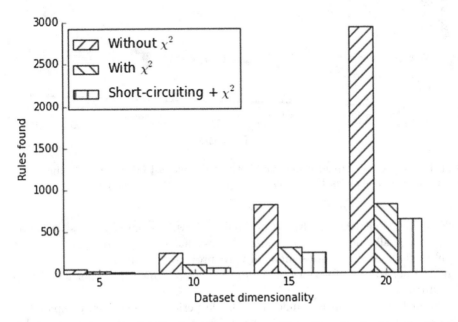

Fig. 2. The amount of rules the algorithm outputs with uniformity checks and without and using the short-circuiting strategy. We reject uniformity if $p \leq 0.05$.

In Fig. 2 we demonstrate how much our checks for uniformity reduce the amount of work the algorithm has to perform. We can see that checking for uniformity drastically reduces the amount of rules we output and, therefore, the amount of computations our algorithm has to perform. Using the short-circuiting strategy further reduces the rules that are generated. We should note, however, that this strategy more than doubled the memory usage of the algorithm.

Overlap. An important characteristic of our algorithm is its capability of finding overlapping regions. To test whether our algorithm is really capable of finding them, we designed a dataset that serves as a demonstration. It consists of several interesting regions that overlap in one or more dimensions. The order in which the algorithms visit the attributes is of great importance for the successful completion of this test. The dataset consists of three sets of records, each of size

10000, along with an additional 10000 items that were added to introduce noise. These record sets are dense in all dimensions they exist in. A summary of this data is shown in Table 1.

Table 1. Summary of the dataset generated to demonstrate overlap. The boundaries the sets in each of the dimensions is shown. Using an approach that visits each dimension separately requires that dimension z is visited first.

Pattern ID	x	y	z
1	[5, 8]	[10, 15]	[10, 15]
2	[10, 18]	[8, 12]	[9, 16]
3	[3, 15]	[5, 15]	[3, 7]

Using this dataset, we came to the following conclusions:

- Discretization does not correctly handle this dataset. That is, while dimension z can be properly discretized, the other dimensions are cut up into chunks. When using confidence as a measure, some of these chunks are discarded as no confident rules can be created from them, leading to information loss.
- QFIMiner is only partially capable of correctly finding all QFIs. QFIMiner uses an Apriori-like build-up of the dimensions that contain dense regions, but does not take into account that the order in which these dimensions are visited can be important. Since it first visits dimension x, then dimension y, and only then dimension z in the last phase, it is only able to detect two regions. QFIMiner's density threshold also causes problems in certain situations. That is, it requires that regions are more dense than the average density of the dataset in the dimensions in which the regions exist. In the case of fairly clean data or data where one region is far more dense than the others, this can lead to regions that go unnoticed.
- DRule is capable of detecting a superset of the rules that QFIMiner finds. That is, since it also visits the dimensions in the same order that QFIMiner does, it finds the same rules. But since it also takes other orderings into account, it is able to detect all three regions and provide rules for them. This ensures that DRule provides more accurate rules compared to competing approaches.

Noise. In Fig. 3 we show how resistant our algorithm is to noise and compare it to the other algorithms (excluding QuantMiner, for reasons mentioned above). Thanks to the first step in our algorithm, the detection of interesting regions, we can quickly filter out dimensions that are not interesting to us. The only work that has to be done on an uninteresting region, is the preliminary 1D clustering, which, depending on the implementation of the algorithm, can be done rather quickly. After this step, uninteresting regions are no longer considered.

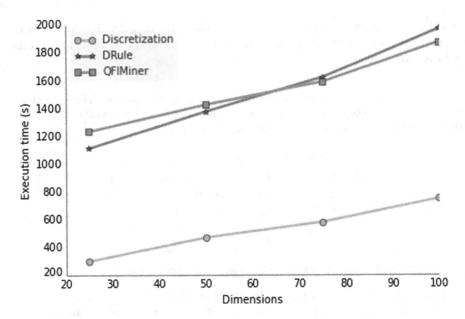

Fig. 3. The algorithm's susceptibility to noise. The dataset contains 10000 items and contains one interesting region.

For this test, we kept the size and amount of interesting regions constant, while introducing additional noisy dimensions with a uniform distribution.

We can see that our algorithm slightly outperforms QFIMiner in terms of efficiency for a lower amount of dimensions, while being slightly outperformed in higher-dimensional data. We outperform QFIMiner in lower-dimensional datasets due to the apriori-like build-up of the dimensions it considers. This approach wastes some time on trying to combine irrelevant dimensions. Due to our reduction step, which efficiently groups dimensions worth exploring together, we can avoid this problem. In higher-dimensional datasets we become less efficient as the amount of patterns we generate increases. Again, as mentioned in the previous test, QFIMiner misses a lot of these patterns.

Real Data

After testing our method on synthetic data to benchmark its efficiency, noise sensitivity, amount of rules it output, and its ability to detect overlap, we move on to real data. We subjectively evaluate the result of our algorithm to illustrate that our algorithm is able to find good rules in the sense that they are very interpretable by the user. To perform this evaluation, we used well-known datasets from the Bilkent University Function Approximation Repository [2], namely the *basketball* and *quake* datasets, and also the *iris* dataset. In Table 2 we present some of the rules found by DRule .

Examining these results, we can immediately see that they are very easy to interpret. Consider for example, the three rules found for the *basketball* dataset. We can immediately conclude that shorter players pass the ball around much more than taller players. Also notable is that the intervals of ASSISTS PER MINUTE actually overlap. While a seemingly small difference, it can prove useful for maximizing your bets.

Our algorithm ran for 1.21, 0.89, and 0.57 s for each of these datasets respectively while QFIMiner ran for 1.02, 0.78, and 0.56 s respectively.

Table 2. Table listing example rules found for real datasets.

Examples (confidence (%))
Iris
SEPAL WIDTH[2.3, 4.2], PETAL LENGTH[1.2, 1.6] → SEPAL LENGTH[4.4, 5.5] (98) CLASS[IRIS-VERSICOLOR] → PETAL LENGTH[3, 5.1] (100) PETAL LENGTH[4.8, 6.9] → PETAL WIDTH[1.4, 2.5] (100)
Quake
LATITUDE[49.83, 50.05], LONGITUDE[78.69, 79.1] → FOCAL DEPTH[0, 9] (100) LONGITUDE[78.69, 79.1], RICHTER[5.9, 6.1] → LATITUDE[49.83, 50.01](95.1) FOCAL DEPTH[0, 24], LONGITUDE[78.69, 79.1] → RICHTER[5.9, 6.1] (71.9)
Basketball
HEIGHT[196, 198] → ASSISTS PER MIN[0.09, 0.14] (72) HEIGHT[183, 185] → ASSISTS PER MIN[0.12, 0.29] (95.5) HEIGHT[191, 193], POINTS PER MIN[0.39, 0.49] → ASSISTS PER MIN[0.07, 0.23] (100)

5 Related Work

Association rule mining was first introduced by Agrawal et al. [1]. Since its introduction, different methods have been extensively studied, as can be demonstrated by the numerous optimizations that have been developed for the Apriori algorithm. However, since these algorithms were developed for discrete values, the so called supermarket basket analysis, they fall short when dealing with numeric attributes. Over the years, many different approaches have been developed to deal with these kinds of attributes. In the following paragraphs, we will give an overview of these approaches and classify them according to how they relate to our approach.

Discretization/Partitioning. The first and classical method of dealing with numeric attributes is to perform a preprocessing step, namely partitioning the data (discretization, binning). This method groups similar items into the same bin. When the data has been partitioned, frequent itemset mining techniques can freely be applied [15,21].

Unfortunately, most of these methods fall short as they perform a univariate discretization, i.e., every dimension is partitioned separately without taking correlations into account. Since quantitative association rules (QAR) generally encompass more than one dimension, these interactions between dimensions remain important. To solve the problem of univariate discretization we can use a multivariate discretization, see for example Bay [4]. These methods attempt to preserve interactions between dimensions. While they improve upon the traditional methods, they still show some shortcomings, such as the assigning each point to a unique bin. This can result in many missed patterns.

A more promising approach to discretization is through the use of fuzzy rules. These allow for some overlap between rules but still have the downside of having to define boundaries beforehand [20].

Clustering-Based Approaches. Another noteworthy attempt at partitioning the data is through the use of clustering algorithms [8]. This approach clusters the data and uses the resulting clusters as boundaries for the bins. Traditional clustering algorithms were designed to handle all attributes simultaneously, which can become a problem as the dimensionality of the data increases. A workaround for the increasing dimensionality of the data has been created through the development of subspace clustering algorithms, e.g. SUBCLU [13] and the FIRES algorithm [12]. A drawback of these techniques is, however, that they are computationally expensive and that they were not designed for rule extraction.

An approach based on the SUBCLU algorithm, which is capable of finding quantitative frequent itemsets, was introduced by Washio et al. [24]. This techniques first generates a partition of the data in each dimension and then uses the anti-monotonicity property of density to combine dimensions. Drawbacks of this technique include: restriction to non-overlapping quantitative predicates, fixed order of processing attributes and thus loss of some rule, and a limitation to quantitative frequent itemsets with no extraction of QARs.

Optimization of Intervals. Other than partitioning approaches, several techniques that allow for non-discretized numeric attributes also exist. These algorithms typically attempt to optimize some criteria in order to interesting rules. The first approach, developed by Fukuda et al. [7] uses techniques from image segmentation to find a region that produces an optimized association rule. This technique was further expanded by Brin et al. [5] to allow for an extra dimension in the rules. A shortcoming is that these techniques are limited in the number of dimensions a rule can contain and that expanding them to higher-dimensional rules is non-trivial. Most recently, Tatti [22] proposes a binarization for real-valued data by selecting sets of thresholds, treating them as random variables and computing the average support.

A wildly different approach was first introduced by Mata et al. [14] and further expanded upon by Salleb-Aouissi et al. [19]. They use a genetic algorithm to find optimal rules according to a rule template. This template has to be passed to the algorithm, so the user has to know which types of rules to look for beforehand.

A last type of techniques operate in a framework of Formal Concept Analysis, and find all possible non-equivalent (or so called closed) vectors of intervals covering a minimal number of points in the data [9,10]. Unfortunately, such techniques are not feasible for large databases.

6   Conclusion

In the present paper, we have introduced a novel method to tackle the problem of finding overlapping quantitative association rules. The rules we find allow for both numeric and categoric attributes and our algorithm is thus more general than many of its competitors. We have demonstrated the algorithm's efficiency and scalability and have shown its resistance to noise. To (partially) solve the problem of increasing rule output, we have also introduced two strategies that decrease the amount of rules the algorithm outputs. Finally, we have demonstrated that DRule is capable of finding association rules showing overlap where other algorithms fail to do so, or only find an incomplete set. Testing DRule on real data leads us to conclude that it has its merit in the field of quantitative association rule mining.

7   Future Work

In the future, we aim to explore different quality measures and different heuristics, other than density, for our rules evaluation and discovery to allow for a more efficient and a more accurate algorithm.

Other than these optimizations, we intend to investigate methods to make our method more robust. That is, currently, as is the case with most ARM algorithms, small changes in parameters can lead to drastic increases in runtime. While a lot of pruning is already done by our algorithm, this can still lead to an increase in the ordering of dimensions that have to be visited.

References

1. Agrawal, R., Imieliński, T., Swami, A.: Mining association rules between sets of items in large databases. ACM SIGMOD **22**(2), 207–216 (1993)
2. Altay Guvenir, H., Uysal, I.: Bilkent university function approximation repository (2000). http://funapp.cs.bilkent.edu.tr
3. Aumann, Y., Lindell, Y.: A statistical theory for quantitative association rules. In: ACM SIGKDD, pp. 261–270 (1999)
4. Bay, S.D.: Multivariate discretization for set mining. Knowl. Inf. Syst. **3**(4), 491–512 (2001)
5. Brin, S., Rastogi, R., Shim, K.: Mining optimized gain rules for numeric attributes. IEEE Trans. Knowl. Data Eng. **15**(2), 324–338 (2003)
6. Ester, M., Kriegel, H.P., Sander, J., Xu, X.: A density-based algorithm for discovering clusters in large spatial databases with noise. In: ACM SIGKDD, pp. 226–231 (1996)
7. Fukuda, T., Morimoto, Y., Morishita, S., Tokuyama, T.: Mining optimized association rules for numeric attributes. J. Comput. Syst. Sci. **58**(1), 1–12 (1999)

148   T. Van Brussel et al.

8. Grzymała-Busse, J.W.: Three strategies to rule induction from data with numerical attributes. In: Peters, J.F., Skowron, A., Dubois, D., Grzymała-Busse, J.W., Inuiguchi, M., Polkowski, L. (eds.) Transactions on Rough Sets II. LNCS, vol. 3135, pp. 54–62. Springer, Heidelberg (2004)
9. Kaytoue, M., Kuznetsov, S.O., Napoli, A.: Revisiting numerical pattern mining with formal concept analysis. International Joint Conference on Artificial Intelligence (IJCAI) arXiv preprint arxiv:1111.5689 (2011)
10. Kaytoue, M., Kuznetsov, S.O., Napoli, A., Duplessis, S.: Mining gene expression data with pattern structures in formal concept analysis. Inf. Sci. 181(10), 1989–2001 (2011)
11. Ke, Y., Cheng, J., Ng, W.: Mic framework: an information-theoretic approach to quantitative association rule mining. In: IEEE ICDE, pp. 112–112 (2006)
12. Kriegel, H.P., Kröger, P., Renz, M., Wurst, S.H.R.: A generic framework for efficient subspace clustering of high-dimensional data. In: IEEE ICDM, pp. 250–257 (2005)
13. Kröger, P., Kriegel, H.P., Kailing, K.: Density-connected subspace clustering for high-dimensional data. In: SIAM SDM, pp. 246–256 (2004)
14. Mata, J., Alvarez, J.L., Riquelme, J.C.: An evolutionary algorithm to discover numeric association rules. In: ACM SAC, pp. 590–594 (2002)
15. Miller, R.J., Yang, Y.: Association rules over interval data. ACM SIGMOD 26(2), 452–461 (1997)
16. Müller, E., Assent, I., Günnemann, S., Seidl, T.: Scalable density-based subspace clustering. In: ACM CIKM, pp. 1077–1086 (2011)
17. Müller, E., Assent, I., Krieger, R., Günnemann, S., Seidl, T.: DensEst: Density estimation for data mining in high dimensional spaces. In: SIAM SDM, pp. 175–186 (2009)
18. Müller, E., Günnemann, S., Assent, I., Seidl, T.: Evaluating clustering in subspace projections of high dimensional data. PVLDB 2(1), 1270–1281 (2009)
19. Salleb-Aouissi, A., Vrain, C., Nortet, C., Kong, X., Rathod, V., Cassard, D.: Quantminer for mining quantitative association rules. J. Mach. Learn. Res. 14(1), 3153–3157 (2013)
20. Serrurier, M., Dubois, D., Prade, H., Sudkamp, T.: Learning fuzzy rules with their implication operators. Data Knowl. Eng. 60(1), 71–89 (2007). http://dx.doi.org/10.1016/j.datak.2006.01.007
21. Srikant, R., Agrawal, R.: Mining quantitative association rules in large relational tables. In: ACM SIGMOD. pp. 1–12 (1996)
22. Tatti, N.: Itemsets for real-valued datasets. In: IEEE ICDM, pp. 717–726 (2013)
23. Vannucci, M., Colla, V.: Meaningful discretization of continuous features for association rules mining by means of a som. In: ESANN, pp. 489–494 (2004)
24. Washio, T., Mitsunaga, Y., Motoda, H.: Mining quantitative frequent itemsets using adaptive density-based subspace clustering. In: IEEE ICDM, pp. 793–796 (2005)
25. Webb, G.I.: Discovering associations with numeric variables. In: ACM SIGKDD, pp. 383–388 (2001)
26. Wijsen, J., Meersman, R.: On the complexity of mining quantitative association rules. Data Min. Knowl. Discov. 2(3), 263–281 (1998)
27. Zhu, F., Yan, X., Han, J., Yu, P.S., Cheng, H.: Mining colossal frequent patterns by core pattern fusion. In: IEEE ICDE, pp. 706–715 (2007)

Semantic Matching Strategies for Job Recruitment: A Comparison of New and Known Approaches

Gábor Rácz[1], Attila Sali[1(✉)], and Klaus-Dieter Schewe[2]

[1] Alfréd Rényi Institute of Mathematics, Hungarian Academy of Sciences,
P.O.Box 127, Budapest 1364, Hungary
gabee33@gmail.com, sali.attila@renyi.mta.hu
[2] Software Competence Center Hagenberg, Softwarepark 21, 4232 Hagenberg, Austria
kd.schewe@scch.at

Abstract. A profile describes a set of skills a person may have or a set of skills required for a particular job. Profile matching aims to determine how well a given profile fits to a requested profile. The research reported in this paper starts from exact matching measure of [21]. It is extended then by matching filters in ontology hierarchies, since profiles naturally determine filters in the subsumption relation. Next we take into consideration similarities between different skills that are not related by the subsumption relation. Finally, a totally different approach, probabilistic matching based on the maximum entropy model is analyzed.

Keywords: Semantic matching · Ontology · Lattice filters · Probabilistic matching · Maximum entropy model

1 Introduction

A profile describes a set of properties and profile matching is concerned with the problem to determine how well a given profile fits to a requested one. Profile matching appears in many application areas such as matching applicants for job requirements, matching system configurations to requirement specifications, etc.

The simplest idea of profile matching is to consider profiles as sets of unrelated items. There are several ways to define distances of sets, such as Jaccard or Sørensen-Dice measures [14] turned out to be useful in ecological applications. However, many dependencies between skills or properties included in profiles exist and need to be taken into account. In the human resources area several taxonomies for skills, competences and education such us DISCO [7], ISCED [9] and ISCO [10] have been set up. Based on these taxonomies a lattice structure

The research reported in this paper has been [partly] supported by the Austrian Ministry for Transport, Innovation and Technology, the Federal Ministry of Science, Research and Economy, and the Province of Upper Austria in the frame of the COMET center SCCH.

© Springer International Publishing Switzerland 2016
M. Gyssens and G. Simari (Eds.): FoIKS 2016, LNCS 9616, pp. 149–168, 2016.
DOI: 10.1007/978-3-319-30024-5_9

of the individual properties can be assumed. Popov and Jebelean [21] exploited this by defining an asymmetric matching measure on the basis of filters in such lattices.

However, there are other relations between skills and properties than the ones given by the ontolgies above. Having some skills imply that the applicant may have some other skills with certain probabilities, or of some (not complete) proficiency level. For example, we may reasonably assume that knowledge of Java implies knowledge of Netbeans up to a grade of 0.7 or with probability 0.7. Our new approach incorporates such relations in the following way. The subsumption hierarchy of the ontology of skills is considered as a directed graph with edge weights 1. A lattice filter generated by a profile corresponds to the set of nodes reachable from the profile's nodes in the directed graph. Then extra edges are added with weights representing the probability/grade of the implication between skills or properties. This may introduce directed cycles. Filters of application profiles are replaced by nodes reachable in the extended graph from the profile's nodes. However, for each vertex x reached a probability/grade is assigned, the largest probability of a path from the profile's nodes to x. Path probability is defined as the product of probabilities of edges of the path. At first sight it seems that determination of the grade of a vertex involves finding a longest path in weighted directed graph, known to be an NP-complete problem. However, in our case the weighting is multiplicative and less than 1, so Dijkstra's Algorithm [5] can be applied. This process results in a set of nodes with grades between zero and one, so it can naturally be interpreted as a fuzzy set. In fact, we prove that it is a fuzzy filter as defined in [8,16].

Considering the grades as probabilities suggests another approach. They can be handled from an information theoretic point of view, with probabilistic logic programs [11] or from set theoretic point of view, with probabilistic models [25], as well. In the present paper the maximum entropy model is used which adds the lowest amount of additional information between single elementary probabilities to the system. In order to apply probabilistic model the information represented by the extended directed graph is translated into sentences over an appropriate measurable space. The matching value of a job offer O and an application A is the result of the probabilistic query obtained from the sentences.

Our paper is organized as follows.

Section 2 contains the description of our novel model of extending the ontology hierarchy with cross relations in the form of weighted directed edges. Two new ranking algorithms are given for job applications and a connection with fuzzy theory is mentioned.

Section 3 is devoted to the comparison of the different approaches. Our findings show that these are basically independent of each other apart from some natural dependences as some of the approaches are extensions of some other ones.

Section 4 discusses related work and how our approach fits into the broad area of semantic matching.

Finally, Sect. 5 contains conclusions.

2 Semantic Matching

Let $S = \{s_1, s_2, \ldots, s_n\}$ be a finite set of skills. Let a job offer $O = \{o_1, o_2, \ldots, o_k\}$ be a subset of S which contains the skills that are required for the job. Then, an application $A = \{a_1, a_2, \ldots, a_l\}$ is also a subset of S which means the applicant possesses these skills. Our task is to find the most suitable applicant for a job offer. Example 2.1 shows a job offer with four applications.

Example 2.1 (A job offer and four applications)

$$\begin{aligned}
Offer_1 &= \{Java, Netbeans, XML\} \\
Application_1 &= \{Java, PHP, Eclipse\} \\
Application_2 &= \{Java, Netbeans, HTML\} \\
Application_3 &= \{C, PHP, XML\} \\
Application_4 &= \{C, Netbeans, XML\}
\end{aligned}$$

Note, that skills are not graded (e.g., basic/medium/expert knowledge in Java) in our examples. If we need such differentiation, we have to handle the grades as separate skills.

2.1 Perfect Matching

A simple idea to decide how much an application fits to a job offer is to compute the number of the matching skills the ones that the applicant possesses and that are required for the job. The result can be normalized with the number of the required skills [21]. Formally,

$$match(O, A) = \frac{|O \cap A|}{|O|}. \tag{1}$$

In the following example, we compute the matching values of the job offer and the applications from Example 2.1.

Example 2.2 (Perfect matching)

$$\begin{aligned}
match(O_1, A_1) &= \frac{|O_1 \cap A_1|}{|O_1|} = \frac{|\{Java\}|}{|\{Java, Netbeans, XML\}|} = \frac{1}{3} \\
match(O_1, A_2) &= \frac{|O_1 \cap A_2|}{|O_1|} = \frac{|\{Java, Netbeans\}|}{|O_1|} = \frac{2}{3} \\
match(O_1, A_3) &= \frac{|O_1 \cap A_3|}{|O_1|} = \frac{|\{XML\}|}{|O_1|} = \frac{1}{3} \\
match(O_1, A_4) &= \frac{|O_1 \cap A_4|}{|O_1|} = \frac{|\{Netbeans, XML\}|}{|O_1|} = \frac{2}{3}
\end{aligned}$$

As it can be seen, this matching function cannot sufficiently distinguish between the applications. It assigned A_2, A_4 and A_1, A_3 the same values, respectively. However, as our goal is to find the most suitable applicants, we want to avoid that two or more candidates get the same values. Therefore, we need extra knowledge to be able to distinguish A_2 from A_4 and A_1 from A_3.

2.2 Matching Using Ontology Edges

Let us suppose that the skills in S form a hierarchy. We can represent a hierarchy several ways, for example, with Description Logic [1], with Resource Description Framework Schema [3] or with Logic programming [17]. We use the description logic approach here, so let the skills corresponds to concepts and we define a specialization relation over them.

Let $\preceq$ be a binary specialization (subsumption) relation over S. Let $s_i, s_j \in S$ be two skills, then $s_i \preceq s_j$ if s_i is a *specialization* of s_j. It means if an applicant possesses the skill s_i, he also possesses the skill s_j as s_i is a more specific skill than s_j (s_j is more general than s_i). Let $\preceq_d$ denote the direct specialization, i.e., $s_i \preceq_d s_j$ if and only if $s_i \preceq s_j$ and $\nexists s_k \in S$ such that $s_k \neq s_i, s_k \neq s_j$ and $s_i \preceq s_k \preceq s_j$. Note, that $\preceq$ is a reflexive, antisymmetric and transitive relation, i.e., it is a partial order over S.

We can always add a top (respectively a bottom) element to the hierarchy that represents all the skills that everybody (nobody) possesses. In addition, let us suppose the concepts define a lattice, i.e., for each pair of skills has unique ancestor (and descendant) that is more general (specific) then both elements of the pair. Let this lattice be denoted by $(S, \preceq)$.

In Fig. 1, the blue edges form a hierarchy among computer science skills.

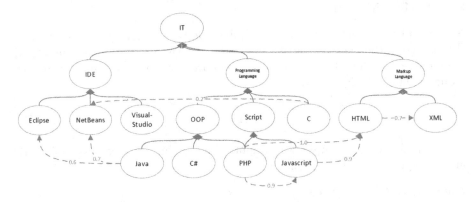

Fig. 1. A hierarchy of skills. The ontology edges are the blue ones (solid) and the extra edges are the orange ones (dashed) (Color figure online).

Definition 2.1. *A filter in a lattice $(S, \preceq)$ is a non-empty subset $F \subseteq S$, such that for all $s, s' \in S$ with $s \preceq s'$ whenever $s \in F$ holds, then also $s' \in F$ holds.*

An $A \subseteq S$ application defines in a natural way an F filter of the $(S, \preceq)$ lattice:

$$F = \{s \in S \mid \exists a \in A, a \preceq s\}.$$

This filter is an extension of the original application with the skills that are more general than the ones in the application. It is reasonable, because if an application possesses a skill, then he must possesses all the skills that are more general by the definition. Note, that a job offer can be extended in the same way as an application. And then, we can apply, for example, the perfect matching function on the extended sets.

The next example shows the filters defined by the job offer and of the applications from Example 2.1, and the values of the perfect matching function applying on the extensions. As can be seen, this method was already able to distinguish A_2 from A_4.

Example 2.3 (Matching based on an ontology)

$$F_{O_1} = \{Java, Netbeans, XML, OOP, PL, IT, IDE, ML\}$$
$$F_{A_1} = \{Java, PHP, Eclipse, OOP, PL, IT, Script, IDE\}$$
$$F_{A_2} = \{Java, Netbeans, HTML, OOP, PL, IT, IDE, ML\}$$
$$F_{A_3} = \{C, PHP, XML, PL, IT, Script, OOP, ML\}$$
$$F_{A_4} = \{C, Netbeans, XML, PL, IT, IDE, ML\}$$

$$match(F_{O_1}, F_{A_1}) = \frac{|\{Java, OOP, PL, IT, IDE\}|}{|F_{O_1}|} = \frac{5}{8}$$

$$match(F_{O_1}, F_{A_2}) = \frac{|\{Java, OOP, PL, IT, Netbeans, IDE, ML\}|}{|F_{O_1}|} = \frac{7}{8}$$

$$match(F_{O_1}, F_{A_3}) = \frac{|\{XML, PL, IT, OOP, ML\}|}{|F_{O_1}|} = \frac{5}{8}$$

$$match(F_{O_1}, F_{A_4}) = \frac{|\{Netbeans, XML, PL, IT, IDE, ML\}|}{|F_{O_1}|} = \frac{6}{8}$$

2.3   Maximal Length Matching

In this section, we are adding extra knowledge to the hierarchy in form of extra edges, and we are investigating how it can be used to find the most suitable application for a job. However, these extra edges can form cycles in the hierarchy, therefore the traditional filters are not applicable in this case. Immediate example for a cycle could be two skills that are connected with two extra edges which are pointing in opposite direction. For this reason, we describe a graph based approach to extend the applications and the offer. Then, we show that this approach corresponds to the definition of fuzzy filters [8, 16].

Let $G = (V, E)$ be a directed weighted graph where V is a finite non-empty set of nodes and $E = \{E_O \cup E_E\} \subseteq V \times V$ is a set of edges. Each node represents a skill and an $e_o = (v_i, v_j) \in E_O$ directed edge is added from the skill v_i to v_j if v_i is a specialization of v_j (this type of edges are called ontology edges). Moreover, an $e_e = (v_i, v_j) \in E_E$ represents a conditional dependency between the skills v_i and v_j. Namely, if a person has the skill v_i then he may have the skill v_j (this type of edges are called extra edges). Let $w : E \to (0, 1]$ be a weighting function that assigns a weight to all edges such that for all ontology edges $e_o \in E_O$ $w(e_o) = 1$ and for all extra edges $e_e \in E_E$ let the weight $w(e_e)$ represents the conditional probability between the start and the end node of the edge. The edge weight can come, for example, from Human Resources experiments or from domain experts.

Definition 2.2. *Let $G = (V, E_O \cup E_E)$ a directed weighted graph with a $w : E \to (0, 1]$ weighting function and let $(S, \preceq)$ be a lattice. We say that G is built on $(S, \preceq)$ if $V = S$ and for all $v_i, v_j \in V$ $(v_i, v_j) \in E_O$ if and only if $v_i \preceq_d v_j$.*

Let $s, t \in V$ be two nodes, then denote by $p_E(s, t)$ all the directed paths between s and t, i.e.,

$$p_E(s, t) = \{(s = v_1, v_2, \ldots, v_n = t) \mid v_i \in V \text{ and } (v_i, v_{i+1}) \in E\}.$$

The extra edges and the directed paths are used to extend the applications with skills that the applicant possibly possesses. Since both sets and uncertainty occur, fuzzy sets [28] are suitable to model the extended applications. A fuzzy set assigns a value for each element that expresses the certainty of that the element is in the set or not.

Definition 2.3. *A fuzzy set in S is a mapping $f : S \to [0, 1]$. A fuzzy set is called empty if f is identically zero on S. Let t be a real number such that $t \in [0, 1]$, then the set $f_t = \{s \in S \mid f(s) \geq t\}$ is called a level subset of f.*

Note that, as S is a finite set, we can define a fuzzy set by enumerating all elements in S with their assigned values (called the grade of membership) if that value is greater than zero, i.e.,

$$f = \{(s, f(s)) \mid s \in S \text{ and } f(s) > 0\}.$$

The intersection and the union of tho fuzzy sets can be defined axiomatically with t-norms and t-conorms [20]. For clarity we use *min* and *max* as t-norm and t-conorm, i.e., for fuzzy sets f, g in S, we define the intersection and the union operations as $(f \cap g)(s) := \min\{f(s), g(s)\}$ and $(f \cup g)(s) := \max\{f(s), g(s)\}$ for all $s \in S$, respectively.

We extend the job offer and the applications to fuzzy sets in the following way. Let $O \subseteq V$ be a set of skills. The extension of O w.r.t. E_O is defined as the set of all the skills that are available from O via directed paths containing edges only from E_O. We assign 1.0 to each element of the extension to create a fuzzy set, that is

$$extend_{E_O}(O) = \hat{O} = \{(v, 1.0) \mid v \in V \text{ and } \exists o \in O : |p_{E_O}(o, v)| \geq 1\}.$$

Let $A \subseteq V$ be a set of skills. The extension of A w.r.t. E is defined as the set of all the skills that are available from A via directed paths containing ontology or extra edges, and we assign the length of the longest path between the node and the elements of A to each element of the extended set, namely

$$extend_E(A) = \hat{A} = \{(v, \mu_v) \mid v \in V \text{ and } \exists a \in A : |p_E(a, v)| \geq 1 \text{ and}$$
$$\mu_v = \max_{a \in A, p \in p_E(a, v)} length(p)\},$$

where $length(p) = \prod_{i=1}^{n-1} w((v_i, v_{i+1}))$ is the product of the weights of the edges on the path p. The μ_v is the length of the longest path from A to v. If the edge weights mean uncertainty or probability, the length of longest path means the joint probability of the applicant possessing all the skills on the path (if we assume some independence), which seems a rational decision.

Note that finding the longest path between two nodes in a graph is generally a hard problem. However, in our case the length of a path is defined as the product of the weight of the edges on the path. Therefore, because of the strict monotonicity of the logarithm function, we can apply the following transformations:

$$
\max_{p \in P_E(v,a)} \prod_{i=1}^{n-1} w((v_i, v_{i+1})) \iff \max_{p \in P_E(v,a)} \log \Big\{ \prod_{i=1}^{n-1} w((v_i, v_{i+1})) \Big\} \iff
$$

$$
\max_{p \in P_E(v,a)} \sum_{i=1}^{n-1} \log w((v_i, v_{i+1})) \iff - \min_{p \in P_E(v,a)} \sum_{i=1}^{n-1} - \log w((v_i, v_{i+1})).
$$

Moreover, the weighting function w assigns a weight from the $(0, 1]$ interval to each edge, thus $- \log w((v_i, v_{i+1})) \in [0, +\inf)$. With these transformation, we got a single-source shortest path problem with non-negative edge weights. That problem can be solved with Dijkstra's algorithm in $O(|E| + |V| \log |V|)$ time [5].

Definition 2.4. Let $(S, \preceq)$ be a lattice, and let f be a fuzzy set in S. f is called a fuzzy filter if for all $t \in [0, 1]$, f_t is either empty or a filter of S.

Next, we show that the extensions presented above define fuzzy filters in S.

Lemma 2.1. Let $(S, \preceq)$ be a lattice, let $G = (V, E_O \cup E_E)$ be a directed weighted graph built on the lattice with the weighting function w and let $s, s' \in S$ be a skill pair such that $s \preceq s'$. Then, there is a $p \in P_E(s, s')$ path such that $length(p) = 1$.

Proof. Since $s \preceq s'$, an $s = s_1, \ldots, s_k = s'$ sequence of skills exists such that $s_i \preceq_d s_{i+1}$ holds for $i = 1, \ldots, k - 1$. As G is built on the lattice, each skill is represented by a node in the graph, and $(s_i, s_{i+1}) \in E_O$ and $w((s_i, s_{i+1})) = 1$ for $i = 1, \ldots, k-1$. Let $p = (s_1, \ldots, s_k)$ be a path between v_1 and v_n. Consequently, $length(p) = \prod_{i=1}^{k-1} w((s_i, s_{i+1})) = 1$ □

Theorem 2.1. Let $(S, \preceq)$ be a lattice, let $G = (V, E_O \cup E_E)$ be a directed weighted graph built on the lattice with weighting function w, let $A \subseteq S$ be a non-empty application, and let $\widehat{A}$ be the extension of A w.r.t. E. Then, $\widehat{A}$ is a fuzzy filter in S.

Proof. For $t \in [0, 1]$, $\widehat{A}_t = \{s \in S | \widehat{A}(s) \geq t\}$. $\widehat{A}_t$ is a filter in S if for all $s, s' \in S$ with $s \preceq s'$ whenever $s \in \widehat{A}_t$ holds, then also $s' \in \widehat{A}_t$ holds. It means if $\widehat{A}(s) \geq t$, then $\widehat{A}(s') \geq t$.

Let s be in $\widehat{A}_t$ and let $p_{a,s} = (a = s_{i_1}, s_{i_2}, \ldots, s_{i_k} = s)$ be one of the maximal length path between A and s. We have to show that if an $s' \in S$ is a generalization of s, i.e., $s \preceq s'$, then a $p_{a,s'}$ path exists such that $length(p_{a,s'}) \geq length(p_{a,s})$.

Lemma 2.1 states that a $p_{s,s'} = (s = s_{j_1}, s_{j_2}, \ldots, s_{j_l} = s')$ path exists such that $length(p_{s,s'}) = 1$. If $p_{a,s}$ and $p_{s,s'}$ are disjoint, so they do not have any node in common except s, then we can concatenate them to $p_{a,s'} = (a = s_{i_1}, \ldots, s_{i_k}, s_{j_1}, \ldots, s_{j_l} = s')$ and its length is $length(p_{a,s'}) = length(p_{a,s}) * 1 \geq$

$length(p_{a,s})$. Otherwise recursively, let $t = \min_{x \in [1,k]} \{s_{i_x} | \exists y \in [1, l] : s_{i_x} = s_{j_y}\}$ be the first common node. Then, consider the $p_{a,t} = (a = s_{i_1}, '\ldots, s_{i_x} = t)$ and the $p_{t,s'} = (t = s_{j_y}, s_{j_{y+1}}, \ldots, s_{j_l} = s')$ paths. Where $length(p_{a,t}) \geq length(p_{a,s})$ as $w(e) \leq 1$ for all $e \in E$ and $length(p_{t,s'}) = 1$ as it contains ontology edges only. If these paths are disjoint except t then we can concatenate them, otherwise repeat this step. Since the number of nodes that are contained in the paths are limited and every step reduces that number, the iteration will stop in finite step. □

Let $\widehat{O}$ be an extended job offer w.r.t. E_O and let $\widehat{A}$ be an extended application w.r.t. E, both are fuzzy sets. We can define a matching value between $\widehat{O}$ and $\widehat{A}$ in a similar way as we did in Eq. 1.

$$match(\widehat{O}, \widehat{A}) = \frac{||\widehat{O} \cap \widehat{A}||}{||\widehat{O}||}, \tag{2}$$

where $||.||$ denotes the sum of the grades of membership of the elements in a fuzzy set, formally $||\widehat{A}|| = \sum_{(a,\mu_a) \in A} \mu_a$. Note that, besides the sigma cardinality used here, many other options are available on the cardinality of a fuzzy set [27].

Algorithm 1 shows how can we find the most suitable applicant for a job using the extensions defined above. The algorithm works as follows:

Algorithm 1. MaximalLengthMatching

Input: a graph $G = (V, E = \{E_O \cup E_E\})$, a job offer $O \subseteq V$, and a set of applications
     $A = \{A_0, A_2, \ldots, A_n\}$.
Output: the most suitable application A_m for the job
1: $O_e \leftarrow extend_{E_O}(O)$
2: $A_m \leftarrow A[0], max \leftarrow extend_E(A[0])$
3: **for** $i = 1 \rightarrow n$ **do**
4:     $A_e \leftarrow extend_E(A[i])$
5:     **if** $match(O_e, A_e) > match(O_e, max)$ **then**
6:         $A_m \leftarrow A[i], max \leftarrow A_e$
7:     **end if**
8: **end for**
9: **return** A_m

First, we extend the job offer O with the skills that are available from O via ontology edges (line 1). It is a reasonable extension because the ontology edges represent specialization relation between to skills. And if an applicant possesses a more general skill than the required one, then he could specialize faster than an applicant that does not possess even that general skill. As everyone possesses the skill at the top of the hierarchy, that does not distinguish one applicant from another.

Next, we also extend each application (line 2, 4). In this step, however, we take into consideration the extra edges as well. This is because, if the skill v_i is a specialization (or conditionally depends, respectively) of another skill v_j, and a person possesses the skill v_i, then he also possesses (may possess) the skill v_j. The matching value of the extended job offer and the extended application is then computed using Eq. 2 (line 5). The most suitable application is stored in the variable A_m (line 6).

The next example shows the extension of the $Offer_1$ and the $Application_1$ from Example 2.1, and their matching value.

Example 2.4 (Maximal length matching).

$$\widehat{O_1} = \{(Java, 1.0), (Netbeans, 1.0), (XML, 1.0), (OOP, 1.0),$$
$$(PL, 1.0), (IT, 1.0), (IDE, 1.0), (ML, 1.0)\}$$

$$\widehat{A_1} = \{(Java, 1.0), (PHP, 1.0), (Eclipse, 1.0), (OOP, 1.0), (PL, 1.0),$$
$$(IT, 1.0), (Script, 1.0), (IDE, 1.0), (Netbeans, 0.7),$$
$$(Javascript, 0.9), (HTML, 1.0), (ML, 1.0), (XML, 0.7)\}$$

$$\widehat{O_1} \cap \widehat{A_1} = \{(Java, 1.0), (OOP, 1.0), (PL, 1.0), (IT, 1.0),$$
$$(IDE, 1.0), (Netbeans, 0.7), (XML, 0.7), (ML, 1.0)\}$$

$$m(\widehat{O_1}, \widehat{A_1}) = \frac{||\widehat{O_1} \cap \widehat{A_1}||}{||\widehat{O_1}||} = \frac{7.4}{8}$$

As one can see, OOP, PL, IT, IDE and ML appeared in the offer as they are available from the originally specified skills (Java, Netbeans, XML) via ontology edges (the solid ones) as Fig. 1 shows. In addition, the $extend_{E_O}$ method assigned 1.0 to each skills in the offer and transformed $\widehat{O_1}$ to fuzzy set. The application was also extended but the $extend_E$ method used the extra edges (the dashed ones) as well during the extension. The intersection of the two extended sets consists of those elements that appeared in both sets and the assigned values are computed with the min function as t-norm. Finally, the ratio of the sum cardinalities is calculated.

2.4 Probabilistic Matching

The ontology and the extra edges can be handled from an information theoretic point of view, with probabilistic logic programs [11] or from set theoretic point of view, with probabilistic models [25] as well. In this paper, we use the latter, the set theoretical approach and we apply the maximum entropy model to give a probabilistic matching method. The following definitions were presented in [25].

2.4.1 Preliminaries

Let Θ be a finite set. Let $R := \{a_1, \ldots, a_l\}$ be a set of subsets of the power set $\mathcal{P}(\Theta)$ of Θ, namely $a_i \in \mathcal{P}(\Theta)$, $i = 1, \ldots, l$. The elements of R are called *events*.

Definition 2.5. *Let X be some set. Let $\mathcal{A}$ be a subset of $\mathcal{P}(X)$. Then, $\mathcal{A}$ is a σ-algebra over X, denoted by $\mathcal{A}(X)$, if*

- *$X \in \mathcal{A}$;*
- *if $Y \in \mathcal{A}$, then $(X \setminus Y) \in \mathcal{A}$; and*
- *if $Y_1, Y_2, \ldots$ is a countable collection of sets in $\mathcal{A}$, then their union $\bigcup\limits_{n=1}^{\infty} Y_n$ is in $\mathcal{A}$ as well.*

The set of full conjunction over R is given by

$$\Omega := \left\{ \bigcap_{i=1}^{l} e_i \mid e_i \in \{a_i, \neg a_i\} \right\},$$

where $a_i \in \mathcal{P}(\Theta)$, $i = 1, \ldots, l$, and $\neg a_i = \Theta \setminus a_i$. It is well known that the 2^l elements of Ω are mutually disjoint and span the set R (any a_i can be expressed by a disjunction of elements of Ω). Therefore the smallest (σ-) algebra $\mathcal{A}(R)$ that contains R is identical to $\mathcal{A}(\Omega)$. For that reason we restrict the set of *elementary events* (set of possible worlds) to Ω instead of the underlying Θ.

Definition 2.6. *Over a set $R := \{a_1, \ldots, a_l\}$, a measurable space $(\Omega, \mathcal{A})$ is defined by*

- *$\Omega := \left\{ \bigcap\limits_{i=1}^{l} e_i \mid e_i \in \{a_i, \neg a_i\} \right\}$; and*
- *$\mathcal{A} = \mathcal{A}(\Omega) = \mathcal{P}(\Omega)$.*

Definition 2.7. *Let $(\Omega, \mathcal{A})$ be a measurable space over R with $\Omega = \{\omega_1, \ldots, \omega_n\}$. A discrete probability measure P or a probability model (P-model) is an assignment of non-negative numerical values to the elements of Ω, which sum up to unity. Formally,*

$$p_i := P(\omega_i) \geq 0, \ i = 1, \ldots, n \text{ and } \sum a_i = 1.$$

The n-tuple $\boldsymbol{p} = (p_1, \ldots, p_n)$ is called a probability vector (P-vector). W_Ω (respectively, V_Ω) denotes the set of all possible P-models (P-vectors) for $(\Omega, \mathcal{A})$.

Definition 2.8. *For given $(\Omega, \mathcal{A})$, $P \in W_\Omega$ $a, b \in \mathcal{A}$, $P(a) > 0$ and $[l, u] \subseteq [0, 1]$ the term*[1]

$$\langle P(b|a) = \delta, \ \delta \in [l, u] \rangle \text{ or } P(b|a)[l, u]$$

is called a sentence in $(\Omega, \mathcal{A})$, where $P(b|a) = P(a \cap b)/P(a)$ denotes the conditional probability of the event b given a. The sentence given above is called true in $P \in W_\Omega$, if and only if $P(b|a) \in [l, u]$. Otherwise it is called false.

---

[1] $P(a) = P(a|\Omega)$.

A sentence $P(b|a)[l, u]$ defines two inequalities, namely

- $P(b|a) \leq u$ (be less than the upper bound); and
- $P(b|a) \geq l$ (be greater than the lower bound).

These inequalities can be further transformed in the following way:

$$P(b|a) \leq u \Leftrightarrow P(a \cap b) \leq u \cdot P(a) \quad \Leftrightarrow P(a \cap b) \leq u \cdot (P(a \cap b) + P(a \cap \neg b))$$
$$P(b|a) \geq l \Leftrightarrow P(a \cap b) \geq l \cdot P(a) \quad \Leftrightarrow P(a \cap b) \geq l \cdot (P(a \cap b) + P(a \cap \neg b))$$

Rearranging the inequalities and using the elementary probabilities $p_i, i = 1, \ldots, n$ yields

$$P(b|a) \leq u \Leftrightarrow (1-u) \cdot \sum_{i:w_i \in a \cap b} p_i + u \cdot \sum_{j:w_j \in a \cap \neg b} p_j \geq 0$$

$$P(b|a) \geq l \Leftrightarrow (1-l) \cdot \sum_{i:w_i \in a \cap b} p_i - l \cdot \sum_{j:w_j \in a \cap \neg b} p_j \geq 0$$

Note, that, if $u = 1$ (respectively, $l = 0$), then the first (second) inequality is always satisfied as $p_i \geq 0$.

Definition 2.9. *Let $DB := \{c_1, \ldots, c_m\}$ be a set of m sentences in $(\Omega, \mathcal{A})$. W_{DB} is defined as the set of all P-models $P \in W_\Omega$ in which $c_1, \ldots, c_m$ are true. We call $c_1, \ldots, c_m$ constraints on W_Ω, and W_{DB} denotes the set of all elements of W_Ω that are consistent with the constraints in DB.*

If W_{DB} is empty, the information in DB is inconsistent. If W_{DB} contains more than one element, the information in DB is incomplete for determining a single P-model.

In the next section, we discuss how the maximum entropy model copes with incomplete information.

2.4.2 Maximum Entropy Model

If W_{DB} contains more than one element, the information in DB is incomplete for determining a single P-model. Therefore, we must add further constraints to the system to get a unique model.

It was shown in [25] that the maximum entropy model adds the lowest amount of additional information between single elementary probabilities to the system. Moreover, the maximum entropy model also satisfies the principle of indifference and the principle of independence. The principle of indifference states that if we have no reason to expect one event rather than another, all the possible events should be assigned the same probability. The principle of independence states the if the independence of two events a and b in a P-model ω is given, any knowledge about the event a does not change the probability of b (and vice verse) in ω, formally $P(b|a) = P(b)$.

To get the consistent P-model to a DB that has the maximum entropy, we have to solve the following linear optimization problem:

Definition 2.10. *Let* $DB := \{c_1, \ldots, c_m\}$ *be a set of* m *sentences in* $(\Omega, \mathcal{A})$ *with* $\Omega = \{\omega_1, \ldots, \omega_n\}$. *Let* W_{DB} *(respectively* V_{DB}*) be the set of all P-models* $P \in W_\Omega$ *(P-vectors* $\boldsymbol{p} \in V_\Omega$*) in which* $c_1, \ldots, c_m$ *are true. The maximum entropy problem is*

$$\max_{\boldsymbol{v}=(p_1,\ldots,p_n)\in[0,1]^n} -\sum_{i=1}^{n} p_i \log p_i$$

subject to

$$\sum_{i=1}^{n} p_i = 1$$

$$(1-u) \cdot \sum_{i:w_i \in a \cap b} p_i + u \cdot \sum_{j:w_j \in a \cap \neg b} p_j \geq 0 \; (\text{for all } c = P(b|a)[l,u] \in DB, l > 0)$$

$$(1-l) \cdot \sum_{i:w_i \in a \cap b} p_i - l \cdot \sum_{j:w_j \in a \cap \neg b} p_j \geq 0 \; (\text{for all } c = P(b|a)[l,u] \in DB, u < 0)$$

$$p_i \geq 0 (i = 1, \ldots, n)$$

Denote by $me[DB]$ the P-model that solves the maximum entropy problem if such model exists.

Definition 2.11. *Let* DB *be a set of sentences and let* $c = \langle P(b|a)[l,u] \rangle$ *be a sentence. We say that* c *is a maximum entropy consequence of* DB, *denoted by* $DB \, \|\sim^{me} c$, *if and only if either*

– *DB is inconsistent, or*
– $me[DB](b|a) \in [l,u]$.

Definition 2.12. *A probabilistic query is an expression* $QP_{DB}(b|a)$ *where* a *and* b *are two events, i.e.,* $a, b \in \mathcal{A}$, *and* DB *is a set of sentences. The query means, what is the probability of* b *given* a *with respect to* DB.

Definition 2.13. *Let* DB *be a set of sentences and let* $QP(b|a)$ *be a probabilistic query. Then, the answer* δ *to the query is*

$$\delta := me[DB](b|a) = \frac{me[DB](a \cap b)}{me[DB](a)}$$

if $DB \, \|\sim^{me} P(a)(0,1]$. *Otherwise,* $\delta := -1$ *means that the set* $DB \cup \{P(a)(0,1]\}$ *is inconsistent.*

The next section shows how an ontology can be transformed into a set of sentences, and how the semantic matching problem can be expressed with probabilistic queries.

2.4.3   Probabilistic Matching

Let $G = (V, \{E_O \cup E_E\})$ be a directed weighted graph as it was defined in Sect. 2.3. We construct a DB from G in the following way:

- We assign a new set (event) a_v to each node v of G which contains the applicants who possess the skill v.
- Next, for each ontology edge $(v_i, v_j) \in E_O$ we add a new sentence s_{ij} to the DB in the form of $P(a_{v_j}|a_{v_i})[1,1]$. The sentence means if an applicant possesses the skill v_i (is an element of a_{v_i}), then the applicant possesses v_j (is an element of a_{v_j}) as well.
- Then, for every extra edge $(v_i, v_j) \in E_E$, we also add a new statement in the form of $P(a_{v_j}|a_{v_i}) = [l, u]$. The weight of an edge can be handled in two different ways. In the first approach, let the lower bound of the interval l be equal to the weight of the edge $w(v_i, v_j)$ and let the upper bound of the interval u be equal to 1. In the second approach, let $l = u = w(v_i, v_j)$. The latter is the stricter approach as it adds constraints to the upper bounds as well.

An application $A = \{v_1, \ldots, v_n\}$ is translated into the event $a = a_{v_1} \cap \cdots \cap a_{v_n}$. The conjunction means that the applicant possesses all the skills $v_1, \ldots, v_n$ at the same time. A job offer $O = \{v_1, \ldots, v_n\}$ is translated into $o = a_{v_1} \cap \ldots \cap a_{v_n}$. It represents the skills that required for the job. The matching value of an job offer O and an application A is the result of the probabilistic query $QP(o|a)$:

$$match(O, A) = QP_{DB}(o|a). \qquad (3)$$

The formula gives the probability of that the applicant possesses the skills that required for the job (supposed that the constructed DB is consistent).

Example 2.5 shows a part of the transformed ontology from Fig. 1 and the matching value of the $Offer_1$ and $Application_1$.

Example 2.5 (Probabilistic matching).

$DB = \{$

   # ontology edges   extra edges $(l = w, u = 1.0)$

   $(it(R)|ide(R))[1, 1],$ $(eclipse(R)|java(R))[0.6, 1.0],$
   $(ide(R)|eclipse(R))[1, 1],$ $(nb(R)|java(R))[0.7, 1.0],$
   $\vdots$ $\vdots$
   $(ml(R)|xml(R))[1, 1],$ $(xml(R)|html(R))[0.7, 1.0]$ $\}$

$Q_{O_1, A_1} = QP(java(a) \wedge nb(a) \wedge xml(a)|java(a) \wedge php(a) \wedge eclipse(a)). \; \delta = 0.51$

The next algorithm shows how the probabilistic model and the probabilistic matching can be used to find the most suitable applicant for a job. It works similarly to the *MaximalLengthMatching* algorithm, but it construct a DB from G first (line 1). Then, instead of extending the offer and the applications, it translates them to probabilistic sentences (line 2, 3, 5) as described above. Next, the algorithm computes the matching values by solving the corresponding probabilistic queries (line 7), and it stores the most suitable application in A_m (line 8).

Algorithm 2. ProbabilisticMatching

Input: a graph $G = (V, E = \{E_O \cup E_E\})$, a job offer $O \subseteq V$, and a set of applications
$\boldsymbol{A} = \{A_1, A_2, ..., A_n\}$.

Output: the most suitable application A_m for the job

1: $DB \leftarrow constructDBFrom(G)$
2: $O' \leftarrow translate(O)$
3: $A_m \leftarrow \boldsymbol{A}[0], max \leftarrow translate(\boldsymbol{A}[0])$
4: **for** $i = 1 \rightarrow n$ **do**
5:      $A' \leftarrow translate(\boldsymbol{A}[i])$
6:      $Q_{O',A'} = \exists(\phi_{O'}(a')|\phi_{A'})[X, Y]$
7:      **if** $solve(Q_{O',A'}, P) > solve(Q_{O',max}, P)$ **then**
8:          $A_m \leftarrow \boldsymbol{A}[i], max \leftarrow A'$
9:      **end if**
10: **end for**
11: **return** A_m

---

3   Comparison

We compared the presented methods on the job offer and applications from
Example 2.1. The skills, their hierarchy, and the conditional dependencies are
shown in Fig. 1. The matching values of the applications to the job can be seen
in the value columns and the order of the applications based on the different
methods can be seen in the order columns in Table 1, where PM, UO, ML, PrM
denote the Perfect Matching, the Matching using ontology edges, the Maximal
Length Matching and the Probabilistic Matching, respectively. The probabilistic
matching values were computed with SPIRIT [23] and PIT [24].

We investigated the matching values of the different algorithms from multiple
aspects see [30–32]. We examined whether there is some kind of regularity among
the values of the algorithms. We compared how the methods sort the applicants,
and how the methods distinguish the applicants from each other. We tried to
abstract from the concrete examples and to describe general observation about
the algorithms.

Table 1. Comparison of the different matching values on the offer {*Java, Netbeans, XML*} and the applications from Example 2.1

Application	PM		UO		ML		PrM_{E_O}		$PrM_{l=w,u=1}$		$PrM_{l=u=w}$	
	value	order	value	order	value	order	value	order	value	order	value	order
A_1	0.33	3,4	0.63	3,4	0.93	2	0.20	4	0.51	2	0.51	3
A_2	0.66	1,2	0.88	1	0.96	1	0.50	1	0.70	1	0.70	1
A_3	0.33	3,4	0.63	3,4	0.68	4	0.22	3	0.32	4	0.14	4
A_4	0.66	1,2	0.75	2	0.75	3	0.43	2	0.49	3	0.54	2

3.1 Perfect Matching vs. Matching Using Ontology Edges

As we saw, the Perfect Matching method assigns the same values to too many applicants as it defines too strictly the matching between an offer and an application. In Sect. 2.2 we introduced a specialization relation among the skills which forms a hierarchy from the skills. We can use that additional knowledge to find the most suitable applicant for a job. We extended the job offer and the applications using the edges of the ontology, and then we applied the perfect matching on the extended sets. This method was called the matching using ontology edges. If only the applications were extended, the matching values on the extended sets would always be greater or equal than on the original sets. Formally, let O an job offer and let A an application, then

$$PM(O, A) \leq UO_{apps}(O, A), \tag{4}$$

where UO_{apps} means that only the applications are extended. It is because the application appears only in the numerator in Eq. 1. However, if the job offer is also extended, the inequality above generally does not hold as the offer appears in the denominator as well. For this reason, the two methods can give different order between the applicants. Note, that the extension of the job offer is reasonable when the skills in the job offer are more specific as in the applications. In this case, we would not get different result from the result of the perfect matching if only the applications were extended.

3.2 Matching Using Ontology Edges vs. Maximal Length Matching

The maximal length matching is a generalization of the matching that uses the ontology edges to extend the applications and the job offer. It also extends the original sets, however, it takes into account the so called extra edges as well. The extra edges could form cycles in the hierarchy graph but it does not affect the computation. Note, that the following connection immediately follows from the definitions:

$$UO(O, A) \leq ML(O, A). \tag{5}$$

Furthermore, if no extra edge is given, then $UO = ML$. However, the order of the applicants can be changed using the two methods as it can be seen in Table 1. For example, A_1 has the lowest matching value in UO, but in ML it has the second greatest value because of the $(Java, Netbeans)$ and $(PHP, HTML, XML)$ paths.

3.3 Probabilistic Matchings

The PrM method uses a totally different strategy to compute the matching value. We tried three different versions; the results are shown in Table 1. In of PrM_{E_O}, only the ontology edges were translated into probabilistic sentences while in the other two cases the extra edges were also translated; in $PrM_{l=w,u=1}$, the lower bound of a sentence was equal to the weight of the edge that the constraint was generated from and the upper bound of the sentence was 1; and in $PrM_{l=u=w}$, both the lower and the upper bound of a sentence were also equal to the weight of the edge.

Generally, the matching values of the three versions are not comparable because of the selection of the underlying probabilistic model. Each algorithm selects a probabilistic model that satisfies the set of sentences generated from the edges, and that has the maximum entropy value among such models. However, the sets of the sentences of the three versions are different from each other. I.e., $PrM_{l=w,u=1}$ added extra constraints to PrM_{E_O} that were generated from the extra edges, and $PrM_{l=u=w}$ added further constraints that came from the limitation of the upper bounds as well. As expected, the entropy of the maximum entropy model of the first version was the greatest (81.68), the entropy of the second model was the next (16.48), and the entropy of third model was the lowest (16.37).

The different probabilistic models give different matching values and it is not guaranteed that the models preserve the order of the applicants. For example, the applicant A_1 is the forth best in PrM_{E_O}, is the second in $PrM_{l=w,u=1}$ and is the third in $PrM_{l=u=w}$ in Table 1.

However, adding extra constraints results in that the algorithms assign the same value to fewer applicants, therefore they distinguish them from each other. It is because while we have no reason to expect one event rather than another, all the possible events should be assigned the same probability as the principle of indifference states.

3.4   Matching Using Ontology Edges vs Probabilistic Matching

When the skills form a hierarchy and no extra edges are given, then the matching using ontology edges and the probabilistic matching PrM_{E_O} can be used.

Unfortunately, there is no connection between the orders of the applicants that the two methods give as Table 1 shows. Although the table suggests that the values of ML are always greater than the values of PrM_{E_O}, but it is generally not true. In that example, all the required skills in the offer and the skills in the applications were selected from the bottom of the hierarchy, therefore the extensions covered large parts of the ontology and the intersections contained many elements. Table 2 shows how the same applications match to another offer which is $Offer_2 = \{IDE, OOP, XML\}$. This offer contains skills from the inner nodes of the ontology as well. It can be seen in Table 2 that all the probabilistic matching methods gave higher values for A_3 and A_4 than the ML method.

3.5   Maximal Length Matching vs. Probabilistic Matching

When the skills form a hierarchy and there are extra edges given as well, then the maximal length matching and the probabilistic matchings $PrM_{l=w,u=1}$ and $PrM_{l=u=w}$ can be used. As we saw the maximal length matching is a generalization of the matching using ontology edges which uses the extra edges too.

Table 2. Comparison of the different matching values on the offer $\{IDE, OOP, XML\}$ and the applications from Example 2.1

Application	PM		UO		ML		PrM_{E_O}		$PrM_{l=w,u=1}$		$PrM_{l=u=w}$	
	value	order	value	order	value	order	value	order	value	order	value	order
A_1	0	3,4	0.67	4	0.95	1,2	0.40	4	0.70	3,4	0.70	3,4
A_2	0	3,4	0.83	1,2,3	0.95	1,2	0.50	3	0.70	3,4	0.70	3,4
A_3	0.33	1,2	0.83	1,2,3	0.87	3	0.89	1	0.92	1	0.85	2
A_4	0.33	1,2	0.83	1,2,3	0.83	4	0.87	2	0.86	2	0.88	1

The $PrM_{l=w,u=1}$ and $PrM_{l=u=w}$ methods also use the extra edges in the transformations, and $PrM_{l=u=w}$ generates additional constraints to the linear optimization problem of $PrM_{l=w,u=1}$ that comes from the limitation of the upper bounds.

As it can be seen in Table 1 and in Table 2, these methods can give totally different orders, and the matching values are incomparable. However, our experiments suggests that the probabilistic matching algorithms can distinguish more applicants from each other than the ML method.

4   Related Work

Semantic matchmaking has become a widely investigated topic recently, due to broad applicability in todays competitive business environment. Its origins go back to Vague query answering, proposed by Motro [19] that was an initial effort to overcome limitations of relational databases, using weights attributed to several search variables. More recent approaches along these lines aim at extending SQL with "preference" clauses (Kießling [12]).

Our main focus in the present paper is facilitating the management of available human resources' competencies. Fully or partially automated techniques were developed (see Colucci et al. [4], Bizer et al. [2], Malinowski et al. [18]).

Several matchmaking approaches exist in the literature that could be applied for matching job applications for job offers. Text based information retrieval techniques such as database querying and similarity between weighted vectors of terms were used by Veit et al. [26]. Ontology based skill profile matching was considered in many papers. Lau and Sure [13] propose an ontology based skill management system for eliciting employee skills and searching for experts within an insurance company. Liu and Dew [15] gives a system that integrates the accuracy of concept search with the flexibility of keyword search to match expertise within academia. Colucci et al. [4] proposes a semantic based approach to the problem of skills finding in an ontology based framework. They use description logic inferences to handle background knowledge and deal with incomplete information. They use profile descriptions sharing a common ontology, our approach is based on this, as well. A fundamental difference between the aforementioned works and our paper is that they facilitate search for matching profiles, while we focus on ranking already given applications.

Di Noia et al. [6] places matchmaking on a consistent theoretical foundation using description logic. They define matchmaking as "an information retrieval task whereby queries (also known as demands) and resources (also known as supplies) are expressed using semi-structured data in the form of advertisements, and task results are ordered (ranked) lists of those resources best fulfilling the query." They also introduce match types and rank individual profiles using penalty function. However, they do not apply the filter approach used in our paper.

Fuzzy techniques are introduced in Ragone et al. [22] where they consider a peer-to-peer e-marketplace of used cars. Also a form of logic programming is applied using fuzzy extension of Datalog. Papers [6,22] contain extensive lists of further references concerning practical algorithms, related areas of multiobjective decision making, logic programming, description logic, query reformulation and top-k query answering.

The manuscript [29] uses exact match in ontologies extended with Euclidean-like distance or similarity measure. They apply different levels of given skills, furthermore the job offer may contain "nice-to-have requirements." In our paper we do not apply different grade levels of skills, instead, we consider them separate skills, so we "blow up" the ontology.

Finally, the use of filters in the ontology hierarchy lattice was initiated by Popov et al. [21].

5   Conclusion

In this paper we described the problem of the semantic matching by examples from the field of human resources management, namely matching job offers with applications. However, the presented methods can be used for other fields as well. We investigated the problem from different aspects with different models.

First, we represented the offers and the applications with set of skills, and we introduced the perfect matching. It is a naive approach that computes the matching value of a job offer and an application based on the intersection of two sets. However, it could not be able to sufficiently distinguish the applicants from each other because of its simplicity.

Next, we defined a specialization relation on the skills and built an ontology over them which was represented with directed graph. Then, we presented a method that can use this additional knowledge to find the best applicant for a job. It extends the set of skills of both the job offer and the applications with the more general skills that are available from the original sets on the edges of the ontology. And on the extended sets the perfect matching is already applicable. Beside the ontology edges, we introduced extra edges as well that express conditional dependencies between skills. And then, we generalized the extension of the sets of skills of the applications to use the extra edges too.

In Sect. 2.4, we presented the probabilistic models and we showed how the ontology edges and extra edges can be translated into probabilistic sentences, and how the problem of the semantic matching can be translated into probabilistic queries. Two different approaches were discussed as the edges can be handled. We used the maximum entropy model to answer the probabilistic queries as it adds the lowest amount of additional information to the system when the given information is incomplete.

Finally, we compared the presented methods from various aspects. We investigated whether there is any connection of the matching values that give the methods. Furthermore, we examined how the methods sort the applicants and how they can distinguish the applicants form each other. We showed that the $PM(O, A) \leq UO_{apps}(O, A)$ and the $UO(O, A) \leq ML(O, A)$ connections hold between the matching values of the perfect matching PM the matching using ontology edges UO and the maximal length matching for arbitrary job offer O and application A. However, our results revealed that the algorithms can give totally different order among the same applicants matching to the same job offer. Therefore, it highly depends on the field which algorithm gives the most suitable order.

References

1. Baader, F., Nutt, W.: Basic description logic. In: Description Logic Handbook, pp. 43–95 (2003)
2. Bizer, C., Heese, R., Mochol, M., Oldakowski, R., Tolksdorf, R., Eckstein, R.: The impact of semantic web technologies on job recruitment processes. In: Proceedings of the 7th International Conference Wirtschaftsinformatik (2005)
3. Brickley, D., Ramanathan, V.G.: RDF Schema 1.1: W3C Recommendation 25, February 2014
4. Colucci, S., Di Noia, T., Di Sciascio, E., Donini, F., Mongiello, M.: Concept abduction and contraction in description logics. In: Proceedings of the 16th International Workshop on Description Logics (DL 2003). CEUR Workshop Proceedings, vol. 81 (2003)
5. Dijkstra, E.W.: A note on two problems in connexion with graphs. Numer. Math. 1(1), 269–271 (1959)
6. Di Noia, T., Di Sciascio, E., Donini, F.M.: Semantic matchmaking as nonmonotonic reasoning: a description logic approach. J. Artif. Intell. Res. 29, 269–307 (2007)
7. European Dictionary of Skills and Competences. http://www.disco-tools.eu
8. Hájek, P.: Mathematics of Fuzzy Logic. Kluwer Academic Publishers, Dordrecht (1998)
9. International Standard Classification of Education. http://www.uis.unesco.org/Education/Pages/international-standard-classification-of-education.aspx
10. International Standard Classification of Occupations (2008)
11. Kern-Isberner, G., Lukasiewicz, T.: Combining probabilistic logic programming with the power of maximum entropy. Artif. Intell. 157(1), 139–202 (2004)
12. Kießling, W.: Foundations of preferences in database systems. In: Proceedings of the 28th International Conference on Very Large Data Bases, pp. 311–322 (2002)
13. Lau, T., Sure, Y.: Introducing ontology-based skills management at a large insurance company. In: Proceedings of the Modellierung, pp. 123–134 (2002)
14. Levandowsky, M., Winter, D.: Distance between sets. Nature 234(5), 34–35 (1971)

15. Liu, P., Dew, P.: Using semantic web technologies to improve expertise matching within academia. In: Proceedings of I-KNOW 2004, pp. 370–378 (2004)
16. Liu, L., Li, K.: Fuzzy filters of BL-algebras. Inf. Sci. **173**(1), 141–154 (2005)
17. Lloyd, J.W.: Foundations of Logic Programming. Springer, Heidelberg (2012)
18. Malinowski, E., Zimányi, E.: Hierarchies in a multidimensional model: from conceptual modeling to logical representation. Data Knowl. Eng. **59**(2), 348–377 (2006)
19. Motro, A.: VAGUE: a user interface to relational databases that permits vague queries. ACM Trans. Off. Inf. Syst. **6**(3), 187–214 (1988)
20. Pedrycz, W., Gomide, F.: An Introduction to Fuzzy Sets: Analysis and Design. MIT Press, Cambridge (1998)
21. Popov, N., Jebelean, T.: Semantic Matching for Job Search Engines: A Logical Approach. Technical report 13–02, Research Institute for Symbolic Computation, JKU Linz (2013)
22. Ragone, A., Straccia, U., Di Noia, T., Di Sciascio, T., Donini, F.M.: Fuzzy matchmaking in e-marketplaces of peer entities using Datalog. Fuzzy Sets Syst. **160**(2), 251–268 (2009)
23. Rödder, W., Meyer, C.-H.: Coherent knowledge processing at maximum entropy by SPIRIT. In: Proceedings of the 12th International Conference on Uncertainty in Artificial Intelligence. Morgan Kaufmann Publishers Inc. (1996)
24. Schramm, M., Ertel, W.: Reasoning with probabilities, maximum entropy: the system PIT and its application in LEXMED. In: Inderfurth, K., Schwödiauer, G., Domschke, W., Juhnke, F., Kleinschmidt, P., Wäscher, G. (eds.) Operations Research Proceedings, vol. 1999, pp. 274–280. Springer, Heidelberg (2000)
25. Schramm, M., Greiner, M.: Non-Monotonic Reasoning on Probability Models: Indifference, Independence. Inst. für Informatik (1995)
26. Veit, D., Müller, J., Schneider, M., Fiehn, B.: Matchmaking for autonomous agents in electronic marketplaces. In: Proceedings of the International Conference on Autonomous Agents 2001, pp. 65–66 (2001)
27. Wygralak, M.: Cardinalities of Fuzzy Sets. Springer, Heidelberg (2003)
28. Zadeh, L.A.: Fuzzy sets. Inf. Control **8**(3), 338–353 (1965)
29. https://users.soe.ucsc.edu/~darrell/tmp/search_subm06.pdf
30. http://www.renyi.hu/~sali/sms/ide_oop_xml.xlsx
31. http://www.renyi.hu/~sali/sms/java_nb_xml.xlsx
32. http://www.renyi.hu/~sali/sms/queries.xlsx

The Challenge of Optional Matching in SPARQL

Shqiponja Ahmetaj, Wolfgang Fischl, Markus Kröll, Reinhard Pichler[✉],
Mantas Šimkus, and Sebastian Skritek

Database and Artificial Intelligence Group, Faculty of Informatics,
TU Wien, Vienna, Austria
{ahmetaj,wfischl,kroell,pichler,simkus,skritek}@dbai.tuwien.ac.at

Abstract. Conjunctive queries are arguably the most widely used querying mechanism in practice and the most intensively studied one in database theory. Answering a conjunctive query (CQ) comes down to matching all atoms of the CQ simultaneously into the database. As a consequence, a CQ fails to provide any answer if the pattern described by the query does not exactly match the data. CQs might thus be too restrictive as a querying mechanism for data on the web, which is considered as inherently incomplete. The semantic web query language SPARQL therefore contains the OPTIONAL operator as a crucial feature. It allows the user to formulate queries which try to match parts of the query over the data if available, but do not destroy answers of the remaining query otherwise. In this article, we have a closer look at this optional matching feature of SPARQL. More specifically, we will survey several results which have recently been obtained for an interesting fragment of SPARQL – the so-called well-designed SPARQL graph patterns.

1 Introduction

Conjunctive queries (or, equivalently, SELECT-FROM-WHERE queries in SQL) are arguably the most widely used querying mechanism in practice and the most intensively studied one in database theory. Answering a conjunctive query (CQ) comes down to matching all atoms of the CQ simultaneously into the database. As a consequence, a CQ fails to provide any answer if the pattern described by the query does not exactly match the data. CQs might thus be too restrictive as a querying mechanism for data on the web, which is considered as inherently incomplete. The semantic web query language SPARQL therefore contains the OPTIONAL operator (abbreviated as OPT henceforth) as a crucial feature. It allows the user to formulate queries which try to match parts of the query over the data if available, but do not destroy answers of the remaining query otherwise. It thus corresponds to the left outer join in the relational algebra. The following example from [24] presents a simple SPARQL query using this feature.

Example 1. Consider the following SPARQL query Q which is posed over a database that stores information about movies[1]:

---

[1] We use here the algebraic-style notation from [29] rather than the official SPARQL syntax of [33]. In particular, we explicitly use an AND operator (rather than comma-separated lists) to denote conjunctions.

© Springer International Publishing Switzerland 2016
M. Gyssens and G. Simari (Eds.): FoIKS 2016, LNCS 9616, pp. 169–190, 2016.
DOI: 10.1007/978-3-319-30024-5_10

$$Q = \Big(\big((?x, \texttt{directed\_by}, ?y) \, \text{AND} \, (?x, \texttt{released}, "\texttt{before\_1980}") \big)$$

$$\text{OPT} \, (?x, \texttt{oscars\_won}, ?z) \Big) \, \text{OPT} \, (?y, \texttt{first\_movie}, ?z').$$

This query retrieves all pairs (m, d) such that movie m is directed by d and released before 1980. This is specified by the pattern $(?x, \texttt{directed\_by}, ?y)$ AND $(?x, \texttt{released}, "\text{before\_1980}")$. Furthermore, whenever possible, this query also retrieves (one or both of) the following pieces of data: the number n of Academy Awards won by movie m and the first movie m' directed by d. In other words, in addition to (m, d) we also retrieve n and/or m' if the information is available in the database. This is specified by the triples $(?x, \texttt{oscars\_won}, ?x)$ and $(?y, \texttt{first\_movie}, ?z')$ following the respective OPT operators.   ◇

Apart from AND and OPT used in Example 1, SPARQL also provides the operators UNION and FILTER. SPARQL 1.1 [18] introduces many further operators, which we ignore for the time being. *Projection* is realized by wrapping a SPARQL graph pattern into a SELECT statement where we may explicitly specify the variables of interest. For instance, in Example 1, we could wrap the query Q into a statement of the form SELECT $?x$, $?z$ WHERE $\{Q\}$ to project out the information on directors and their first movie.

As far as the expressive power of SPARQL is concerned, it was shown in [3,32] that SPARQL is relational complete. Not surprisingly, the SPARQL query evaluation problem (i.e., given an RDF graph G, a SPARQL query Q, and a set μ of variable bindings, check if μ is a solution) is PSPACE-complete (combined complexity) [29,35]. The OPT operator was identified as one of the main sources of complexity. Indeed, it was shown in [35] that the PSPACE-completeness of SPARQL query evaluation holds even if we restrict SPARQL to the AND and OPT operator. The reason for this high complexity is the unrestricted use of variables inside and outside an OPT expression. Therefore, in [29], the class of *well-designed* SPARQL graph patterns was introduced. The restriction imposed there is that if a variable occurs on the right-hand side of an OPT expression and anywhere else in the SPARQL graph pattern, then it must also occur on the left-hand side of the OPT expression. It was shown that the complexity of the evaluation problem for the well-designed fragment drops to coNP-completeness [29].

In [29], many further interesting properties of well-designed SPARQL graph patterns were shown. At this point, we mention only one, namely the efficient transformation into so-called OPT *normal form*: a SPARQL graph pattern using AND and OPT operators is in this normal form, if the OPT operator does not occur in the scope of an AND operator. It was shown in [29] that this can always be achieved efficiently by exploiting the equivalence $(P_1 \text{ AND } (P_2 \text{ OPT } P_3)) \equiv ((P_1 \text{ AND } P_2) \text{ OPT } P_3)$, which holds for well-designed SPARQL graph patterns. Moreover, such graph patterns allow for a natural tree representation, formalized by so-called *well-designed pattern trees* (wdPTs, for short) in [25].

Intuitively, the nodes in a wdPT correspond to CQs while the tree structure represents the optional extensions. For instance, the wdPT corresponding to the

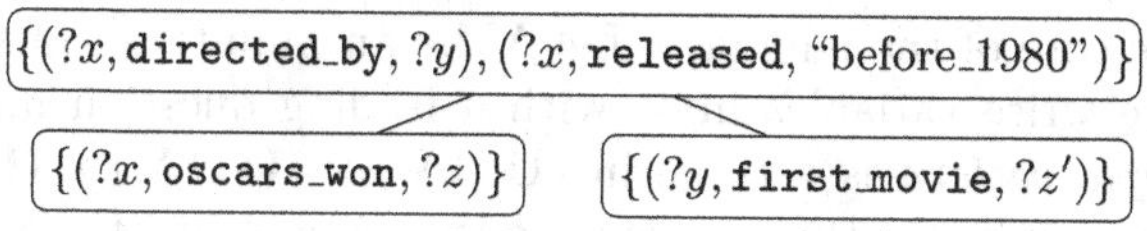

$$\{(?x, \texttt{directed\_by}, ?y), (?x, \texttt{released}, \text{``before\_1980''})\}$$

$$\{(?x, \texttt{oscars\_won}, ?z)\} \qquad \{(?y, \texttt{first\_movie}, ?z')\}$$

Fig. 1. The wdPT $\mathcal{T}$ representing the query Q from Example 1.

query in Example 1 is displayed in Fig. 1. As with SPARQL graph patterns, we can add projection to wdPTs by indicating the distinguished variables to which the result shall be projected. Well-designed pattern trees then yield a natural extension of conjunctive queries. Indeed, a CQ corresponds to a wdPT consisting of the root node only. It turns out that the extension of CQs to wdPTs can have a significant effect on various computational tasks. For instance, while query evaluation and query containment are both NP-complete for CQs, these tasks become Σ_2^P-complete [25] or even undecidable [30], respectively, for wdPTs with projection. Actually, it is even questionable if the definition of containment via set inclusion is appropriate in case of optional matching. Also the semantics definition of answering CQs in the presence of ontologies requires rethinking for wdPTs [2].

In this article, we survey these and several further results which have recently been obtained for well-designed SPARQL graph patterns or, equivalently, for well-designed pattern trees. We shall thus mainly concentrate on algorithms and complexity results obtained for the most fundamental computational problems in this area, namely *query evaluation* (see Sect. 3) and basic *static query analysis* tasks such as testing containment and equivalence of two queries (see Sect. 4). Finally, we shall also recall results on the evaluation of wdPTs in the presence of ontologies from the DL-Lite family [10] and briefly discuss some unintuitive behavior of SPARQL entailment regimes according to the the recently released W3C recommendation [14] (see Sect. 5).

2   RDF, SPARQL, and Pattern Trees

RDF. The data model designed for the *Semantic Web* is the Resource Description Framework (RDF) [13]. We focus here on ground RDF graphs and assume them to be composed of URIs only. Formally, let **U** be an infinite set of URIs. An *RDF triple* (s, p, o) is a tuple in $\mathbf{U} \times \mathbf{U} \times \mathbf{U}$, whose components are referred to as subject, predicate, and object, respectively. An *RDF graph* is a finite set of RDF triples. Note that a set of triples (s, p, o) can be seen as an edge-labeled graph, where s and o denote vertices and p denotes an edge label. The active domain $\mathsf{dom}(G) \subseteq \mathbf{U}$ of an RDF graph G is the set of URIs actually appearing in G.

SPARQL Syntax. SPARQL [33], which was later extended to SPARQL 1.1 [18], is the standard query language for RDF data. Following the presentation in [29], we next recall the formalization of its graph pattern matching facility, which forms the core of the language. Let $\mathbf{V}$ be an infinite set of variables with $\mathbf{U} \cap \mathbf{V} = \emptyset$. We write variables in $\mathbf{V}$ with a leading question mark, as in $?x$. A *SPARQL triple pattern* is a tuple in $(\mathbf{U} \cup \mathbf{V}) \times (\mathbf{U} \cup \mathbf{V}) \times (\mathbf{U} \cup \mathbf{V})$. More complex patterns studied in this article are constructed using the operators AND, OPT, and UNION. We omit here further operators specified by [33] and [18], including the FILTER operator. Formally, *SPARQL graph patterns* (or simply *graph patterns*, for short) are thus recursively defined as follows. (1) a triple pattern is a graph pattern, and (2) if P_1 and P_2 are graph patterns, then $(P_1 \circ P_2)$ is a graph pattern for $\circ \in \{\text{AND}, \text{OPT}, \text{UNION}\}$. Let P be a graph pattern or a set of graph patterns; then we write $\text{vars}(P)$ to denote the set of variables occurring in P.

SPARQL Semantics. For defining the semantics of SPARQL graph patterns, we again follow closely the definitions proposed in [29]. A *mapping* is a function $\mu \colon A \to \mathbf{U}$ for some $A \subset \mathbf{V}$. For a triple pattern t with $\text{vars}(t) \subseteq \text{dom}(\mu)$, we write $\mu(t)$ to denote the triple obtained by replacing the variables in t according to μ. Two mappings μ_1 and μ_2 are called compatible (written $\mu_1 \sim \mu_2$) if $\mu_1(?x) = \mu_2(?x)$ for all $?x \in \text{dom}(\mu_1) \cap \text{dom}(\mu_2)$. A mapping μ_1 is subsumed by μ_2 (written $\mu_1 \sqsubseteq \mu_2$) if $\mu_1 \sim \mu_2$ and $\text{dom}(\mu_1) \subseteq \text{dom}(\mu_2)$. In this case, we also say that μ_2 is an extension of μ_1. Subsumption is naturally extended to sets of mappings, e.g., $\mu \sqsubseteq M$ for a set M of mappings, if $\mu \sqsubseteq \mu'$ for some $\mu' \in M$.

We formalize the evaluation of graph patterns over an RDF graph G as a function $[\![\cdot]\!]_G$ that, given a graph pattern P, returns a set of mappings (i.e., the "solutions" or "answers" of P over G). It is defined recursively as follows [29]:

1. $[\![t]\!]_G = \{\mu \mid \text{dom}(\mu) = \text{vars}(t) \text{ and } \mu(t) \in G\}$ for a triple pattern t.
2. $[\![P_1 \text{ AND } P_2]\!]_G = \{\mu_1 \cup \mu_2 \mid \mu_1 \in [\![P_1]\!]_G, \mu_2 \in [\![P_2]\!]_G, \text{ and } \mu_1 \sim \mu_2\}$.
3. $[\![P_1 \text{ OPT } P_2]\!]_G = [\![P_1 \text{ AND } P_2]\!]_G \cup \{\mu_1 \in [\![P_1]\!]_G \mid \forall \mu_2 \in [\![P_2]\!]_G \colon \mu_1 \not\sim \mu_2\}$.
4. $[\![P_1 \text{ UNION } P_2]\!]_G = [\![P_1]\!]_G \cup [\![P_2]\!]_G$.

Note that, as in [29], we assume set semantics, while the W3C Recommendation specifies bag-semantics [33].

Well-Designed SPARQL. In [29], the authors identify several classes of graph patterns. One of these classes, which is at the heart of this survey, is formed by the so-called *well-designed* SPARQL graph patterns. A graph pattern P built only from AND and OPT is *well-designed* if there does not exist a subpattern $P' = (P_1 \text{ OPT } P_2)$ of P and a variable $?x$, such that $?x$ occurs in P_2 and in P outside P', but not in P_1. A graph pattern $P = P_1 \text{ UNION} \ldots \text{UNION } P_n$ is well-designed if each subpattern P_i is UNION-free and well-designed. Thus, as in [29], when including the UNION operator, we only allow it to appear outside the scope of other operators.

Well-Designed Pattern Trees. We have already mentioned above the OPT *normal form* [29], which forbids occurrences of the OPT operator in the scope of an AND operator. Well-designed graph patterns in OPT normal form allow for a natural tree representation, formalized by so-called *well-designed pattern trees (wdPTs)* in [25]. A wdPT $\mathcal{T}$ is a pair $(T, \mathcal{P})$ where $T = (V, E, r)$

is a rooted, unordered, tree and $\mathcal{P} = (P_n)_{n \in V}$ is a labeling of the nodes in V, s.t. P_n is a non-empty set of triple patterns (or, equivalently, a conjunction of triple patterns) for every $n \in V$. The *well-designedness* condition requires that, for every variable $?x$ occurring in $\mathcal{T}$, the nodes $\{n \in V \mid ?x$ occurs in $P_n\}$ must induce a connected subgraph of $\mathcal{T}$. For instance, the graph pattern Q in Example 1 is in OPT normal form. Its corresponding pattern tree $\mathcal{T}$ is shown in Fig. 1.

Graph patterns in OPT normal form consist of conjunctive parts (represented by the nodes of the pattern tree) that are located in a structure of nested OPT operators (modeled by the tree-structure). Note that the order of child nodes in such a tree does not matter. This is due to equivalence $((P_1 \text{ OPT } P_2) \text{ OPT } P_3) \equiv ((P_1 \text{ OPT } P_3) \text{ OPT } P_2)$, which holds for well-designed graph patterns [29]. This is why wdPTs are defined as *unordered* trees.

Components of a Pattern Tree. Let $\mathcal{T} = ((V, E, r), \mathcal{P})$ be a wdPT. We call a wdPT $\mathcal{T}' = ((V', E', r'), (P_n)_{n \in V'})$ a *subtree* of $\mathcal{T}$ if (V', E', r') is a subtree of $\mathcal{T}$. $\mathcal{T}'$ is a *subtree of $\mathcal{T}$ containing the root* if $r' = r$. Throughout this article, unless explicitly specified otherwise, we always consider subtrees containing the root, and will thus refer to them simply as "subtrees", omitting the phrase "containing the root". An *extension* $\hat{\mathcal{T}}'$ of a subtree $\mathcal{T}'$ of $\mathcal{T}$ is a subtree $\hat{\mathcal{T}}'$ of $\mathcal{T}$, s.t. $\mathcal{T}'$ is in turn a subtree of $\hat{\mathcal{T}}'$. A subtree or extension is *proper* if some node of the bigger tree is missing in the smaller tree.

Given a wdPT $\mathcal{T} = ((V, E, r), \mathcal{P})$, we write $V(\mathcal{T})$ to denote the set V of vertices. We sometimes refer to the set P_n of triple patterns at vertex $n \in V$ as $pat(n)$ and we denote by $pat(\mathcal{T})$ the set $\bigcup_{n \in V(\mathcal{T})} P_n$ of triple patterns occurring in $\mathcal{T}$. We write $\text{vars}(\mathcal{T})$ (resp. $\text{vars}(n)$) as an abbreviation for $\text{vars}(pat(\mathcal{T}))$ (resp. $\text{vars}(pat(n))$). These notions extend naturally to sets of nodes. For nodes $n, \hat{n} \in V(\mathcal{T})$, s.t. $\hat{n}$ is the parent of n, let $\text{newvars}(n) = \text{vars}(n) \setminus \text{vars}(\hat{n})$. A wdPT $\mathcal{T}$ is said to be in *NR normal form*, if $\text{newvars}(n) \neq \emptyset$ for every $n \in V(\mathcal{T})$ [25]. It was shown in [25], that every wdPT can be transformed efficiently into an equivalent wdPT in NR normal form. We therefore assume w.l.o.g. that all wdPTs dealt with here are in NR normal form.

Semantics of Pattern Trees. Analogously to graph patterns, the result of evaluating a wdPT $\mathcal{T}$ over some RDF graph G is denoted by $[\![\mathcal{T}]\!]_G$. In [25], the set $[\![\mathcal{T}]\!]_G$ of solutions was defined via a translation to graph patterns. However, for wdPTs in NR normal form, the set of solutions $[\![\mathcal{T}]\!]_G$ has a nice direct characterization in terms of maximal subtrees of $\mathcal{T}$:

Lemma 1 ([25]). *Let $\mathcal{T}$ be a wdPT in NR normal form and G an RDF graph. Then $\mu \in [\![\mathcal{T}]\!]_G$ iff there exists a subtree $\mathcal{T}'$ of $\mathcal{T}$, s.t. (1) $\text{dom}(\mu) = \text{vars}(\mathcal{T}')$, and (2) $\mathcal{T}'$ is the maximal subtree of $\mathcal{T}$, s.t. $\mu \sqsubseteq [\![pat(\mathcal{T}')]\!]_G$.*

It can be easily checked that $\mathcal{T}'$ is uniquely defined by $\text{dom}(\mu)$. We refer to this tree as $\mathcal{T}_\mu$. We illustrate the evaluation of graph patterns or, equivalently, of wdPTs) by revisiting Example 1.

Example 2. Consider the following RDF graph G:

$$G = \{ (\text{"American\_Graffiti"}, \texttt{directed\_by}, \text{"George\_Lucas"}),$$
$$(\text{"American\_Graffiti"}, \texttt{released}, \text{"before\_1980"}),$$
$$(\text{"Star\_Wars"}, \texttt{directed\_by}, \text{"George\_Lucas"}),$$
$$(\text{"Star\_Wars"}, \texttt{released}, \text{"before\_1980"}),$$
$$(\text{"Star\_Wars"}, \texttt{oscars\_won}, \text{"6"})\}.$$

The evaluation of the query Q from Example 1 (or, equivalently of the wdPT T in Fig. 1) over G, yields the partial mappings μ_1 and μ_2 defined on the variables $?x, ?y, ?t$, and $?z'$ such that: (1) $\mathsf{dom}(\mu_1) = \{?x, ?y\}$ with $\mu_1 = \{?x \leftarrow$ "American_Graffiti", $?y \leftarrow$ "George_Lucas"$\}$ and (2) $\mathsf{dom}(\mu_2) = \{?x, ?y, ?z\}$ with $\mu_2 = \{?x \leftarrow$ "Star_Wars", $?y \leftarrow$ "George_Lucas", $?z \leftarrow$ "6"$\}$.   ◇

Projection. Recall that projection is not considered as part of a graph pattern [33]; instead, it is realized by the SELECT result modifier on top of a graph pattern. For a mapping μ and a set $\mathcal{X}$ of variables, let $\mu_{|\mathcal{X}}$ denote the projection of μ to the variables in $\mathcal{X}$, that is, the mapping μ' defined as $\mathsf{dom}(\mu') := \mathcal{X} \cap \mathsf{dom}(\mu)$ and $\mu'(?x) := \mu(?x)$ for all $?x \in \mathsf{dom}(\mu')$.

It is convenient to denote a graph pattern P or a wdPT T with projection to $\mathcal{X}$ as $(P, \mathcal{X})$ and $(T, \mathcal{X})$, respectively. The evaluation of such a graph pattern or wdPT over an RDF graph G is then defined as $[\![(P, \mathcal{X})]\!]_G = \{\mu_{|\mathcal{X}} \mid \mu \in [\![P]\!]_G\}$ and $[\![(T, \mathcal{X})]\!]_G = \{\mu_{|\mathcal{X}} \mid \mu \in [\![T]\!]_G\}$, respectively. We refer to the pair $(T, \mathcal{X})$ as a *wdPT with projection* or simply a *wdPT*, for short. Moreover, we refer to $\mathsf{vars}(T) \cap \mathcal{X}$ as the free variables ($\mathsf{fvars}(T)$) and to $\mathsf{vars}(T) \setminus \mathsf{fvars}(T)$ as the existential variables in T ($\mathsf{evars}(T)$). Analogously, we write $\mathsf{fvars}(n)$ and $\mathsf{evars}(n)$, respectively, for nodes $n \in V(T)$. Moreover, for $n \in V(T)$, let $\mathsf{newfvars}(n) = \mathsf{newvars}(n) \cap \mathsf{fvars}(n)$. W.l.o.g., we assume that existential variables in wdPTs with projection are always renamed apart, i.e., $\mathsf{evars}(T_1) \cap \mathsf{evars}(T_2) = \emptyset$ for any two distinct wdPTs T_1 and T_2.

A wdPT $(T, \mathcal{X})$ is in NR normal form if T is. For wdPTs with projection, a similar characterization of solutions as Lemma 1 exists.

Lemma 2 ([25]). *Let $(T, \mathcal{X})$ be a wdPT with projection in NR normal form, G an RDF graph and μ a mapping with $\mathsf{dom}(\mu) \subseteq \mathcal{X}$. Then $\mu \in [\![(T, \mathcal{X})]\!]_G$ iff there exists a subtree T' of T, s.t. (1) $\mathsf{dom}(\mu) = \mathsf{fvars}(T')$, and (2) there exists a mapping $\lambda: \mathsf{evars}(T') \to \mathsf{dom}(G)$, s.t. $\mu \cup \lambda \in [\![T]\!]_G$.*

SPARQL allows the use of blank nodes in graph patterns (see [18] for details), which we do not consider here. This is however no restriction, since every well-designed graph pattern with blank nodes is equivalent to a well-designed graph pattern with projection but without blank nodes.

Union. Recall that we allow the UNION operator only to be applied "top-level", i.e., well-designed SPARQL graph patterns involving the UNION operator are of the form $P = P_1$ UNION … UNION P_k, such that each P_i is a UNION-free well-designed graph pattern. Analogously, we consider a set $\{T_1, \ldots, T_k\}$ of wdPTs (i.e., a *well-designed pattern forest (wdPF)*) with the intended meaning that it stands for the union of the wdPTs. All notions introduced for wdPTs extend naturally to wdPFs, e.g., a subtree T' of a wdPF $\mathcal{F}$ is a subtree for some wdPT $T \in \mathcal{F}$. We define the set of solutions of a wdPF $\mathcal{F}$ without projection and of a wdPF $(\mathcal{F}', \mathcal{X})$ with projection over an RDF graph G as $[\![\mathcal{F}]\!]_G := \bigcup_{T \in \mathcal{F}} [\![T]\!]_G$ and $[\![(\mathcal{F}', \mathcal{X})]\!]_G := \bigcup_{(T, \mathcal{X}) \in \mathcal{F}'} [\![(T, \mathcal{X})]\!]_G$, respectively.

3 Query Evaluation

In this section we have a closer look at the evaluation of well-designed SPARQL graph patterns or, equivalently, of wdPTs. To this end, we first revisit the semantics definition of SPARQL graph patterns or wdPTs from Sect. 2. Recall that the semantics $[\![\cdot]\!]_G$ is inductively defined over the structure of SPARQL graph patterns. In terms of wdPTs, a direct implementation of this semantics definition corresponds to a bottom-up traversal of the tree. Clearly, it may thus happen that one computes big intermediate results for some subtree (not containing the root) of the wdPT, which ultimately have to be deleted since these intermediate results cannot be extended to mappings up to the root of the wdPT. For instance, suppose that in Example 2, the graph G is augmented by triples $(d_1, \texttt{first\_movie}, m_1)$, $(d_2, \texttt{first\_movie}, m_2)$, etc. Then the evaluation of the pattern $\{(?y, \texttt{first\_movie}, ?z')\}$ at the right leaf node of T yields the mappings $\nu_1 = \{?y \leftarrow d_1, ?z' \leftarrow m_1\}$, $\nu_2 = \{?y \leftarrow d_2, ?z' \leftarrow m_2\}$, etc. Obviously, none of these mappings can be extended further up to the root node.

In [28] the authors therefore proposed a top-down evaluation method for well-designed SPARQL graph patterns, which avoids the computation of "useless" intermediate results, i.e.: every partial mapping produced by this evaluation method is indeed a solution or can be extended to a solution. Below we illustrate this top-down evaluation for wdPTs [25].

Top-Down Evaluation. Lemma 1 essentially states that the solutions of a wdPT over some graph G are exactly those mappings which map all triples in some subtree T' of T into G, and which cannot be extended to some bigger subtree T'' of T. This characterization inspires the following procedural semantics that is obtained by evaluating the pattern tree via a top-down traversal. Given a label P_n of node n in T and a graph G, we denote by $[\![P_n]\!]_G$ the set of mappings μ that send all triples in P_n into G, i.e., $[\![P_n]\!]_G = \{\mu \mid \mu(t) \in G \text{ for all } t \in P_n\}$.

Definition 1. *Consider an RDF graph G, a wdPT $T = ((V, E, r), \mathcal{P})$ with $\mathcal{P} = (P_n)_{n \in V}$, and a set M of mappings. For $n \in V$, we define the evaluation of T_n (the complete subtree of T rooted at n) given M over G, denoted by $\mathrm{ext}(M, n, G)$ as follows. If n is a leaf, then*

$$\mathrm{ext}(M, n, G) = M \bowtie [\![P_n]\!]_G,$$

and, otherwise, if $n_1, \ldots, n_k$ are the child nodes of n, then

$$\mathrm{ext}(M, n, G) = M_1 \bowtie M_2 \bowtie \cdots \bowtie M_k,$$

where $M_i = (M \bowtie [\![P_n]\!]_G) \bowtie \text{ext}(M \bowtie [\![P_n]\!]_G, n_i, G)$. We define the top-down evaluation of T over G, denoted by $[\![T]\!]_G^{td}$, as

$$[\![T]\!]_G^{td} = \text{ext}(\{\mu_\emptyset\}, r, G),$$

where $\mu_\emptyset$ is the mapping with the empty domain.

The above definition can be also seen in a more procedural way: Given some wdPT with root r and some RDF graph G, first get the set M of all mappings that map P_r into G. For each mapping $\mu \in M$ property (1) of Lemma 1 is satisfied. Now in order to test property (2), it suffices to check for each such mapping μ if it can be extended to some child n of r, i.e. to some mapping $\mu' : \text{vars}(P_r) \cup \text{vars}(P_n) \to \text{dom}(G)$ with $\mu'(P_n) \subseteq G$. If this is possible, replace μ by μ'. Note that μ' again satisfies property (1) of Lemma 1. Hence one way to think of this evaluation method is to maintain a set of partial solutions together with a subtree T' of the input wdPT rooted at r for each of them. In order to determine whether the mapping can be extended, it suffices to check if it can be extended to a child node of the leaf nodes of T'.

The following theorem states that the top-down evaluation defined above coincides with the semantics of pattern trees recalled in Sect. 2.

Theorem 1 ([25]). *Let T be a wdPT and G an RDF graph. Then $[\![T]\!]_G = [\![T]\!]_G^{td}$.*

Complexity of Evaluation Without Projection. We now look at the complexity of the EVALUATION problem of SPARQL graph patterns or, equivalently, of wdPTs. We thus study the following decision problem: Given a wdPT T, an RDF graph G, and a mapping μ, check if μ is a solution. For wdPTs without projection, it was shown in in [29] that this problem is coNP-complete. For our representation of SPARQL graph patterns as wdPTs in NR normal form, a coNP test can work as follows. Let $T = ((V, E, r), \mathcal{P})$ be a wdPT. By using the characterization of the evaluation of wdPTs provided in Lemma 1, in order to check whether μ is a solution of T over G, the coNP-algorithm can first find a subtree T' of T rooted at r s.t. $\text{dom}(\mu) = \text{vars}(T')$. Notice that if this subtree exists, then it is unique (since T is in NR normal form), and thus, this step can be done in polynomial time. Then the algorithm checks that T' is a maximal subtree such that $\mu \sqsubseteq [\![pat(T')]\!]_G$. The latter test requires coNP-power since we have to check that μ cannot be extended to any of the sets of triple patterns at nodes in T, which are "below" the leaf nodes of T'. However, it is sufficient to check this for every child node of T' (i.e., for every child in T of a leaf node of T') individually: if μ can be extended to any child node, this immediately proves that it is not maximal. Note that this simple coNP-algorithm heavily relies on the NR normal form; the coNP-algorithm provided in [29] is considerably more involved.

Now consider the relationship with CQs. Clearly, if all sets of triple patterns are from tractable fragments of CQ evaluation, then the problem of checking whether μ is a solution of T over G also becomes tractable. This follows immediately from the algorithm sketched above: instead of coNP-power to test whether μ cannot be extended to any "child" node of T', this is now feasible in polynomial time. Note that tractability is required for each set P_n individually, hence for different nodes n and n', the sets P_n and $P_{n'}$ may belong to different tractable fragments.

Complexity of Evaluation with Projection. For CQs without existentially quantified variables, the decision problem corresponding to EVALUATION is tractable. However, it becomes NP-complete for CQs with existentially quantified variables. The next result shows that a similar behavior can be observed for well-designed SPARQL graph patterns as well. I.e., the complexity increases by one level in the polynomial hierarchy if projection is added.

Theorem 2 ([25]). *The* EVALUATION *problem of wdPTs with projection is* Σ_2^P-*complete.*

The membership is shown by devising a simple "guess and check" algorithm that tests whether the solution candidate μ satisfies Lemma 2. Given a wdPT T, a mapping μ, a set $\mathcal{X}$ of free variables, and an RDF graph G, the witness that must be guessed by the algorithm consists of

1. the subtree T' of T rooted at r and
2. the mapping λ on evars(T').

For the "check" part, it remains to test whether fvars(T') = dom(μ) and whether $\mu \cup \lambda \in [\![T]\!]_G$. The first test can be obviously done in polynomial time, while the second test is in coNP [29].

Tractable Evaluation. A condition that has been shown to help identifying relevant tractable fragments of wdPTs is *local tractability* [25]. This refers to restricting the CQ defined by each node in a wdPT to belong to a tractable class. The classes of CQ patterns which admit an efficient evaluation include classes of bounded *treewidth* [12], *hypertreewidth*, [15] (generalizing acyclic CQs [36]), *fractional* hypertreewidth, [17], etc. We concentrate on the first two. From now on, we denote by TW(k) (resp., HW(k)), for $k \geq 1$, the class of CQs of treewidth (resp., hypertreewidth) at most k. A wdPT $((V, E, r), (P_n)_{n \in V})$ is *locally in* $\mathcal{C}$, if for each node $n \in V$ the CQ $ANS \leftarrow P_n$ is in $\mathcal{C}$. We write ℓ-$\mathcal{C}$ for the set of all wdPTs that are locally in $\mathcal{C}$. Moreover we denote by EVAL($\mathcal{C}$) the evaluation problem of wdPTs restricted to the class $\mathcal{C}$.

As already mentioned before, local tractability leads to tractability of evaluation for projection-free wdPTs. On the other hand, this result does not hold in the presence of projection, even when $\mathcal{C}$ is of bounded treewidth. Formally, EVAL(ℓ-TW(k)) and EVAL(ℓ-HW(k)) are NP-complete for every $k \geq 1$ [25].

This raises the question of which further restrictions on wdPTs are needed to achieve tractability. In [8], a natural such restriction is introduced, called *bounded interface*. Intuitively, this restricts the number of variables shared between a node in a wdPT and its children. We say that a wdPT has *c-bounded interface*, for $c \geq 1$, if for each node n of the wdPT, the number of variables that appear both in n and its children is at most c. We denote by BI(c) the set of wdPTs of

c-bounded interface. It can be shown that local tractability and bounded interface yield tractability of the EVALUATION problem of wdPTs with projection:

Theorem 3 ([8]). *Let C be $\mathsf{TW}(k)$ or $\mathsf{HW}(k)$ and $c \geq 1$. Then $\mathrm{EVAL}(\ell\text{-}C \cap \mathsf{BI}(c))$ is in PTIME.*

Notice that CQs are a special case of wdPTs consisting of the root node only. Hence, $\mathsf{TW}(k) \subseteq \ell\text{-}\mathsf{TW}(k) \cap \mathsf{BI}(c)$ and $\mathsf{HW}(k) \subseteq \ell\text{-}\mathsf{HW}(k) \cap \mathsf{BI}(c)$ hold for each $c \geq 1$. Therefore, Theorem 3 tells us that $\ell\text{-}\mathsf{TW}(k) \cap \mathsf{BI}(c)$ and $\ell\text{-}\mathsf{HW}(k) \cap \mathsf{BI}(c)$ define relevant extensions of $\mathsf{TW}(k)$ and $\mathsf{HW}(k)$, respectively, that preserve tractability of evaluation.

Partial Evaluation of wdPTs. Given the nature of wdPTs, it is also interesting to check whether a mapping μ is a *partial* solution of the wdPT T over G [29], i.e., whether μ can be extended to some solution μ' of T over G. This gives rise to the partial evaluation problem PARTIAL-EVAL(C) for a class C of wdPTs defined as follows: Given a graph G and a wdPT $T \in C$, as well as a partial mapping $\mu : X \to U$, where X is the set of variables mentioned in T, is there a $\mu' \in \llbracket T \rrbracket_G$ such that μ' extends μ?

Partial evaluation is tractable for the class of projection-free wdPTs [29]. On the other hand, if projection is allowed, then partial evaluation is NP-complete even under local tractability, i.e., even for the classes $\ell\text{-}\mathsf{TW}(k)$ and $\ell\text{-}\mathsf{HW}(k)$, for each $k \geq 1$ [25].

It is easy to modify the proof of Theorem 3 to show that adding bounded interface to local tractability yields efficient partial evaluation; that is, PARTIAL-EVAL($\ell\text{-}\mathsf{TW}(k) \cap \mathsf{BI}(c)$) and PARTIAL-EVAL($\ell\text{-}\mathsf{HW}(k) \cap \mathsf{BI}(c)$) are in PTIME. However, partial evaluation is seemingly easier than exact evaluation. Hence, the question naturally arises if tractability of partial evaluation of wdPTs can be ensured by a weaker condition. Indeed, we give a positive answer to this question below. This condition will be referred to as *global tractability*. Intuitively, it states that there is a bound on the treewidth (resp., hypertreewidth) of the CQs defined by the different subtrees of a wdPT T rooted in r. Formally, let C be $\mathsf{TW}(k)$ or $\mathsf{HW}(k)$, for $k \geq 1$. A wdPT T is *globally in C*, if for each subtree T' of T rooted in r it is the case that the CQ $ANS \leftarrow pat(T')$ is in C. We denote by $g\text{-}C$ the set of all wdPTs that are globally in C.

Theorem 4 ([8]). *PARTIAL-EVAL($g\text{-}\mathsf{TW}(k)$) and PARTIAL-EVAL($g\text{-}\mathsf{HW}(k)$) are in PTIME for every $k \geq 1$.*

It remains to answer the question if global tractability also suffices to ensure tractability of (exact) evaluation for wdPTs. It turns out that this is not the case.

Proposition 1 ([8]). *EVAL($g\text{-}\mathsf{TW}(k)$) and EVAL($g\text{-}\mathsf{HW}(k)$) are NP-complete for all $k \geq 1$.*

Semantics Based on Maximal Mappings. The semantics of projection-free wdPTs is only based on *maximal* mappings, i.e., mappings that are not subsumed by any other mapping in the answer. This is no longer the case in the presence of projection [25]. As we will see in Sect. 5, for query answering of SPARQL under entailment regimes, it will turn out advantageous to define a semantics for wdPTs that is based on maximal mappings. This semantics is formalized as follows. Assume G is an RDF Graph and T is a wdPT. The *evaluation of T over G under maximal mappings*, denoted $[\![T]\!]_G^m$, corresponds to the restriction of $[\![T]\!]_G$ to those mappings μ which are not extended by any other mapping $\mu' \in [\![T]\!]_G$. This naturally leads to the decision problem MAX-EVAL($\mathcal{C}$) defined as follows: Given an RDF graph G and a wdPT $T \in \mathcal{C}$, as well as a partial mapping $\mu : \mathcal{X} \to \mathbf{U}$, where $\mathcal{X}$ is the set of variables mentioned in T, is $\mu \in [\![T]\!]_G^m$?

MAX-EVAL($\mathcal{C}$) is clearly intractable for the class $\mathcal{C}$ of all wdPTs. Analogously to PARTIAL-EVAL, local tractability is not sufficient to ensure tractability of MAX-EVAL:

Proposition 2 ([8]). *For every $k \geq 1$ the problems* MAX-EVAL(ℓ-TW(k)) *and* MAX-EVAL(ℓ-HW(k)) *are* NP-*hard.*

To obtain tractability in this case it is however sufficient to impose global tractability, which is exactly the same condition that yields tractability of partial evaluation for wdPTs (as stated in Theorem 4):

Theorem 5 ([8]). MAX-EVAL(g-TW(k)) *and* MAX-EVAL(g-HW(k)) *are in* PTIME *for every $k \geq 1$.*

4 Static Query Analysis

Static query analysis is a fundamental task in query optimization. Two of the most important problems in this context are query containment and query equivalence, which are very well understood for a variety of query languages [1]. For instance, since by Trakhtenbrot's theorem both problems are undecidable for the full relational calculus, they have been studied for several interesting fragments of relational calculus, including CQs and several extensions thereof [11,21,34].

Since SPARQL has the same expressive power as the relational calculus and queries from one language can be effectively transformed into equivalent queries of the other language [3,32], containment and equivalence are undecidable for full SPARQL. Hence, analogously to the relational calculus, both problems have been studied for fragments of SPARQL, with well-designed graph patterns being the core fragment.

We use the notation wd-SPARQL[S] to refer to the different classes of SPARQL queries reviewed in this section, where $S \subseteq \{\cup, \pi\}$. I.e., we consider well-designed SPARQL queries which use the AND and OPT operator and which may be extended by UNION (if $\cup \in S$) and/or projection (if $\pi \in S$). We will consider the problems CONTAINMENT[S_1, S_2] and EQUIVALENCE[S_1, S_2], which take as input queries $Q_1 \in$ wd-SPARQL[S_1], $Q_2 \in$ wd-SPARQL[S_2] and ask if for all RDF graphs G it is the case that $[\![Q_1]\!]_G \subseteq [\![Q_2]\!]_G$ or $[\![Q_1]\!]_G = [\![Q_2]\!]_G$, respectively, holds.

It was argued that in the presence of optional matching, the classical notion of query containment via the subset relation ($\subseteq$) might be too restrictive for certain applications. The reason for this is illustrated by the following example.

Example 3. Consider the two SPARQL queries $Q_1 = (?x, \texttt{directed\_by}, ?y)$ and $Q_2 = (?y, \texttt{directed\_by}, ?y) \, \mathsf{OPT} \, (?x, \texttt{oscars\_won}, ?z)$ which are simplified variants of the query Q from Example 1, and an RDF graph $G = \{(\text{"Star\_Wars"}, \texttt{directed\_by}, \text{"George\_Lucas"}), (\text{"Star\_Wars"}, \texttt{oscars\_won}, \text{"6"})\}$. Then $[\![Q_1]\!]_G = \{\mu\}$ with $\mu = \{?x \leftarrow \text{"Star\_Wars"}, (?y \leftarrow \text{"George\_Lucas"}\}$ and $[\![Q_2]\!]_G = \{\mu'\}$ with $\mu' = \mu \cup \{?z \leftarrow \text{"6"}\}$. Hence $Q_1 \not\subseteq Q_2$. This might be, however, unintuitive or even unintended, since answers to Q_2 always contain at least the same amount of information as those to Q_1.   ◇

One way to address this concern is to resort to the subsumption relation mentioned in Sect. 2. This gives rise to the problem SUBSUMPTION$[S_1, S_2]$ which, given two queries $Q_1 \in$ wd-SPARQL$[S_1]$ and $Q_2 \in$ wd-SPARQL$[S_2]$, asks if $[\![Q_1]\!]_G \sqsubseteq [\![Q_2]\!]_G$ holds for all RDF graphs G. Clearly, for CQs, the notions of containment and subsumption coincide. Subsumption has already been used in the past as a meaningful way of testing containment of queries with incomplete answers over semistructured data [20], and it has been convincingly argued that it is also a suitable notion for comparing the result of SPARQL queries containing the OPT operator [5]. It has also been considered in foundational work on SPARQL to compare the evaluation of two patterns containing OPT operators [5, 29].

Subsumption. It turns out that in the presence of optional matching not only the semantics of subsumption is more robust than that of containment, but also its complexity is much more stable for the different fragments of SPARQL. In fact, the subsumption problem is Π_2^P-complete in all of the cases considered in this survey.

For CQs, the containment problem $Q_1 \subseteq Q_2$ is equivalent to deciding if there exists a homomorphism h from Q_2 to Q_1. Recall that the main intuition behind this is that h allows one to "translate" solutions of Q_1 to solutions of Q_2.

The subsumption problem for well-designed SPARQL queries can be decided in a similar way, and can essentially be reduced to a (possibly exponential) number of containment tests between CQs: An immediate consequence of Lemma 2 and Lemma 1 is that for every wdPF $(\mathcal{F}, \mathcal{X})$ and RDF graph G, every solution $\mu \in [\![(\mathcal{F}, \mathcal{X})]\!]_G$ is witnessed by some subtree T' of $\mathcal{F}$ and some extension μ' of μ s.t. (a) $\mathsf{dom}(\mu') = \mathsf{vars}(T')$, (b) $\mathsf{dom}(\mu) = \mathsf{fvars}(T')$, and (c) μ' maps all triple patterns in T' into G. Thus, subsumption between two wdPFs $(\mathcal{F}_1, \mathcal{X}), (\mathcal{F}_2, \mathcal{X})$ holds if and only if every such mapping μ' for $(\mathcal{F}_1, \mathcal{X})$ can be "translated" (in the same sense as for CQs) to a corresponding mapping on $(\mathcal{F}_2, \mathcal{X})$. This allows for the following characterization of subsumption between wdPFs.

Lemma 3 ([30]). *Let $(\mathcal{F}_1, \mathcal{X})$ and $(\mathcal{F}_2, \mathcal{X})$ be two wdPFs. Then $(\mathcal{F}_1, \mathcal{X}) \sqsubseteq (\mathcal{F}_2, \mathcal{X})$ iff for every subtree T_1' of $\mathcal{F}_1$, there exists a subtree T_2' of $\mathcal{F}_2$, s.t.*

(1) fvars$(T_1') \subseteq$ fvars(T_2') *and*
(2) *there exists a homomorphism* $h\colon pat(T_2') \to pat(T_1')$ *with* $h(?x) = ?x$ *for all* $?x \in$ fvars(T_1').

Assuming this characterization is satisfied, given some $\mu_1 \in [\![(\mathcal{F}_1, \mathcal{X})]\!]_G$ witnessed by some subtree T_1' and mapping μ_1', we get a corresponding mapping μ_2' as $\mu_2'(\cdot) = \mu_1'(h(\cdot))$, where h is the homomorphism guaranteed to exist by the characterization. Observe, however, that the lemma only guarantees $\mu_2' \sqsubseteq [\![\mathcal{F}_2]\!]_G$, and not $\mu_2' \in [\![\mathcal{F}_2]\!]_G$, since μ_2' need not be maximal. Thus the characterization guarantees $\mu_2 \sqsubseteq [\![(\mathcal{F}_2, \mathcal{X})]\!]_G$ as required, but not $\mu_2 \in [\![(\mathcal{F}_2, \mathcal{X})]\!]_G$.

This characterization can be immediately turned into a Π_2^P-algorithm for deciding subsumption. On the other hand, Π_2^P-hardness was shown to already hold for SUBSUMPTION$[\emptyset, \emptyset]$ in [25]. We thus get the following result.

Theorem 6 ([25,30]). *The problem* SUBSUMPTION$[S_1, S_2]$ *is* Π_2^P-*complete for all* $S_1, S_2 \subseteq \{\cup, \pi\}$.

Table 1. Complexity of the Containment and Equivalence problem, [25,30].

$\downarrow S_1 / S_2 \to$	CONTAINMENT$[S_1, S_2]$				EQUIVALENCE$[S_1, S_2]$			
	$\emptyset$	$\{\cup\}$	$\{\pi\}$	$\{\cup, \pi\}$	$\emptyset$	$\{\cup\}$	$\{\pi\}$	$\{\cup, \pi\}$
$\emptyset$	NP-c.	Π_2^P-c.	undec.	undec.	NP-c.	–	–	–
$\{\cup\}$	NP-c.	Π_2^P-c.	undec.	undec.	Π_2^P-c	Π_2^P-c	–	–
$\{\pi\}$	NP-c.	Π_2^P-c.	undec.	undec.	Π_2^P-c	Π_2^P-h	undec	–
$\{\cup, \pi\}$	NP-c.	Π_2^P-c.	undec.	undec.	Π_2^P-c	undec	undec	undec

Containment. The complexity of the CONTAINMENT problem is summarized in Table 1. Beside ranging from NP-completeness to even undecidability, it also displays a surprising asymmetry: For instance, CONTAINMENT$[\{\pi\}, \emptyset]$ is NP-complete, while CONTAINMENT$[\emptyset, \{\pi\}]$ is undecidable.

Recall that for subsumption $(\mathcal{F}_1, \mathcal{X}) \sqsubseteq (\mathcal{F}_2, \mathcal{X})$, the crucial property for the characterization in Lemma 3 to be correct is that it is irrelevant whether the subtrees of $\mathcal{F}_2$ are maximal or not. However, for containment, this is no longer the case since now it must be guaranteed that for every solution to $(\mathcal{F}_1, \mathcal{X})$, the exact same mapping (and not an extension of it) is also a solution to $(\mathcal{F}_2, \mathcal{X})$. While homomorphisms are too weak to directly express such a property, for the decidable cases in Table 1, it is possible to express this in an indirect way. We demonstrate this idea for the problem CONTAINMENT$[\{\pi\}, \emptyset]$:

Lemma 4 ([30]). *Let* $(T_1, \mathcal{X})$ *and* T_2 *be wdPTs. Then* $(T_1, \mathcal{X}) \subseteq T_2$ *iff for every subtree* T_1' *of* T_1,

(1) *either there exists a child node* n *of* T_1' *and a homomorphism* $h\colon pat(n) \to pat(T_1')$ *with* $h(?x) = ?x$ *for all* $?x \in$ vars$(n) \cap$ vars(T_1')

(2) or there exists a subtree T_2' of T_2, s.t. (a) $\mathsf{fvars}(T_1') = \mathsf{vars}(T_2')$, (b) $pat(T_2') \subseteq pat(T_1')$, and (c) for all extensions $\hat{T}_2'$ of T_2' there exists an extension $\hat{T}_1'$ of T_1' and a homomorphism $h\colon pat(\hat{T}_1') \to pat(T_1') \cup pat(\hat{T}_2')$ with $h(?x) = ?x$ for all $?x \in \mathsf{vars}(T_1')$.

The intuition of this characterization is as follows: Property (1) is a technical detail dealing with subtrees that can always be extended. This case was implicitly covered for subsumption but must now be made explicit. Property (2a) and (2b) are the adaptations of the properties (1) and (2) in Lemma 3: (2a) follows immediately from the fact that extensions are not allowed. For (2b), observe that in the present case, looking for a homomorphism as in property (2) in Lemma 3 means to look for a homomorphism that is the identity on all variables in its domain, hence degenerating to a subset inclusion test. Finally, property (2c) implicitly ensures that the mapping on T_2 cannot be extended. Intuitively, it expresses the following: Assume some RDF graph G, a mapping $\mu \in [\![(T_1, \mathcal{X})]\!]_G$ witnessed by $\mu_1 \in [\![T_1]\!]_G$ and T_1'. Assuming further the properties of the lemma to be satisfied, we know from (2a) and (2b) that $\mu \sqsubseteq [\![T_2]\!]_G$. Thus assume to the contrary that $\mu \notin [\![T_2]\!]_G$ because of some $\mu' \in [\![T_2]\!]_G$ with $\mu \sqsubseteq \mu'$. Let T_2' be the subtree of T_2 corresponding to μ and let μ' be witnessed by some subtree $\hat{T}_2'$. Then $\hat{T}_2'$ must be an extension of T_2'. But then the subtree $\hat{T}_1'$ of T_1 and homomorphism h according to property (2c) provide a contradiction to $\mu_1 \in [\![T_1]\!]_G$, since $\mu_1 \sqsubset \mu_1'$ where $\mu_1'(\cdot) = \mu'(h(\cdot))$, and $\mu_1' \in [\![T_1]\!]_G$. I.e., the characterization guarantees the maximality on T_2 implicitly by making sure that if the mapping is not maximal on T_2, then it is not on T_1 either. This characterization can be easily extended to $\text{CONTAINMENT}[\{\cup, \pi\}, \emptyset]$.

A direct implementation of this characterization would lead to a Π_2^P-algorithm for deciding $\text{CONTAINMENT}[\{\cup, \pi\}, \emptyset]$. However, it is in fact not necessary to perform the test for all subtrees of T_1, but it suffices to just test a linear number of them. This pushes the complexity down to NP. Allowing for UNION on the right hand side requires some non-trivial extension of property (2c) for the characterization to still work in such a setting. As a result, the complexity rises by one level in the polynomial hierarchy.

Once projection is allowed to occur in the containing query (i.e. the query on the right hand side), this approach no longer works, and in fact the problem becomes undecidable. To get an idea of why this is the case, observe that an alternative way to look at property (2c) is that we create a canonical RDF graph over which μ – the mapping of interest – is guaranteed not to be a solution. Without projection, such a canonical graph can always be found since we get the following property: μ is not a solution if it can be extended to a bigger solution. It thus basically suffices to just add one such extension to the canonical graph. In the presence of projection, however, we get the following situation: μ is not a solution to T_2, if for every subtree T_2' of T_2 with $\mathsf{fvars}(T_2') = \text{dom}(\mu)$ *and every mapping μ' on $\mathsf{evars}(T_2')$,* there exists an extension of $\mu \cup \mu'$ that is a solution to T_2. Thus we have to provide an extension for all possible mappings μ' in the canonical graph. Adding these extensions may give rise to new mappings μ' on the existential variables, which in turn require new extensions to be provided in the canonical RDF graph, and it is not clear when this can be stopped.

This behavior is reminiscent of the chase termination (cf. [9,16,26] and related problems, and in fact the undecidability of the containment problem was shown by reduction from the following problem: Given a set Σ of tuple-generating dependencies, a database instance I and a Boolean CQ Q, is Q true in every (finite) model of Σ and I (cf. [9,30])?

Equivalence. The complexity of the EQUIVALENCE problem is also depicted in Table 1. Of course, for two SPARQL queries Q_1 and Q_2 it holds that $Q_1 \equiv Q_2$ iff $Q_1 \subseteq Q_2$ and $Q_2 \subseteq Q_1$. Thus an upper-bound on the complexity of the containment problem also provides an upper-bound for the equivalence problem. In addition, it was shown in [25] that for queries in wd-SPARQL[$\emptyset$] it is also the case that $Q_1 \equiv Q_2$ iff $Q_1 \sqsubseteq Q_2$ and $Q_2 \sqsubseteq Q_1$. However, as soon as we add union or projection on either side, this property does no longer hold.

Of course, the EQUIVALENCE$[S_1, S_2]$ problem is symmetric in S_1 and S_2. Hence, only the lower triangle of the table has been filled in. The reason that EQUIVALENCE$[\{\cup, \pi\}, \emptyset]$ is decidable in Π_2^P, while CONTAINMENT$[\emptyset, \{\cup, \pi\}]$ is undecidable is that in order to decide equivalence, it actually suffices to test containment in one, and only subsumption in the other direction. We would like to point out that not only is the exact complexity of EQUIVALENCE$[\{\pi\}, \{\cup\}]$ still open, but it is even unknown if the problem is decidable or not.

5   Ontology-Based Query Answering

In the recently released recommendation [14], the W3C has defined various SPARQL entailment regimes to allow users to specify implicit knowledge about the vocabulary in an RDF graph. The theoretical underpinning of query answering under entailment regimes is provided by the big body of work on *ontology-based query answering*, notably in the area of description logics (DLs) [6]. However, the semantics of query answering under SPARQL entailment regimes is defined in a simpler and less expressive way than the *certain answer semantics* usually adopted in the DL and database literature.

Example 4. Consider an RDF graph G containing a single triple ("Star_Wars", rdf:type,movie) – stating that "Star_Wars" is a movie – and an ontology $\mathcal{O}$ containing the triple (movie, rdfs:subClassOf, $\exists$has_actor). – stating that every movie has some actor who acts in it. Now consider the following simple graph pattern $(P, \{?x\})$ with $P = (?x, \text{has\_actor}, ?y)$, where $?x$ is the only output variable. Following the SPARQL entailment regimes standard [14], this query yields an empty result. $\diamond$

By the concept inclusion (movie, rdfs:subClassOf, $\exists$has_actor), we know for certain that there is some actor who acts in "Star_Wars". Hence, the result in the above example is rather unintuitive. The reason for this behavior is that the standard for SPARQL entailment regimes [14] requires that all values assigned to any variable must come from the RDF graph. In other words, distinguished variables (which are ultimately output) and non-distinguished variables (which are eventually projected out) are treated in the same way. In contrast, the certain answer semantics retrieves all mappings on the distinguished variables that allow to satisfy the query in every possible model of the database and the ontology – yielding the certain answer $\mu = \{?x \leftarrow$ "Star_Wars"$\}$ in the above example. The certain answer semantics has been extensively studied in the database and DL literature for CQs [1, 10]. However, in the presence of optional matching the usual certain answer semantics (i.e., something is a certain answer if it is present in every model) turns out to be unsatisfactory:

Example 5. Consider the graph pattern $(P, \{?x, ?z\})$ with P: $((?x, \texttt{has\_actor}, ?y)\texttt{OPT}(?y, \texttt{was\_born}, ?z))$ over the graph $G = \{(\text{"Star\_Wars"},$ $\texttt{has\_actor}, \text{"Harrison\_Ford"})\}$ and empty ontology $\mathcal{O}$. The query yields a unique solution $\mu = \{?x \leftarrow$ "Star_Wars"$\}$. Clearly, also the extended graph $G' = G \cup \{(\text{"Harrison\_Ford"}, \texttt{was\_born}, \text{"1942"})\}$ is a model of $(G, \mathcal{O})$. But in G', μ is no longer a solution since μ can be extended to solution $\mu' = \{?x \leftarrow$ "Star_Wars", $?z \leftarrow$ "1942"$\}$. Hence, there exists no mapping which is a solution in every possible model of $(G, \mathcal{O})$.   ◇

As Example 5 illustrates, a literal adoption of the certain answer semantics in the presence of the OPT operator leads to having no solutions, even though there is a solution that can be extended to a solution in every model. In order to tackle this and further problems, the definition of certain answers for the class of wdPTs has to be suitably modified [2]. This modification of the semantics also requires an adaptation and extension of known query answering algorithms established in the area of description logics. We mention two such modified algorithms for query evaluation under DL-Lite [10]. It turns out that the additional expressive power due to the certain answers comes without an increase of the complexity.

OWL 2 QL. RDF has been enhanced by the OWL 2 Web Ontology Language [27], a World Wide Web Consortium (W3C) recommendation to enable the specification of background knowledge about the application domain, and to enrich query answers with implicit information. The logical underpinning of OWL 2 and its sub-languages are description logics. One such sub-language is OWL 2 QL which is based on DL-Lite$_\mathcal{R}$, a member of the DL-lite family [10]. Its fundamental building blocks are *constants* c, *atomic concepts* A and *atomic roles* R, which are countably infinite and mutually disjoint subsets of a set $\mathbf{U}$ of URIs. From these we can build *basic roles* R and R^-, and *basic concepts* B and $\exists Q$, where Q is a basic role. Using the above, DL-Lite$_\mathcal{R}$ allows one to express the following kind of statements: Membership assertions $(c, \texttt{rdf:type}, B)$ or (c, Q, c'), concept inclusions $(B_1, \texttt{rdfs:subClassOf}, B_2)$, role inclusions $(Q_1, \texttt{rdfs:subPropertyOf}, Q_2)$ as well as concept and role disjointness (where c, c' are constants and B_i, Q_i are basic concepts resp. basic roles). In the following, an ontology $\mathcal{O}$ is any set of such expressions, excluding membership assertions, which we assume to be part of the RDF graph. A *knowledge base (KB)* $\mathcal{G} = (G, \mathcal{O})$ consists of an RDF graph G and an ontology $\mathcal{O}$.

Certain Answers of wdPTs. Before providing our definition of certain answers, we need to introduce two additional notions. Let P be a well-designed graph pattern. Following [29], we say that P' is a reduction of P (denoted as $P' \trianglelefteq P$) if P' can be constructed from P by replacing in P sub-patterns of the form $(P_1 \text{ OPT } P_2)$ by P_1. Note that, in terms of wdPTs, a reduction corresponds to a subtree containing the root node of the wdPT. Moreover, for a mapping μ and some property A, we shall say that μ is $\sqsubseteq$-*maximal w.r.t.* A if μ satisfies A, and there is no μ' such that $\mu \sqsubseteq \mu'$, $\mu' \not\sqsubseteq \mu$, and μ' satisfies A.

Definition 2. *Let* $\mathcal{G} = (G, \mathcal{O})$ *be a KB and* $Q = (P, \mathcal{X})$ *a well-designed graph pattern. A mapping* μ *is a* certain answer *to* Q *over* $\mathcal{G}$ *if it is a* $\sqsubseteq$-*maximal mapping with the following properties: (1)* $\mu \sqsubseteq [\![Q]\!]_{G'}$ *for every model G' of $\mathcal{G}$, and (2)* $\mathsf{vars}(P') \cap \mathcal{X} = \mathsf{dom}(\mu)$ *for some* $P' \trianglelefteq P$. *We denote by* $\mathsf{cert}(P, \mathcal{X}, \mathcal{G})$ *the set of all certain answers to* Q *over* $\mathcal{G}$.

The reason for restricting the set of certain answers to $\sqsubseteq$-maximal mappings is that queries with projection and/or UNION may have "subsumed" solutions, i.e. solutions s.t. also some proper extension is a solution. But then – with set semantics – we cannot recognize the reason why some subsumed solution may be not a solution in some possible world, as illustrated in Example 6.

Example 6. Let us revisit the graph pattern $(P, \{?x, ?z\})$ of Example 5 with $P = (?x, \mathsf{has\_actor}, ?y)$. Consider the following RDF graph G:

$$G = \{\, (\text{``Star\_Wars''}, \mathsf{has\_actor}, \text{``Harrison\_Ford''}),$$
$$(\text{``Star\_Wars''}, \mathsf{has\_actor}, \text{``Mark\_Hamill''}),$$
$$(\text{``Harrison\_Ford''}, \mathsf{was\_born}, \text{``1942''})\}.$$

and empty ontology $\mathcal{O}$. As possible models of $(G, \mathcal{O})$ we have all graphs containing G. Hence, $\mu = \{?x \leftarrow \text{``Star\_Wars''}, ?z \leftarrow \text{``1942''}\}$ and $\mu' = \{?x \leftarrow \text{``Star\_Wars''}\}$ are both solutions of $(P, \{?x, ?z\})$ over G and can be extended to solutions in every possible model.

Next consider the RDF graph:

$$G' = \{\, (\text{``Star\_Wars''}, \mathsf{has\_actor}, \text{``Harrison\_Ford''}),$$
$$(\text{``Harrison\_Ford''}, \mathsf{was\_born}, \text{``1942''})\}.$$

If we take as certain answers all mappings that can be extended to some solution in every possible model, then μ' from above is still a certain answer, which is clearly undesired. ◇

A key idea to solve the problem illustrated in Example 6 is to allow only "maximal" solutions. In addition, Property (2) in Definition 2 ensures that the domain of such an answer adheres to the structure of nested OPT operators in the query. However, we can show that this property need not be considered during the computation of the certain answers, but can be enforced in a simple post-processing step. We call such answers that satisfy Definition 2 except

Property (2) *certain pre-answers*, and use $\mathsf{certp}(P, \mathcal{X}, \mathcal{G})$ to denote the set of all certain pre-answers. The same is true for projection, which can also be performed in a simple post-processing step. Thus, it suffices to compute $\mathsf{certp}(P, \mathcal{G})$, which can be done via a universal solution (referred to as *canonical model* in the area of DLs) as follows.

Theorem 7 ([2]). *Let $\mathcal{G} = (G, \mathcal{O})$ be a KB and P a well-designed graph pattern. Then, $\mathsf{certp}(P, \mathcal{G}) = \mathrm{MAX}(\llbracket P \rrbracket_{\mathsf{univ}(G)}{\downarrow})$, where $\mathrm{MAX}(M)$ is the set of $\sqsubseteq$-maximal mappings in M, $M{\downarrow} := \{\mu{\downarrow}\mid \mu \in M\}$ ($\mu{\downarrow}$ is the restriction of μ to those variables which are mapped by μ to the active domain of G), and $\mathsf{univ}(G)$ is a universal solution of $\mathcal{G}$.*

However, computing the certain answers via a universal solution is not always practical, since universal solutions can be infinite. As a result, query rewriting algorithms have been developed. These algorithms take the input query and the ontology, and rewrite them into a single query that can be evaluated over the input database without considering the ontology. By introducing several adaptations and extensions of the rewriting-based CQ evaluation for DL-Lite from [10], we developed two different approaches to compute the certain answers for well-designed SPARQL graph patterns (or, equivalently, of wdPTs) under OWL 2 QL entailment [2].

The first one proceeds in a modular way by rewriting the pattern P_n at each node n in a wdPT individually. It thus follows the general philosophy of SPARQL entailment regimes [14]. One possible disadvantage of this modular approach is that it requires to maintain additional data structures to ensure consistency when combining the partial solutions for the patterns of different nodes. As a consequence, the complete algorithm has to be implemented from scratch because the standard tools cannot handle these additional data structures.

The goal of the second approach is to make use of standard technology as much as possible. The idea is to transform the OWL 2 QL entailment under our new semantics into SPARQL query evaluation under RDFS entailment [14], for which strong tools are available. Unlike the first – modular – approach, this rewriting proceeds in a holistic way, i.e. it always operates on the whole query.

Based on these rewriting algorithms, we can show that the complexity of query answering and of several static query analysis tasks does not increase despite the additional power of OWL 2 QL entailment under our new semantics.

Recall from Sect. 3 the two variants PARTIAL-EVAL and MAX-EVAL of the EVALUATION problem of wdPTs, where we have to decide for a graph G, wdPT $(\mathcal{T}, \mathcal{X})$, and mapping μ, if μ can be extended to a solution or is a maximal solution, respectively, of $(\mathcal{T}, \mathcal{X})$ over G. Note that we cannot directly compare the EVALUATION problem of wdPTs under OWL 2 QL entailment (with our certain answer semantics) and the EVALUATION problem of wdPTs without entailment regimes. This is due to the fact that our certain answer semantics (for reasons explained above) only allows maximal solutions. Hence, PARTIAL-EVAL and MAX-EVAL are the right problems to look at. The PARTIAL-EVAL problem is NP-complete [25] and the MAX-EVAL problem is DP-complete [2], and it was shown in [2] that these complexities remain unchanged under OWL 2 QL entailment with our certain answer semantics.

As far as static query analysis is concerned, we now have to redefine the problems SUBSUMPTION, CONTAINMENT, and EQUIVALENCE so as to take a given ontology into account. For instance, for the SUBSUMPTION problem, we are now given two wdPTs T_1, T_2 plus an ontology $\mathcal{O}$; we have to decide if for every RDF graph G, the relationship $\mathsf{cert}(T_1, \mathcal{X}_1, \mathcal{G}) \sqsubseteq \mathsf{cert}(T_2, \mathcal{X}_2, \mathcal{G})$ holds, where $\mathcal{G}$ denotes the knowledge base $\mathcal{G} = (G, \mathcal{O})$. Recall that SUBSUMPTION without entailment regimes is Π_2^P-complete in all settings considered here (i.e., with or without projection; with our without the UNION operator). It can be shown that the complexity remains the same also for the SUBSUMPTION problem under OWL2 QL entailment [2]. Note that the subsumption relation between two queries only depends on the *maximal* solutions over an arbitrary graph. Hence, we did not need to define yet another variant of SUBSUMPTION, which takes only the maximal solutions into account. In contrast, for CONTAINMENT and EQUIVALENCE, a comparison between the settings with and without ontologies only makes sense if we check for containment (resp. equivalence) of the maximal solutions only. In [2], the resulting problems were shown Π_2^P-complete both for the settings with and without OWL2 QL entailment.

6   Conclusion and Future Work

We have recalled some recent results on an interesting fragment of SPARQL, the so-called well-designed SPARQL graph patterns or, equivalently, well-designed pattern trees (wdPTs). Such queries can be seen as a natural extension of conjunctive queries (CQs) by the optional matching feature. It has turned out that this feature makes virtually all relevant computational tasks more complex: the complexity of the EVALUATION problem raises from NP-completeness to Σ_2^P-completeness. The CONTAINMENT and EQUIVALENCE problems even become undecidable unless we forbid projection. In [29], SUBSUMPTION has been proposed as an interesting variant of CONTAINMENT, which is computationally better behaved and which may be more intuitive in the presence of optional matching. Its complexity is Π_2^P-complete in all settings considered here. Finally, we have seen that an additional extension of wdPTs by entailment under OWL2 QL (which corresponds to DL-Lite$_\mathcal{R}$) does not increase the complexity anymore.

Note that many further aspects of well-designed SPARQL graph patterns have been studied, which were not recalled in this survey. In response to the intractabilty of the EVALUATION problem of wdPTs, works on the approximation of CQs [7] were extended to wdPTs in [8]. The COUNTING problem of wdPTs (i.e., given a wdPT T and a graph G, how many solutions does T have over G) was studied in [31]. The COUNTING problem turned out to be more complex than the EVALUATION problem in the sense that the restrictions guaranteeing tractability of EVALUATION do not suffice to achieve tractability of COUNTING. In [24], various aspects of the ENUMERATION problem of wdPTs (i.e., given a wdPT T and a graph G, output all solutions of T over graph G) are studied.

As has been recalled in Sect. 3, tractability of the MAX-EVAL problem is easier to achieve than for the EVALUATION problem. Interestingly, for the ENUMERATION problem, outputting the *maximal* solutions may become harder than outputting *all* solutions.

Recall that projection is realized in SPARQL by wrapping a SPARQL graph pattern into a SELECT statement. Another query form provided by the SPARQL standard [33] is to wrap a SPARQL graph pattern into a CONSTRUCT statement. The result of applying a CONSTRUCT query to an RDF graph is again an RDF graph (rather than a set of mappings). In [23], several interesting properties of CONSTRUCT queries were presented. For instance, for CONSTRUCT queries with well-designed SPARQL graph patterns, it was shown that they correspond to positive first order queries. An important extension provided by SPARQL 1.1 [18] is the possibility to formulate queries which have to be evaluated over different endpoints. Various aspects of this federation extension were studied in [4] – including the extension of well-designed SPARQL to federated SPARQL queries. Most of the computational problems mentioned here (EVALUATION, CONTAINMENT, EQUIVALENCE, SUBSUMPTION) were studied in [22] for another important extension of SPARQL 1.1 [18] – the so-called property paths. This extension introduces the ability to navigate in RDF graphs. Property paths thus resemble regular path queries. However, as is shown in [22], the interaction with the other SPARQL operators – in particular, with the OPT operator – requires new techniques.

Despite the great variety of results obtained for wdPTs, many questions have remained open. First, in most of the complexity analyses carried out so far, some cases could not be fully classified. For instance, in Table 1, the exact complexity (even the question of decidability) of EQUIVALENCE$[\{\pi\}, \{\cup\}]$ is still open. Closing these gaps is a natural task for future work. Strongly related to such complexity analyses is the quest for tractable fragments of the various problems studied so far. For instance, we have recalled here that the EVALUATION problem of wdPTs becomes tractable if wdPTs are restricted to the class ℓ-$\mathcal{C} \cap \mathsf{BI}(c)$ where $\mathcal{C}$ is $\mathsf{TW}(k)$ or $\mathsf{HW}(k)$ and $c \geq 1$. The same restriction guarantees tractability of the ENUMERATION problem [24], which is not the case for the COUNTING problem [31]. For all these problems, further approaches have to explored to find (further) "natural" tractable classes of wdPTs.

Finally, the language fragments studied so far should be extended in several directions. For instance, we have recalled here some results obtained for the evaluation of wdPTs under OWL2 QL entailment (or, equivalently, under DL-Lite$_{\mathcal{R}}$). This work should be extended to more expressive entailment regimes. Another important extension is concerned with extending well-designed SPARQL itself. We have recalled several favorable features of this fragment of SPARQL. For instance, the complexity of the EVALUATION problem drops from PSPACE-completeness (for the AND/OPT-fragment without well-designedness restriction) to coNP (without projection) or Σ_2^P (with projection), respectively. However, the restriction to well-designedness may be too strong. Hence, very recently [19], the extension of well-designed SPARQL to *weakly well-designed* SPARQL has been presented, by allowing some typical uses of non-well-designedness. On the one hand, it is shown that the extension of well-designed SPARQL to weakly well-designed SPARQL does not make the EVALUATION problem harder. On the other hand, the authors give evidence that the resulting fragment of SPARQL is practically highly relevant by observing that in DBpedia query logs, almost all queries containing the OPT operator are weakly well-designed. Of course, the study of the various computational tasks mentioned here should be extended to yet further (and bigger) fragments of SPARQL.

Acknowledgments. This work was supported by the Vienna Science and Technology Fund (WWTF), project ICT12-15 and by the Austrian Science Fund (FWF): P25207-N23 and W1255-N23.

References

1. Abiteboul, S., Hull, R., Vianu, V.: Foundations of Databases. Addison Wesley, Boston (1995). http://www.bibsonomy.org/bibtex/224822a68f8997f802f178bb7b146c6ff/algebradresden
2. Ahmetaj, S., Fischl, W., Pichler, R., Simkus, M., Skritek, S.: Towards reconciling SPARQL and certain answers. In: Proceedings of the WWW 2015, pp. 23–33. ACM (2015)
3. Angles, R., Gutierrez, C.: The expressive power of SPARQL. In: Sheth, A.P., Staab, S., Dean, M., Paolucci, M., Maynard, D., Finin, T., Thirunarayan, K. (eds.) ISWC 2008. LNCS, vol. 5318, pp. 114–129. Springer, Heidelberg (2008)
4. Aranda, C.B., Arenas, M., Corcho, Ó., Polleres, A.: Federating queries in SPARQL 1.1: syntax, semantics and evaluation. J. Web Sem. **18**(1), 1–17 (2013)
5. Arenas, M., Pérez, J.: Querying semantic web data with SPARQL. In: Proceedings of the PODS 2011, pp. 305–316. ACM (2011)
6. Baader, F., Calvanese, D., McGuinness, D.L., Nardi, D., Patel-Schneider, P.F. (eds.): The Description Logic Handbook: Theory, Implementation, and Applications. Cambridge University Press, Cambridge (2003)
7. Barceló, P., Libkin, L., Romero, M.: Efficient approximations of conjunctive queries. SIAM J. Comput. **43**(3), 1085–1130 (2014)
8. Barceló, P., Pichler, R., Skritek, S.: Efficient evaluation and approximation of well-designed pattern trees. In: Proceedings of the PODS 2015, pp. 131–144. ACM (2015)
9. Calì, A., Gottlob, G., Kifer, M.: Taming the infinite chase: Query answering under expressive relational constraints. In: Proceedings of the KR 2008, pp. 70–80. AAAI Press (2008)
10. Calvanese, D., De Giacomo, G., Lembo, D., Lenzerini, M., Rosati, R.: Tractable reasoning and efficient query answering in description logics: The DL-Lite family. J. Autom. Reason. **39**(3), 385–429 (2007)
11. Chandra, A.K., Merlin, P.M.: Optimal implementation of conjunctive queries in relational data bases. In: Proceedings of the STOC 1977, pp. 77–90. ACM (1977)
12. Chekuri, C., Rajaraman, A.: Conjunctive query containment revisited. Theor. Comput. Sci. **239**(2), 211–229 (2000)
13. Cyganiak, R., Wood, D., Lanthaler, M.: RDF 1.1 concepts and abstract syntax. W3C Recommendation, W3C (2014). http://www.w3.org/TR/rdf11-concepts
14. Glimm, B., Ogbuji, C.: SPARQL 1.1 Entailment Regimes. W3C Recommendation, W3C, March 2013. http://www.w3.org/TR/sparql11-entailment

15. Gottlob, G., Leone, N., Scarcello, F.: Hypertree decompositions and tractable queries. J. Comput. Syst. Sci. **64**(3), 579–627 (2002)
16. Greco, S., Spezzano, F., Trubitsyna, I.: Checking chase termination: Cyclicity analysis and rewriting techniques. IEEE Trans. Knowl. Data Eng. **27**(3), 621–635 (2015)
17. Grohe, M., Marx, D.: Constraint solving via fractional edge covers. ACM Trans. Algor. **11**(1), 4 (2014)
18. Harris, S., Seaborne, A.: SPARQL 1.1 Query Language. W3C Recommendation, W3C, March 2013. http://www.w3.org/TR/sparql11-query
19. Kaminski, M., Kostylev, E.V.: Beyond well-designed SPARQL. In: Proceedings of the ICDT 2016 (to appear, 2016)
20. Kanza, Y., Nutt, W., Sagiv, Y.: Querying incomplete information in semistructured data. J. Comput. Syst. Sci. **64**(3), 655–693 (2002)
21. Klug, A.C.: On conjunctive queries containing inequalities. J. ACM **35**(1), 146–160 (1988)
22. Kostylev, E.V., Reutter, J.L., Romero, M., Vrgoč, D.: SPARQL with Property Paths. In: Arenas, M., et al. (eds.) The Semantic Web - ISWC 2015. LNCS, vol. 9366, pp. 3–18. Springer, Heidelberg (2015)
23. Kostylev, E.V., Reutter, J.L., Ugarte, M.: CONSTRUCT queries in SPARQL. In: Proceedings of the ICDT 2015. LIPIcs, vol. 31, pp. 212–229 (2015)
24. Kröll, M., Pichler, R., Skritek, S.: On the complexity of enumerating the answers to well-designed pattern trees. In: Proceedings of the ICDT 2016 (to appear, 2016)
25. Letelier, A., Pérez, J., Pichler, R., Skritek, S.: Static analysis and optimization of semantic web queries. ACM Trans. Database Syst. **38**(4), 25 (2013)
26. Meier, M.: On the termination of the chase algorithm. Ph.D. Thesis, University of Freiburg (2010). http://www.freidok.uni-freiburg.de/volltexte/7590/
27. Motik, B., Grau, B.C., Horrocks, I., Wu, Z., Fokoue, A., Lutz, C.: Owl 2 web ontology language: Profiles. W3C working draft, W3C, October 2008. http://www.w3.org/TR/2008/WD-owl2-profiles-20081008/
28. Pérez, J., Arenas, M., Gutierrez, C.: Semantics and complexity of SPARQL. In: Cruz, I., Decker, S., Allemang, D., Preist, C., Schwabe, D., Mika, P., Uschold, M., Aroyo, L.M. (eds.) ISWC 2006. LNCS, vol. 4273, pp. 30–43. Springer, Heidelberg (2006)
29. Pérez, J., Arenas, M., Gutierrez, C.: Semantics and complexity of SPARQL. ACM Trans. Database Syst. **34**(3), 1–45 (2009)
30. Pichler, R., Skritek, S.: Containment and equivalence of well-designed SPARQL. In: Proceedings of the PODS 2014, pp. 39–50. ACM (2014)
31. Pichler, R., Skritek, S.: On the hardness of counting the solutions of SPARQL queries. In: Proceedings of the AMW 2014. CEUR Workshop Proceedings, vol. 1189. CEUR-WS.org (2014)
32. Polleres, A.: From SPARQl to rules (and back). In: Proceedings of the WWW 2007, pp. 787–796. ACM (2007)
33. Prud'hommeaux, E., Seaborne, A.: SPARQL Query Language for RDF. W3C Recommendation, W3C (2008). http://www.w3.org/TR/rdf-sparql-query/
34. Sagiv, Y., Yannakakis, M.: Equivalences among relational expressions with the union and difference operators. J. ACM **27**(4), 633–655 (1980)
35. Schmidt, M., Meier, M., Lausen, G.: Foundations of SPARQL query optimization. In: Proceedings of the ICDT 2010, pp. 4–33. ACM (2010)
36. Yannakakis, M.: Algorithms for acyclic database schemes. In: Proceedings of the VLDB 1981, pp. 82–94. IEEE Computer Society (1981)

Maintenance of Queries Under Database Changes: A Unified Logic Based Approach

Elena V. Ravve[✉]

Ort Braude College, Karmiel, Israel
elena_v_m@hotmail.com

Abstract. This contribution deals with one single theme, the exploitation of logical reduction techniques in database theory. Two kinds of changes may be applied to databases: structural changes, known also as restructuring or schema evolution, and data changes. We present both of them in the terms of syntactically defined translation schemes.

At the same time, we have application programs, computing different queries on the database, which are oriented on some specific generation of the database. Systematically using the technique of translation scheme, we introduce the notion of Φ-sums and show how queries, expressible in extensions of First Order Logic (FOL) may be handled over different generations of the Φ-sums. Moreover, using the technique of translation scheme, we introduce the notions of an incremental view recomputations. We prove when queries expressible in extensions of FOL allow incremental view recomputations.

Our approach covers uniformly the cases we have encountered in the literature and can be applied to all existing query languages.

1 Introduction

Over time, databases undergo two kinds of changes: structural changes (i.e., changes in the schema), known also as restructuring or schema evolution, and data changes (i.e., insertion, deletions and modifications of tuples). Data changes are usually referred to as updates. In the same way, one may talk about selection queries or simply queries (non-modification queries) and updates.

Non-modification query usually must be answered lots of time. That is why, as a rule, such queries are maintained as auxiliary relations, called in the context of databases as *materialized views*. Non-materialized views are called *virtual views* and, as a rule, are not updatable. We consider only materialized views in the paper. Moreover, in this paper, we consider only <u>relational</u> databases.

In this paper, we are concentrated on two problems: handling queries under restructuring of databases and under database updates.

Handling Queries Under Restructuring of Databases: Database, during its life cycle, may be restructured several times. At the same time, we have several application programs, oriented on some specific generation of the database. The problem under investigation is:

© Springer International Publishing Switzerland 2016
M. Gyssens and G. Simari (Eds.): FoIKS 2016, LNCS 9616, pp. 191–208, 2016.
DOI: 10.1007/978-3-319-30024-5_11

Given: There are two different generations of the same database, g and $g+1$. There is an application, running on the gth generation: Q_g.

Find: An application Q_{g+1}, running on the $(g+1)^{th}$ generation with the same results.

Let us consider a toy example. The gth generation of the database contains only one relation P, while the $(g+1)^{th}$ generation contains two relations R and S, such that $P = (R \bowtie S)$. The application, running on the gth generation Q_g is a simple modification query on P, which deletes tuples from P, according to some condition θ, expressed in terms of P. The set of deleted tuples is defined by $\nabla_\theta P$ rather than given by enumeration.

We have problems with this kind of rules like deletion over join. In fact: in $\nabla_\theta (R \bowtie S)$, we deal with formula θ that can be complicated. When we use the substitution of $(R \bowtie S)$ instead of P in θ, we receive a new formula in terms of R and S that contains a mix of attributes from both relations: R and S. In order to evaluate θ, we must first produce $(R \bowtie S)$ and then delete ∇_θ from the join, while we are mostly interested to derive (if possible) from θ some formulae: $\theta_1^R, \ldots, \theta_i^R$ over R and $\theta_1^S, \ldots, \theta_j^S$ over S, which we will apply to R and S respectively in order to obtain the same desired result.

In logical notation, the formulae: $\theta_1^R, \ldots, \theta_i^R$ over R and $\theta_1^S, \ldots, \theta_j^S$ over S are Feferman-Vaught reduction sequences (or simply, reductions), cf. [19]. The sequences are sets of queries such that each such a query can be evaluated on the components: R and S. Next, from the local answers, and possibly some additional information, we compute the answer. In this paper, we generalize the notion of Feferman-Vaught reduction sequences to handling queries over Φ–sums.

Handling queries under database updates: Materialized views contain some derived portion of the database information and storing as new relations. In order to reflect the changes, made on the source relations, the views should be modified by adding or deletions tuples without total re-computation from the database.

Given: A materialized view and a database update.

Find: A set of view updates that uses the old content of the view and delete from and inserts in the view some set of tuples defined on the source database.

In the case of the incremental view maintenance, we try to find some effective way to refresh the content of the view by some updates on it. The updates should be derived from the update on the source database, without the total view re-computation. In many case, it permits to simplify the maintenance procedure.

Unfortunately, as a rule, the derived view contains only some small part of the database information, and it is just impossible to obtain the desired results as a map over only the view. Using extension of the logical machinery of syntactically defined *translations schemes*, first introduced in [29] and recently used in [21] in the context of the database theory, we give precise definition of *incremental view re-computation* and prove that every query expressible in several extensions of First Order Logic (FOL) allows the incremental view re-computation.

In general, this contribution deals with exploitation of logical reduction techniques in database theory. This approach unifies different aspects, related to both schema and data evolution in databases, into a single framework. It is assumed that the reader is familiar with database theory as presented in [1] and has logical background as described in [15].

The used logical reduction techniques come in the form of Feferman-Vaught reduction sequences and translation schemes, known also in model theory as interpretations. The interpretations give rise to two induced maps, translations and transductions. Transductions describe the induced transformation of database instances and the translations describe the induced transformations of queries. Translation schemes appear naturally in the context of databases: The first example is the vertical decomposition of a relation scheme into two relation schemes with overlapping attribute sets. Also the reconstruction of the original scheme of the vertical decomposition can be looked at as translation scheme. The same is true for horizontal decompositions and the definition of views. More surprisingly, also updates can be cast into this framework. Finally, translation schemes describe also the evolution of one database scheme over different generations of database designs.

The paper is structured in the following way. Section 2 presents short review of the related works. Section 3 provides the definitions and main results, related to syntactically defined translation schemes. Section 4 is dedicated to handling of queries under restructuring of databases. Section 5 is dedicated to handling of queries under database updates. Section 6 summarizes the paper.

2   Related Works

Maintaining dynamic databases, have a long history, cf. [9–11,13]. One of the most recent paper is [21], inspired by [31]. Like [21], we are also *"interested in some arbitrary but fixed query on a finite structure, which is subject to an ongoing sequence of local changes, and after each change the answer to the query should remain available."*

In [21], the local changes of the database were limited to elements, which are constantly inserted in and deleted from the database. We take the following verbatim from the Conclusion and future work section of [21]: *We think that it is interesting to consider updates that are induced by first-order formulae. On the one hand one can consider formulae which induce updates directly to the structure, i.e. consider updates that change all tuples with the property defined by the formula. On the other hand one can perform canonical updates to one structure and consider the changes that are induced on a first-order interpreted structure.* In this paper, we propose a unified logic based approach to maintenance of queries under database changes. We show how this approach works not only for FOL but also for different extensions of it, used in the database theory.

In [26], the incremental view maintenance problem was investigated from an algebraic perspective. The author constructed a ring of databases and used it as the foundation of the design of a query calculus that allowed to express powerful aggregate queries. In this framework, a query language needed to be closed under computing an additive inverse (as a generalization of the union operation on relations to support insertions and deletions) and the join operation had to be distributive over this addition to support normalization, factorization, and the taking of deltas of queries.

Some propagation techniques for view updates may be found in [3]. In [24], the complexity of testing the correctness of an arbitrary update to a database view is analyzed, coming back to constant-complement approach of Bancilhon and Spyratos, cf. [5]. We must mention the recent exciting works [16–18,20,23], which use propagation techniques for view updates as well. However, no one of them considers the question in comparable generality. In fact, we do not need most of usually used additional assumptions. For example, we do not need the structures to be ordered. Moreover, we allow both restructuring of the database and insertion, deletion or set operations under the same logical framework. In addition, we do not restrict ourselves to the use of FOL but rather different its extensions.

3   Translation Schemes

In this section, we introduce the general framework for syntactically defined translation schemes in terms of databases. We assume that the reader is familiar with precise definitions of extensions of FOL, cf. [25]. The notion of abstract translation schemes comes back to Rabin, cf. [29]. The translation schemes are also known in model theory as interpretations, as described in particular in [25]. The definition is valid for a wide class of logics or query languages, including Datalog or Second Order Logic (SOL) as well as FOL, $MSOL$, TC, n-TC, LFP or n-LFP. However, we start from Relational Calculus in the form of FOL. Occasionally, we use Relational Algebra expressions when they are more convenient to readers.

We follow Codd's notations, cf. [7]. Database systems should present the user with tables called *relations* $(R_1, R_2, \ldots)$ and their columns are headed by *attributes* $(A_1, A_2, \ldots)$ for a relation is called the *schema* for that relation $(R_1[\bar{A}], R_2[\bar{B}], \ldots)$. The set of schema for the relations $(\mathbf{R}, \mathbf{S}, \ldots)$ is called a *relational database schema*, or just *database schema*. The row of a relation (t) are called *tuples*. A tuple has one *component* $(t[A_1], t[A_2], \ldots)$ for each attribute of the relation. We shall call a set of tuples for a given relation an *instance* $(I(R_1),$ $I(R_2), \ldots)$ of that relation.

Definition 1 (Translation Schemes Φ). *Let $\mathbf{R}$ and $\mathbf{S}$ be two database schemes. Let $\mathbf{S} = (S_1, \ldots, S_m)$ and let $\rho(S_i)$ be the arity of S_i. Let $\Phi = \langle \phi, \phi_1, \ldots, \phi_m \rangle$ be FOL formulae over $\mathbf{R}$. Φ is k–feasible for $\mathbf{S}$ over $\mathbf{R}$ if ϕ has exactly k distinct free FOL variables and each ϕ_i has $k\rho(S_i)$ distinct free first order variables. Such a $\Phi = \langle \phi, \phi_1, \ldots, \phi_m \rangle$ is also called a k-$\mathbf{R}$-$\mathbf{S}$-translation scheme or, in short, a translation scheme, if the parameters are clear in the context.*

If $k = 1$ we speak of **scalar** *or* **non–vectorized** *translation schemes. If ϕ is a tautology, then the translation scheme is* **non–relativized.** *Otherwise, ϕ defines* **relativization** *of the new database domain.*

The formulae $\phi, \phi_1, \ldots, \phi_m$ can be thought of as queries. ϕ describes the new domain, and the ϕ_i's describe the new relations. Vectorization creates one attribute out of a finite sequence of attributes. The use of vectorized translation schemes in the context of databases is shown in particular in [2] and [27]. We shall discuss concrete examples after we have introduced the induced transformation of database instances.

A (partial) function Φ^* from **R** instances to **S** instances can be directly associated with a translation scheme Φ.

Definition 2 (Induced Map Φ^*). *Let $I(\mathbf{R})$ be a* **R** *instance and Φ be k–feasible for* **S** *over* **R**. *The instance $I(\mathbf{S})_\Phi$ is defined as follows:*

1. *The universe of $I(\mathbf{S})_\Phi$ is the set $I(\mathbf{S})_\Phi = \{\bar{a} \in I(\mathbf{R})^k : I(\mathbf{R}) \models \phi(\bar{a})\}$.*
2. *The interpretation of S_i in $I(\mathbf{S})_\Phi$ is the set*

$$I(\mathbf{S})_\Phi(S_i) = \{\bar{a} \in I(\mathbf{S})_\Phi{}^{\rho(S_i)} : I(\mathbf{R}) \models (\phi_i(\bar{a}))\}.$$

Note that $I(\mathbf{S})_\Phi$ is a **S** *instance of cardinality at most $\mid \mathbf{R} \mid^k$.*

3. *The partial function $\Phi^* : I(\mathbf{R}) \rightarrow I(\mathbf{S})$ is defined by $\Phi^*(I(\mathbf{R})) = I(\mathbf{S})_\Phi$. Note that $\Phi^*(I(\mathbf{R}))$ is defined iff $I(\mathbf{R}) \models \exists \bar{x}\phi$.*

Φ^* maps **R** instances into **S** instances, by computing the answers to the queries $\phi_1, \ldots, \phi_m$ over the domain of **R** specified by ϕ, see Fig. 1. The definition of Φ^* can be extended on the case of sub-sets of **R** instances in the regular way.

Next we want to describe the way formulae (query expressions) are transformed when we transform databases by Φ^*. For this a function $\Phi^\#$ from $\mathcal{L}_1$–formulae over **S** $\mathcal{L}_2$–formulae over **R** can be directly associated with a translation scheme Φ, see Fig. 1.

Definition 3 (Induced map $\Phi^\#$). *Let θ be a* **S**–*formula and Φ be k–feasible for* **S** *over* **R**. *The formula θ_Φ is defined inductively as follows:*

1. *For each $S_i \in \mathbf{S}$ and $\theta = S_i(x_1, \ldots, x_l)$ let $x_{j,h}$ be new variables with $j \leq l$ and $h \leq k$ and denote by $\bar{x}_j = \langle x_{j,1}, \ldots, x_{j,k} \rangle$. We make $\theta_\Phi = \phi_i(\bar{x}_1, \ldots, \bar{x}_l)$.*
2. *For the boolean connectives, the translation distributes, i.e. if $\theta = (\theta_1 \vee \theta_2)$ then $\theta_\Phi = (\theta_{1\Phi} \vee \theta_{2\Phi})$ and if $\theta = \neg\theta_1$ then $\theta_\Phi = \neg\theta_{1\Phi}$, and similarly for $\wedge$.*
3. *For the existential quantifier, we use relativization, i.e., if $\theta = \exists y\theta_1$, let $\bar{y} = \langle y_1, \ldots, y_k \rangle$ be new variables. We make $\theta_\Phi = \exists \bar{y}(\phi(\bar{y}) \wedge (\theta_1)_\Phi)$.*
4. *For infinitary logics: if $\theta = \bigwedge \Psi$ then $\theta_\Phi = \bigwedge \Psi_\Phi$.*
5. *For second order variables U of arity ℓ and $\bar{a}$ a vector of length ℓ of first order variables or constants we translate $V(\bar{a})$ by treating V like a relation symbol and put $\theta_\Phi = \exists V(\forall \bar{v}(V(\bar{v}) \rightarrow (\phi(\bar{v}_1) \wedge \ldots \phi(\bar{v}_\ell) \wedge (\theta_1)_\Phi)))$.*
6. *For LFP, if $\theta = n\text{-}LFP\bar{x}, \bar{y}, \bar{u}, \bar{v}\theta_1$ then $\theta_\Phi = nk\text{-}LFP\bar{x}, \bar{y}, \bar{u}, \bar{v}\theta_{1\Phi}$.*
7. *For TC: if $\theta = n\text{-}TC\bar{x}, \bar{y}, \bar{u}, \bar{v}\theta_1$ then $\theta_\Phi = nk\text{-}TC\bar{x}, \bar{y}, \bar{u}, \bar{v}\theta_{1\Phi}$.*

8. *The function* $\Phi^\# : \mathcal{L}_1$ *over* $\mathbf{S} \to \mathcal{L}_2$ *over* $\mathbf{R}$ *is defined by* $\Phi^\#(\theta) = \theta_\Phi$.

9. *For a set of* $\mathbf{S}$*-formulae* Σ *we define*

$$\Phi^\#(\Sigma) = \{\theta_\Phi : \theta \in \Sigma \text{ or } \theta = \forall \bar{y}(S_i \leftrightarrow S_i)\}$$

This is to avoid problems with Σ *containing only quantifier free formulae, as* $\Phi^\#(\Sigma)$ *need not be a set of tautologies even if* Σ *is. If* Σ *contains only quantifier free formulae, we can reflect effect of relativization.*

Observation 1. *1.* $\Phi^\#(\theta) \in FOL$ (SOL, TC, LFP) *if* $\theta \in FOL(SOL, TC, LFP)$*, even for vectorized* Φ.

2. $\Phi^\#(\theta) \in MSOL$ *provided* $\theta \in MSOL$*, but only for scalar* Φ.

3. $\Phi^\#(\theta) \in nk\text{-}TC(nk\text{-}LFP)$ *provided* $\theta \in n\text{-}TC(n\text{-}LFP)$ *and* Φ *is a* k*-feasible.*

4. $\Phi^\#(\theta) \in TC^{kn}(LFP^{kn}, L^{kn}_{\infty\omega})$ *provided* $\theta \in TC^n(LFP^n, L^n_{\infty\omega})$ *and* Φ *is a* k*-feasible.*

The following fundamental theorem is folklore and establishes the correctness of the translation, cf. [15]. Figure 1 illustrates the fundamental theorem.

Theorem 1. *Let* $\Phi = \langle \phi, \phi_1, \ldots, \phi_m \rangle$ *be a* k*-$\mathbf{R}$-$\mathbf{S}$-translation scheme,* $I(\mathbf{R})$ *be a* $\mathbf{R}$*-instance and* θ *be a* FOL*-formula over* $\mathbf{S}$*. Then* $I(\mathbf{R}) \models \Phi^\#(\theta)$ *iff* $\Phi^*(I(\mathbf{R})) \models \theta$.

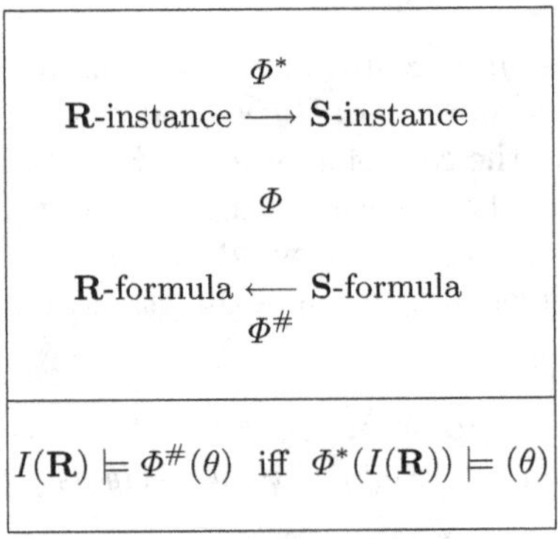

$$
\begin{array}{c}
\Phi^* \\
\mathbf{R}\text{-instance} \longrightarrow \mathbf{S}\text{-instance} \\
\\
\Phi \\
\\
\mathbf{R}\text{-formula} \longleftarrow \mathbf{S}\text{-formula} \\
\Phi^\#
\end{array}
$$

$$I(\mathbf{R}) \models \Phi^\#(\theta) \text{ iff } \Phi^*(I(\mathbf{R})) \models (\theta)$$

Fig. 1. Components of translation schemes the fundamental property

Now, we can define the composition of translation schemes:

Definition 4 (Composition of Translation Schemes). *Let* $\Psi = \langle \psi, \psi_1, \ldots, \psi_{m_1} \rangle$ *be a* k_1*-$\mathbf{R}$-$\mathbf{S}$-translation scheme, and let* $\Phi = \langle \phi, \phi_1, \ldots, \phi_{m_2} \rangle$ *be a* k_2*-$\mathbf{S}$-$\mathbf{T}$-translation scheme. Then we denote by* $\Psi \circ \Phi$ *the* $(k_1 \cdot k_2)$*-$\mathbf{R}$-$\mathbf{T}$-translation scheme given by* $\langle \Psi^\#(\phi), \Psi^\#(\phi_1), \ldots, \Psi^\#(\phi_{m_1}) \rangle$*.* $\Psi(\Phi)$ *is called the composition of* Φ *with* Ψ.

One can easily check that the syntactically defined composition of translation schemes has the following semantic property: $\Psi \circ \Phi(I(\mathbf{R})) = \Psi(\Phi(I(\mathbf{R})))$.

Now, we give a line of examples of translation schemes, relevant to the field of database theory. Assume that in all the examples, we a given database scheme $\mathbf{R} = (R_1, R_2, \ldots, R_n)$.

Example 1 (Restriction of the Domain). Assume we want to restrict the domain of $\mathbf{R}$ by allowing only elements, which satisfy some condition, defined by formula $\phi(x)$ in the chosen language (FOL, relation calculus, etc.). The corresponding translation scheme $\Phi_{Restriction}$ is:

$$\Phi_{Restriction} = \langle \phi, R_1, R_2, \ldots, R_n \rangle.$$

Example 2 (Deletion of a Definable set of Tuples from a Relation). Assume we want to delete from a relation, say, R_i of the $\mathbf{R}$ a set of tuples, which do not satisfy some condition, defined by formula θ. The corresponding translation scheme Φ_{DT} is:

$$\Phi_{DT} = \langle x \approx x, R_1, \ldots, R_{i-1}, R_i \wedge \neg\theta, R_{i+1}, \ldots, R_n \rangle.$$

Example 3 (Insertion of a Tuple into a Relation). Assume we want to insert a tuple into a relation, say, $R_i[A_1, \ldots, A_{k_i}]$ of the $\mathbf{R}$, where R_i contains k_i attributes. The corresponding translation scheme Φ_{InT} is a parametrized translation scheme with k_i parameters $a_1, \ldots, a_{k_i}$, which can be expressed, for example, in FOL in the following way:

$$\Phi_{InT} = \langle x \approx x, R_1, R_2, \ldots, R_{i-1}, (R_i \bigvee (\bigwedge_{1 \leq j \leq k_i} x_j \approx a_j)), R_{i+1}, \ldots, R_n \rangle.$$

Example 4 (Vertical Decomposition (Projections)). The *vertical decomposition,* given by a translation scheme in FOL notation, is:

$$\Phi_{VD} = \langle x \approx x, \phi_1, \ldots, \phi_n \rangle,$$

where each ϕ_i is of the form $\phi_i(\bar{x}_i) = \exists \bar{y}_i R_i(\bar{x}_i, \bar{y}_i)$. R_i is a relation symbol from $\mathbf{R}$ and $\bar{x}_i$ is a vector of free variables. In relational algebra notation, this amounts to $\phi_i(\bar{x}_i) = \pi_{\bar{x}_i} R_i$.

Example 5 (Vertical Composition (Join)). The *vertical composition,* given by a translation scheme in FOL notation, is:

$$\Phi_{VC} = \langle x \approx x, \phi_1, \ldots, \phi_n \rangle,$$

where each ψ_i is of the form $\phi_i(\bar{x}_i) = \bigwedge_{l=1}^{k} R_{i_l}(\bar{x}_{i_l})$, R_{i_l} is a relation symbol from $\mathbf{R}$ and $\bar{x}_i$ is a vector of free variables. Furthermore $\cup_j \bar{x}_{i_j} = \bar{x}_i$ and for all $\bar{x}_{i_{j_1}}$ there is $\bar{x}_{i_{j_2}}$ such that $\bar{x}_{i_{j_1}} \cap \bar{x}_{i_{j_2}} \neq \emptyset$. In relational algebra notation, this amounts to $\phi_i(\bar{x}_i) = \bowtie_{l=1}^{k} R_{i_l}$. If there are no common free variables, this just defines the Cartesian product.

Example 6 (Horizontal Decomposition (Exceptions)). Assume we want to decompose a relation, say, $R_i[A_1, \ldots, A_{k_i}]$ into two parts $R_i^1[A_1, \ldots, A_{k_i}]$ and $R_i^2[A_1, \ldots, A_{k_i}]$ such that all tuples of the first part satisfy some definable condition (formula) θ and all tuples of the second part do not. Such a transformation is called *horizontal decomposition of R_i along θ* and in *FOL* notation is:

$$\Phi_{HD} = \langle x \approx x, R_1, R_2, \ldots, R_{i-1}, R_i \wedge \theta, R_i \wedge \neg\theta, R_{i+1}, \ldots, R_n \rangle.$$

Example 7 (Horizontal Composition (Union)). Assume we want to compose a new relation, say, $R_{n+1}[A_1, \ldots, A_{k_{n+1}}]$ from two given relations $R_{i_1}[A_1, \ldots, A_{k_{n+1}}]$ and $R_{i_2}[A_1, \ldots, A_{k_{n+1}}]$. Such a transformation is called *horizontal composition of R_{n+1}* and in *FOL* notation is:

$$\Phi_{HC} = \langle x \approx x, R_1, R_2, \ldots, R_n, R_{i_1} \vee R_{i_2} \rangle.$$

The translation Φ_{HC} is called the *horizontal composition (union)* of R_{i_1} and R_{i_2}.

Example 8 (Definition of a View). Assume we are given a database scheme that contains four relations: $\mathbf{R} = (R_1, R_2, R_3, R_4)$. Assume that we want to define a view of a snapshot that is derived from the database by applying the following query, given in the format of relational algebra: $\phi_{View} = (\pi_A R_1 \cup R_2) \bowtie (R_3 - \sigma_\xi R_4)$. In this case, the corresponding translation scheme is:

$$\Phi_{View} = \langle x = x, \phi_{View} \rangle.$$

4   Handling Queries Under Restructuring of Databases

In terms of translation schemes, the problem of handling queries under restructuring of databases may be paraphrased in the following way, see Fig. 2:

Given: Two different generations of the same database, say, $\mathbf{R}^g$ and $\mathbf{R}^{g+1}$. Additionally, we have two maps: Φ_g and Ψ_g, where Φ_g produces $\mathbf{R}^{g+1}$ from $\mathbf{R}^g$ and Ψ_g is the corresponding reconstruction map. Finally, there is an application (translation scheme) Φ_g^{app} on the gth generation.

Find: An application (translation scheme) Φ_{g+1}^{app} on the $(g+1)^{th}$ generation, such that: $\Phi_{g+1}^{app*}(\Phi_g^*(\mathbf{R}^g)) = \Phi_g^{app*}(\mathbf{R}^g)$.

Example 9. Assume that we are given database scheme $\mathbf{R^g} = (R^g)$ and two restructurings, defined by the following pair of translation schemes:

1. $\mathbf{R^{g+1}} = (R_1^{g+1}, R_2^{g+1})$, $\Psi_g = (\psi^g)$ and $\psi^g = (R_1^{g+2} \bowtie R_2^{g+2})$.
2. $\mathbf{R^{g+2}} = (R_1^{g+2}, R_2^{g+2}, R_3^{g+2}, R_4^{g+2})$, $\Psi_{g+1} = (\psi_1^{g+1}, \psi_2^{g+1})$ and $\psi_1^{g+1} = (\pi_A R_1^{g+2} \cup R_2^{g+2})$, $\psi_2^{g+1} = (R_3^{g+2} - \sigma_\zeta R_4^{g+2})$.

Assume that Q_g is a simple modification query on $\mathbf{R}^g$, which deletes tuples from R^g, according to some condition θ, expressed in terms of R^g. The set of deleted tuples is defined by $\bigtriangledown_\theta R^g$ rather than given by enumeration. In such a

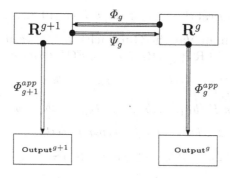

Fig. 2. Query on two different generations of database

case, we want to understand which tuples of which relations from $\mathbf{R}^{g+2}$ must be deleted, or moreover not only deleted, in order to produce the same output. Using substitutions, we obtain over $\mathbf{R}^{g+2}$:

$$\triangledown_{\Psi_{g+1}^{\#}(\Psi_g^{\#}(\theta))}((\pi_A R_1^{g+2} \cup R_2^{g+2}) \bowtie (R_3^{g+2} - \sigma_\zeta R_4^{g+2})).$$

From Example 9, we observe that the derived set of tuples, defined by $\Psi_{g+1}^{\#}(\Psi_g^{\#}(\theta))$, seems to be already in terms of $\mathbf{R}^{g+2}$. However, the corresponding modification procedure can not be directly presented in terms of updates of **relations** from $\mathbf{R}^{g+2}$.

4.1   Handling of Queries over Disjoint Unions and Shufflings

The *Disjoint Union (DJ)* is the simplest example of juxtaposition, where none of the components are linked to each other. Assume we have a set of database schemes $\mathbf{R}_i$'s and we want to define a database scheme that represents their *DJ*. In this case, we add an, so called, *index scheme* $\mathbf{R}_I$, which specifies the parameters of the composition of the database schemes. The index scheme is a database scheme, whose instances are used in combining disjoint databases into a single database.

Definition 5 (Disjoint Union). *Let* $\mathbf{R}_I$ *be a database scheme chosen as an index scheme* $\mathbf{R}_I = (R_1^I, \ldots, R_{j^I}^I)$ *with domain* I *and* $\mathbf{R}_i = (R_1^i, \ldots, R_{j^i}^i)$ *be a database scheme with domain* D_i. *In the general case, the resulting database scheme* $\mathbf{R} = \bigsqcup_{i \in I} \mathbf{R}_i$ *with the domain* $I \cup \bigcup_{i \in I} D_i$ *will be*

$$\mathbf{R} = (P(i,x), Index(x), R_j^I(1 \le j \le j^I), R_{j^i}^i(i \in I, 1 \le j^i \le j^i)) \text{ for all } i \in I,$$

- *the instance of* $P(i,x)$ *in* $\mathbf{R}$ *contains a tuple* (i,x) *iff* x *came from* R_i;
- *the instance of* $Index(x)$ *in* $\mathbf{R}$ *contains* x *iff* x *came from* I;
- $R_j^I(1 \le j \le j^I)$ *are from* $\mathbf{R}_I$ *and*
- $R_{j^i}^i(i \in I, 1 \le j^i \le j^i)$ *are from* $\mathbf{R}_i$

Now, we give the classical theorem for the DJ, cf. [19,22].

Theorem 2 (Feferman-Vaught-Gurevich). *Let $\mathbf{R}_I$ be an index scheme with domain of size k and let $\mathbf{R}=\bigsqcup_{i \in I}\mathbf{R}_i$. For any FOL formula φ over $\mathbf{R}$ there are:*

1. *formulae of FOL $\psi_{1,1},\ldots,\psi_{1,j_1},\ldots,\psi_{k,1},\ldots,\psi_{k,j_k}$*
2. *a formula of MSOL ψ_I*
3. *a boolean function $F_\varphi(b_{1,1},\ldots,b_{1,j_1},\ldots,b_{k,1},\ldots,b_{k,j_k},b_I)$*

with the formulae in 1-2 having the following property:

$$I(\mathbf{R}_i) \models \psi_{i,j} \text{ iff } b_{i,j} = 1 \text{ , and } I(\mathbf{R}_I) \models \psi_I \text{ iff } b_I = 1$$

and, for the boolean function of 3, we have

$$I(\mathbf{R}) \models \varphi \text{ iff } F_\varphi(b_{1,1},\ldots,b_{1,j_1},\ldots,b_{k,1},\ldots,b_{k,j_k},b_I) = 1.$$

Note that we require that F_φ and the $\psi_{i,j}$'s depend only on φ, k and $\mathbf{R}_1,\ldots,\mathbf{R}_k$ but not on the instances involved.

For the case of the DJ, we assume that domains of databases in each site are disjoint. However, as a rule, the values of certain attributes may appear at several sites. We can assume that the domain of the index scheme is fixed and known, however we can not (without additional assumption) fix finite number of one place predicates. This puts the main limitation on the use of Theorem 2. Moreover, even if $\phi_=$ exists for some fixed database instance, it must be independent upon the current content of database and must be formulated ahead syntactically. In addition, it must be relatively small, as otherwise it causes explosion in size of other formulae. Now, we apply logical machinery.

Definition 6 (Partitioned Index Structure). *Let $\mathcal{I}$ be an index structure over τ_{ind}. $\mathcal{I}$ is called finitely partitioned into ℓ parts if there are unary predicates I_α, $\alpha < \ell$, in the vocabulary τ_{ind} of $\mathcal{I}$ such that their interpretation forms a partition of the universe of $\mathcal{I}$.*

In addition to the DJ, one may produce a new structure by shuffling.

Definition 7 (Shuffle over Partitioned Index Structure). *Let $\mathcal{A}_i, i \in I$ be a family of structures such that for each $i \in I_\alpha$: $\mathcal{A}_i \cong \mathcal{B}_\alpha$. In this case, we say that $\biguplus^{\mathcal{I}}_{\alpha < \beta}\mathcal{A}_\alpha$ is the shuffle of $\mathcal{B}_\alpha$ along the partitioned index structure $\mathcal{I}$.*

We generalize Theorem 2 by introducing abstract preservation properties in the following way:

Definition 8 (Preservation Properties with Fixed Index Set). *For two logics $\mathcal{L}_1$ and $\mathcal{L}_2$ we define Preservation Property for Disjoint Union*

Input of operation: *Indexed set of structures;*
Preservation Property: *if for each $i \in I$ (index set) $\mathcal{A}_i$ and $\mathcal{B}_i$ satisfy the same sentences of $\mathcal{L}_1$ then the disjoint unions $\bigsqcup_{i \in I}\mathcal{A}_i$ and $\bigsqcup_{i \in I}\mathcal{B}_i$ satisfy the same sentences of $\mathcal{L}_2$.*
Notation: $DJ\text{-}PP(\mathcal{L}_1,\mathcal{L}_2)$

Definition 9 (Preservation Properties with Variable Index Structures). *For two logics $\mathcal{L}_1$ and $\mathcal{L}_2$ we define Preservation Properties for Shuffle*

Input of operation: *A family of structures $\mathcal{B}_\alpha : \alpha < \beta$ and a (finitely) partitioned index structure $\mathcal{I}$ with I_α a partition.*

Preservation Property: *Assume that for each $\alpha < \beta$ the pair of structures $\mathcal{A}_\alpha, \mathcal{B}_\alpha$ satisfy the same sentences of $\mathcal{L}_1$, and $\mathcal{I}, \mathcal{I}$ satisfy the same MSOL-sentences. Then the schuffles $\biguplus_{\alpha<\beta}^{\mathcal{I}} \mathcal{A}_\alpha$ and $\biguplus_{\alpha<\beta}^{\mathcal{I}} \mathcal{B}_\alpha$ satisfy the same sentences of $\mathcal{L}_2$.*

Notation: $Shu\text{-}PP(\mathcal{L}_1, \mathcal{L}_2)$ $(FShu\text{-}PP(\mathcal{L}_1, \mathcal{L}_2))$

Now, we list which Preservation Properties hold for which logics.

Theorem 3. *Let $\mathcal{I}$ be an index structure and $\mathcal{L}$ be any of FOL, $FOL^{m,k}$, $L^\omega_{\omega_1,\omega}$, $L^k_{\omega_1,\omega}$, $MSOL^m$, MTC^m, $MLFP^m$, or $FOL[\mathbf{Q}]^{m,k}$ $(L_{\omega_1,\omega}[\mathbf{Q}]^k)$ with unary generalized quantifiers. Then $DJ\text{-}PP(\mathcal{L}, \mathcal{L})$ and $FShu\text{-}PP(\mathcal{L}, \mathcal{L})$ hold. Note that this includes $DJ\text{-}PP(FOL^{m,k}, FOL^{m,k})$ and $FShu\text{-}PP(FOL^{m,k}, FOL^{m,k})$ with the same bounds for both arguments, and similarly for the other logics.*

Proof.

<u>FOL and $FOL^{m,k}$:</u> The proofs for FOL and $MSOL$ are classical, see in particular [6]. Extension for $FOL^{m,k}$ can be done directly from the proof for FOL.
<u>$MLFP$ and $MLFP^m$:</u> The proof for $MLFP$ was given in [4].
<u>$L_{\omega_1,\omega}(\mathbf{Q})^k$:</u> The proof was given in [8].
<u>MTC^m:</u> The proof was given in [30].

Now, we recall that analyzing Example 9, we decided that we are interested to derive from θ of $\bigtriangledown_{\Psi^\#_{g+1}(\Psi^\#_g(\theta))}((\pi_A R_1^{g+2} \cup R_2^{g+2}) \bowtie (R_3^{g+2} - \sigma_\zeta R_4^{g+2}))$ some formulae: $\theta_1^R, \ldots, \theta_i^R$ over R and $\theta_1^S, \ldots, \theta_j^S$ over S, which we will apply to R and S respectively. Now, we formulate the requirement more formally:

Definition 10 (Reduction Sequence). *Let $\mathcal{I}$ be a finitely partitioned τ_{ind}-index structure and $\mathcal{L}$ be logic.*

Let $\mathcal{A} = \biguplus_{\alpha<\beta}^{\mathcal{I}} \mathcal{B}_\alpha$ be the τ-structure which is the finite shuffle of the τ_α-structures $\mathcal{B}_\alpha$ over $\mathcal{I}$ or another combination of the components. A $\mathcal{L}_1$-reduction sequence for shuffling for $\phi \in \mathcal{L}_2(\tau_{shuffle})$ is given by

1. a boolean function $F_\phi(b_{1,1}, \ldots, b_{1,j_1}, \ldots, b_{\beta,1}, \ldots, b_{\beta,j_\beta}, b_{I,1}, \ldots, b_{I,j_I})$
2. set Υ of $\mathcal{L}_1$-formulae $\Upsilon = \{\psi_{1,1}, \ldots, \psi_{1,j_1}, \ldots, \psi_{\beta,1}, \ldots, \psi_{\beta,j_\beta}\}$
3. MSOL-formulae $\psi_{I,1}, \ldots, \psi_{I,j_I}$

and has the property that for every $\mathcal{A}, \mathcal{I}$ and $\mathcal{B}_\alpha$ as above with $\mathcal{B}_\alpha \models \psi_{\alpha,j}$ iff $b_{\alpha,j} = 1$ and $\mathcal{B}_I \models \psi_{I,j}$ iff $b_{I,j} = 1$ we have

$$\mathcal{A} \models \phi \text{ iff } F_\phi(b_{1,1}, \ldots, b_{1,j_1}, \ldots, b_{\beta,1}, \ldots, b_{\beta,j_\beta}, b_{I,1}, \ldots, b_{I,j_I}) = 1.$$

Note that we require that F_ϕ and the $\psi_{\alpha,j}$'s depend only on ϕ, β and $\tau_1, \ldots, \tau_\beta$ but not on the structures involved.

The following theorem partially answers the question of Example 9.

Theorem 4. *Let $\mathcal{L}$ be any of FOL, $FOL^{m,k}$, $L^{\omega}_{\omega_1,\omega}$, $L^{k}_{\omega_1,\omega}$ $MSOL^m$, MTC^m, $MLFP^m$, or $FOL[\mathbf{Q}]^{m,k}$ with unary generalized quantifiers. There is an algorithm, which for given $\mathcal{L}$, τ_{ind}, τ_α, $\alpha < \beta$, $\tau_{shuffle}$ and $\phi \in \mathcal{L}(\tau_{shuffle})$ produces a reduction sequence for ϕ for $(\tau_{ind}, \tau_{shuffle})$-shuffling. However, F_ϕ and the $\psi_{\alpha,j}$ are tower exponential in the quantifier rank of ϕ. Furthermore, F depends on the MSOL–theory of the index structure restricted to the same quantifier rank as ϕ.*

Proof. By analyzing the proof of Theorem 3.

Note that Theorem 4 is not true for all logics as shown in [30].

4.2   Handling Queries Over Φ–Sum

Combining Disjoint Unions and Shuffles with translation schemes, we can reach a very large set of useful structures. In this section, we present our new results in the field. We expend the classical Theorem 2 and more recent Theorems 3 and 4 to the cases, when translation schemes are involved in process of construction of the desired structure from the Disjoint Unions and Shuffles.

Definition 11 (Φ–Sum for extensions of FOL). *Let $\mathcal{I}$ be a finitely partitioned index structure and $\mathcal{L}$ be any of FOL, MSOL, MTC, MLFP, or FOL with unary generalized quantifiers. Let $\mathcal{A} = \bigsqcup_{i \in I} \mathcal{A}_i$ or $\mathcal{A} = \biguplus^{\mathcal{I}}_{\alpha < \beta} \mathcal{B}_\alpha$ be a τ–structure, where each $\mathcal{A}_i$ is isomorphic to some $\mathcal{B}_1, \ldots, \mathcal{B}_\beta$ over the vocabularies $\tau_1, \ldots, \tau_\beta$, in accordance with the partition.*

For a Φ be a scalar (non–vectorized) τ–σ $\mathcal{L}$–translation scheme, the Φ–sum of $\mathcal{B}_1, \ldots, \mathcal{B}_\beta$ over I is the structure $\Phi^*(\mathcal{A})$, or rather any structure isomorphic to it.

Theorem 5. *Let $\mathbf{R}_I$ be a finitely partitioned index database scheme, $\mathcal{L}$ be any of FOL, MSOL, MTC, MLFP, MSOL or FOL with unary generalized quantifiers. Let $\mathbf{R}$ be the Φ–sum of $\mathbf{R}_{\mathcal{B}_1}, \ldots, \mathbf{R}_{\mathcal{B}_\beta}$ over I, as above. For every $\varphi \in \mathcal{L}(\tau)$ there are*

1. *a boolean function $F_{\Phi,\varphi}(b_{1,1}, \ldots, b_{1,j_1}, \ldots, b_{\beta,1}, \ldots, b_{\beta,j_\beta}, b_{I,1}, \ldots, b_{I,j_I})$*
2. *$\mathcal{L}$–formulae $\psi_{1,1}, \ldots, \psi_{1,j_1}, \ldots \psi_{\beta,1}, \ldots, \psi_{\beta,j_\beta}$*
3. *and MSOL–formulae $\psi_{I,1}, \ldots, \psi_{I,j_I}$*

such that for every $\mathbf{R}$, $\mathbf{R}_I$ and $\mathbf{R}_{\mathcal{B}_i}$ as above with $\mathbf{R}_{\mathcal{B}_i} \models \psi_{i,j}$ iff $b_{i,j} = 1$ and $\mathbf{R}_I \models \psi_{I,j}$ iff $b_{I,j} = 1$ we have

$$\mathbf{R} \models \varphi \text{ iff } F_{\Phi,\varphi}(b_{1,1}, \ldots, b_{1,j_1}, \ldots, b_{\beta,1}, \ldots, b_{\beta,j_\beta}, b_{I,1}, \ldots, b_{I,j_I}) = 1.$$

Moreover, $F_{\Phi,\varphi}$ and the $\psi_{i,j}$ are computable from $\Phi^\#$ and φ, but are tower exponential in the quantifier depth of φ[1].

---

[1] Note that in most real applications, F_ϕ and the $\psi_{\alpha,j}$ are single exponential in the quantifier rank of ϕ.

Proof. By analyzing the proof of Theorem 4 and using Theorem 1.

Finally, we receive our main result, concerning handling of queries under restructuring of databases:

Theorem 6. *Let I be an index, $\mathcal{L}$ be FOL (or rather any language for which Theorem 5 holds), and let $\mathbf{R}^{g+1}$ be the generalized sum of $\mathbf{R}_1^{g+1'},\ldots,\mathbf{R}_\ell^{g+1'}$ over I, as usual. Let Φ_g, Ψ_g and Φ_g^{up} of the logic $\mathcal{L}$ be as above. Any query Φ_g^{app} over $\mathbf{R}^g$ gives the corresponding query Φ_{g+1}^{app} over $\mathbf{R}^{g+1}$, where $\Phi_{g+1}^{app} = \Phi_g^{app}(\Psi_g)$ and each $\varphi_{g+1,i}^{app}$ in Φ_{g+1}^{app} may be computed with the help of the corresponding boolean function $F_{\{\Phi_g,\Psi_g,\Phi_g^{app}\},\varphi_{g+1,i}^{app}}(b_{1,1},\ldots,b_{1,j_1},\ldots,b_{\ell,1}, \ldots,b_{\ell,j_\ell},b_{I,1},\ldots,b_{I,j_I})$ as in Theorem 5.*

5 Handling Queries Under Database Updates

Assume that we have a database scheme $\mathbf{R}$ and a query (translation scheme) Φ_{View}, which defines the view. Assume that $\mathbf{R}$ was updated by translation scheme Φ^{up}. In terms of translation schemes, we obtain the following formulation:

Given: Translation scheme Φ_V, and the database update Φ_{DB}^{up}.

Find: A set of view updates Φ_V^{incr} that uses the old content of the view and delete from and inserts in the view some set of tuples defined on the source database.

This leads to the situation on Fig. 3, where Φ_V^{incr} uses both: database and the old view. For the case of queries defined in relational algebra and for updates given as deletion and insertion of a (undefined) set of tuples, the question was investigated in [28]. For the case of Datalog, the answer for the same kind of updates is given in [12]. However, the techniques were defined for the specific languages. Moreover, the update operations, used in both cases are *data changes*. It means that sets of tuples, which we insert in relations or delete from relations are not defined, but given by enumeration.

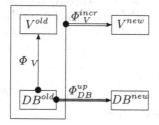

Fig. 3. Incremental view maintenance under update.

Recently, in [21], *dynamic problem* was introduced in the following way. For a sequence $w = \sigma_1,\ldots,\sigma_m \in \triangle_{can}^*(\tau)$ of operations (update translation schemes like Φ_{DB}^{up} in our formulation) and a structure $\mathcal{U}$, $w(\mathcal{U})$ is the result of subsequently applying the operations to $\mathcal{U}$ (DB^{new}), and $\mathcal{U}$ (DB^{old}) if $w = \epsilon$.

Definition 12 ([21]). *Let S be a Boolean query on τ-structures. The dynamic problem $\mathcal{D}(S)$ associated with S is the set of pairs $\mathcal{D} = (\mathcal{U}, w)$ where $\mathcal{U} \in Fin(\tau)$ and $w \in \triangle_{can}(\tau)$ is an update sequence with $w(\mathcal{U}) \in S$. The query S is called the underlying static problem of $\mathcal{D}(S)$.*

The dynamic problems are handled by *incremental evaluation systems*. These systems allow underlined relations over the universe of the input structure $\mathcal{U}$. Incremental Evaluation System (IES) for a dynamic problem $\mathcal{D}(S)$ consists of a set of logical interpretations (translation schemes) and an additional logical sentence φ. Given an initial structure $\mathcal{D}$, the IES defines auxiliary relations over the universe of $\mathcal{U}$ by an interpretation called the initial interpretation.

In practice, we are interested in update operations, which we call *relational updates*, that means definable updates. Indeed, as a rule, a regular query that deletes (inserts) data from (to) a database looks like: delete from relation R all tuples, such that ...

Let us use one example from [28] for our purposes and paraphrase it the following way:

Example 10. Given database scheme $\mathbf{R} = (R_1, R_2, R_3, R_4)$ and $\Phi_V = (\phi)$, where

$$\phi = (\pi_A R_1 \cup R_2) \bowtie (R_3 - \sigma_\zeta R_4).$$

Suppose a database update causes a set of tuples $\triangledown_\theta R_4$ to be deleted, where θ is a formula that defines the set of tuples to be deleted.

The update changes only one relation and its translation scheme is: $\Phi_{DB}^{up} = (R_1(x_1, \ldots, x_{n_1}), R_2(x_1, \ldots, x_{n_2}), R_3(x_1, \ldots, x_{n_3}), R_4(x_1, \ldots, x_{n_4}) \wedge \neg\theta(x_1, \ldots, x_{n_4}))$, where θ, in general, contains parameters.

In terms of FOL, the query that defines the view is:
$$\phi(x_1, \ldots, x_{n_3}) = ((\exists x_2 \ldots \exists x_{n_1} R_1(x_1, \ldots, x_{n_1}) \vee R_2(x_1)) \wedge$$
$$(R_3(x_1, \ldots, x_{n_3}) \wedge \neg(R_4(x_1, \ldots, x_{n_3}) \wedge \zeta(x_1, \ldots, x_{n_3})))).$$

After the update, made by Φ_{DB}^{up}, the query is:
$$\Phi_{DB}^{up\#}(\phi(x_1, \ldots, x_{n_3})) = ((\exists x_2 \ldots \exists x_{n_1} R_1(x_1, \ldots, x_{n_1}) \vee R_2(x_1)) \wedge$$
$$(R_3(x_1, \ldots, x_{n_3}) \wedge \neg((R_4(x_1, \ldots, x_{n_3})$$
$$\wedge \neg\theta(x_1, \ldots, x_{n_3})) \wedge \zeta(x_1, \ldots, x_{n_3})))).$$

First, we show:

$$((R_4(x_1, \ldots, x_{n_3}) \wedge \neg\theta) \wedge \zeta) =$$
$$(R_4(x_1, \ldots, x_{n_3}) \wedge \zeta \wedge \neg\zeta) \vee (R_4(x_1, \ldots, x_{n_3}) \wedge \zeta \wedge \neg\theta) =$$
$$(R_4(x_1, \ldots, x_{n_3}) \wedge \zeta) \wedge (\neg\theta \vee \neg\zeta) =$$
$$(R_4(x_1, \ldots, x_{n_3}) \wedge \zeta \wedge \neg R_4(x_1, \ldots, x_{n_3})) \vee ((R_4(x_1, \ldots, x_{n_3}) \wedge \zeta) \wedge (\neg\theta \vee \neg\zeta)) =$$
$$(R_4(x_1, \ldots, x_{n_3}) \wedge \zeta) \wedge (\neg R_4(x_1, \ldots, x_{n_3}) \vee \neg\zeta \vee \neg\theta) =$$
$$(R_4(x_1, \ldots, x_{n_3}) \wedge \zeta) \wedge \neg(R_4(x_1, \ldots, x_{n_3}) \wedge \zeta \wedge \theta).$$

Now, we use the equivalence, obtained above, for $\Phi_{DB}^{up\#}(\phi(x_1, \ldots, x_{n_3}))$:

$$\Phi_{DB}^{up\#}(\phi) = ((\exists x_2 \ldots \exists x_{n_1} R_1(x_1, \ldots, x_{n_1}) \vee R_2(x_1)) \wedge$$
$$(R_3(x_1, \ldots, x_{n_3}) \wedge \neg((R_4(x_1, \ldots, x_{n_3}) \wedge \neg\theta) \wedge \zeta))) =$$
$$((\exists x_2 \ldots \exists x_{n_1} R_1(x_1, \ldots, x_{n_1}) \vee R_2(x_1)) \wedge$$
$$(R_3(x_1, \ldots, x_{n_3}) \wedge \neg((R_4(x_1, \ldots, x_{n_3}) \wedge \zeta) \wedge$$
$$\neg(R_4(x_1, \ldots, x_{n_3}) \wedge \zeta \wedge \theta)) =$$
$$((\exists x_2 \ldots \exists x_{n_1} R_1(x_1, \ldots, x_{n_1}) \vee R_2(x_1)) \wedge$$
$$((R_3(x_1, \ldots, x_{n_3}) \wedge \neg(R_4(x_1, \ldots, x_{n_3}) \wedge \zeta)) \vee$$
$$(R_3(x_1, \ldots, x_{n_3}) \wedge (R_4(x_1, \ldots, x_{n_3}) \wedge \zeta \wedge \theta)))) =$$
$$((\exists x_2 \ldots \exists x_{n_1} R_1(x_1, \ldots, x_{n_1}) \vee R_2(x_1)) \wedge (R_3 \wedge \neg(R_4 \wedge \zeta))) \vee$$
$$((\exists x_2 \ldots \exists x_{n_1} R_1(x_1, \ldots, x_{n_1}) \vee R_2(x_1)) \wedge (R_3 \wedge (R_4 \wedge \zeta \wedge \theta)))) =$$
$$\phi(x_1, \ldots, x_{n_3}) \vee$$
$$((\exists x_2 \ldots \exists x_{n_1} R_1(x_1, \ldots, x_{n_1}) \vee R_2(x_1)) \wedge (R_3(x_1, \ldots, x_{n_3}) \wedge$$
$$(R_4(x_1, \ldots, x_{n_3}) \wedge \zeta(x_1, \ldots, x_{n_3}) \wedge \theta(x_1, \ldots, x_{n_3})))).$$

The second part of $\Phi_{DB}^{up\#}(\phi(x_1, \ldots, x_{n_3}))$ is exactly

$$((\pi_A R_1 \cup R_2) \bowtie (R_3 \cap \sigma_\zeta \nabla_\theta R_4)),$$

if written in relational algebra notation.

Example 10 shows that the only tools, which we really used in order to obtain the new propagation rules, were logical equivalences. Note additionally that, in general, any update translation scheme $\Phi^{up} = (\phi_1, \ldots, \phi_i, \ldots, \phi_n)$, which deletes (inserts) tuples, according to condition θ, from (to) relation R_i of database scheme $\mathbf{R} = (R_1, \ldots, R_i, \ldots, R_n)$ is in the form: $\phi_j = R_j$ if $i \neq j$ and $\phi_i = (R_i \wedge \neg\theta)$ (or $\phi_i = (R_i \vee \theta)$ for insertion of tuples, described by θ), without relativization but parametrized.

Now, the following proposition generalizes the example and gives the following answer:

Proposition 1. *For any formula ξ of FOL, MSOL or SOL and for any update translation scheme Φ^{up} of the same logic, it holds: $\Phi^{up\#}(\xi) = \xi$ or there is a set of formulae ξ'_i, $1 \leq i \leq n$ of the same logic, such that $\Phi^{up\#}(\xi) = (\ldots((\xi \circ_1 \xi'_1) \circ_2 \xi'_2)\ldots \circ_n \xi'_n)$, where $\circ_i \in \{\wedge, \vee\}$.*

Proof. By induction on ξ.

To show the same fact for LFP, IFP and TC, we use:

Theorem 7. *Given $\psi_1(\bar{x}, X, \bar{y})$ and $\psi_2(\bar{x}, X, \bar{z})$, it holds:*

$LFP\bar{x}, X, \bar{u}(\psi_1(\bar{x}, X, \bar{y}) \vee \psi_2(\bar{x}, X, \bar{z})) = LFP\bar{x}, X, \bar{u}(LFP\bar{x}, X, \bar{u}(\psi_1(\bar{x}, X, \bar{y})) \vee \psi_2(\bar{x}, X, \bar{z}));$
$IFP\bar{x}, X, \bar{u}(\psi_1(\bar{x}, X, \bar{y}) \vee \psi_2(\bar{x}, X, \bar{z})) = IFP\bar{x}, X, \bar{u}(IFP\bar{x}, X, \bar{u}(\psi_1(\bar{x}, X, \bar{y})) \vee \psi_2(\bar{x}, X, \bar{z}));$
$TC\bar{x}, X, \bar{u}(\psi_1(\bar{x}, X, \bar{y}) \vee \psi_2(\bar{x}, X, \bar{z})) = TC\bar{x}, X, \bar{u}(TC\bar{x}, X, \bar{u}(\psi_1(\bar{x}, X, \bar{y})) \vee \psi_2(\bar{x}, X, \bar{z})).$

The same holds for $\wedge$ as well.

Proof. The proof follows directly from the semantics of LFP, IFP and TC.

Now, it remains to combine Proposition 1 and Theorem 7 with the following results, proven in [14]:

Theorem 8. *If φ is an LFP-formula and φ' is an IFP-formula then there is a first-order formula ψ, such that φ is equivalent to $\exists(\forall)\bar{u}'LFP\bar{x}, X, \bar{u}\psi$ and there is an existential first-order formula ψ', such that φ' is equivalent to $\exists(\forall)\bar{u}'IFP\bar{x}, X, \bar{u}\psi'$.*

Theorem 9. *Suppose that we have two constant c and d and in our model $c \neq d$. Let φ be an existential pos-TC-formula. Then φ is equivalent to a formula of the form: $TC\bar{x}, \bar{x}', c, d\psi(\bar{x}, \bar{x}')$, where ψ is a first-order quantifier-free formula.*

Finally, we receive our main result, concerning handling of queries under database updates:

Theorem 10. *Every query expressible in FOL, MSOL, SOL, LFP, IFP and existential pos-TC allows incremental view re-computation.*

Proof. Use Proposition 1 with Theorems 7, 8 and 9.

As $I\text{-}DATALOG \equiv IFP$ and on ordered databases $LFP(TC)$ covers polynomial time (logarithmic space) computations, we, in particular, have:

Corollary 1. *1. Every I-DATALOG program allows incremental re-computation.*

2. On ordered databases every program, computable in polynomial time or logarithmic space, allows incremental re-computation.

6   Discussion and Conclusions

The paper introduces a unified logic based approach to maintenance of queries under database changes and shows how known results in translations schemes transfer can be applied to particular problems in database maintenance. This approach unifies different aspects, related to both schema and data evolution in databases, into a single framework. The basic underlying notion of a *logical translation scheme* and its *induced maps*, is based on the classical syntactic notion of interpretability from logic, made explicit by M. Rabin in [29].

Analyzing computations on different generations of databases, using our general technique, we encountered several problems with some kinds of rules, for example, deletion over join. Systematically using the technique of translation scheme, we introduced the notion of Φ-sums and showed how queries, expressible in different extensions of FOL may be handled over different generations of the Φ-sums.

Moreover, using the technique of translation scheme, we introduced the notions of an *incremental view re-computations*. We proved that every query expressible in FOL, $MSOL$, SOL, LFP, IFP and existential $pos\text{-}TC$ allows incremental view re-computations. The last results lead to the corollary that every $I\text{-}DATALOG$ program allows incremental re-computation. Moreover, it follows from our main results that on ordered databases every program, computable in polynomial time or logarithmic space, allows incremental re-computation.

References

1. Abiteboul, S., Hull, R., Vianu, V.: Foundations of Databases. Addison-Wesley, Boston (1995)
2. Benedikt, M., Koch, C.: From XQuery to relational logics. ACM Trans. Database Syst. **34**(4), 25:1–25:48 (2009)
3. Buneman, P., Khanna, S., Tan, W.-C.: On propagation of deletions and annotations through views. In: Proceedings of the 21st ACM SIGMOD-SIGACT-SIGART Symposium on Principles of Database Systems, Madison, WI, 3–6 June 2002, pp. 150–158 (2002)
4. Bosse, U.: Ehrenfeucht-Fraïssé Games for Fixed Point Logic. Ph.D. thesis. Department of Mathematics, University of Freiburg, Germany (1995)
5. Bancilhon, F., Spyratos, N.: Update semantics of relational views. ACM Trans. Database Syst. **6**(4), 557–575 (1981)
6. Chang, C.C., Keisler, H.J.: Model Theory. Studies in Logic, 3rd edn. vol. 73. North-Holland, Amsterdam (1990)
7. Codd, E.F.: A relational model of large shared data banks. Commun. ACM **13**(2), 377–387 (1970)
8. Dawar, A., Hellat, L.: The expressive power of finitely many genaralized quantifiers. Technical report CSR 24–93. Computer Science Department, University of Wales, University College of Swansea, UK (1993)
9. Dong, G., Libkin, L., Wong, L.: Incremental recomputation in local languages. Inf. Comput. **181**(2), 88–98 (2003)
10. Dong, G., Su, J.: Deterministic FOIES are strictly weaker. Ann. Math. Artif. Intell. **19**(1–2), 127–146 (1997)
11. Dong, G., Su, J.: Arity bounds in first-order incremental evaluation and definition of polynomial time database queries. J. Comput. Syst. Sci. **57**(3), 289–308 (1998)
12. Dong, G., Topor, R.: Incremental evaluation of Datalog queries. In: Hull, R., Biskup, J. (eds.) ICDT 1992. LNCS, vol. 646, pp. 282–296. Springer, Heidelberg (1992)
13. Dong, G., Zhang, L.: Separating auxiliary arity hierarchy of first-order incremental evaluation systems using $(3k + 1)$-ary input relations. Int. J. Found. Comput. Sci. **11**(4), 573–578 (2000)
14. Ebbinghaus, H.D., Flum, J.: Finite Model Theory. Perspectives in Mathematical Logic. Springer, Berlin (1995)
15. Ebbinghaus, H.D., Flum, J., Thomas, W.: Mathematical Logic. Undergraduate Texts in Mathematics, 2nd edn. Springer, New York (1994)
16. Franconi, E., Guagliardo, P.: On the translatability of view updates. In: Freire, J., Suciu, D. (eds.) AMW. CEUR Workshop Proceedings, vol. 866, pp. 154–167 (2012)
17. Franconi, E., Guagliardo, P.: The view update problem revisited. CoRR abs/1211.3016 (2012)
18. Franconi, E., Guagliardo, P.: Effectively updatable conjunctive views. In: Proceedings of the 7th Alberto Mendelzon International Workshop on Foundations of Data Management, Puebla/Cholula, Mexico, 21–23 May 2013
19. Feferman, S., Vaught, R.: The first order properties of products of algebraic systems. Fundam. Math. **47**, 57–103 (1959)
20. Guagliardo, P., Pichler, R., Sallinger, E.: Enhancing the updatability of projective views. In: Proceedings of the 7th Alberto Mendelzon International Workshop on Foundations of Data Management, Puebla/Cholula, Mexico, 21–23 May 2013

21. Grädel, E., Siebertz, S.: Dynamic definability. In: Proceedings 15th International Conference on Database Theory ICDT, Berlin, Germany, 26–29 March 2012, pp. 236–248 (2012)
22. Gurevich, Y.: Modest theory of short chains, I. J. Symbolic Logic **44**, 481–490 (1979)
23. Guagliardo, P., Wieczorek, P.: Query processing in data integration. In: Kolaitis, P.G., Lenzerini, M., Schweikardt, N. (eds.) Data Exchange, Information, and Streams. Dagstuhl Follow-Ups, vol. 5, pp. 129–160. Schloss Dagstuhl, Leibniz-Zentrum für Informatik (2013)
24. Hegner, S.J.: The relative complexity of updates for a class of database views. In: Seipel, D., Turull-Torres, J.M. (eds.) FoIKS 2004. LNCS, vol. 2942, pp. 155–175. Springer, Heidelberg (2004)
25. Immerman, N.: Descriptive Complexity. Graduate Texts in Computer Science. Springer, New York (1999)
26. Koch, C.: Incremental query evaluation in a ring of databases. In: Proceedings of the 29th ACM SIGMOD-SIGACT-SIGART Symposium on Principles of Database Systems, Indianapolis, IN, 6–11 June 2010, pp. 87–98 (2010)
27. Makowsky, J.A., Ravve, E.V.: BCNF via attribute splitting. In: Düsterhöft, A., Klettke, M., Schewe, K.-D. (eds.) Conceptual Modelling and Its Theoretical Foundations. LNCS, vol. 7260, pp. 73–84. Springer, Heidelberg (2012)
28. Quan, X., Wiederhold, G.: Incremental recomputation of active relational expressions. IEEE Trans. Knowl. Data Eng. **3**(3), 337–341 (1991)
29. Rabin, M.O.: A simple method for undecidability proofs and some applications. In: Bar Hillel, Y. (ed.) Logic, Methodology and Philosophy of Science II. Studies in Logic, pp. 58–68. North Holland, (1965)
30. Ravve, E.V., Volkovich, Z., Weber, G.-W.: A uniform approach to incremental reasoning on strongly distributed systems. In: Proceedings of GCAI2015 (2015, to appear)
31. Weber, V., Schwentick, T.: Dynamic complexity theory revisited. In: Proceedings 22nd Annual Symposium on Theoretical Aspects of Computer Science, Stuttgart, Germany, 24–26 February 2005, pp. 256–268 (2005)

Dealing with Knowledge

Selected Results and Related Issues of Confidentiality-Preserving Controlled Interaction Execution

Joachim Biskup[✉]

Technische Universität Dortmund, Dortmund, Germany
`joachim.biskup@cs.tu-dortmund.de`

Abstract. Controlled Interaction Execution has been developed as a security server for inference control shielding an isolated, logic-oriented information system when interacting over the time with a client by means of messages, in particular for query and transaction processing. The control aims at preserving confidentiality in a formalized sense, intuitively and simplifying rephrased as follows: Even when having (assumed) a priori knowledge, recording the interaction history, being aware of the details of the control mechanism, and unrestrictedly rationally reasoning, the client should never be able to infer the validity of any sentence declared as a potential secret in the security server's confidentiality policy. To enforce this goal, for each of a rich variety of specific situations a dedicated censor has been designed. As far as needed, a censor distorts a functionally expected reaction message such that suitably weakened or even believably incorrect information is communicated to the client. In this article, we consider selected results of recent and ongoing work and discuss several issues for further research and development. The topics covered range from the impact of the underlying logic, whether propositional or first-order or for non-monotonic beliefs or an abstraction from any specific one, to the kind of the interactions, whether only queries or also view publishing or updates or revisions or even procedural programs.

Keywords: A priori knowledge · Belief · Censor · Client state · Completeness · Confidentiality · Constraint satisfaction · Distortion · Evaluated secrecy · First-order logic · Guarded commands · Inference control · Information system · Information flow control · Interaction history · Knowledge · Lying · Model theory · Monitoring · Non-monotonic reasoning · Policy · Possibilistic secrecy · Proof theory · Program execution · Query answering · Rational reasoning · Refusal · Relational database · Security automaton · Security invariant · Theorem proving · Update processing · View publishing · Weakening

1 Introduction

As surveyed in [11,12,17,34], Controlled Interaction Execution, CIE, has been developed as a *security server* for *inference control* [9,54] shielding an isolated

© Springer International Publishing Switzerland 2016
M. Gyssens and G. Simari (Eds.): FoIKS 2016, LNCS 9616, pp. 211–234, 2016.
DOI: 10.1007/978-3-319-30024-5_12

information system when *interacting* over the time with a *client* by means of *messages*. Controlled interactions might comprise *query* answering, *update* processing complemented with refreshment notifications, *revision* processing, more generally *transaction* processing, even more generally execution of a *procedural program* with guarded commands, and *view publishing*, in each case based on logic-based *formal semantics* [1,63], like for relational databases.

Following the spirit of many other works on secrecy [28,61], a CIE-control aims at provably *preserving confidentiality* in a fully formalized sense, intuitively and simplifying rephrased as follows: Even when having (assumed) *a priori knowledge*, recording the *interaction history*, being aware of the *details of the control mechanism*, and unrestrictedly *rationally reasoning*, the client should never be able to infer the validity of any sentence declared as a potential secret in the server's *confidentiality policy*. In other words, the client should always believe in the *possibility* that such a sentence is not valid in the underlying information system, or at least not plausible. If interactions may modify the instance of the information system, this requirement refers to either the current instance only or to previous instances as well. Moreover, the *notion of validity* might depend on the kind of the underlying information system, e.g., whether seen as providing a formal and either complete or incomplete representation of an outside "real world", or whether treated as formally reflecting somebody's internal *belief* under non-monotonic reasoning.

To enforce this goal, for each of a rich variety of specific situations a dedicated *censor* has been designed. Basically, on a client's request or triggered by a spontaneous activity of the information system, such a censor first inspects the functionally expected interaction behavior, whether it would preserve confidentiality in the strong sense sketched above. If it does, the *expected* reaction message is sent to the client. Otherwise, the censor determines a *distorted* reaction message that first of all preserves confidentiality and additionally should be as informative as possible for the sake of the conflicting goal of availability. Distortions will lead the system to communicate suitably *weakened* or even believably *incorrect* data to the client, depending on the basic enforcement strategy of the chosen censor. In particular, the choice has to consider whether reaction messages containing lies are seen as socially acceptable for the concrete application.

In principle, at any point of time, the decision taken and the distortions made by the censor not only have to consider the *past* of the interactions but also have to ensure the option to continue with interactions in the *future*. Accordingly, the effectiveness of each censor is based on maintaining a suitably formed *security invariant*. In a sense, the control instantiated by a censor is proceeding like a *security automaton* [7,51,74], which *monitors* an unlimited *stream* of messages built from a client's requests and the corresponding reactions.

Typically, checking the pertinent invariant for a tentative reaction message requires to solve one or several *entailment problems* in the formal logics on which the underlying information system is based or by which the client's reasoning is assumed to be captured, respectively. Hence, from an algorithmic point of view, in general the censor has to be supported by applicable *theorem provers*.

For special cases, however, we would prefer to exploit more dedicated procedures to enhance the runtime efficiency.

In this article, we consider a selection of the results of recent and ongoing work about CIE and discuss several issues for further research and development:

- In Sect. 2, within a simple framework based on finite classical propositional logic, we introduce into the main concepts of CIE for controlling sequences of queries and in particular present the basic approaches to construct censors employing refusals as the strongest form of weakening, lying, and a combination of refusal and lying, respectively.
- In Sect. 3 we abstract from using a specific logic, in particular to compare the basic approaches and to determine their inherent complexity.
- In Sect. 4 we examine the problems arising from essentially increasing the expressiveness of the underlying logic, more specifically of using first-order logic as a foundation of relational databases, in particular enabling to deal with open queries with the need to control completeness sentences.
- In Sect. 5 we describe a static alternative to dynamic query processing, namely to publish a controlled view, basically expanding on two fundamental strategies, an intensionally working one based on sufficiently exhaustive querying and an extensionally working one based on removing violations of constraints stemming from a priori knowledge and the confidentiality policy.
- In Sect. 6 we examine the impact of a more advanced information system, in particular handling incompleteness and belief rather than complete knowledge and the client's corresponding possibilities of inferences.
- In Sect. 7 we extend the interactions to also process updates or revisions, and even to execute a procedural program, finally leading to combine CIE with language based information flow control with declassification.

When considering formal theorems presented in previous work, we will often neglect technical details and omit precise suppositions in order to focus on the main assertion in hopefully intuitive terms. Accordingly, we will refer to such a rephrasement as a "Result", and the reader is kindly advised to find the missing technicalities in the original publications. Moreover, we do not repeat technically elaborated examples. Furthermore, we will summarize an outlook to future work as an "Issue", also mostly in simplifying terms and leaving open the exact status.

As a guideline for reading the remainder, the *overall conclusion* will be the following. A computing agent's reasoning about its *own* knowledge or belief has been a successfully explored research topic, the results of which are first of all used for the information system agent in our context. This research has also been extended to the considerations of one agent, in our context the client, about the *internal* knowledge or belief of *another* agent, in our context the information system, based on *observable* communication data. Now, the goal of inference control adds a further challenge: How can the latter agent, the information system, *minimally distort* communication data in order to confine the achievements in reasoning by the former agent, the client, according to a declarative *confidentiality policy* to be enforced by the information system's security server?

2   A Simple Propositional Framework

We start our considerations with a simple *logic-oriented* information system: A *query* is expressed as a sentence of the language $\mathcal{L}_{pl}$ of classical propositional logic over a finite set of propositional atoms, and an *instance* of the information system is just a (semantic) model represented by a complete truth assignment to the atoms. The information system stores a fixed instance *db* and then would grant the right to submit queries to the client without, however, permitting any direct access to the instance. Moreover, initially the client is assumed to only know a priori that a set *prior* of sentences is satisfied by (valid under) *db*.

Further on, at each point in time i, the client submits a query request with a discretionarily specified sentence φ_i, and – without any control – the information system would then return either φ_i or $\neg\varphi_i$, depending on the truth evaluation $eval(db, \varphi_i)$ of the i-th query sentence regarding the fixed instance, i.e., whether or not $db \models \varphi_i$. Accordingly, after the i-th interaction, the client would be able to infer that *db* satisfies the elements of the current *"syntactic" view* $synView_i :=$ *prior* $\cup \{eval(db, \varphi_1), \dots, eval(db, \varphi_i)\}$, together with all sentences entailed by that set. For any other sentence ψ, from the point of view of the client, it would appear to be possible that ψ is not satisfied.

Thus, the closure of that set under entailment, in this context treated as the current *"semantic" view* denoted by $semView_i$, would constitute the client's current knowledge about the stored instance. Clearly, without control, the client could obtain complete knowledge about the instance, just by submitting a suitable sequence of queries. Accordingly, the (owner of the) information system would potentially share all information about the instance with the (human user of the) client.

Though *sharing* information would be the main goal of permitting the client to submit any sequence of queries, and thus in effect to learn their actual truth evaluations, the information system's owner might nevertheless want to enforce some *exceptions* for certain sentences seen as being too sensitive and in this context referred to as *potential secrets*. For such a sentence, independently of the actual truth value, from the point of view of the client it should *always* appear to be possible that the sentence is *not* satisfied. This goal could be achieved in two steps. In a first *declarative* step, the owner defines a *confidentiality policy* *psec* containing all sentences to be treated as a potential secret.

In a second *enforcing* step, the original information system is shielded by a security server for inference control by a censor, which gets the policy as an input parameter. The control then *intercepts* each query request and, only as far as needed, the censor *distorts* the correct truth evaluation $eval(db, \varphi_i)$ of the query sentence into a controlled answer sentence ans_i, in order to confine the information content of the reaction message returned to the client appropriately, as required by the policy. Consequently, the syntactic material available to the client becomes $synView_i := prior \cup \{ans_1, \dots, ans_i\}$.

Now it is important to observe that the distortions might have broken the straightforward relationship between a syntactic view, literally extracted from the messages of the interactions, and the corresponding semantic view:

- Purely functionally, without control and according to common usage of formal logic, the semantic view is obtained by applying the closure of the syntactic view under entailment.
- With inference control, however, facing potential distortions, the semantic view can only be determined by considering the details of the censor.

More specifically, in the case of inference control, the client has to investigate questions of the following kind. Why did the censor require to return the "verbatim" answer ans_i to the query about the truth evaluation of φ_i? Which possible instances of the information system do lead to that verbatim answer? Which of the two possible truth evaluations of φ_i do cause that verbatim answer? Thus, in most general mathematical terms (see, e.g., Sect. 4 of [9] for further exposition), the semantic view has to be determined by inverting the function that describes the censor on the function values observed as verbatim answers. If the *inverted function* happens to map a verbatim answer to a singleton pre-image containing *exactly one* element, and the client can actually compute this element, then this element contributes full knowledge to the semantic view of the client; the distortions might have changed the syntactic form of the correct answer, but the "real" information content has been preserved. Otherwise, if the pre-image has *at least two* elements, the distortions have not only changed the syntactic form of correct answers but also introduced *uncertainty* about them.

Having the distinction between a syntactic view and the corresponding semantic view in mind, one can construct a concrete censor following three guidelines:

1. Let the censor express any answer, whether correct or distorted, *as a sentence* of the underlying language $\mathcal{L}_{pl}$ (or as a convenient abbreviation of such a sentence) such that the answer looks like "being informative" and the syntactic view $synView_i$ remains a consistent subset of $\mathcal{L}_{pl}$.
2. Let the censor maintain a suitable *security invariant*, also to be ensured as a *precondition* for $synView_0 := prior$, which in particular expresses that none of the potential secrets in the policy *psec* is ever entailed by the *syntactic view* $synView_i$:

$$\text{for all } \psi \in psec : synView_i \not\models \psi . \qquad (1)$$

   Since for propositional logic the semantic notion of *entailment*, $\models$, is equivalent to a syntactic notion of *derivability (formal provability)*, $\vdash$, given a tentative answer the censor can *computationally* check whether the invariant would be maintained.
3. For each query, in general also dependent on the history and thus on the current view, let the censor *computationally* check this *derivation problem* expressed in the logic – and possibly further or more general ones –, to determine the need of a distortion regarding the *semantic* view (for which the inverted censor function is involved). Then, as indicated by the outcomes of the checks, let the censor form the answer sentence such that, from the client's point of view, it remains *indistinguishable* what the correct answer would have been, i.e., the inversion of the answer would show a pre-image containing both possibilities.

Obviously, the third guideline is the most difficult one to handle, since it is directed to capture the crucial relationship between the syntactic view (what has been shown to the client) and the semantic view (what was the cause of what has been shown). The basic approaches to the wanted construction successfully handle this difficulty by proceeding as sketched in the following.

A censor following the basic *refusal* approach [8,10,13,14,17,36,76] first checks whether the correct answer could already be inferred from the current view; if this is not the case, then – in particular to ensure indistinguishability by instance-independence – the censor inspects both the query sentence φ_i and its negation $\neg\varphi_i$: if returning any of them would lead to a direct violation of the confidentiality policy, then the answer sentence is formed by *weakening* the correct answer into a tautology expressing "tertium non datur" for the query sentence (which w.l.o.g. can be abbreviated by a keyword like mum, interpreted as a refusal notification):

$$ans_i :=$$ (2)
$$\textbf{if } synView_{i-1} \models eval(db, \varphi_i)$$
$$\textbf{then } eval(db, \varphi_i) \qquad \text{\%the correct answer}$$
$$\textbf{else if } (exists\ \psi)[\psi \in psec \text{ and}$$
$$(synView_{i-1} \cup \{\varphi_i\} \models \psi \text{ or } synView_{i-1} \cup \{\neg\varphi_i\} \models \psi)]$$
$$\textbf{then } (eval(db, \varphi_i) \vee \neg eval(db, \varphi_i))\% \text{ a tautology, or mum}$$
$$\textbf{else } eval(db, \varphi_i) \qquad \text{\% the correct answer}$$

A censor following the basic *lying* approach [8,13,14,17,36,39] only inspects the correct truth evaluation $eval(db, \varphi_i)$ of the query sentence φ_i but – in particular to ensure consistent answers – regarding a stronger violation condition, namely whether the disjunction of all policy elements would be entailed:

$$ans_i :=$$ (3)
$$\textbf{if } synView_{i-1} \cup \{eval(db, \varphi_i)\} \models \bigvee_{\psi \in psec}\psi$$
$$\textbf{then } \neg eval(db, \varphi_i) \text{ \% a lie}$$
$$\textbf{else } eval(db, \varphi_i) \quad \text{\% the correct answer}$$

A censor following the basic *combined* approach [14,15,36] first inspects the correct truth evaluation $eval(db, \varphi_i)$ of the query sentence φ_i; if it would lead to a direct violation then – in particular to ensure consistent answers – the censor additionally inspects the negation of the correct truth evaluation: if that negation would also lead to a violation, then the answer sentence is formed by *weakening* the correct answer into a tautology (or mum); otherwise the negation is returned as a lie:

$$ans_i :=$$ (4)
$$\textbf{if } (exists\ \psi)[\psi \in psec \text{ and } synView_{i-1} \cup \{eval(db, \varphi_i)\} \models \psi]$$
$$\textbf{then if } (exists\ \psi)[\psi \in psec \text{ and } synView_{i-1} \cup \{\neg eval(db, \varphi_i)\} \models \psi]$$
$$\textbf{then } (eval(db, \varphi_i) \vee \neg eval(db, \varphi_i)) \text{ \% a tautology, or mum}$$
$$\textbf{else } \neg eval(db, \varphi_i) \qquad \text{\% a lie}$$
$$\textbf{else } eval(db, \varphi_i) \qquad \text{\% the correct answer}$$

Result 1 (Effectiveness of Basic Censors for Query Sequences). *For the propositional framework (and any similar ones) used for controlling* sequences of queries, *each of the basic censors for* refusal, lying, *or the* combination *of refusal and lying, respectively,* preserves confidentiality, *i.e.,*

> *for each actual instance, for each confidentiality policy, for each potential secret in that policy, for each assumed a priori knowledge, and for each sequence of query sentences, there exists an alternative instance that satisfies the a priori knowledge as well, generates the same controlled answer sentences, but does not satisfy the potential secret.*

The proofs are based on a structurally quite simple argument outlined as follows. We consider any potential secret $\psi \in psec$. First, at each point in time i, the applicable security invariant ensures the existence of an "alternative instance" that satisfies the current syntactic view but not ψ. Second, a more or less sophisticated induction up to i shows that the actual instance and the alternative instance generate the same controlled answers, and thus are indistinguishable from the client's point of view. Hence, the "alternative instance" is a witness for the *possibility* that the potential secret ψ is *not* valid.

Similarly, as already observed above, a client could gain some kind of best achievable knowledge about the actual instance by submitting an *exhaustive sequence of queries* consisting of all possible queries (up to equivalences). Clearly, the security server can use the same approach for controlled *view publishing*: on request or discretionarily, the censor just generates the final (syntactic) view as the limit of the intermediate views and then sends it to the client. So we have the following corollary to the preceding result.

Result 2 (Effectiveness of Basic Censors for Published Views). *For the propositional framework (and any similar ones) used for controlled* view publishing, *for each of the basic censors for* refusal, lying, *or the* combination *of refusal and lying, respectively, the limit of the controlled answers of any exhaustive query sequence* preserves confidentiality.

The simple framework suggests several dimensions of elaborating more sophisticated and more comprehensive approaches. In fact, many works on CIE have been motivated this way. In the remainder, we will review and discuss some of these dimensions, as announced in the introduction. Besides considering any of these dimensions in isolation, it would be worthwhile to explore which instantiations of the dimensions are compatible, or could be smoothly composed by suitable constructions.

Issue 1 (Compositionality). *Identify* composition *guidelines for suitably combining features of different dimensions, and establish the corresponding formal assurances regarding preservation of confidentiality.*

To actually design and implement a control mechanism as sketched so far, we could employ an architecture as roughly visualized by Fig. 1, which is built from at least the following components:

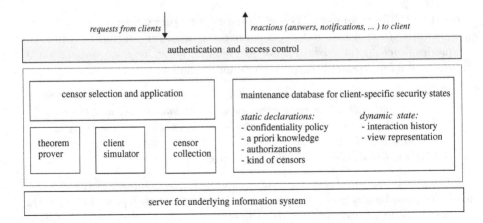

requests from clients   *reactions (answers, notifications, ...) to client*

authentication and access control

censor selection and application

| theorem prover | client simulator | censor collection |

maintenance database for client-specific security states

static declarations:
- confidentiality policy
- a priori knowledge
- authorizations
- kind of censors

dynamic state:
- interaction history
- view representation

server for underlying information system

Fig. 1. Rough architecture of Controlled Interaction Execution

- a *functional server* for the underlying information system for storing the instance and (correct) interaction processing;
- a *theorem prover* for solving entailment problems for the logics involved;
- a collection of *censors*, each of which has been verified to meet the confidentiality requirement;
- a *maintenance* database which stores for each authorized client a *client state*, in particular comprising *statically* declared parameters including
  1. the wanted client-specific *confidentiality policy*,
  2. the assumed *a priori knowledge*,
  3. the *authorizations* for interactions, and
  4. the kind of *censors* that could be applied,
  as well as *dynamic* information about
  5. the *interaction history* including the actually applied censors and
  6. a *view representation* of the client's views according to previously returned messages (in the simplest case just *synView*);
- a client *simulator* that determines that representation (in the simplest case just by adding the answer sentences to a log file, which has been initialized with the a priori knowledge).

Issue 2 (Comprehensive System Architecture). *Refine the roughly sketched architecture to an extendible software package which– given suitable parameters for each of the dimensions – can uniformly be configured and then employed as a comprehensive implementation of CIE, and of related and compatible security techniques as well.*

3   An Abstract Framework

In any logic-oriented framework, whether a simple propositional one as sketched in Sect. 2 or a suitably extended one, two aspects are combined:

- an underlying classical (or even non-classical) *logic* comprising an intuitively expressive syntax and a formal notion of either entailment based on models or, if applicable even equivalently, of computational derivability, and
- a *censor function* together with its inverted function, where we deal with reasoning about employing that or a somehow related logic.

Since in general the latter aspect appears to be not directly expressible in the respective logic, one could attempt to bridge the gap between the two aspects by dealing with both of them in a purely functional manner.

Such a unifying treatment could be useful for several purposes, in particular for identifying features that are common to several logics and for separating the computational complexity stemming from the underlying logic and the computational complexity of inference control as essentially encoded in the inverted censor function. These and further considerations have motivated the abstract framework presented in [17]. This framework is inspired by the model-theoretic approach to semantics of formal logics, but without dealing with any concrete syntax.

More specifically, an *information system* is thought to be given by the set $\mathcal{I}$ of its possible abstract *instances* (or data sources), which are functionally treated like (semantic) models in a logic. An *abstract query* φ is then identified with its meaning, namely with a subset of $\mathcal{I}$, such that the evaluation $eval(db, \varphi)$ for an instance db just checks whether or not $db \in \varphi$ (which corresponds to $db \models \varphi$ in a logic framework) and then returns either $\varphi \subseteq \mathcal{I}$ or $(\mathcal{I} \setminus \varphi) \subseteq \mathcal{I}$. Accordingly, if an (abstract) user wants to learn about the *conjunction* of two queries φ_1 and φ_2 and submits them accordingly, he would get $\varphi_1 \cap \varphi_2$ and, similarly, $\varphi_1 \cup \varphi_2$ for the *disjunction*. Thus, *refusing* an answer to a query φ by weakening the correct answer to a *tautology* corresponds to returning $\varphi \cup (\mathcal{I} \setminus \varphi) = \mathcal{I}$, i.e., by saying that the actual instance might be any one, which a client is assumed to know anyhow. Furthermore *lying* on φ by *negation* corresponds to returning the complement $\mathcal{I} \setminus \varphi$ of the correct answer. Hence, as by the model theory of a logic, the intuitive meanings of phrases like "conjunction", "disjunction", "negation" or "tautology" are reflected by set-theoretic operations on sets of instances, i.e., by some algebra over the powerset of $\mathcal{I}$ (or a suitable subset of that powerset). Finally, an *abstract potential secret* is just given by an abstract query.

Result 3 (Effectiveness of Basic Censors in the Abstract Framework). *For the abstract framework used for controlling sequences of queries or controlled view publishing, respectively, the application of each of the basic censors for refusal, lying, or the combination of refusal and lying, respectively, preserves confidentiality.*

Result 4 (Refusal as Normal Form). *For the abstract framework used for controlled view publishing, the achievements of any effective censor can equivalently be described in terms of the basic refusal approach.*

Result 5 (Limits of Refusals are not Refinable). *For the abstract framework used for controlled view publishing, the limit (under intersection) of the* *controlled answers under the basic* refusal *approach of any exhaustive query sequence cannot be refined, and in this sense it is* optimal.

Notably, for the basic *lying* approach a corresponding result does *not* hold, essentially due to the need of protecting disjunctions of potential secrets.

Result 6 (Inherent Computational Complexity of Optimal Censoring). *For the abstract framework used for controlled view publishing (and thus, in a sense, for any sufficiently expressive framework), under suitable assumptions about the finiteness of the situation and the encoding of censors and their inversions, the following problem is* coNP-complete: *given a confidentiality policy and a censor, decide whether for each instance of the information system the censor generates a published view that is both confidentiality preserving and optimal w.r.t. the policy.*

Issue 3 (Notions of Optimality and Related Approximations). *Define and investigate meaningful notions of* optimality, *capturing suitable intuitions of "best availability", together with convincing notions of* approximation *to overcome the inherently high computational complexity.*

So far, in the simple propositional framework as well as in its extensions and in the abstract framework, the works on CIE have considered a *possibilistic* notion of confidentiality, which only requires the *existence* of at least one witness of the required property regarding an alternative instance that is both indistinguishable and "harmless". However, one might be interested in a more refined notion which treats *degrees of confidentiality* based on an *evaluation* of all indistinguishable instances regarding being either "harmful" or "harmless" [60,61,69]. For example, such an evaluation might count the *cardinalities* of the two classes and then relate the cardinalities according to a declared threshold or, if an a priori probability distribution over the set of all instances is known, determine and relate the respective *probabilities*.

Issue 4 (Generalized Abstract Framework). *Generalize the abstract framework so far dealing with possibilistic confidentiality towards kinds of* evaluated confidentiality, *in particular probabilistic confidentiality.*

4   A Relational Framework

On the one hand, classical propositional logic over a finite set of propositional atoms, as considered in Sect. 2, enjoys many nice computational properties, including computationally solvable decision problems with theorem provers like SAT(isfiability)-solvers and C(onstraint)S(atisfaction)P(roblem)-solvers which are usually highly efficient [42,62,67,83] (despite the intractable worst-case complexity). But on the other hand, that logic lacks expressiveness to capture many features needed for more advanced applications. In contrast, classical *first-order logic* is often expressive enough for such needs but suffers from essential restrictions regarding general decidability and from potentially unaffordable computational efficiency of decidable fragments [43] or practical theorem proving [71,79,80].

Serving as a foundation of *relational databases* [1], first-order logic provides formal means to interpret a stored relational instance as a (semantic) *model* satisfying the *integrity constraints* declared in the schema and to deal with *open queries*, intuitively of the kind "give me all $x, y, \ldots$ such that the property $\ldots$ holds". In this context, an open query is expressed by a formula containing free occurrences of one or more variables and expected to return those sentences that result from substituting the free occurrences with constants and then, as a (closed) sentence, are evaluated to *true* regarding the stored relational instance.

However, a closer look reveals that we have to take care about several subtle details. Classical model theory for first-order logic deals with universes (sets over which interpretations are formed and variables are ranging) and with interpretations of relation symbols of *any cardinality* [68, 75]. Accordingly, a formula with free variables might return infinitely many "true" sentences by substitution, even worse, without any further specification by typing, from any universe. But a database relation is a stored *finite* object, and an open query should always return a *finite* object, too. The latter property is guaranteed if the query formula is *domain-independent* and in particular *safe* (see, e.g., [1]), i.e., intuitively, whenever a *negation* occurs in the query formula – in principle evaluated by taking a set-theoretic complement w.r.t. to some previously determined and possibly infinite set – or a variable occurs – in principle ranging over a possibly infinite set – then the possibility of dealing with an infinite set does not actually occur, since the pertinent sets can be bounded to a finite subset.

Tentatively, all these problems could be avoided by employing only models with a *finite universe* and thus finite interpretations [52, 64]. But then at least the following problems occur: applications often suggest not to define a cardinality bound on the type of an attribute in a relation scheme, and inference control often wants to avoid *combinatorial inferences* based on a fixed and known finite cardinality of some set, like applying the pigeon hole principle, in particular when the application does not justify such a bound.

Seeing neither the classical model-theoretic semantics nor the finite-model semantics as appropriate for general inference control of advanced applications, all works of CIE dealing with relational databases [16, 18–20, 22, 23, 25–27, 35, 38] are based on so-called *DB-semantics*: interpretations are restricted to Herbrand-like ones over a fixed *infinite* universe of *constant symbols*, which are constraint by *unique names* axioms, with only *finitely many positively evaluated ground* facts. This feature has some unusual consequences, e.g., for each safe open query formula $\varphi(x)$, the sentence $(\exists x)[\neg\varphi(x)]$ is a tautology. But a comprehensive exploration of the exact relationship between classical semantics, finite-model semantics and DB-semantics appears to be not available, but see, e.g., [1, 3, 16, 63, 81].

Issue 5 (Logical Foundation of the Relational Model). *Reconsider the theory of relational databases in terms of first-order logic with DB-semantics, postulating infinite domains of constants used as unique names but considering only finite relational instances.*

Employing DB-semantics and restricting a priori knowledge including integrity constraints, confidentiality policies, (closed) query sentences and (open) query formulas such that all DB-entailment problems to be considered by a censor will be in a suitable decidable fragment of first-order logic, the basic approaches to construct a censor for sequences of queries, originally designed for closed queries only, can be extended to include also open queries [16,18]. More specifically, the extension is based on the decidability of the universal validity problem of the *Bernays-Schoenfinkel class* of sentences in prenex normal form having an $\forall^*\exists^*$ prefix, which not only holds for classical semantics and finite-model semantics, but also for DB-semantics.

The extensions of the basic approaches are then based on the following features, the first and the second of which are supported by DB-semantics:

- An open query can be evaluated by systematically *enumerating all substitutions* of the free occurrences of variables in the query formula by constants taking from the fixed universe, and handling the resulting sentences as *closed queries* to be controlled.
- Such an in principle infinite enumeration can be *terminated* after a finite number of rounds by suitably inspecting pertinent *completeness sentences* that basically state that in all further rounds the considered closed queries will be answered negatively, basically capturing a *closed world assumption* for the answers generated before. As far as needed, and at least after termination, the *controlled* truth evaluation of such a completeness sentence is explicitly added to the current view, and thus any implicit knowledge provided by the closed-world assumption is under effective inference control.
- The pertinent completeness sentences are expressible in first-order logic such that their usage in the entailment problems inspected by the censor remains in the decidable fragment.
- Besides others, statically *fixing* the enumeration sequence in advance ensures the kind of indistinguishability required by the formal notion of preservation of confidentiality, even if that enumeration is known to the client.

Result 7 (Effectiveness of Basic Approaches for Query Sequences). *For the relational framework under* DB-semantics *of first-order logic used for controlling sequences of queries including open ones, each of the basic approaches of refusal, lying, or the combination of refusal and lying, respectively, can be extended to open queries. In each case, the extended censor controls sufficiently many closed sentences obtained by a substitution in a fixed sequence and inspects suitably formed completeness sentences in a controlled way, such that each controlled answer processing terminates and preserves confidentiality.*

Issue 6 (Entailment Problems with Completeness Sentences). *Explore efficient computational approaches to decide entailment problems of first-order logic under DB-semantics when relational completeness sentences are involved.*

5   Static View Publishing

Research on confidentiality-preserving *view publishing* [57,58] spans a broad range of frameworks, including pioneering work on distortions of *statistical databases* [50,82], *value generalization* and *row-suppressing* for achieving *k-anonymity* and *l-diversity* of tables [46,66], and *database fragmentation and encryption* for cloud computing [2,45,48,59]. View publishing has also been studied for CIE for several frameworks and approaches to censor construction, guided by three different strategies as discussed below:

1. for the abstract framework using any of the basic approaches, by taking the limit of controlled answers to an exhaustive sequence of all queries [17];
2. somehow implicitly, for the relational framework with DB-semantics following any of the basic approaches, by controlling those open queries that would return a whole relation, based on a fixed exhaustive sequence of all closed and elementary queries each of which is about just one tuple [16];
3. for a specific description logics framework [4] using a variant of the basic approaches, by iteratively enumerating all possible atoms of the logic [40,41];
4. for both the propositional and the relational framework with DB-semantics following the lying approach, by iteratively modifying a given instance while also aiming at a minimum number of distortions [19,37,38];
5. for an XML-approach following a weakening approach by iteratively suppressing harmful parts [24];
6. for the relational framework with DB-semantics following a weakening approach that refines the refusal approach by globally determined value generalization [26]; and
7. for the relational framework with DB-semantics following a weakening approach by globally determined fragmentation and encryption [25,27].

The first kind of a strategy [16,17,40,41], items 1–3 above, treats a view in an *intensional* way, seeing a view as being fully characterized by its relevant properties. In this context, the relevant properties are the *controlled answers* to an *exhaustive sequence of queries* evaluated regarding the actual instance. In the abstract framework [17], see Sect. 3, due to the lack of any internal structure of instances, exhaustiveness requires to include *all* queries. In the relational framework [16], see Sect. 4, where an instance is built from tuples, exhaustiveness can be accomplished by including all *elementary* queries about just one tuple. Similarly, in the description logics framework [40,41] all atoms are employed. In more procedural terms, the view to be published is iteratively approximated "from above", starting with full ignorance (or with the assumed a priori knowledge) and then stepwise adding information to narrow it down towards the final limit. And in computational terms, the iteration should terminate in finite time to come up with a final view.

The second kind of a strategy [19,24,37,38], items 4–5 above, works in an *extensional* way, starting with the extension of the actual instance and treating both the elements of the a priori knowledge and the potential secrets in the confidentiality policy as *constraints*, employed for iteratively *modifying* the original instance: as long as any of the given or dynamically derived constraints is still violated, a violating constraint is selected and the currently considered instance is minimally modified to comply with the selected constraint. So, the view to be published is approximated "from below", starting with the actual instance and then stepwise distorting it. Again, the crucial point is termination: a modification to satisfy one constraint might cause to newly violate another one. Clearly, if the framework is expressive enough, the *constraint satisfaction problem* becomes undecidable, and thus we have to suitably restrict the expressiveness.

A third kind of strategy [25–27], items 6–7 above, also works in an *extensional* way, but in a sense more globally than iteratively. Regarding [26], without giving details here, in a first instance-independent step only considering the potential secrets, some kind of constraints on "admissible" weakenings are generated, which then, at least conceptually, are "globally solved" in a minimal way (where the actually used solver might work sequentially). Only in a second step, the actual relational instance is weakened by converting each harmful tuple (in logic terms, each ground fact) into an admissible disjunction. This two step procedure ensures that undistorted parts of the view remain isolated from weakened parts, and thus any harmful inferences are blocked. A related guarantee by isolation is employed in [25, 27]. Again, computability and efficiency is a problem, demanding for suitable restrictions.

In all strategies, while giving precedence to preserve confidentiality, availability is considered as an important secondary goal. Accordingly, the "difference" between the actual instance and the view to be published should be at least "minimal" in the sense that discarding any single distortion would lead to a violation of confidentiality. More ambitiously, however, we might even aim at finding a view that has a minimum number of distortions among the set of all confidentiality-preserving views.

So far, adding such an overall *numerical optimization problem* to the problem of preserving confidentiality has only been thoroughly treated for the relational framework following the lying approach with the extensionally working strategy [19, 37, 38]. Though this attempt has required to combine the *satisfaction problem* for sentences in an expressive fragment of first-order logic with a numerical *optimization problem*, it has been proved to be conceptually successful [38]. But this attempt appears to be not practically feasible in general, and thus often requires to relax the optimization requirement by allowing an approximation or to suitably restrict the constraints [19].

Result 8 (Intensional Iterative View Generation by Exhaustive Querying). *Subject to appropriate operations of information manipulation and to termination, an intensionally working and iteratively proceeding generation strategy returns a view that preserves confidentiality.*

Result 9 (Extensional Iterative View Generation by Eliminating Violations). *Subject to appropriate restrictions on expressiveness and to termination, an extensionally working and iteratively proceeding generation strategy returns a view that preserves confidentiality.*

Result 10 (Extensional View Generation by Global Distortions). *Subject to appropriate restrictions on expressiveness, in dedicated cases an exten-sionally working generation strategy that globally determines distortions returns a view that preserves confidentiality.*

Issue 7 (Comparison of Generalized View Generation Strategies). *Generalize and elaborate both the extensionally working and the intensionally working view generation strategy, respectively, and systematically compare their achievements, in particular regarding the availability of information provided by the confidentiality-preserving views.*

6   Advanced Reasoning

In both the propositional framework of Sect. 2 and the relational framework of Sect. 4, the underlying logic-oriented information system is supposed to *com-pletely* describe some outside "real world" by storing a representation of a (semantic) model which assigns a truth value to *all* atomic sentences and thus, by induction, to *all* sentences. In many applications, however, the (owner of the) information system might have only *incomplete knowledge* about the outside world or even only some *fragmentary internal belief.* In the rich literature about knowledge and belief engineering, many approaches to deal with such situations have been proposed and studied in detail, see, e.g., [4,6,44,53,56,63].

For inference control by means of CIE, *incompleteness* has first been treated for an extended propositional framework [36]: now, an *instance db* of the infor-mation system is a consistent set of propositional sentences of the language $\mathcal{L}_{pl}$ of classical propositional logic over a finite set of propositional atoms. While, syntactically, a *query* φ is still a sentence of $\mathcal{L}_{pl}$, semantically its evaluation is now based on the notion of entailment, also denoted by $\models$, rather than directly on truth evaluation with respect to a (semantic) model, tentatively given by

$$eval(db, \varphi) := \begin{cases} true & \text{if } db \models \varphi, \\ false & \text{if } db \models \neg\varphi, \\ undefined & \text{otherwise.} \end{cases} \tag{5}$$

whereas as before the definite results of the first two cases are directly expressible in $\mathcal{L}_{pl}$, the result of the third case is not. So, extending propositional logic, a *knowledge operator* K for a *modal logic* [53] is introduced to enable us to speak about "the information system knows that ..." and, correspondingly, "the information system does not know that ...":

$$eval(db, \varphi) := \begin{cases} K\varphi & \text{if } db \models \varphi, \\ K\neg\varphi & \text{if } db \models \neg\varphi, \\ \neg K\varphi \wedge \neg K\neg\varphi & \text{otherwise.} \end{cases} \tag{6}$$

By this approach, constructing a censor, we can now distinguish whether the information system itself does *not know* the answer to a query or whether the censor merely demands to *refuse* an informative answer. More generally, we now have four possible controlled answers, which provides additional flexibility for distorting answers. This flexibility is exploited by defining so-called *distortion tables* which determine for each combination of a client state in need of a distortion and the correct answer a controlled (possibly distorted) "harmless" answer, based on a finite list of representations of the relevant client states.

Result 11 (Effectiveness of Adapted Basic Censors for Query Sequences to Incomplete Information Systems). *For the extended* propositional *framework with* incomplete *instances used for controlling* sequences of queries *based on modal logic and employing a* distortion table, *all adaptions of each of the basic censors for* refusal, lying, *or the* combination *of refusal and lying, respectively,* preserve confidentiality.

Whereas propositional modal logic evolves from classical propositional logic in a quite natural way, extending classical first-order logic by modalities requires highly sophisticated considerations [55]. So far our attempts to transform and extend the propositional case treated in [36] to the general first-order case, which among others have also been inspired by [63,70], have not been successful.

Issue 8 (First-Order Modal Logic for Censor Construction). *Elaborate the* modal logic approach *to construct censors for* incomplete instances *of an information system based on* first-order logic.

However, restricting the first-order case to a *finite situation*, we could successfully treat a comprehensive *propositionalization* [35].

Result 12 (Propositionalized First-Order Modal Logic for Censor Construction). *The* modal logic approach *to construct confidentiality-preserving censors for* incomplete instances *of an information system can be extended to a first-order logic framework that can be finitely propositionalized.*

An alternative way to deal with inference control for incomplete information systems [40,41] has been based on description logics, which provides efficiently tractable fragments of first-order logic.

Result 13 (Censor Construction for Incomplete Information Systems Based on Description Logics). *For a* description logics *framework of an* incomplete *information system used for controlled* view publishing *following a variant of the basic approaches, the limit of the controlled answers of any exhaustive sequence of atoms* preserves confidentiality.

Though not elaborated in the context of CIE, a further interesting and very flexible approach to censor constructions for sequences of queries evaluated w.r.t. an incomplete information system has been proposed for a Boolean description logics framework [78].

Incompleteness of an instance complicates query answering and view publishing by the information system, and thus also increases the client's challenge to infer confidential information from visible reaction data. Though the client does not know the incomplete instance stored by the information system, the client is still fully aware about the system's reasoning procedure to generate an answer or a view, respectively.

However, the situation is changing, and becoming even more challenging, if the information system represents an internal *belief*, which is not only based on (classical) sentences but also on *conditionals* (also known as default rules or probabilistic rules). In order to form a consistent belief, such a system employs *non-monotonic reasoning* parameterized with an *instantiation* of some *plausibility structure* such as preference orderings, ordinal conditional functions, possibility or plausibility spaces [56]. The client then faces the additional problem of being *uncertain* about the concrete instantiation actually used by the system, and thus also the censor has to appropriately deal with that uncertainty.

Exemplarily for potentially many similar situations, CIE has been conceptually extended for a propositional information system that is based on *ordinal conditional functions* [6,77] – or, more generally, an abstract class of suitable consequence relations – and handles query requests regarding its current belief as well as revision requests [30,33]. In this work, such an abstract class is shown to be obtainable by an "allowed" axiomatization, and the censor construction is directed to preserve confidentiality regarding a client that knows the pertinent class and masters its uncertainty about which instantiation is taken by "accepting" a sentence if and only if the sentence is plausible under *all* instantiations. But other kinds of treating that kind of uncertainty could also be meaningful, for instance credulous reasoning.

Result 14 (Effectiveness of a Refusal Censor for Sequences of Belief Queries and Belief Revisions). *For a non-monotonic propositional framework for belief based on a class of consequence relations having an "allowed" axiomatization used for controlling mixed sequences of queries and revisions, a computational adaption of the basic censor for refusal preserves confidentiality assuming a skeptically reasoning client.*

Issue 9 (General Censor Constructions for Non-monotonic Frameworks). *For further examples of a non-monotonic framework, explore the options to construct a confidentiality-preserving censor, and concisely generalize such constructions.*

A crucially important aspect of any censor construction for an information system based on advanced reasoning is an (at least) two-step reflection of the system's reasoning under mutual uncertainty. As further discussed in [34], such a reflection is needed for the client simulator in the rough architecture of CIE shown in Fig. 1.

Issue 10 (A Censor's Simulation of the Client's Inference of the System's Parameterized Belief). *Given an information system based on advanced parameterized belief reasoning, identify the following: (i) what a client* *can infer about the system's belief from the reactions and (ii) what a censor can computationally determine about what the client can infer, for both cases of whether the reactions are controlled or not, respectively.*

7   Advanced Interactions

Early work about CIE has focused on inference control of query answering and view publishing regarding a *fixed* instance stored by the underlying information system. In general, however, an instance will be *modified* over the time. Then answers to queries become time-dependent, and a simple syntactic view obtained by the client by directly logging the data received might become inconsistent. Moreover, not only the information system can autonomously modify the instance, but the client itself might request a modification. For example, in a multiagent system, after having observed that the outside "real world" has changed, a client agent might inform the information system agent about the observation and request a corresponding *update* of the system's belief. Or a client agent has learnt further aspects about the unchanged "real world" and then suggests a corresponding *revision* of the system's belief.

In general, processing an update or revision request follows a sometimes quite involved protocol, in particular in order to maintain *invariants* declared in the schema of the information system in the form of sentences expressing *integrity constraints*, which are seen as being "unmodifiable" or, in other terms, "unquestionable". If a requested modification would violate a constraint, the request is either totally rejected or at least somehow "corrected". In any case, the information system would externally react by sending a corresponding notification to the client. Moreover, some complicated updates or revisions can only be handled as *transactions* such that only the finally resulting instance is guaranteed to satisfy the constraints, but the auxiliary versions generated during processing are not.

Now, receiving a *notification* about success, correction or failure of a modification request implies getting answers to *implicit queries* regarding the constraints. Thus inference control of interactions that modify the instance has to suitably distort such notifications in order to enforce the required confidentiality. Unfortunately, early research on multilevel databases with mandatory access control has already shown that maintaining integrity on the one hand and enforcing confidentiality on the other hand might be conflicting goals [47,49,65,73]. A proposed resolution has been the concept of *polyinstantiation*, i.e., introducing some kind of cover stories or lies for specific clients.

If inference control considers that a client infers knowledge in a *history-aware* way, as CIE is doing, a further difficulty arises. Observing time-dependent data about different versions of the stored instance, the client might get new options of inferences by reasoning about the causes that led to semantically different reactions on syntactically the same or related requests. Moreover, later reactions might reveal that confidential information has been valid earlier. Thus, if wanted according to the application, *continuous confidentiality preservation* might be required, i.e., to not only confine knowledge about the current instance but also about previous ones.

The problems sketched above have first been studied for the propositional framework presented in Sect. 2, suitably extended to include single updates as well as transactional ones under some simplifying restrictions [21, 29].

Result 15 (Effectiveness of Adapted Basic Censors for Sequences of Queries and Updates). *For the* propositional *framework with* complete *instances used for controlling mixed sequences of* queries *and suitably restricted single or transactional* updates, *adaptions of the basic censors for* refusal *or* lying, *respectively,* preserve confidentiality.

As mentioned before in Result 14, a similar achievement has been obtained for a propositional framework with belief revision [32].

Issue 11 (General Censor Constructions for Classical, Incomplete and Non-monotonic Frameworks with Modification of Instances). *For further examples of a classical, incomplete or non-monotonic framework with updates and, as far as applicable, revisions, explore the options to construct a confidentiality-preserving censor, and concisely generalize these constructions.*

A protocol for processing a modification can be seen as a *procedural application program* or a *stored procedure* that, depending on the client's request,

- generates and submits queries regarding the current instance,
- potentially level-wise branches according to the corresponding answers used as conditions in guarded commands, and
- in each branch
  - actually modifies the instance in a possibly "corrected" way,
  - prepares a corresponding notification and
  - finally sends it to the client.

Clearly, such procedural programs are of interest not only for specific processing of modifications but for reacting on any kind of messages received from a client. So, we would like to elaborate a generic approach to apply inference control for the execution of any such procedural program, in particular for preparing controlled notifications. Accordingly, still under some restrictions, in recent and ongoing work [31, 32] we have designed and verified a combination

- of CIE-like inference control by means of *abstract representations* of the information content of program variables keeping answers to queries regarding the stored instance and of suitably generated *distortion tables*
- with security techniques for language-based *information flow control*, in particular capturing *implicit flows* by guarded commands, by means of *security typing* and of *declassification* [5, 72].

Result 16 (Controlled Mediation of Client Requests Processed by Procedural Programs). *Assuming an integrated* fixed belief instance *obtained from one or more underlying information systems (and thus so far not allowing modifications of that belief and, suitably propagated, of the underlying instances),* *and restricting to guarded commands of the* if-then-else *form (and thus so far not allowing arbitrary repetitions) and to sensitive program variables with manageably small domain extensions, the designed combination of CIE-like inference control following a* weakening *approach with language-based information flow control* preserves confidentiality.

Issue 12 (Generalized Controlled Mediation of Client Requests Processed by Procedural Programs). *Extend and generalize the designed combination of* CIE-like inference control *with* language-based information flow control *for* procedural programs *as expressive as possible.*

Acknowledgements. I would like to sincerely thank all colleagues who have worked together with me on Controlled Interaction Execution, in particular the co-authors of joint publications. Moreover, I am specially indebted to Marcel Preuß and Cornelia Tadros for many helpful comments on an earlier draft. Finally, I gratefully acknowledge the longtime support of the German Research Council, DFG, under grants Bi 311/12 and SFB 876/A5.

References

1. Abiteboul, S., Hull, R., Vianu, V.: Foundations of Databases. Addison-Wesley, Reading (1995)
2. Aggarwal, G., Bawa, M., Ganesan, P., Garcia-Molina, H., Kenthapadi, K., Motwani, R., Srivastava, U., Thomas, D., Xu, Y.: Two can keep a secret: a distributed architecture for secure database services. In: 2nd Biennial Conference on Innovative Data Systems Research, CIDR 2005, pp. 186–199. Online Proceedings (2005)
3. Ailamazyan, A.K., Gilula, M.M., Stolbushkin, A.P., Shvarts, G.F.: Reduction of a relational model with infinite domains to the finite-domain case. Russian version: Dokl. Akad. Nauk SSSR **286**, 308–311; English translation: Sov. Phys. Dokl. **31**(1), 11–13 (1968)
4. Baader, F., Calvanese, D., McGuinness, D.L., Nardi, D., Patel-Schneider, P.F. (eds.): The Description Logic Handbook: Theory, Implementation, and Applications. Cambridge University Press, Cambridge (2003)
5. Balliu, M., Dam, M., Guernic, G.L.: Encover: symbolic exploration for information flow security. In: Chong, S. (ed.) IEEE Computer Security Foundations Symposium, CSF 2012, pp. 30–44. IEEE Computer Society, Los Alamitos (2012)
6. Beierle, C., Kern-Isberner, G.: A conceptual agent model based on a uniform approach to various belief operations. In: Mertsching, B., Hund, M., Aziz, Z. (eds.) KI 2009. LNCS, vol. 5803, pp. 273–280. Springer, Heidelberg (2009)
7. Bell, D.E., LaPadula, L.J.: Secure computer systems: a mathematical model, volume II. J. Comput. Sec. **4**(2/3), 229–263 (1996). Reprint of MITRE Corporation (1974)
8. Biskup, J.: For unknown secrecies refusal is better than lying. Data Knowl. Eng. **33**(1), 1–23 (2000)
9. Biskup, J.: Security in Computing Systems - Challenges. Approaches and Solutions. Springer, Heidelberg (2009)

10. Biskup, J.: Dynamic policy adaption for inference control of queries to a propositional information system. J. Comput. Secur. **20**, 509–546 (2012)
11. Biskup, J.: Inference-usability confinement by maintaining inference-proof views of an information system. Int. J. Comput. Sci. Eng. **7**(1), 17–37 (2012)
12. Biskup, J.: Logic-oriented confidentiality policies for controlled interaction execution. In: Madaan, A., Kikuchi, S., Bhalla, S. (eds.) DNIS 2013. LNCS, vol. 7813, pp. 1–22. Springer, Heidelberg (2013)
13. Biskup, J., Bonatti, P.A.: Lying versus refusal for known potential secrets. Data Knowl. Eng. **38**(2), 199–222 (2001)
14. Biskup, J., Bonatti, P.A.: Controlled query evaluation for enforcing confidentiality in complete information systems. Int. J. Inf. Secur. **3**(1), 14–27 (2004)
15. Biskup, J., Bonatti, P.A.: Controlled query evaluation for known policies by combining lying and refusal. Ann. Math. Artif. Intell. **40**(1–2), 37–62 (2004)
16. Biskup, J., Bonatti, P.A.: Controlled query evaluation with open queries for a decidable relational submodel. Ann. Math. Artif. Intell. **50**(1–2), 39–77 (2007)
17. Biskup, J., Bonatti, P.A., Galdi, C., Sauro, L.: Optimality and complexity of inference-proof data filtering and CQE. In: Kutyłowski, M., Vaidya, J. (eds.) ESORICS 2014, Part II. LNCS, vol. 8713, pp. 165–181. Springer, Heidelberg (2014)
18. Biskup, J., Bring, M., Bulinski, M.: Confidentiality preserving evaluation of open relational queries. In: Morzy, T., Valduriez, P., Bellatreche, L. (eds.) ADBIS 2015. LNCS, vol. 9282, pp. 431–445. Springer, Heidelberg (2015)
19. Biskup, J., Dahn, C., Diekmann, K., Menzel, R., Schalge, D., Wiese, L.: Publishing inference-proof relational data: an implementation and experiments (2015) (submitted for publication)
20. Biskup, J., Embley, D.W., Lochner, J.H.: Reducing inference control to access control for normalized database schemas. Inf. Process. Lett. **106**(1), 8–12 (2008)
21. Biskup, J., Gogolin, C., Seiler, J., Weibert, T.: Inference-proof view update transactions with forwarded refreshments. J. Comput. Secur. **19**, 487–529 (2011)
22. Biskup, J., Hartmann, S., Link, S., Lochner, J.-H.: Efficient inference control for open relational queries. In: Foresti, S., Jajodia, S. (eds.) Data and Applications Security and Privacy XXIV. LNCS, vol. 6166, pp. 162–176. Springer, Heidelberg (2010)
23. Biskup, J., Hartmann, S., Link, S., Lochner, J.-H., Schlotmann, T.: Signature-based inference-usability confinement for relational databases under functional and join dependencies. In: Cuppens-Boulahia, N., Cuppens, F., Garcia-Alfaro, J. (eds.) DBSec 2012. LNCS, vol. 7371, pp. 56–73. Springer, Heidelberg (2012)
24. Biskup, J., Li, L.: On inference-proof view processing of XML documents. IEEE Trans. Dependable Sec. Comput. **10**(2), 99–113 (2013)
25. Biskup, J., Preuß, M.: Database fragmentation with encryption: under which semantic constraints and a priori knowledge can two keep a secret? In: Wang, L., Shafiq, B. (eds.) DBSec 2013. LNCS, vol. 7964, pp. 17–32. Springer, Heidelberg (2013)
26. Biskup, J., Preuß, M.: Inference-proof data publishing by minimally weakening a database instance. In: Prakash, A., Shyamasundar, R. (eds.) ICISS 2014. LNCS, vol. 8880, pp. 30–49. Springer, Heidelberg (2014)
27. Biskup, J., Preuß, M., Wiese, L.: On the inference-proofness of database fragmentation satisfying confidentiality constraints. In: Lai, X., Zhou, J., Li, H. (eds.) ISC 2011. LNCS, vol. 7001, pp. 246–261. Springer, Heidelberg (2011)
28. Biskup, J., Tadros, C.: Policy-based secrecy in the Runs & Systems framework and controlled query evaluation. In: Echizen, I., Kunihiro, N., Sasaki, R. (eds.)

Advances in Information and Computer Security, IWSEC 2010, Short Papers, pp. 60–77. Information Processing Society of Japan (IPSJ) (2010)

29. Biskup, J., Tadros, C.: Inference-Proof View Update Transactions with Minimal Refusals. In: Garcia-Alfaro, J., Navarro-Arribas, G., Cuppens-Boulahia, N., de Capitani di Vimercati, S. (eds.) DPM 2011 and SETOP 2011. LNCS, vol. 7122, pp. 104–121. Springer, Heidelberg (2012)

30. Biskup, J., Tadros, C.: Revising belief without revealing secrets. In: Lukasiewicz, T., Sali, A. (eds.) FoIKS 2012. LNCS, vol. 7153, pp. 51–70. Springer, Heidelberg (2012)

31. Biskup, J., Tadros, C.: Confidentiality enforcement by hybrid control of flows from abstract information states through program execution via declassification (2015) (submitted for publication)

32. Biskup, J., Tadros, C.: Constructing inference-proof belief mediators. In: Samarati, P. (ed.) DBSec 2015. LNCS, vol. 9149, pp. 188–203. Springer, Heidelberg (2015)

33. Biskup, J., Tadros, C.: Preserving confidentiality while reacting on iterated queries and belief revisions. Ann. Math. Artif. Intell. **73**(1–2), 75–123 (2015)

34. Biskup, J., Tadros, C.: On the simulation assumption for controlled interaction processing (to appear, 2016)

35. Biskup, J., Tadros, C., Wiese, L.: Towards controlled query evaluation for incomplete first-order databases. In: Link, S., Prade, H. (eds.) FoIKS 2010. LNCS, vol. 5956, pp. 230–247. Springer, Heidelberg (2010)

36. Biskup, J., Weibert, T.: Keeping secrets in incomplete databases. Int. J. Inf. Secur. **7**(3), 199–217 (2008)

37. Biskup, J., Wiese, L.: Preprocessing for controlled query evaluation with availability policy. J. Comput. Secur. **16**(4), 477–494 (2008)

38. Biskup, J., Wiese, L.: A sound and complete model-generation procedure for consistent and confidentiality-preserving databases. Theoret. Comput. Sci. **412**, 4044–4072 (2011)

39. Bonatti, P.A., Kraus, S., Subrahmanian, V.S.: Foundations of secure deductive databases. IEEE Trans. Knowl. Data Eng. **7**(3), 406–422 (1995)

40. Bonatti, P.A., Petrova, I.M., Sauro, L.: Optimized construction of secure knowledge-base views. In: Calvanese, D., Konev, B. (eds.) International Workshop on Description Logics 2015. CEUR Workshop Proceedings, vol. 1350. CEUR-WS.org (2015)

41. Bonatti, P.A., Sauro, L.: A confidentiality model for ontologies. In: Alani, H., Kagal, L., Fokoue, A., Groth, P., Biemann, C., Parreira, J.X., Aroyo, L., Noy, N., Welty, C., Janowicz, K. (eds.) ISWC 2013, Part I. LNCS, vol. 8218, pp. 17–32. Springer, Heidelberg (2013)

42. Bordeaux, L., Hamadi, Y., Zhang, L.: Propositional satisfiability and constraint programming: a comparative survey. ACM Comput. Surv. **38**(4), 12.1–12.54 (2006)

43. Börger, E., Grädel, E., Gurevich, Y.: The Classical Decision Problem. Perspectives in Mathematical Logic. Springer, Heidelberg (1997)

44. Brachman, R.J., Levesque, H.J.: Knowledge Representation and Reasoning. Elsevier, Amsterdam (2004)

45. Ciriani, V., De Capitani di Vermercati, S., Foresti, S., Jajodia, S., Paraboschi, S., Samarati, P.: Combining fragmentation and encryption to protect privacy in data storage. ACM Trans. Inf. Syst. Secur. **13**(3), 1–33 (2010)

46. Ciriani, V., De Capitani di Vermercati, S., Foresti, S., Samarati, P.: K-anonymity. In: Yu, T., Jajodia, S. (eds.) Secure Data Management in Decentralized Systems. Advances in Information Security, vol. 33, pp. 323–353. Springer, New York (2007)

47. Cuppens, F., Gabillon, A.: Cover story management. Data Knowl. Eng. **37**(2), 177–201 (2001)

48. De Capitani di Vermercati, S., Foresti, S., Jajodia, S., Livraga, G., Paraboschi, S., Samarati, P.: Fragmentation in presence of data dependencies. IEEE Trans. Dependable Sec. Comput. **11**(6), 510–523 (2014)

49. Denning, D.E., Akl, S.G., Heckman, M., Lunt, T.F., Morgenstern, M., Neumann, P.G., Schell, R.R.: Views for multilevel database security. IEEE Trans. Software Eng. **13**(2), 129–140 (1987)

50. Denning, D.E., Schlörer, J.: Inference controls for statistical databases. IEEE Comput. **16**(7), 69–82 (1983)

51. Dolzhenko, E., Ligatti, J., Reddy, S.: Modeling runtime enforcement with mandatory results automata. Int. J. Inf. Secur. **14**(1), 47–60 (2015)

52. Ebbinghaus, H.D., Flum, J.: Finite Model Theory. Springer, Heidelberg (1995)

53. Fagin, R., Halpern, J.Y., Moses, Y., Vardi, M.Y.: Reasoning about Knowledge. MIT Press, Cambridge (1995)

54. Farkas, C., Jajodia, S.: The inference problem: a survey. SIGKDD Explor. **4**(2), 6–11 (2002)

55. Fitting, M., Mendelsohn, R.L.: First-Order Modal Logic, Synthese Library, vol. 277. Kluwer Academic Publishers, Dordrecht (1998)

56. Friedman, N., Halpern, J.Y.: Plausibility measures and default reasoning. J. ACM **48**(4), 648–685 (2001)

57. Fung, B.C.M., Wang, K., Chen, R., Yu, P.S.: Privacy-preserving data publishing: a survey of recent developments. ACM Comput. Surv. **42**(4), 1–53 (2010)

58. Fung, B.C.M., Wang, K., Fu, A.W.C., Yu, P.S.: Introduction to Privacy-Preserving Data Publishing - Concepts and Techniques. Chapman & Hall/CRC, Boca Raton (2010)

59. Ganapathy, V., Thomas, D., Feder, T., Garcia-Molina, H., Motwani, R.: Distributing data for secure database services. Trans. Data Priv. **5**(1), 253–272 (2012)

60. Gray III, J.W.: Toward a mathematical foundation for information flow security. In: IEEE Symposium on Security and Privacy, pp. 21–35 (1991)

61. Halpern, J.Y., O'Neill, K.R.: Secrecy in multiagent systems. ACM Trans. Inf. Syst. Secur. **12**(1), 1–47 (2008)

62. Katebi, H., Sakallah, K.A., Marques-Silva, J.P.: Empirical study of the anatomy of modern sat solvers. In: Sakallah, K.A., Simon, L. (eds.) SAT 2011. LNCS, vol. 6695, pp. 343–356. Springer, Heidelberg (2011)

63. Levesque, H.J., Lakemeyer, G.: The Logic of Knowledge Bases. MIT Press, Cambridge (2000)

64. Libkin, L.: Elements of Finite Model Theory. Springer, Heidelberg (2004)

65. Lunt, T.F., Denning, D.E., Schell, R.R., Heckman, M., Shockley, W.R.: The SeaView security model. IEEE Trans. Software Eng. **16**(6), 593–607 (1990)

66. Machanavajjhala, A., Kifer, D., Gehrke, J., Venkitasubramaniam, M.: L-diversity: privacy beyond k-anonymity. TKDD **1**(1), 3 (2007)

67. Malik, S., Zhang, L.: Boolean satisfiability from theoretical hardness to practical success. Commun. ACM **52**(8), 76–82 (2009)

68. Nerode, A., Shore, R.: Logic for Applications, 2nd edn. Springer, New York (1997)

69. Ray, D., Ligatti, J.: A theory of gray security policies. In: Pernul, G., Ryan, P.Y.A., Weippl, E.R. (eds.) ESORICS 2015. LNCS, vol. 9327, pp. 481–499. Springer, Heidelberg (2015)

70. Reiter, R.: What should a database know? Logic Program. **14**, 127–153 (1992)

71. Robinson, J.A., Voronkov, A. (eds.): Handbook of Automated Reasoning (in 2 volumes). Elsevier, MIT Press, Amsterdam, Cambridge (2001)

72. Sabelfeld, A., Sands, D.: Dimensions and principles of declassification. In: IEEE Computer Security Foundations Workshop, CSFW 2005, pp. 255–269. IEEE Computer Society (2005)
73. Sandhu, R.S., Jajodia, S.: Polyinstantation for cover stories. In: Deswarte, Y., Quisquater, J.-J., Eizenberg, G. (eds.) ESORICS 1992. LNCS, pp. 307–328. Springer, Heidelberg (1992)
74. Schneider, F.B.: Enforceable security policies. ACM Trans. Inf. Syst. Secur. 3(1), 30–50 (2000)
75. Shoenfield, J.R.: Mathematical Logic. Addison-Wesley, Reading (1967)
76. Sicherman, G.L., de Jonge, W., van de Riet, R.P.: Answering queries without revealing secrets. ACM Trans. Database Syst. 8(1), 41–59 (1983)
77. Spohn, W.: Ordinal conditional functions: A dynamic theory of epistemic states. In: Skyrms, B., Harper, W.L. (eds.) Irvine Conference on Probability and Causation. Causation in Decision, Belief Change, and Statistics, vol. II, pp. 105–134. Kluwer, Dordrecht (1988)
78. Studer, T., Werner, J.: Censors for boolean description logic. Trans. Data Priv. 7(3), 223–252 (2014)
79. Sutcliff, G., Suttner, C.: The TPTP problem library for automated theorem proving. Technical report (2015). http://www.tptp.org
80. Sutcliffe, G.: The TPTP problem library and associated infrastructure: The FOF and CNF parts, v3.5.0. J. Autom. Reason. 43(4), 337–362 (2009)
81. Thalheim, B.: Entity-Relationship Modeling - Foundations of Database Technology. Springer, Heidelberg (2000)
82. Traub, J.F., Yemini, Y., Wozniakowski, H.: The statistical security of a statistical database. ACM Trans. Database Syst. 9(4), 672–679 (1984)
83. Weissenbacher, G., Malik, S.: Boolean satisfiability solvers: techniques and extensions. In: Nipkow, T., Grumberg, O., Hauptmann, B. (eds.) Software Safety and Security - Tools for Analysis and Verification, pp. 205–253. IOS Press (2012)

Integrity Constraints for General-Purpose Knowledge Bases

Luís Cruz-Filipe[1(✉)], Isabel Nunes[2], and Peter Schneider-Kamp[1]

[1] Department of Mathematics and Computer Science,
University of Southern Denmark, Odense, Denmark
lcfilipe@gmail.com
[2] Faculdade de Ciências da Universidade de Lisboa, Lisbon, Portugal

Abstract. Integrity constraints in databases have been studied extensively since the 1980s, and they are considered essential to guarantee database integrity. In recent years, several authors have studied how the same notion can be adapted to reasoning frameworks, in such a way that they achieve the purpose of guaranteeing a system's consistency, but are kept separate from the reasoning mechanisms.

In this paper we focus on multi-context systems, a general-purpose framework for combining heterogeneous reasoning systems, enhancing them with a notion of integrity constraints that generalizes the corresponding concept in the database world.

1 Introduction

Integrity constraints in databases have now been around for decades, and are universally acknowledged as one of the essential tools to ensure database consistency [2]. The associated problem of finding out how to repair an inconsistent database – i.e., change it so that it again satisfies the integrity constraints – was soon recognized as an important and difficult one [1], which would unlikely be solvable in a completely automatic way [18].

Since the turn of the century, much focus in research has moved from classical databases to more powerful reasoning systems, where information is not all explicitly described, but may be inferred by logical means. In this setting, an important topic of study is how to combine the reasoning capabilities of different systems, preferrably preserving the properties that make them useful in practice – e.g. consistency, decidability of reasoning, efficient computation. One of the most general frameworks to combine reasoning systems abstractly is that of heterogeneous nonmonotonic multi-context systems [5]. Besides being studied from a theoretical perspective, these have been implemented, and many specialized versions have been introduced to deal with particular aspects deemed relevant in practice [7,15,23,31]. In this work, we will work with relational multi-context systems [20], a first-order generalization of the original, propositional-based systems, which we will refer to simply as multi-context systems, or MCSs.

As a very simple kind of reasoning system, databases can naturally be viewed as particular cases of MCSs. In this paper we propose to define integrity constraints in MCSs in a way that naturally generalizes the usual definitions for

© Springer International Publishing Switzerland 2016
M. Gyssens and G. Simari (Eds.): FoIKS 2016, LNCS 9616, pp. 235–254, 2016.
DOI: 10.1007/978-3-319-30024-5_13

relational databases. Some authors have previously discussed modelling integrity constraints in MCSs, but their approach differs substantially from the typical database perspective, as integrity constraints are embedded *into* the system, thereby becoming part of the reasoning mechanism – unlike the situation in databases, where they form an independent layer that simply signals whether the database is in a consistent state. We argue that integrity constraints for MCSs should also follow this principle, and show how our approach is also in line with investigations on how to add integrity constraints to other reasoning frameworks, namely description logic knowledge bases [19,26]. Due to the richer structure of MCSs, we can define two distinct notions of consistency with respect to integrity constraints, which coincide for the case of databases.

We also address the problem of repairing an MCS that does not satisfy its integrity constraints by moving to managed multi-context systems (mMCSs) [7], which offer additional structure that helps defining the notion of repair.

Contributions. Our main contribution is a uniform notion of integrity constraint over several formalisms. We define integrity constraints over an MCS, together with notions of weak and strong satisfaction of these. We show that the problem of deciding whether an MCS satisfies a set of integrity constraints is polynomial-time reducible to the problem of deciding whether an MCS is logically consistent (i.e., it has a model). We show how our definition captures the traditional notion of integrity constraints over relational databases, and how it naturally generalizes this concept to distributed databases and deductive databases. We also compare our definition with existing proposals for integrity constraints over ontology languages. Finally, we define repairs, and show how our definition again generalizes the traditional concept in databases.

Outline. In Sect. 2 we introduce the framework of multi-context systems. In Sect. 3 we define integrity constraints over MCSs, together with the notions of weak and strong satisfaction. We show how we can encode an MCS with integrity constraints as a different MCS, and obtain decidability and complexity results for satisfaction of integrity constraints by reducing to the problem of logical consistency. In Sect. 4 we justify our definition of integrity constraint, by showing that it generalizes the usual concept in relational databases, as well as other authors' proposals for ontology languages [26] and peer-to-peer systems [9]. We also show that it induces a natural concept of integrity constraint for distributed databases, as well as providing a similar notion for deductive databases that is more expressive than the usual one; and provide complexity results for these concrete cases. In Sect. 5 we recall the notion of a database repair, and show how repairs can be naturally defined in a simple extension of MCSs. We conclude with an overview of our results and future directions in Sect. 6.

1.1   Related Work

The topic of integrity constraints has been extensively studied in the literature. In this section, we discuss the work that we feel to be more directly relevant to the tasks we carry out in this paper.

Integrity constraints and updates – ways of repairing inconsistent databases – were identified as a seminal problem in database theory almost thirty years ago [1]. The case for viewing integrity constraints as a layer on top of the database, rather than as a component of it, has been made since the 1980s. The idea is that the data inconsistencies captured by integrity constraints need to be resolved, but they should not interfere with the ability to continue using the database. In this line, much work has been done e.g. in query answering from inconsistent databases [3,30], by ensuring that the only answers generated are those that hold in minimally repaired versions of the database.

The first authors to consider deductive databases [4,22] also discussed this issue. They identify three ways to look at deductive databases: by viewing the whole system as a first-order theory; by viewing it as an extensional database together with integrity constraints; and a mixed view, where some rules are considered part of the logic theory represented by the database, and others as integrity constraints identifying preferred models. In [4], it is argued that this third approach is the correct one, as it cleanly separates rules that are meant to be used in logic inferencing from those that only specify consistency requirements.

More recently, authors have considered adding integrity constraints to open-world systems such as ontologies. Although integrity constraints can be written in the syntax of terminological axioms, the authors of [26] discuss why they should still be kept separate from the logical theory. Therefore, they separate the axioms in the T-Box (the deductive part of an ontology) into two groups: reasoning rules, which are used to infer new information, and integrity constraints, which only verify the consistency of the knowledge state without changing it.

The setting of multiple ontologies was considered in [19], which considers the problem of combining information from different knowledge sources while guaranteeing the overall consistency, and preserving this consistency when one of the individual ontologies is changed. This is achieved by external integrity constraints, written in a Datalog-like syntax, which can refer to knowledge in different ontologies in order to express relationships between them. Again, the purpose of these rules is uniquely to identify incompatibilities in the data, and not to infer new information.

By contrast, the authors who have discussed multi-context systems have not felt the need to take a similar approach. Integrity constraints appear routinely in examples in e.g. [6,7,17,28], but always encoded within the system, so that their violation leads to logical inconsistency of the global knowledge base. Their work focuses rather on the aspect of identifying the sources of inconsistencies – integrity constraints being only one example, not given any special analysis – and ways in which it can be repaired.

Although we believe this last work to be of the utmost importance, and show how satisfaction of integrity constraints can be reduced to consistency checking (which in turn implies that computing repairs can be reduced to restoring consistency), we strive for the clean separation between integrity constraints and reasoning that is present in other formalisms, and believe our proposal to be an important complement to the analysis of inconsistency in MCSs.

2   Background

We begin this section with a summary of the notion of relational multi-context system [20]. Intuitively, these are a collection of logic knowledge bases – the *contexts* – connected by Datalog-style *bridge rules*. The formal definition proceeds in several layers. The first notion is that of *relational logic*, an abstract notion of a logic with a first-order sublanguage.

Definition 1. *Formally, a relational logic L is a tuple $\langle \mathsf{KB}_L, \mathsf{BS}_L, \mathsf{ACC}_L, \Sigma_L \rangle$, where KB_L is the set of well-formed knowledge bases of L (sets of well-formed formulas), BS_L is a set of possible belief sets (models), $\mathsf{ACC}_L : \mathsf{KB}_L \to 2^{\mathsf{BS}_L}$ is a function assigning to each knowledge base a set of acceptable sets of beliefs (i.e., its models), and Σ_L is a signature consisting of sets P_L^{KB} and P_L^{BS} of predicate names (with associated arity) and a universe U_L of object constants, such that $U_L \cap (P_L^{\mathsf{KB}} \cup P_L^{\mathsf{BS}}) = \emptyset$.*

If $p \in P_L^{\mathsf{KB}}$ has arity k and $c_1, \ldots, c_k \in U_L$, then $p(c_1, \ldots, c_k)$ must be an element of some knowledge base, and if $p \in P_L^{\mathsf{BS}}$, then $p(c_1, \ldots, c_k)$ must be an element of some belief set. Therefore, we can view Σ_L as a first-order signature generating a sublanguage of L. The elements in this sublanguage are called *relational ground elements*, while the remaining elements of knowledge bases or belief sets are called *ordinary*.

Example 1. We can see first-order logic over a first-order signature Σ_{FOL} as a logic $\mathsf{FOL} = \langle \mathsf{KB}_{\mathsf{FOL}}, \mathsf{BS}_{\mathsf{FOL}}, \mathsf{ACC}_{\mathsf{FOL}}, \Sigma_{\mathsf{FOL}} \rangle$, where $\mathsf{KB}_{\mathsf{FOL}}$ is the set of sets of well-formed formulas over Σ_{FOL}, $\mathsf{BS}_{\mathsf{FOL}}$ is the set of first-order interpretations over Σ_{FOL}, and $\mathsf{ACC}_{\mathsf{FOL}}$ maps each set of formulas to the set of its models. This logic only contains relational elements.

Definition 2. *Let $\mathfrak{I}$ be a finite set of indices, $\{L_i\}_{i \in \mathfrak{I}}$ be a set of relational logics, and V be a set of (first-order) variables distinct from predicate and constant names in any L_i. A relational element of L_i has the form $p(t_1, \ldots, t_k)$, where $p \in P_{L_i}^{\mathsf{KB}} \cup P_{L_i}^{\mathsf{BS}}$ has arity k and each t_j is a term from $V \cup U_{L_i}$, for $1 \le j \le k$. A relational k-bridge rule over $\{L_i\}_{i \in \mathfrak{I}}$ and V is a rule of the form*

$$(k : s) \leftarrow (c_1 : p_1), \ldots, (c_q : p_q), \mathsf{not}\ (c_{q+1} : p_{q+1}), \ldots, \mathsf{not}\ (c_m : p_m) \quad (1)$$

such that $k, c_i \in \mathfrak{I}$, s is an ordinary or a relational knowledge base element of L_k and $p_1, \ldots, p_m$ are ordinary or relational beliefs of L_{c_i}.

The notation $(c : p)$ indicates that p is evaluated in context c. These rules intuitively generalize logic programming rules, and as usual in that context we impose a *safety condition*: all variables occurring in $p_{q+1}, \ldots, p_m$ must also occur at least once in $p_1, \ldots, p_q$.

Definition 3. *A relational multi-context system is a collection $M = \{C_i\}_{i \in \mathfrak{I}}$ of contexts $C_i = \langle L_i, \mathsf{kb}_i, \mathsf{br}_i, D_i \rangle$, where L_i is a relational logic, kb_i is a knowledge base, br_i is a set of relational i-bridge rules, and D_i is a set of import domains $D_{i,j}$, with $j \in \mathfrak{I}$, such that $D_{i,j} \subseteq U_j$.*

Import domains define which constants are exported from one context to another: as the underlying logic languages can be different, these sets are essential to allow one context to reason about individuals introduced in another. We will assume that $D_{i,j}$ is the finite domain consisting of the object constants appearing in kb_j or in the head of a relational bridge rule in br_j, unless otherwise stated.

Example 2. Let C_1 and C_2 be contexts over the first-order logic FOL with R and Rt binary predicates in Σ_{FOL}, and let $\mathsf{kb}_1 = \mathsf{kb}_2 = \emptyset$. We can use the following bridge rules in br_2 to define Rt in C_2 as the transitive closure of R in C_1.

$$(2 : \mathsf{Rt}(x,y)) \leftarrow (1 : \mathsf{R}(x,y)) \qquad (2 : \mathsf{Rt}(x,y)) \leftarrow (1 : \mathsf{R}(x,z)), (2 : \mathsf{Rt}(z,y))$$

We will use the MCS $M = \langle C_1, C_2 \rangle$ to exemplify the concepts we introduce.

The semantics of relational MCSs is defined in terms of ground instances of bridge rules: the instances obtained from each rule $r \in \mathsf{br}_i$ by uniform substitution of each variable X in r by a constant in $\bigcap D_{i,j}$, with j ranging over the indices of the contexts to which queries containing X are made in r.

Definition 4. *A belief state for M is a collection $S = \{S_i\}_{i \in \mathfrak{I}}$ where $S_i \in \mathsf{BS}_i$ for each $i \in \mathfrak{I}$ – i.e., a tuple of models, one for each context. The ground bridge rule Eq. (1) is applicable in a belief state S if $p_i \in S_{c_i}$ for $1 \le i \le q$ and $p_i \notin S_{c_i}$ for $q < i \le m$. The set of the heads of all applicable ground instances of bridge rules of context C_i w.r.t. S is denoted by $\mathsf{app}_i(S)$. An equilibrium is a belief state S such that $S_i \in \mathsf{ACC}_i(\mathsf{kb}_i \cup \mathsf{app}_i(S))$.*

Particular types of equilibria (minimal, grounded, well-founded) [5] can be defined for relational MCSs, but we will not discuss them here.

Example 3. In the setting of the previous example, all equilibria of M will have to include the transitive closure of R in S_1 in the interpretation of Rt in S_2. For example, if we take $S = \langle S_1, S_2 \rangle$ with $S_1 = \{\mathsf{R}(a,b), \mathsf{R}(b,c)\}$ and $S_2 = \{\mathsf{Rt}(a,b), \mathsf{Rt}(b,c), \mathsf{Rt}(a,c)\}$, then S is an equilibrium. However, $S' = \langle S_1, S_2' \rangle$ with $S_2' = \{\mathsf{Rt}(a,b), \mathsf{Rt}(b,c)\}$ is not an equilibrium, as it does not satisfy the second bridge rule.

Checking whether an MCS has an equilibrium is known as the *consistency problem* in the literature. We will refer to this property as *logical consistency* (to distinguish from consistency w.r.t. integrity constraints, defined in the next section) throughout this paper. This problem has been studied extensively [6,16,17,35]; its decidability depends on decidability of reasoning in the underlying contexts. The complexity of checking logical consistency of an MCS M depends on the context complexity of M – the highest complexity of deciding consistency in one of the contexts in M (cf. [17] for a formal definition and known results).

3   Integrity Constraints on Multi-context Systems

In their full generality, integrity constraints in databases can be arbitrary first-order formulas, and reasoning with them is therefore undecidable. For this reason, it is common practice to restrict their syntax in order to regain decidability;

our definition follows the standard approach of writing integrity constraints in denial clausal form.

Definition 5. *Let* $M = \langle C_1, \ldots, C_n \rangle$ *be an MCS. An* integrity constraint *over an MCS* M *(in denial form) is a formula*

$$\leftarrow (i_1 : P_1), \ldots, (i_m : P_m), \mathsf{not}\ (i_{m+1} : P_{m+1}), \ldots, \mathsf{not}\ (i_\ell : P_\ell) \qquad (2)$$

where $M = \langle C_1, \ldots, C_n \rangle$, $i_k \in \{1, \ldots, n\}$, *each* P_k *is a relational element of* C_{i_k}, *and the variables in* $P_{m+1}, \ldots, P_\ell$ *all occur in* $P_1, \ldots, P_m$.

Syntactically, integrity constraints are similar to "headless bridge rules". However, we will treat them differently: while bridge rules influence the semantics of MCSs, being part of the notion of equilibrium, integrity constraints are meant to be checked at the level of equilibria.

Example 4. Continuing the example from the previous section, we can write an integrity constraint over M stating that the relation R (in context C_1) is transitive.

$$\leftarrow (2 : \mathsf{Rt}(x, y)), \mathsf{not}\ (1 : \mathsf{R}(x, y)) \qquad (3)$$

The restriction on variables again amounts to the usual Logic Programming requirement that bridge rules be safe. To capture general tuple-generating dependencies we could relax this constraint slightly, and allow $P_{m+1}, \ldots, P_\ell$ to introduce new variables, with the restriction that they can be used only once in the whole rule. This generalization poses no significant changes to the theory, but makes the presentation heavier, and we will therefore assume safety.

Definition 6. *Let* $M = \langle C_1, \ldots, C_n \rangle$ *be an MCS and* $S = \langle S_1, \ldots, S_n \rangle$ *be a belief state for* M. *Then* S satisfies *the integrity constraint Eq. (2) if, for every instantiation* θ *of the variables in* $P_1, \ldots, P_m$, *either* $P_k\theta \notin S_k$ *for some* $1 \leq k \leq m$ *or* $P_k\theta \in S_k$ *for some* $m < k \leq \ell$.

In other words: equilibria must satisfy all bridge rules (if their body holds, then so must their heads), but they may or may not satisfy all integrity constraints. In this sense, integrity constraints express preferences among equilibria.

Example 5. The equilibrium S from Example 3 does not satisfy the integrity constraint (3), thus M does not strongly satisfy this formula. However, M weakly satisfies (3), as seen by the equilibrium $S'' = \langle S_1', S_2' \rangle$ where S_2' is as above and $S_1' = \{\mathsf{R}(\mathsf{a}, \mathsf{b}), \mathsf{R}(\mathsf{b}, \mathsf{c}), \mathsf{R}(\mathsf{a}, \mathsf{c})\}$.

Definition 7. *Let* M *be an MCS and* η *be a set of integrity constraints.*

1. M strongly satisfies η, $M \models_s \eta$, *if: (i)* M *is logically consistent and (ii) every equilibrium of* M *satisfies all integrity constraints in* η.
2. M weakly satisfies η, $M \models_w \eta$, *if there is an equilibrium of* M *that satisfies all integrity constraints in* η.

We say that M is (strongly/weakly) *consistent* w.r.t. a set of integrity constraints η if M (strongly/weakly) satisfies η. These two notions express different interpretations of integrity constraints. Strong satisfaction views them as necessary requirements, imposing that all models of the MCS to satisfy them. Examples of these are the usual integrity constraints over databases, which express semantic connections between relations that must always hold. Weak satisfaction views integrity constraints as expressing preferences: the MCS may have several equilibria, and we see those that do satisfy the integrity constraints as "better".

The distinction is also related to the use of brave (credulous) or cautious (skeptical) reasoning. If M strongly satisfies a set of integrity constraints η, then any inferences we draw from M using brave reasoning are guaranteed to hold in some equilibrium that also satisfies η. If, however, M only weakly satisfies η, then this no longer holds, and we can only use cautious reasoning if we want to be certain that any inferences are still compatible with η.

Both strong and weak satisfaction require M to be logically consistent, so $M \models_s \eta$ implies $M \models_w \eta$. This implies that deciding whether $M \models_s \eta$ and $M \models_w \eta$ are both at least as hard as deciding whether M has an equilibrium – thus undecidable in the general case.[1] When logical consistency of M is decidable and its set of equilibria is enumerable, weak satisfaction is semi-decidable (if there is an equilibrium that satisfies η, we eventually encounter it), while strong satisfaction is co-semi-decidable (if there is an equilibrium that does not satisfy η, we eventually encounter it). The converse also holds.

Theorem 1. *Weak satisfaction of integrity constraints is reducible to logical consistency.*

Proof. To decide whether $M \models_w \eta$, construct M' by extending M with a context C_0 where $\mathsf{KB}_0 = \wp(\{*\})$, $\mathsf{kb}_0 = \emptyset$, $\mathsf{ACC}_0(\emptyset) = \{\emptyset\}$, $\mathsf{ACC}_0(\{*\}) = \emptyset$, and the bridge rules obtained by adding $(0 : *)$ to the head of the rules in η. Then M' has an equilibrium iff $M \models_w \eta$: any equilibrium of M not satisfying η corresponds to a belief state of M' where $\mathsf{app}_0(S) = \{*\}$, which is never an equilibrium of M'; but equilibria of M satisfying η give rise to equilibria of M' taking $S_0 = \emptyset$. □

Theorem 2. *Strong satisfaction of integrity constraints is reducible to logical inconsistency.*

Proof. Construct M' as before, but now defining $\mathsf{ACC}_0(\emptyset) = \emptyset$, $\mathsf{ACC}_0(\{*\}) = \{\{*\}\}$. If M is inconsistent, then $M \not\models_s \eta$. If M is consistent, then any equilibrium of M satisfying η corresponds to a belief state of M' where $\mathsf{app}_0(S) = \emptyset$, which can never be an equilibrium of M'; in turn, equilibria of M not satisfying η give rise to equilibria of M' taking $S_0 = \{*\}$. So if M is consistent, then $M \models_s \eta$ iff M' is inconsistent. □

Combining the two above results with the well-known complexity results for consistency checking (Table 1 in [17]), we directly obtain the following results.

---

[1] If consistency of one of M's contexts is undecidable, then clearly the question of whether M has an equilibrium is also undecidable.

Table 1. Complexity of integrity checking of an MCS in terms of its context complexity.

$\mathcal{CC}(M)$	P	NP	Σ_i^p	PSPACE	EXPTIME
$M \models_w \eta$	NP	NP	Σ_i^p	PSPACE	EXPTIME
$M \models_s \eta$	Δ_2^p	Δ_2^p	Δ_{i+1}^p	PSPACE	EXPTIME

Corollary 1. *The complexity of deciding whether $M \models_w \eta$ or $M \models_s \eta$, depending on the context complexity of M, $\mathcal{CC}(M)$, is given in* Table 1.

These results suggest an alternative way of modelling integrity constraints in MCSs: adding them as bridge rules whose head is a special atom interpreted as inconsistency. This approach was taken in e.g. [16]. However, we believe that integrity constraints should be kept separate from the data, and having them as a separate layer achieves this purpose. In this way, we do not restrict the models of MCSs, and we avoid issues of logical inconsistency. Furthermore, violation of integrity constraints typically is indicative of some error in the model or in the data, which should result in an alert and not in additional inferences.

These considerations are similar to those made in Sect. 2.7 of [26] and in [19], in the (more restricted) context of integrity constraints over description logic knowledge bases. Likewise, the approach taken for integrity constraints in databases is that inconsistencies should be brought to the users' attention, but not affect the semantics of the database [1,18]. In particular, it may be meaningful to work with reasoning systems not satisfying integrity constraints (see [30] for databases and [28] for description logic knowledge bases). Our approach is also in line with [7], where it is argued that in MCSs it is important to "distinguish data from additional operations on it".

4   Applications of ICs for MCSs

In this section we look at particular cases of MCSs with integrity constraints. We begin by showing that our notion generalizes the usual one for standard databases. Then we look into other types of databases and show how we obtain interesting notions for these systems.

4.1   Relational Databases

Integrity constraints in relational databases can be written as first-order formulas in denial clausal form [21] – which are essentially equivalent in form to bridge rules with no head.

Definition 8. *Let DB be a database. The context generated by DB, $\mathsf{Ctx}(DB)$, is defined as follows.*

- *The underlying logic is first-order logic.*
- *Belief sets are sets of ground literals.*

– *The knowledge base is DB.*
– *For all* kb, *the only belief set compatible with* kb *is* ACC(kb) = kb$^\vdash$ = kb$\cup\{\neg a \mid a \notin$ kb$\}$.
– *The set of bridge rules is empty.*

We can see any database DB as a single-context MCS consisting of exactly the context Ctx(DB); we will also denote this MCS by Ctx(DB), as this poses no ambiguity. The only equilibrium for Ctx(DB) is $DB^\vdash$ itself, corresponding to the usual closed-world semantics of relational databases. Previous work (cf. [7,17]) implicitly treats databases in this way, although Ctx is not formally defined.

Let DB be a database and r be an integrity constraint over DB in denial clausal form. We can rewrite r as an integrity constraint over Ctx(DB): if r is $\forall(A_1 \wedge \ldots \wedge A_k \wedge \neg B_1 \wedge \ldots \wedge \neg B_m \rightarrow \perp)$, then br($r$) is

$$\leftarrow (1 : A_1), \ldots, (1 : A_k), \text{not } (1 : B_1), \ldots, \text{not } (1 : B_m).$$

Note that general tuple-generating dependencies require allowing singleton variables in the B_is, as discussed earlier. The following result is straightforward to prove. If we assume first-order logic with equality, we can also write equality-generating constraints, thus obtaining the expressivity used in databases.

Theorem 3. *Let DB be a database and η be a set of ICs over DB. Then DB satisfies all ICs in η iff* Ctx(DB) $\models_s$ br(η) *iff* Ctx(DB) $\models_w$ br(η), *where* br *is extended to sets in the standard way.*

In this setting, weak and strong satisfaction of integrity constraints coincide, as every database has exactly one equilibrium. Furthermore, deciding whether Ctx(DB) $\models$ br(η) can be done in time $O(|DB| \times |\eta|)$, where $|DB|$ is the number of elements in DB and $|\eta|$ is the total number of literals in all integrity constraints in η. This means that the data complexity [34] of this problem is linear, as we can query the database using the open bridge rules in η, rather than considering the set of all ground instances of those rules.

Theorem 3 could be obtained by adding integrity constraints as bridge rules with a special inconsistency atom, as discussed earlier, and done in [16]). This would significantly blur the picture, though, as in principle nothing would prevent us from writing integrity constraints referencing the inconsistency atom in their body, potentially leading to circular reasoning. Our approach guarantees that there is no such internalization of inconsistencies into the database.

Our results show that the notion of integrity constraint we propose directly generalizes the traditional notion of integrity constraints over databases [1].

4.2   Distributed DBs

Distributed databases are databases that store their information at different sites in a network, typically including information that is duplicated at different nodes [33] in order to promote resilience of the whole system.

A distributed database consisting of individual databases $DB_1, \ldots, DB_n$ can be modeled as an MCS with n contexts $\mathsf{Ctx}(DB_1), \ldots, \mathsf{Ctx}(DB_n)$. The internal consistency of the database, in the sense that tables that occur in different DB_is must have the same rows, can be specified as integrity constraints over this MCS as follows. For each relation p, let $\gamma(p)$ be the number of columns of p and $\delta(p)$ be the set of indices of the databases containing p. Then

$$\{\leftarrow (i : p(x_1, \ldots, x_{\gamma(p)})), \mathsf{not}\ (j : p(x_1, \ldots, x_{\gamma(p)})) \mid i, j \in \delta(p), p \text{ is a relation}\}$$

logically specifies the integrity of the system. Different strategies for fixing inconsistencies in distributed databases (e.g. majority vote or siding with the most recently updated node) correspond to different preferences for choosing repairs in the sense of the next section.

Again, such integrity constraints can be written as bridge rules in the form

$$(j : p(x_1, \ldots, x_{\gamma(p)})) \leftarrow (i : p(x_1, \ldots, x_{\gamma(p)})) .$$

but these significantly change the semantics of the database: instead of describing preferred equilibria, they impose a flow of information between nodes.

Example 6. Consider a country with a central person register (CPR), mapping a unique identifying number to the name and current address of each citizen using a relation person, e.g. person($1111111118, old\_lady, gjern$). Furthermore, each electoral district keeps a local voter register using a relation voter, e.g. voter(1111111118), and a list of addresses local to the given electoral district using a relation address, e.g. address($gjern$). Then the integrity constraints

$$\leftarrow \mathsf{Skborg} : \mathsf{voter}(Id), \mathsf{not}\ (\mathsf{CPR} : \mathsf{person}(Id)) \tag{4}$$

$$\leftarrow \mathsf{Skborg} : \mathsf{voter}(Id), \mathsf{CPR} : \mathsf{person}(Id, Add), \mathsf{not}\ (\mathsf{Skborg} : \mathsf{address}(Add)) \tag{5}$$

ensure that all voters registered in the Silkeborg electoral district are registered in the central person register, and that they are registered with an address that is local to the Silkeborg electoral district. Here, we are implicitly assuming that the database is closed under projection, and overload the person relation for the sake of simplicity. In addition, the following set of integrity constraints models the fact that each person registered in the Silkeborg electoral district is not registered in any other electoral districts from the set ED.

$$\{\leftarrow \mathsf{Skborg} : \mathsf{voter}(Id), C_i : \mathsf{voter}(Id) \mid C_i \in ED \setminus \{\mathsf{Skborg}\}\}$$

This assumption of closure under projection is meaningful from a practical point of view, and has been implemented e.g. in [12]. Alternatively, we could define the projections as bridge rules of the MCSs, in line with the idea of encoding views of deductive databases presented in the next section.

This section's treatment of distributed databases is equivalent to considering their disjoint union as a database. Consequently, there is no need to use MCSs for distributed databases, but this mapping shows that our notion of integrity constraints abstracts the practice in this field. Furthermore, results in previous work [11] indicate that the processing of integrity constraints can be efficiently parallelized in this disjoint scenario, given suitable assumptions.

4.3   Deductive DBs

We now address the case of deductive databases. These consist of two different components: the (extensional) fact database, containing only concrete instances of relations, and the (intensional) rule database, containing Datalog-style rules defining new relations. Every relation must be either intensional or extensional, unlike in e.g. full-fledged logic programming.

One standard way to see the intensional component(s) of deductive databases is as *views* of the original database. The instances of the new relations defined by rules are generated automatically from the data in the database, and these relations can thus be seen as content-free, having a purely presentational nature. For simplicity of presentation, we consider the case where there is one single view.

Definition 9. *Let Σ_E and Σ_I be two disjoint first-order signatures. A deductive database over Σ_E and Σ_I is a pair $\langle DB, R \rangle$, where DB is a relational database over Σ_E and R is a set of rules of the form $p \leftarrow q_1, \ldots, q_n$, where p is an atom of Σ_I and $q_1, \ldots, q_n$ are atoms over $\Sigma_E \cup \Sigma_I$.*

More precisely, this definition corresponds to the definite deductive databases in [22]; we do not consider the case of indefinite databases in this work. We can view deductive databases as MCSs.

Definition 10. *Let $\langle DB, R \rangle$ be a deductive database over Σ_E and Σ_I. The MCS induced by $\langle DB, R \rangle$ is $M = \langle C_E, C_I \rangle$, where $C_E = \mathsf{Ctx}(DB)$ defined as above and $C_I = \mathsf{Ctx}(R)$ is a similar context where:*

- *The knowledge base is $\emptyset$.*
- *For each rule $p \leftarrow q_1, \ldots, q_n$ in R there is a bridge rule $(I : p) \leftarrow (i_1 : q_1), \ldots, (i_n : q_n)$ in $\mathsf{Ctx}(R)$, where $i_k = E$ if q_k is an atom over Σ_E and $i_k = I$ otherwise.*

Integrity constraints over such MCSs correspond precisely to the definition of integrity constraints over deductive databases from [4]. By combining this with the adequate notion of repair, we capture the typical constraints of deductive databases – that consistency can only be regained by changing extensional predicates – in line with the traditional view-update problem. More modern works [8] restrict the syntax of integrity constraints, allowing them to use only extensional relations; in the induced MCS, this translates to the additional requirement that only relational elements from C_E appear in the body of integrity constraints.

Example 7. Consider a deductive database for class diagrams, where information about direct subclasses is stored in the extensional database using a relation isa, e.g. isa(*list, collection*) and isa(*array, list*). Intensionally, we model the transitive closure of the subclass relation using a view created by the two rules sub$(A, B) \leftarrow$ isa(A, B) and sub$(A, C) \leftarrow$ isa(A, B), sub(B, C), thus allowing us to find out that in our example sub(*array, collection*). The integrity constraint

$$\leftarrow \mathsf{sub}(A, A)$$

can then be used to state the acyclicity of the subclass relation. Integrity constraints restricted to the extensional database could not express this, as there would be no way to define a fixpoint. The only (incomplete) solution would be to add n integrity constraints disallowing cycles of length up to n. This example illustrates our gain of expressive power compared to the approach in [8].

We can also consider databases with several, different views, each view generating a different context. Integrity constraints over the resulting MCS can then specify relationships between relations in different views.

Yet again, the complexity of verifying whether an MCS induced by a deductive database satisfies its integrity constraints is lower than the general case. In particular, consistency checking is reducible to query answering (all integrity constraints are satisfied iff there are no answers to the queries expressed in their bodies). If we do not allow negation in the definition of the intensional relations, then there is only one model of the database as before, and consistency checking w.r.t. a fixed set of integrity constraints is PTIME-complete [29]. In the general case, weak and strong consistency correspond, respectively, to brave and cautious reasoning for Datalog programs under answer set semantics, which are known to be co-NP-complete and NP-complete, respectively.

4.4   Peer-to-Peer Systems

Peer-to-peer (P2P) networks are distributed systems where each node (the peer) has an identical status in the hierarchy, i.e., there is no centralized control. Queries can be posed to each peer, and peers communicate amongst themselves in order to produce the desired answer. For a general overview see e.g. [27].

A particularly interesting application are P2P systems, which integrate features of both distributed and deductive databases. We follow [9], which also addresses the issue of integrity constraints. In this framework, P2P systems consist of several nodes (the peers), each of them a deductive database of its own, connected via *mapping rules* that port relations from one peer to another.

Definition 11. *A peer-to-peer system* $\mathcal{P}$ *is a set of peers* $\mathcal{P} = \{P_i\}_{i=1}^{n}$. *Each peer is a tuple* $\langle \Sigma^i, DB_i, R_i, M_i, IC_i \rangle$, *where:*

- Σ^i *is the disjoint union of three signatures* Σ_E^i, Σ_I^i *and* Σ_M^i;
- $\langle DB_i, R_i \rangle$ *is a deductive database over signatures* Σ_E^i *and* Σ_I^i, *where the rules in* R_i *may also use relations from* Σ_M^i;
- M_i *is a set of mapping rules of the form* $p \leftarrow_j q_1, \ldots, q_m$ *with* $j \neq i$, *where* p *is an atom over a signature* Σ_M^i *and each* q_k *is an atom over* Σ^j;
- IC_i *is a set of integrity constraints over* Σ^i.

Intuitively, relations can be defined either extensionally (those in Σ_E), intensionally (those in Σ_I) or as mappings from another peer (those in Σ_M), and these definitions may not be mixed. Observe that, with these definitions, negations may only occur in the bodies of the integrity constraints.

We can view a P2P system as a MCS with integrity constraints. To simplify the construction, we adapt the definition from the case of deductive databases slightly, so that there is a one-to-one correspondence between peers and contexts.

Definition 12. *Let $\mathcal{P} = \{P_i\}_{i=1}^n$ be a P2P system. The MCS induced by $\mathcal{P}$ is defined as follows.*

- *There are n contexts, where C_i is constructed as $\mathsf{Ctx}(DB_i)$ together with the following set of bridge rules:*
  - $(i : p) \leftarrow (i : q_1), \ldots, (i : q_m)$ *for each rule $p \leftarrow q_1, \ldots, q_m \in R_i$;*
  - $(i : p) \leftarrow (j : q_1), \ldots, (j : q_m)$ *for each rule $p \leftarrow_j q_1, \ldots, q_m \in M_i$.*
- *Each integrity constraint $\leftarrow q_1, \ldots, q_m$ in IC_i is translated to the integrity constraint $\leftarrow (i : q_1), \ldots, (i : q_m)$, where we take $(i : \neg q)$ to mean $\mathsf{not}\ (i : q)$.*

The definition of the bridge rules from R_i is identical to what one would obtain by constructing the context $\mathsf{Ctx}(R_i)$ described in the previous section.

This interpretation does not preserve the semantics for P2P systems given in [9,10]. Therein, mapping rules can only be applied if they do not generate violations of the integrity constraints. This is directly related to the real-life implementation of these systems, where this option represents a "cheap" strategy to ensure local enforcement of integrity constraints; as discussed in [35], the underlying philosophy of P2P systems and MCSs is significantly different.

We now show that, while the semantics differ, there is a correspondence between P2P systems and their representation as an MCS, and the "ideal" models of both coincide. When no such models exist, the MCS formulation can be helpful in identifying the problematic mapping rules.

The semantics of P2P systems implicitly sees them as logic programs.

Definition 13. *Let $\mathcal{P} = \{P_i\}_{i=1}^n$ be a P2P system and I be a Herbrand interpretation over $\bigcup \Sigma^i$. The program $\mathcal{P}^I$ is obtained from $\mathcal{P}$ by (i) grounding all rules and (ii) removing the mapping rules whose head is not in I.*

Let $\mathcal{MM}(P)$ denote the minimal model of a logic program. A weak model for $\mathcal{P}$ is an interpretation I such that $I = \mathcal{MM}(\mathcal{P}^I)$.

Since integrity constraints are rules with empty head, this definition implicitly requires weak models to satisfy them. Interpretations over a P2P system and equilibria over the induced MCS are trivially in bijection, as the latter simply assign each atom to the right context, and we implicitly identify them hereafter. We can relate the "perfect" models in both systems.

Theorem 4. *Let $\mathcal{P}$ be a P2P system, I an interpretation for $\mathcal{P}$, and M the induced MCS. Then $I = \mathcal{MM}(\mathcal{P}) = \mathcal{MM}(\mathcal{P}^I)$ iff I is an equilibrium for M satisfying all the integrity constraints.*

Proof. Since $\mathcal{P}$ corresponds to a positive program, the only equilibrium of M is $\mathcal{MM}(\mathcal{P})$ (see [14]). Furthermore, for any I, $\mathcal{MM}(\mathcal{P}^I)$ includes the facts in all extensional databases and satisfies all rules in R_i and all integrity constraints. Thus, it also corresponds to a belief state satisfying their counterparts in M.

Suppose that $\mathcal{MM}(\mathcal{P}) = \mathcal{MM}(\mathcal{P}^I)$. Since mapping rules are the only ones that can add information about relations in Σ_M^i to I, the second equality implies that no mapping rules are removed in $\mathcal{P}^I$. Therefore $I = \mathcal{MM}(\mathcal{P})$ satisfies all bridge rules of M obtained from the mapping rules in $\mathcal{P}$, whence I is an equilibrium of M satisfying all integrity constraints.

Conversely, if I is an equilibrium of M and r is a mapping rule, then either I does not satisfy the body of r or I contains its head. Since no other rules can infer instances of relations in Σ_M^i, this implies that $\mathcal{MM}(\mathcal{P}) = \mathcal{MM}(\mathcal{P}^I)$, and being an equilibrium implies that $I = \mathcal{MM}(\mathcal{P})$. $\qquad\square$

The MCS representation has an interesting connection with the notion of weak model in general, though: if there are integrity constraints in M that are not satisfied by $\mathcal{MM}(\mathcal{P})$, then repairing M by removing mapping rules is equivalent to finding a weak model for $\mathcal{P}$. This is again reminescent of the view-update problem.

The MCS representation allows us to write seemingly more powerful integrity constraints over a P2P system, as we can use literals from different contexts in the same rule. However, this does not give us more expressive power: for example, the integrity constraint $\leftarrow (1 : a), (2 : b)$ can be written as $\leftarrow (1 : a), (1 : b_2)$ adding the mapping rule $(1 : b_2) \leftarrow (2 : b)$, where b_2 is a fresh relation in peer 1.

4.5   Description Logic Knowledge Bases

We now discuss the connection between our work and results on adding integrity constraints to description logic knowledge bases, namely OWL ontologies.

Description logics differ from databases in their rejection of the closed-world assumption, thereby contradicting the semantics of negation-by-failure. For this reason, encoding ontologies as a context in an MCS is a bit different than the previous examples. We follow the approach from [13], refering the reader to the discussion therein of why the embeddding from e.g. [5] is not satisfactory.

Definition 14. *A description logic $\mathcal{L}$ is represented as the relational logic $L_{\mathcal{L}} = \langle \mathsf{KB}_{\mathcal{L}}, \mathsf{BS}_{\mathcal{L}}, \mathsf{ACC}_{\mathcal{L}}, \Sigma_{\mathcal{L}} \rangle$ defined as follows:*

- *$\mathsf{KB}_{\mathcal{L}}$ contains all well-formed knowledge bases (including a T-Box and an A-Box) of $\mathcal{L}$;*
- *$\mathsf{BS}_{\mathcal{L}}$ is the set of all possible A-Boxes in the language of $\mathcal{L}$;*
- *$\mathsf{ACC}_{\mathcal{L}}(\mathsf{kb})$ is the singleton set containing the set of kb's known consequences (positive and negative);*
- *$\Sigma_{\mathcal{L}}$ is the signature underlying $\mathcal{L}$.*

Regarding the choice of acceptable belief sets (the elements of $\mathsf{BS}_{\mathcal{L}}$), the possible A-Boxes correspond to (partial) models of $\mathcal{L}$, seen as a first-order theory: they contain concepts and roles applied to particular known individuals, or negations thereof. However, they need not be categorical: they may contain neither $C(a)$ nor $\neg C(a)$ for particular C and a. This reflects the typical open-world semantics of ontologies. In particular, the only element of $\mathsf{ACC}_{\mathcal{L}}(\mathsf{kb})$ may not be a model of kb in the classical sense of first-order logic. This is in contrast with [5], where $ACC_{\mathcal{L}}(kb)$ contains all models of kb; as discussed in [13], this is essential to model e.g. default reasoning correctly.

Definition 15. *An ontology $\mathcal{O}$ based on description logic $\mathcal{L}$ induces a context with underlying logic $L_{\mathcal{L}}$, knowledge base $\mathcal{O}$, and an empty set of bridge rules.*

Like in the database scenario, ontologies viewed as MCSs always have one equilibrium, as long as they are logically consistent. Therefore, the notions of weak and strong satisfaction of integrity constraints again coincide, and we get the same notion of consistency w.r.t. a set of integrity constraints as that defined in [26]; however, our syntax is more restricted, as we do not allow general formulas as integrity constraints. Observe that, as in that work, our integrity constraints only apply to named individuals (explicitly mentioned in the ontology's A-Box), which is a desirable consequence that yet again can only be gained from keeping integrity constraints separate from the knowledge base.

Example 8. We illustrate the construction in this section with a classical example. We assume that we have an ontology O including a concept person and a role hasCPR, which associates individuals with their CPR number. (So we are essentially resetting Example 6 to use an ontology, rather than a distributed database.) We can add the integrity constraint

$$\leftarrow (O : \mathsf{person}(x)), \mathsf{not}\ (O : \mathsf{hasCPR}(x, y))$$

requiring each person to have a CPR number. Due to the semantics of ontologies, this actually requires each person's CPR number to be explicitly present in the ontology: the presence of an axiom such as person $\sqsubseteq$ ($\exists$person.hasCPR) does not yield any instance hasCPR(x, y) in the set of the ontology's known consequences. This also justifies our definition of $ACC_{\mathcal{L}}$: if we take the model-based approach of [5], then this integrity constraint no longer demands the actual presence of such a fact in the A-Box.

This integrity constraint is an example of one that does not satisfy the safety condition (the variable y occurs only in a negated literal), but as discussed in Sect. 3 our theory is easily extended to cover this case, as y only occurs once in the formula.

Our scenario is also expressive enough to model the distributed ontology scenario of [19], which defines integrity constraints as logic programming-style rules with empty head whose body can include atoms from different ontologies: we can simply consider the MCS obtained from viewing each ontology as a separate context, and the integrity constraints as ranging over the joint system.

5   Repairs and Managed Multi-context Systems

The definitions in the previous section allow us to distinguish between acceptable and non-acceptable equilibria w.r.t. a set of integrity constraints, but they do not help with the analog of the problem of database repair [1] – namely, given an inconsistent equilibrium for a given MCS, how do we change it into a consistent one. In order to address this issue, we turn our attention to *managed* multi-context systems (mMCS) [7].

Definition 16. *A managed multi-context system is a collection of managed contexts* $\{C_i\}_{i\in\mathcal{J}}$, *with each* $C_i = \langle L_i, \mathsf{kb}_i, \mathsf{br}_i, D_i, OP_i, \mathsf{mng}_i \rangle$ *as follows.*

- L_i *is a relational logic,* kb_i *is a knowledge base, and* D_i *is a set of import domains, as in standard MCSs.*
- OP_i *is a set of operation names.*
- br_i *is a set of managed bridge rules, with the form of* Eq. (1), *but where s is of the form* $o(p)$ *with* $o \in OP_i$ *and* $p \in \bigcup \mathsf{KB}_i$.
- $\mathsf{mng}_i : \wp(OP_i \times \bigcup \mathsf{KB}_i) \times \mathsf{KB}_i \to \mathsf{KB}_i$ *is a management function.*

The intuition is as follows: the heads of bridge rules can now contain arbitrary actions (identified by the labels in OP_i), and the management function specifies the semantics of these labels – see [7] for a more detailed discussion. Our definition is simplified from those authors', as they allow the management function to change the semantics of the contexts and return several possible effects for each action. This simplification results in a less flexible concept of mMCS, which is however more useful for the purposes of defining repairs.

Example 9. The management function can perform several manipulations of the knowledge base in one update action. For example, considering the setting of Example 6, we could include an operation replace $\in OP_{\mathsf{CPR}}$ such that $\mathsf{mng}(\{\langle \mathsf{replace}, \mathsf{person}(Id, Name, Add)\rangle\}, \mathsf{kb})$ inserts the tuple $(Id, Name, Add)$ into the person table and removes any other tuple $(Id, Name', Add')$ from that table.

Every MCS (in the sense of the previous section) can be seen as an mMCS by taking every context to have exactly one operation add with the natural semantics of adding its argument (the head of the rule) to the belief set associated with the context in question. We will therefore discuss integrity constraints over mMCS in the remainder of this section. The motivation of generalizing database tradition also suggests that we include another operation remove that removes an element from the specified context.

Definition 17. *Let* $M = \{C_i\}_{i\in\mathcal{I}}$ *be an mMCS. An* update action *for* M *is of the form* $(i : o(p))$, *with* $i \in \mathcal{J}$, $o \in OP_i$ *and* $p \in \bigcup \mathsf{KB}_i$.

Given a set of update actions $\mathcal{U}$ *and an mMCS* M, *the result of applying* $\mathcal{U}$ *to* M, *denoted* $\mathcal{U}(M)$, *is computed by replacing each* kb_i *(in context* C_i) *by* $\mathsf{mng}_i(\mathcal{U}_i, \mathsf{kb}_i)$, *where* $\mathcal{U}_i$ *is the set of update actions of the form* $(i : o(p))$.

Updates differ from applying (managed) bridge rules, as they actually change one or more knowledge bases in M's contexts *before* any evaluation of bridge rules takes place. This is similar to database updates, which change the database before and independent of the query processing. Based on this notion of update, we can define (weak) repairs as follows.

Definition 18. *Let M be an mMCS, η be a set of ICs over M, and assume that M is inconsistent w.r.t. η. A set of update actions $\mathcal{U}$ is a* weak repair *for M and η if $\mathcal{U}(M)$ is consistent w.r.t. η. If there is no subset $\mathcal{U}'$ of $\mathcal{U}$ that is also a weak repair for M and η, then $\mathcal{U}$ is a* repair.

Example 10. Again in the setting of Example 6, suppose that the CPR database contains the record person(1111111118, *old_lady*, *odense*) and the Silkeborg electoral database contains the records voter(1111111118) and address(*gjern*), but not the record address(*odense*) as Odense is not in Silkeborg. The induced mMCS is inconsistent w.r.t. the integrity constraint Eq. (5), and a possible repair is $\{(\text{CPR} : \text{add}(\text{person}(1111111118, old\_lady, gjern)))\}$. The semantics of the management function guarantee that only the new record will persist in the mMCS.

As is the case in databases, it can happen that a set of integrity constraints is inconsistent, in the sense that no MCS can satisfy it. However, this inconsistency can also arise from incompatibility between integrity constraints and bridge rules – consider the very simple case where there is a bridge rule $(B : b) \leftarrow (A : a)$ and an integrity constraint $\leftarrow (A : a), \text{not}\ (B : b)$. Since our notion of update does not allow one to change bridge rules, this inconsistency is unsurmountable.

In general, this interaction between integrity constraints and bridge rules makes the problem of finding repairs for inconsistent MCSs more complex than in the database world. However, Theorems 1 and 2 show that the problem of finding a repair for an MCS that is inconsistent w.r.t. a set of integrity constraints can be reduced to finding a set of update actions that will make a logically inconsistent MCS have equilibria. The results on diagnosing and repairing logical inconsistency in multi-context systems [16,17] can therefore be used to tackle this problem. By considering deductive databases as MCSs, we also see the problem of repairing an inconsistent MCS as a generalization of the view-update problem [24,25,32].

Another issue is how to choose between different repairs: as in the database case, some repairs are preferable to others. Consider the following toy example.

Example 11. Let M be the MCS induced by a deductive database with one extensional relation p and one intensional relation q, both 0-ary, connected by the rule q $\leftarrow$ p, and consider the integrity constraint $(I : \text{q})$.

Assume the usual operations add and remove. There are two repairs for M, namely $\{(E : \text{add}(\text{p}))\}$ and $\{(I : \text{add}(\text{q}))\}$, but only the former is valid from the perspective of deductive databases.

The usual consensus in databases is that, in general, deciding which repair to apply is a task that needs human intervention [18]. However, several formalisms also include criteria to help automate such preferences. In our setting, a simple way to restrict the set of possible repairs would be to restrict the update actions to use only a subset of the OP_is – in the case of deductive databases, we could simply restrict them to the operations over C_E. An alternative that offers more fine-tuning capabilities would be to go in the direction of active integrity constraints [21], which require the user to be explicit about which update actions can be used to repair the integrity constraints that are not satisfied. We plan to pursue the study of such formalisms to discuss repairs of MCSs with integrity constraints in future work. We also intend to study generalizations of repairs to include the possibility of changing bridge rules.

6   Conclusions and Future Work

In this paper, we proposed a notion of integrity constraint for multi-context systems, a general framework for combining reasoning systems. We showed that our notion generalizes the well-studied concept of integrity constraint over databases, and studied its relation to similar notions in other formalisms. Satisfaction of integrity constraints comes in two variants, weak and strong, related to the usual concepts of brave and cautious reasoning.

By showing how to encode integrity constraints within the syntax of MCSs, we obtained decidability and complexity results for the problem of whether a particular MCS weakly or strongly satisfies a set of integrity constraints, and of repairing it in the negative case. We argued however that by keeping integrity constraints as an added layer on top of an MCS we are able to separate intrinsic logical inconsistency from inconsistencies that may arise e.g. from improper changes to an individual context, which we want to detect and fix, rather than propagate to other contexts. Our examples show that we indeed capture the usual behaviour of integrity constraints in several existing formalisms.

We also defined a notion of repair, consistent with the tradition in databases, and identified new research problems related to which repairs should be preferred that arise in the MCS scenario. We intend to pursue this study further by developing a theory of active integrity constraints, in the style of [21].

Acknowledgements. We would like to thank Graça Gaspar for introducing us to the exciting topic of integrity constraints and for many fruitful discussions. We also thank the anonymous referees for many valuable suggestions that improved the overall quality of this paper. This work was supported by the Danish Council for Independent Research, Natural Sciences, and by FCT/MCTES/PIDDAC under centre grant to BioISI (Centre Reference: UID/MULTI/04046/2013).

References

1. Abiteboul, S.: Updates, a new frontier. In: Gyssens, M., Paredaens, J., van Gucht, D. (eds.) ICDT'88. LNCS, vol. 326, pp. 1–18. Springer, Heidelberg (1988)
2. Abiteboul, S., Hull, R., Vianu, V.: Foundations of Databases. Addison Wesley, Reading (1995)
3. Arenas, M., Bertossi, L., Chomicki, J.: Consistent query answers in inconsistent databases. In: Vianu, V., Papadimitriou, C. (eds.) PODS 1999, pp. 68–79. ACM Press (1999)
4. Asirelli, P., Santis, M.D., Martelli, M.: Integrity constraints for logic databases. J. Log. Program. **2**(3), 221–232 (1985)

5. Brewka, G., Eiter, T.: Equilibria in heterogeneous nonmonotonic multi-context systems. In: AAAI 2007, pp. 385–390. AAAI Press (2007)
6. Brewka, G., Eiter, T., Fink, M.: Nonmonotonic multi-context systems: a flexible approach for integrating heterogeneous knowledge sources. In: Balduccini, M., Son, T.C. (eds.) Logic Programming, Knowledge Representation, and Nonmonotonic Reasoning. LNCS, vol. 6565, pp. 233–258. Springer, Heidelberg (2011)
7. Brewka, G., Eiter, T., Fink, M., Weinzierl, A.: Managed multi-context systems. In: Walsh, T. (ed.) IJCAI 2011, IJCAI/AAAI, pp. 786–791 (2011)
8. Caroprese, L., Trubitsyna, I., Truszczyński, M., Zumpano, E.: The view-update problem for indefinite databases. In: del Cerro, L.F., Herzig, A., Mengin, J. (eds.) JELIA 2012. LNCS, vol. 7519, pp. 134–146. Springer, Heidelberg (2012)
9. Caroprese, L., Zumpano, E.: Consistent data integration in P2P deductive databases. In: Prade, H., Subrahmanian, V.S. (eds.) SUM 2007. LNCS (LNAI), vol. 4772, pp. 230–243. Springer, Heidelberg (2007)
10. Caroprese, L., Zumpano, E.: Dealing with incompleteness and inconsistency in P2P deductive databases. In: Desai, B., Almeida, A., Bernardino, J., Ferreira Gomes, E. (eds.) IDEAS 2014, pp. 124–131. ACM (2014)
11. Cruz-Filipe, L.: Optimizing computation of repairs from active integrity constraints. In: Beierle, C., Meghini, C. (eds.) FoIKS 2014. LNCS, vol. 8367, pp. 361–380. Springer, Heidelberg (2014)
12. Cruz-Filipe, L., Franz, M., Hakhverdyan, A., Ludovico, M., Nunes, I., Schneider-Kamp, P.: repAIrC: a tool for ensuring data consistency by means of active integrity constraints. In: Fred, A., Dietz, J., Aveiro, D., Liu, K., Filipe, J. (eds.) KMIS, pp. 17–26. SciTePress (2015)
13. Cruz-Filipe, L., Gaspar, G., Nunes, I.: Information flow within relational multi-context systems. In: Janowicz, K., Schlobach, S., Lambrix, P., Hyvönen, E. (eds.) EKAW 2014. LNCS, vol. 8876, pp. 97–108. Springer, Heidelberg (2014)
14. Cruz-Filipe, L., Henriques, R., Nunes, I.: Description logics, rules and multi-context systems. In: McMillan, K., Middeldorp, A., Voronkov, A. (eds.) LPAR-19 2013. LNCS, vol. 8312, pp. 243–257. Springer, Heidelberg (2013)
15. Dao-Tran, M., Eiter, T., Fink, M., Krennwallner, T.: Distributed nonmonotonic multi-context systems. In: Lin, F., Sattler, U., Truszczynski, M. (eds.) KR 2010. AAAI Press (2010)
16. Eiter, T., Fink, M., Ianni, G., Schüller, P.: The IMPL policy language for managing inconsistency in multi-context systems. In: Tompits, H., Abreu, S., Oetsch, J., Pührer, J., Seipel, D., Umeda, M., Wolf, A. (eds.) INAP/WLP 2011. LNCS, vol. 7773, pp. 2–25. Springer, Heidelberg (2013)
17. Eiter, T., Fink, M., Schüller, P., Weinzierl, A.: Finding explanations of inconsistency in multi-context systems. Artif. Intell. **216**, 233–274 (2014)
18. Eiter, T., Gottlob, G.: On the complexity of propositional knowledge base revision, updates, and counterfactuals. Artif. Intell. **57**(2–3), 227–270 (1992)
19. Fang, M., Li, W., Sunderraman, R.: Maintaining integrity constraints among distributed ontologies. In: CISIS 2011, pp. 184–191. IEEE (2011)
20. Fink, M., Ghionna, L., Weinzierl, A.: Relational information exchange and aggregation in multi-context systems. In: Delgrande, J.P., Faber, W. (eds.) LPNMR 2011. LNCS, vol. 6645, pp. 120–133. Springer, Heidelberg (2011)
21. Flesca, S., Greco, S., Zumpano, E.: Active integrity constraints. In: Moggi, E., Scott Warren, D. (eds.) PPDP 2004, pp. 98–107. ACM (2004)
22. Gallaire, H., Minker, J., Nicolas, J.M.: Logic and databases: a deductive approach. ACM Comput. Surv. **16**(2), 153–185 (1984)

23. Gonçalves, R., Knorr, M., Leite, J.: Evolving multi-context systems. In: Schaub, T., Friedrich, G., O'Sullivan, B. (eds.) ECAI 2014, Frontiers in Artificial Intelligence and Applications, vol. 263, pp. 375–380. IOS Press (2014)

24. Kakas, A., Mancarella, P.: Database updates through abduction. In: McLeod, D., Sacks-Davis, R., Schek, H.J. (eds.) VLDB 1990, pp. 650–661. Morgan Kaufmann (1990)

25. Mayol, E., Teniente, E.: Consistency preserving updates in deductive databases. Data Knowl. Eng. **47**(1), 61–103 (2003)

26. Motik, B., Horrocks, I., Sattler, U.: Bridging the gap between OWL and relational databases. Web Seman. Sci. Serv. Agents World Wide Web **7**(2), 74–89 (2011)

27. Pourebrahimi, B., Bertels, K., Vassiliadis, S.: A survey of peer-to-peer networks. In: ProRISC 2005 (2005)

28. Pührer, J., Heymans, S., Eiter, T.: Dealing with inconsistency when combining ontologies and rules using DL-programs. In: Aroyo, L., Antoniou, G., Hyvönen, E., ten Teije, A., Stuckenschmidt, H., Cabral, L., Tudorache, T. (eds.) ESWC 2010, Part I. LNCS, vol. 6088, pp. 183–197. Springer, Heidelberg (2010)

29. Schlipf, J.: Complexity and undecidability results for logic programming. Ann. Math. Artif. Intell. **15**(3–4), 257–288 (1995)

30. Staworko, S., Chomicki, J.: Consistent query answers in the presence of universal constraints. Inf. Syst. **35**(1), 1–22 (2010)

31. Tasharrofi, S., Ternovska, E.: Generalized multi-context systems. In: Baral, C., de Giacomo, G., Eiter, T. (eds.) KR 2014. AAAI Press (2014)

32. Teniente, E., Olivé, A.: Updating knowledge bases while maintaining their consistency. VLDB J. **4**(2), 193–241 (1995)

33. Ullman, J.: Principles of Database and Knowledge-Base Systems, Volume I. Computer Science Press, Cambridge (1988)

34. Vardi, M.: The complexity of relational query languages (extended abstract). In: Lewis, H., Simons, B., Burkhard, W., Landweber, L. (eds.) STOC 1982, pp. 137–146. ACM (1982)

35. Weinzierl, A.: Advancing multi-context systems by inconsistency management. In: Bragaglia, S., Damásio, C., Montali, M., Preece, A., Petrie, C., Proctor, M., Straccia, U. (eds.) RuleML2011@BRF Challenge, CEUR Workshop Proceedings, vol. 799, CEUR-WS.org (2011)

A Knowledge Based Framework for Link Prediction in Social Networks

Pooya Moradian Zadeh[(⊠)] and Ziad Kobti

School of Computer Science, University of Windsor, Windsor, ON, Canada
{moradiap,kobti}@uwindsor.ca

Abstract. Social networks have a dynamic nature so their structures change over time. In this paper, we propose a new evolutionary method to predict the state of a network in the near future by extracting knowledge from its current structure. This method is based on the fact that social networks consist of communities. Observing current state of a given network, the method calculates the probability of a relationship between each pair of individuals who are not directly connected to each other and estimate the chance of being connected in the next time slot. We have tested and compared the method on one synthetic and one large real dataset with 117 185 083 edges. Results show that our method can predict the next state of a network with a high rate of accuracy.

Keywords: Social networks · Link prediction · Cultural algorithm · Evolutionary algorithm · Knowledge · Community detection

1 Introduction

People use social networks to interact with others. Regardless of the content, these interactions can reveal valuable information about real societies and individuals. This information can be useful to identify the structure and topology of these networks, which makes it possible to track their evolutions and predict the next state. Naturally, these networks are extremely dynamic and their rate of evolution is very high. Consequently, their structure changes frequently. Since these networks reflect real life events, having knowledge about their next state can be applied to various domains such as recommendation systems, decision making, marketing and risk analysis [1–4,6,9]. In the field of social network analysis, this problem is known as Link Prediction, which can be defined as estimating the likelihood of a connection between two disconnected entities in a network in the near future [2,4,6].

The main idea behind this problem is, the future state of a network is not random and has a dependency on the current state. Therefore, the target is to find the level of dependency and the main factors affecting it.

Social networks, as a subset of complex networks have some particular characteristics such as power-law distribution and high value of cluster coefficiency. Having a high level of cluster co-efficiency in the network indicates the tendency

© Springer International Publishing Switzerland 2016
M. Gyssens and G. Simari (Eds.): FoIKS 2016, LNCS 9616, pp. 255–268, 2016.
DOI: 10.1007/978-3-319-30024-5_14

of users to join communities is high. Accordingly, in this paper we propose a knowledge-based evolutionary framework based on these properties to estimate the state of a network in the near future just by having one snapshot of the network.

Our proposed model is defined based on the similarity approach with two main assumptions. The first is that an individual in a network tends to join a community. The second is that, according to the homophily phenomenon in social network, each individual joins a community through their friends. Hence the similarity measurement here is defined as having a common community. For example, if a person in a network has 6 friends and 5 of them are members of a community with 30 people. The probability of a friendship between this person and members of the community in the near future is higher than other cases and it can be estimated approximately.

To estimate this likelihood, a knowledge-based structure which is called belief space has been adapted from the evolutionary cultural algorithm which has been proposed for the community detection problem in [13]. Cultural algorithms are a specific type of evolutionary algorithm that use knowledge to enhance the search process to find near optimal solutions for a problem [10,13]. As shown in Fig. 1, a cultural algorithm consists of Population and Belief spaces. In fact, the population space is a list of probable solutions for the community detection problem and the belief space is a knowledge-based structure which guides the population generation process in each iteration and it is evolved by extracting information from the population space [10,13].

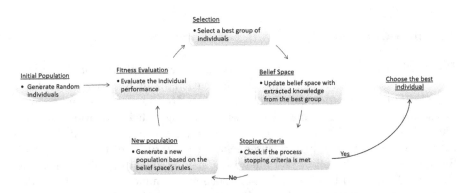

Fig. 1. A cultural algorithm process

In this paper, by focusing on the belief space as a great source of knowledge, we propose an algorithm to determine the level of dependency between each pair of users and estimate their tendency to communicate with each other. The main structure of our proposed algorithm is a directed weighted graph which is generated from the belief space data and demonstrates levels of relationships between all neighbor nodes. For predictions, a mathematical formula is proposed to estimate the likelihood of having relationship between each unconnected pair. This formula has been defined based on two main concepts, number of paths between each unconnected pair and length of these paths. Generally having more paths and shorter lengths implies higher chance of connection in the next timeslot. Finally, our algorithm calculates the probability of a relation between pairs of nodes which are not connected together directly and ranks them.

In this research, we present a novel concept of observing the quality of links between pairs of nodes. We also introduce a method to extract information from structure of the network as a similarity index.

The rest of the paper is organized as follows: In the next section, the problem definition and related works will be reviewed. In Sect. 3, we present our model and, after that, the evaluation of the model will be discussed. Conclusions are presented in the last section.

2   Problem Definition and Related Works

If a network maps to a graph, $G(V, E)$, where V is a fixed number of nodes and E represents links between each pair of nodes, an edge is defined as $e = (u, v) \in E$, where $u, v \in V$, at a particular timeslot (t). Predicting a state of the graph at time $t + 1$ by having a snapshot of it at time t, is defined as the Link Prediction Problem in social networks. In other words, given a network G_t at time t, the output of a link prediction algorithm will be a list of edges which are not in G_t and have high probability of appearing in G_{t+1} [2,4,6]. See also Fig. 2.

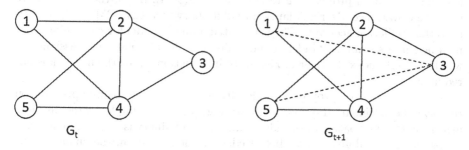

Fig. 2. Predicting the state of a network at time $t + 1$, given a snapshot of it at time t

To solve this problem many studies have been carried out. A similarity-based approach is one of them [4,6]. In fact, having a high number of common features among pair of users increases their chance for making a link in the near future. Therefore, these types of algorithms calculate the level of similarity between each pair of nodes x and node y and assign a score to them. After ranking them, they select the pairs which have higher scores as they have more likelihood to be linked in the near future.

One of the most famous approaches in this field is an unsupervised method which is based on the similarity of the nodes' structure [3,4,6]. To calculate the similarity, many indexes have been proposed, such as the Jaccard similarity co-efficient, Katz, Common Neighbors, Leicht-Holme-Newman, etc. These indexes are mainly based on the number of common neighbors [4,6].

The Common Neighbors index is defined as $C(x, y) = |\Gamma(x) \cap \Gamma(y)|$, where $\Gamma(x)$ and $\Gamma(y)$ are the lists of neighbors of nodes x and y, respectively. This index counts the number of shared neighbors of nodes x and y.

The Jaccard similarity co-efficient is an important index in this field which is defined as $J(x, y) = |\Gamma(x) \cap \Gamma(y)|/|\Gamma(x) \cup \Gamma(y)|$. It measures the number of shared neighbors between two nodes over number of their all unique neighbors.

Leicht-Holme-Newman is also an index which measures the similarity by calculating the number of common neighbors between nodes x and y relative to the product of their degrees: $L(x, y) = |\Gamma(x) \cap \Gamma(y)|/d(x)d(y)$, where $d(x)$ and $d(y)$ are the degrees of nodes x and y, respectively.

The Resource Allocation Index is another index which performs well on real networks. Consider the situation where node x sends some resources to node y through its mutual neighbors. The similarity between x and y is then defined as the amount of resources received by node y: $RA(x, y) = \sum_{z \in \Gamma(x) \cap \Gamma(y)} 1/|\Gamma(z)|$.

However, these indexes are not suitable for all types of networks, their performance is varied based on the structures of different networks [4,6].

In addition, maximum likelihood and probabilistic models approaches which are supervised methods are also used to solve the link prediction problem. However, by increasing the size of the network ($|\text{network}| > 10^4$), these models become impractical because of their time complexity [4,6].

Evolutionary and swarm-based approaches are also used to solve the problem that have been proposed in recent years [1,2,9,11]. In [1], the authors have used the Covariance Matrix Adaption Evolutionary Strategy (CMA-ES) to optimize the prediction accuracy. They suggested a linear model for combining common neighbor's similarity indexes and nodes specific information by assigning a weight to each index. In their model, prior information about the network is not required.

In [2], the authors proposed an algorithm based on ant colony optimization to solve the problem. Random walk strategy has been implemented in their algorithm to select paths. In this algorithm, the probability is assigned to an edge to help an artificial ant select a better edge. In each iteration, the quality of the paths are evaluated to update the probabilities for the next iterations. Finally, the path with higher quality is selected as a link which has more likelihood to appear.

On the other hand, since the future actually is not predictable, to test the accuracy of the algorithm, a network must be randomly divided into two subsets, the training set, E^T, and the probe set, E^P. Here, E^T can be considered as the observed known interactions and E^P as the set of links that must be predicted for testing. In the prediction process, information from E^T must not be used. As a result of this division, $E^T \cup E^P = E$ (the set of the network's edges) and $E^T \cap E^P = \emptyset$.

To evaluate the performance of these algorithms, two main methods are commonly used, the Area Under the Receiver Operating Characteristic Curve (AUC) and Precision [4,6].

For the former, $AUC = (n'+0.5n'')/n$, where n is the number of independent comparisons and n' denotes the number of times a randomly chosen missing link (a link in E^P) had a higher score than a randomly chosen nonexistent link (a link in $U - E$, where U denotes the universal set containing all possible links, of which there are $|V|(|V| - 1)/2$, with $|V|$ the number of nodes in the network). Furthermore, n'' denotes the number of times that their score is the same [4,6].

For the latter, if the ranked non-observed links are given, Precision is defined as the number of relevant items selected divided by the total number of items selected. In the case that the top-L links from the predicted links are chosen, and L_r denotes the number of these links which are in E^P, then Precision can be defined as L_r/L [4,6].

3 Proposed Evolutionary Model

As we mentioned before, community is the core of our model. Thus, in our model we adapt outputs of the evolutionary cultural algorithm which has been proposed to detect communities on social networks in [13]. While the output of this algorithm is the list of communities, the focus of this research is on the belief space. This belief space can be visualized as a probability matrix which estimates the quality of relationships between each pair of nodes in the network which are directly connected together. Using this belief space which is updated by the extracted information from populations in each iteration, the cultural algorithm limits the search space and enhances the individual evolutions. In our model, we propose using this knowledge repository as a source of information. As shown in Fig. 3, the belief space will map to a directed weighted graph. The weights indicate the level of dependency between each connected pair of nodes. After that, we propose a method to estimate the likelihood of relationships between two unlinked nodes of the graph. Ranking them will be the last process of this model.

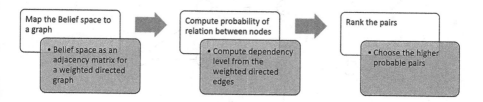

Fig. 3. Components of the proposed model

3.1   Making the Weighted Graph

First we briefly describe the mentioned cultural algorithm [13]. In this algorithm, an individual is represented as a probable solution based on a particular locus-based adjacency method [8] stored in an array structure. The length of this array is equal to number of nodes in the graph. Each cell of this array is addressed from 1 to n (length of the array) which determines a node in the graph with the same number. E.g., cell #10 corresponds to node #10. For each cell #i, the algorithm will choose an address of a node from the list of neighbors of node #i.

For example, as shown in Fig. 4, if a network has 7 nodes, one sample individual can be defined as an array of nodes, shown in Fig. 5, and illustrated in Fig. 6, which shows two communities in this graph (nodes #1, 5, 6, and 7 in one community and nodes #2, 3, and 4 in another).

As mentioned before and presented in Fig. 1, in each iteration, specific number of individuals are generated by the algorithm (to make a population)

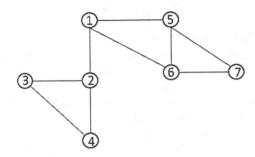

Fig. 4. A sample network

1	2	3	4	5	6	7
5	3	2	2	7	7	6

Fig. 5. A random individual

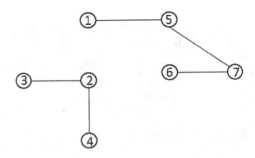

Fig. 6. Illustration of the individual in Fig. 5 which clearly shows two separate communities (1,5,6,7) and (2,3,4)

according to the rules which are set in the belief space. The quality of these individuals is evaluated based on a fitness function. As a result, these individuals can be compared with each other. After sorting them, a group of them that have better fitness values are selected to enter the belief space and if they meet some conditions they can update the belief space.

To update the belief space, each cell of these individuals adds its value to the n by n belief space matrix, where n is the number of nodes, and the algorithm will calculate the relative frequency of it and store it in the matrix as shown in Fig. 7. With this method, the belief space can be considered as an alternative adjacency matrix for the graph, because it is a weighted sub-graph of the main network that shows the level of dependency between nodes according to the community index.

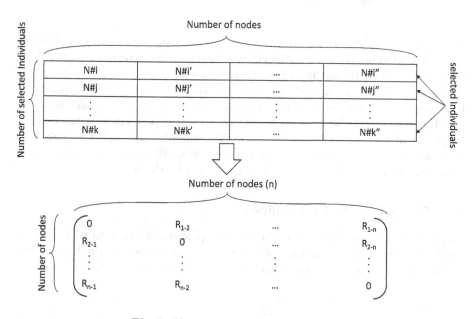

Fig. 7. The structure of the belief space

The belief space plays a key role by setting some rules for generating new generations of individuals. This space collects and saves normative knowledge of the best group of individuals. The assumption is that best individuals are close to an optimal solution, thus the final solution can be generated by combining components of them. In fact, the belief space defines a new state space for the network by storing best individuals. In the subsequent iterations, new generations of individuals are produced mostly based on this state space.

Our main assumption here is, if the number of iterations approaches infinity, the belief space matrix can accurately represent some information about the level of dependency between the connected nodes. Consequently, these relative frequencies can be used as the probability of a relation in the next timeslot

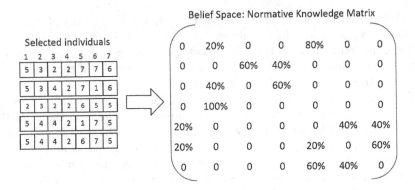

Fig. 8. Belief space formed by 5 selected individuals

based on the community function. By processing a snap shot of an undirected and unweighted network, a weighted directed graph is made which reveals hidden information about the quality of relations in the network.

Figure 8 shows an example for updating the belief space. Five individuals have been selected to update the belief space of the same network shown in Fig. 4. If the matrix had been empty before, then it is populated by the relative frequency of nodes and their neighbors. For example, node #5 was linked to node #1, 20 % of times (once out of 5 times). If we illustrate this belief space, as shown in Fig. 9, a directed weighted graph will be the result.

3.2   Computing the Probabilities

To compute the probabilities of relations of a pair of disconnected nodes in this weighted graph, two criteria have been considered to propose a formula. The first is the number of paths between each pairs of disconnected nodes. The second

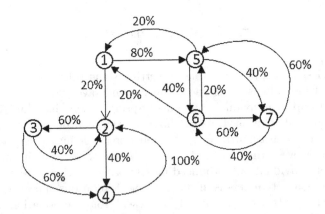

Fig. 9. Illustration of the belief space in Fig. 8

is the length of these paths. To reduce the complexity, we assume the length of the paths is always 1, which means that the probability is computed for those pairs of disconnected nodes that have only one node between themselves. Let $G(V, E, W)$ denote the input weighted graph, where V is a set of nodes and E a set of edges between each pair of nodes (hence, each edge e is of the form (i, j), with $i, j \in V$. Furthermore W is a set of weights of edges, with $0 \le W(i,j) \le 1$ for all edges (i, j). For each pair of disconnected nodes (i, k), where $i, k \in V$ and $(i, k) \notin E$, if there is a node j with $j \in V$ and $(i, j), (j, k) \in E$, the estimated weight between i and k is computed as follows:

$$\forall j \in V \to (i, j), (j, k) \in E, (i, k) \notin E,$$
$$W'(i, j, k) = \max(W(i, j), W(j, i)) \times \max(W(j, k), W(k, j)). \quad (1)$$

If there were a link between two nodes i and k in the absence of node j, then $W'(i, j, k)$ can be interpreted as the estimated weight of that link.

For each similar path this weight must be computed accordingly, and, finally, the probability of a relation between nodes i and k is computed as follows:

$$P(i, k) = 1 - \frac{1}{2(n + \sum_1^n W'(i, j, k))}, \quad (2)$$

where n is the number of paths between i and k.

For example, in Fig. 9, a direct link does not exist between node#1 and #7 but there are 2 paths of length 1 between them. Therefore, $n = 2$, and the nodes #5 and #6 represent j. We have $W'(1, 5, 7) = 0.8 \times 0.6 = 0.48$ and $W'(1, 6, 7) = 0.2 \times 0.6 = 0.12$, and $P(1, 7) = 1 - (1/(2 \times (2 + 0.6)) = 0.6153$.

3.3   Ranking the Probabilities

After calculating all the probabilities, the predicted pairs must be ranked based on their probabilities. Finally, the top-L of them will be selected as the final predicted edges. This process is shown in the following algorithm:

```
Algorithm CA-LP (G,A,B,L)

Input:
G: an undirected and unweighted graph, G(V,E)
A: adjacency matrix of G
B: Belief Space matrix
L: desired number of top predicted links

Output:
O: n*n matrix of L probabilities where
   O(i,j)=P(i,j), ( i,j are members of V)
```

```
Main:
1: Map Belief space to a weighted directed Graph
2: Compute
P(i,k) by extracting weights from B according to
    (1) and (2), for all pairs where A(i,k)=0 and A(i,j), A(j,k)=1
3: Store probabilities in a array
4: Sort the array
5: Choose the top-L and store in O where O(i,k)=P(i,k)
```

4   Evaluation

To evaluate the performance of the proposed algorithm, we have used one synthetic network and one real large social network dataset. For the synthetic network, 10 graphs were generated randomly based on Newman's method in [7]. Each of these graphs has 128 nodes with degree 16, therefore the graph has 1024 edges. It consists of 4 same-sized communities where each community has 32 members. Each of these members have Z_{in} links to other members who are inside its own community and Z_{out} links to members from other communities ($Z_{in} + Z_{out} = 16$). The range of Z_{out} in these 10 graphs were set from 3 to 5.

As shown in Table 1, we selected 90 % of the graph as E^T and the rest as E^P to evaluate the performance. The belief space which was imported to the algorithm was obtained from the result of running the community detection algorithm proposed in [13]. We tested the effectiveness of the algorithm according to both AUC and Precision methods. The results are illustrated in Table 2 and Fig. 10. Tests were implemented 100 times independently on the top-100 instances. We also compared the results of AUC with three other similarity metrics, Common Neighbors (CN), Jaccard (JC) and Leicht-Holme-Newman (LH).

The results clearly show that the proposed algorithm has better performance in comparison with other metrics on synthetic networks. Another interesting observation is that, by increasing the complexity of the network ($Z_{out} > 4$) the performance of the algorithm reduced significantly. We believe the cause to be the increasing rate of errors in the community detection algorithm when Z_{out} becomes larger.

In addition to Precision, we also compared the top-102 predicted links calculated by the algorithm with the probe set, E^P (|predicted links in E^P|/|E^P|). As a result, in average 78.28 % of the predicted links were among the probe set, which means that the algorithm could predict the correct links by an accuracy of more than 75 %.

Table 1. Description of the synthetic network

#Nodes	#Edges	E^T	E^P	U
128	1024	922	102	8128

Table 2. Comparision between different methods

Z_{out}	AUC				Precision
	CA-LP	CN	JC	LH	CA-LP
3	0.901	0.756	0.696	0.899	0.57
3	0.934	0.780	0.754	0.796	0.59
3	0.930	0.893	0.890	0.943	0.63
4	0.963	0.772	0.771	0.957	0.56
4	0.993	0.723	0.623	0.803	0.78
4	0.979	0.801	0.692	0.967	0.70
4	0.955	0.882	0.802	0.940	0.73
5	0.912	0.902	0.800	0.912	0.67
5	0.856	0.834	0.870	0.884	0.65
5	0.895	0.722	0.704	0.809	0.65

We also tested the performance of our proposed algorithm on a big real dataset, Orkut, with 117 185 083 edges [12]. The dataset obtained from the Stanford Large Network Dataset repository [5] is a benchmark dataset used by most researchers in social network analysis. Another reason for selecting this dataset is that it is a network with ground-truth communities which make us possible to validate our results. Information about this dataset is represented in Table 3. The procedure for running the experiment is similar to the procedure described before in experimental setup for synthetic networks. The network was divided

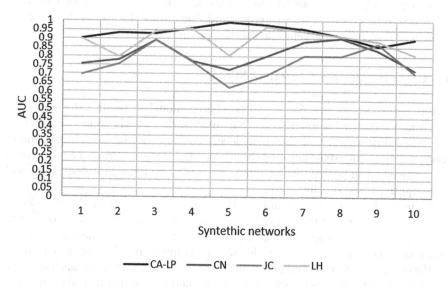

Fig. 10. Comparision of the algorithms based on AUC

Table 3. Orkut dataset specification

#Nodes	#Edges	Cluster coefficiency	E^T	E^P	U
3072441	117185083	0.1666	105466575	11718508	4719945313020

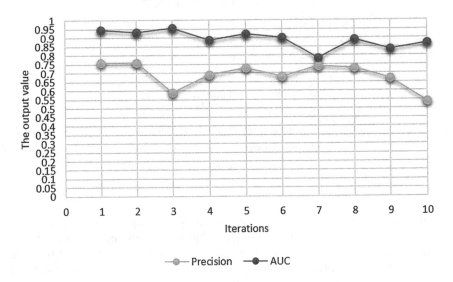

Fig. 11. The results from the orkut dataset

into two sets, the training set (90 %) and the probe set (10 %). After 10 iterations of independent experiments, the AUC and the precision were calculated. As shown in Fig. 11, the algorithm could estimate the correct links with over 68 % success based on the precision method. Regarding the size of the network, we believe that it is an acceptable rate for prediction.

5   Conclusion and Future Work

In this paper, we proposed a knowledge-based model to predict the state of a network in the near future. The key part of this model is the belief space which is a probability matrix that shows the level of dependency between linked nodes. Assuming it as an adjacency matrix, a weighted directed graph can be made. Consequently, the probability of relation between two disconnected nodes will be computed based on this graph.

Estimating the quality of links between a pair of nodes in the network is the first contribution of the algorithm. The second one is defining the concept of community as a similarity index. Finally, the third one is using the cultural algorithm as a knowledge-based evolutionary algorithm to predict the near future.

We evaluated the performance of our algorithm on one synthetic and one large real dataset and compared it with three other metrics. Regarding the results, the algorithm can predict the state of a network with a high accuracy. According to this issue that the objective of evolutionary algorithms is to find near optimal solutions, we believe that by increasing the number of iterations, the quality of prediction will improve. Meanwhile, since the size of the belief space is fixed to the number of nodes, the complexity of the algorithm will not change based on the number of iterations or the number of edges.

In the future, we would like to observe the performance of the algorithm in different type of social networks and extend our work to multiple networks. In addition, currently we have tested the algorithm using the common standard procedure of dividing the training and probe set in the ratio of 90 % and 10 %, in the future we would like to test the performance on different ratios to find the optimal training size.

Acknowledgments. This work is partially supported by a Cross-Border Institute (CBI) Research Grant.

References

1. Bliss, C.A., Frank, M.R., Danforth, C.M., Dodds, P.S.: An evolutionary algorithm approach to link prediction in dynamic social networks. J. Comput. Sci. **5**(5), 750–764 (2014)
2. Chen, B., Chen, L.: A link prediction algorithm based on ant colony optimization. Appl. Intell. **41**(3), 694–708 (2014)
3. Fire, M., Tenenboim, L., Lesser, O., Puzis, R., Rokach, L., Elovici, Y.: Link prediction in social networks using computationally efficient topological features. In: 2011 IEEE Third Inernational Conference on Social Computing (SocialCom), PAS-SAT/SocialCom 2011, Privacy, Security, Risk and Trust (PASSAT), Boston, MA, USA, pp. 73–80, 9–11 October 2011
4. Hasan, M.A., Zaki, M.J.: A survey of link prediction in social networks. In: Aggarwal, C.C. (ed.) Social Network Data Analytics, pp. 243–275. Springer, USA (2011)
5. Leskovic, J., Krevl, A.: SNAP datasets. In: SNAP Datasets: Stanford Large Network Dataset Collection. http://snap.stanford.edu/data
6. Lü, L., Zhou, T.: Link prediction in complex networks: a survey. Physica A **390**(6), 1150–1170 (2011)
7. Newman, M.: Detecting community structure in networks. Eur. Phys. J. B **38**(2), 321–330 (2004)
8. Park, Y., Song, M.: A genetic algorithm for clustering problems. In: Koza, J.R., Banzhaf, W., Chellapilla, K., Deb, K., Dorigo, M., Fogel, D.B., Garzon, M.H., Goldberg, D.E., Iba, H., Riolo, R. (eds.) Proceedings of the Third Annual Conference on Genetic Programming, pp. 568–575. Morgan Kaufmann, University of Wisconsin, Madison, Wisconsin, 22–25 July 1998
9. Qiu, B., He, Q., Yen, J.: Evolution of node behavior in link prediction. In: Burgard, W., Roth, D. (eds.) Proceedings of the Twenty-Fifth AAAI Conference on Artificial Intelligence, AAAI 2011, San Francisco, California, USA, 7–11 August 2011
10. Reynolds, R.G.: An introduction to cultural algorithms. In: Sebald, A.V., Fogel, L.J. (eds.) Proceedings of the Third Annual Conference Evolutionary Programming, pp. 131–139. World Scientific Press, San Diego, CA, 24–26 February 1994
11. Sherkat, E., Rahgozar, M., Asadpour, M.: Structural link prediction based on ant colony approach in social networks. Physica A **419**, 80–94 (2015)

12. Yang, J., Leskovec, J.: Defining and evaluating network communities based on ground-truth. Knowl. Inf. Syst. **42**(1), 181–213 (2015)
13. Zadeh, P.M., Kobti, Z.: A multi-population cultural algorithm for community detection in social networks. Procedia Comput. Sci. **52**, 342–349 (2015). Shakshuki, E.M. (ed.) Proceedings of the 6th International Conference on Ambient Systems, Networks and Technologies (ANT 2015), the 5th International Conference on Sustainable Energy Information Technology (SEIT-2015), London, UK, 2–5 June 2015

Logics and Complexity

Approximation and Dependence via Multiteam Semantics

Arnaud Durand[1], Miika Hannula[2], Juha Kontinen[2(✉)], Arne Meier[3], and Jonni Virtema[3]

[1] Institut de Mathématiques de Jussieu - Paris Rive Gauche, CNRS UMR 7586, Université Paris Diderot, Paris, France
durand@math.univ-paris-diderot.fr
[2] Department of Mathematics and Statistics, University of Helsinki, Helsinki, Finland
{miika.hannula,juha.kontinen}@helsinki.fi
[3] Leibniz Universität Hannover, Institut für Theoretische Informatik, Hanover, Germany
{meier,virtema}@thi.uni-hannover.de

Abstract. We define a variant of team semantics called *multiteam semantics* based on multisets and study the properties of various logics in this framework. In particular, we define natural probabilistic versions of inclusion and independence atoms and certain approximation operators motivated by approximate dependence atoms of Väänänen.

1 Introduction

Dependence logic was introduced by Väänänen in 2007 [36]. It extends first-order logic with dependence atomic formulas (dependence atoms) $=\!(x, y)$ with the intuitive meaning that the value of the variable y is functionally determined by the values of the variables x. The notion of dependence has real meaning only in plurals. Thus, in contrast to the usual Tarskian semantics, in dependence logic the satisfaction of formulas is defined not via single assignments but via sets of assignments. Such sets are called *teams* and the semantics is called *team semantics*. In this article we take a further step of replacing structures and teams by their multiset analogues. Multiteams have been considered in some earlier works [20,21,38] but so far no systematic study of the subject in the team semantics context has appeared. In the temporal logic setting (in the context of computation tree logic) multiteam semantics have been introduced and studied recently [28]. In this article we define the so-called *lax* and *strict* multiteam semantics and study properties of various logics under these semantics. Moreover we show how the shift from sets to multisets naturally gives rise to probabilistic and approximate versions of dependence logic.

The idea of team semantics goes back to Hodges [19] whose aim was to define compositional semantics for *independence-friendly logic* [18]. The introduction of dependence logic and its many variants has evinced that team semantics is a very interesting and versatile semantical framework. In fact, team semantics

M. Gyssens and G. Simari (Eds.): FoIKS 2016, LNCS 9616, pp. 271–291, 2016.
DOI: 10.1007/978-3-319-30024-5_15

has natural propositional, modal, and temporal variants. The study of *modal dependence logic* was initiated by Väänänen [37] in 2008. Shortly after, *extended modal dependence logic* was introduced by Ebbing et al. [7] and *modal independence logic* by Kontinen et al. [27]. In purely propositional context the study was initiated by Yang and Väänänen [42] and further studied, e.g., by Hannula et al. [17]. One of the most important developments in the area of team semantics was the introduction of *independence logic* [13] in which dependence atoms of dependence logic are replaced by *independence atoms* $y \perp_x z$. The intuitive meaning of the independence atom $y \perp_x z$ is that, when the value of x is fixed, knowing the value of z does not tell us anything new about the value of y. Soon after the introduction of independence logic, Galliani [9] showed that independence atoms can be further analysed, and alternatively expressed, in terms of inclusion and exclusion atoms. The inclusion atom $x \subseteq y$ expresses that each value taken by x in a team X appears also as a value of y in X. The meaning of the exclusion atom $x|y$ is that x and y have no common values in X.

Independence, inclusion, and exclusion atoms have very interesting properties in the team semantics setting. For example, inclusion atoms give rise to a variant of dependence logic that corresponds to the complexity class PTIME over finite ordered structures [10]. In fact, the complexity theoretic aspects of these atoms in propositional, modal, and first-order setting have been studied extensively during the past few years (see the survey of Durand et al. [6] and the references therein).

A team X over variables $x_1, \ldots, x_n$ can be viewed as a database table with $x_1, \ldots, x_n$ as its attributes. Under this interpretation, dependence, inclusion, exclusion, and independence atoms correspond exactly to functional, inclusion, exclusion, and embedded multivalued dependencies, respectively. These dependencies have been studied extensively in database theory. The close connection between team semantics and database theory has already led to fruitful interactions between these areas [15,16,26]. It is worth noting that multiset semantics (also known as bag semantics) is widely used in databases [1,24,29]. On the other hand, independence atoms, embedded multivalued dependencies, and the notion of conditional independence $Y \perp Z | X$ in statistics have very interesting connections, see, e.g., [14,40]. In this article we establish that, in the multiteam semantics setting, independence atoms can be naturally interpreted exactly as statistical conditional independence. Probabilistic versions of dependence logic have been previously studied by Galliani and Mann [8,11].

In practice dependencies such as functional dependence do not hold absolutely but with a small margin of error. In order to logically model such scenarios, Väänänen introduced approximate dependence atoms [38]. The corresponding approximate functional dependencies have been studied in the context of data mining [22]. In this article we define a general approximation operator which, in particular, can be used to express approximate dependence atoms. In the last sections of the article, we study the computational aspects of logics extended by the approximation operator.

Previous work on multisets in team semantics. The idea of generalising team semantics by the use of multisets has been discussed in several articles. Hyttinen et al. [20] study multiteams, and their generalisations called *quantum teams*, which they use to give semantics to a propositional logic called *quantum team logic* that can be used for the logical analysis of phenomena in quantum physics. Moreover Hyttinen et al. [21] define a notion of a *measure team* and *measure team logic*. The latter is a logic for making inferences about probabilities of first-order formulas in measure teams. Furthermore Krebs et al. introduced team semantics with multisets for the temporal logic CTL [28]. Finally the fact that under multiteam semantics approximate dependence atoms have the locality property (compare to Proposition 11) is discussed by Väänänen [38].

Organisation. This article is organised as follows. Section 2 briefly discusses the basic concepts and definitions. The generalisation of team semantics to multisets is presented in Sect. 3. Section 4 defines the approximation operators, and in Sect. 5 the complexity theoretic aspects of logics with the approximation operators are studied.

2   Preliminaries

We assume familiarity with standard notions in computational complexity theory and logic. We will make use of the complexity classes NP and PTIME. For an introduction into this topic we refer to the good textbook of Papadimitriou [34].

2.1   Team Semantics

Vocabularies τ are finite sets of relation symbols with prescribed arities. For each $R \in \tau$, let $ar(R) \in Z_+$ denote the arity of R. A τ-structure is a tuple $\mathfrak{A} = \big(A, (R_i^{\mathfrak{A}})_{R_i \in \tau}\big)$, where A is a set and each $R_i^{\mathfrak{A}}$ is an $ar(R_i)$-ary relation on A (i.e., $R_i^{\mathfrak{A}} \subseteq A^{ar(R_i)}$). We use $\mathfrak{A}$, $\mathfrak{B}$, etc. to denote τ-structures and A, B, etc. to denote the corresponding domains. In this article we restrict attention to finite structures.

Let D be a finite set of first-order variables and A be a nonempty set. A function $s: D \rightarrow A$ is called an *assignment*. The set D is the *domain* of s, and the set A the *codomain* of s. For a variable x and $a \in A$, the assignment $s(a/x): D \cup \{x\} \rightarrow A$ is obtained from s as follows:

$$s(a/x)(y) := \begin{cases} a & \text{if } y = x, \\ s(y) & \text{otherwise.} \end{cases}$$

A *team* is a finite set of assignments with a common domain and codomain. Let X be a team, A a finite set, and $F: X \rightarrow \mathcal{P}(A) \setminus \{\emptyset\}$ a function. We denote by $X[A/x]$ the modified team $\{s(a/x) \mid s \in X, a \in A\}$, and by $X[F/x]$ the team $\{s(a/x) \mid s \in X, a \in F(s)\}$. Let $\mathfrak{A}$ be a τ-structure and X a team with codomain A, then we say that X is a team of $\mathfrak{A}$.

Let τ be a set of relation symbols. The syntax of first-order logic $\mathsf{FO}(\tau)$ is given by the following grammar, where $R \in \tau$, x is a tuple of variables, and x and y are variables. Note that in the definition the scope of negation is restricted to atomic formulae.

$$\varphi ::= x = y \mid x \neq y \mid R(x) \mid \neg R(x) \mid (\varphi \wedge \varphi) \mid (\varphi \vee \varphi) \mid \exists x \varphi \mid \forall x \varphi.$$

Let x, y be tuples of variables and φ a formula. We write $\mathrm{Var}(\varphi)$ for the set of variables that occur in φ, and $\mathrm{Var}(x)$ for the set of variables listed in x. We also write xy for the concatenation of x and y, $x \cap y$ for any tuple listing the variables that occur both in x and y, and $x \setminus y$ for any tuple listing the variables that occur in x but not in y. For an assignment s, we write $s(x)$ to denote the sequence $(s(x_1), \ldots, s(x_n))$.

Next we define the lax and strict team semantics of first-order logic. It is worth noting that the disjunction has a non-classical interpretation. The classical (or intuitionistic) disjunction is usually denoted by $\varovee$ in the team semantics framework. However, as exemplified by Proposition 1, the non-classical disjunction of team semantics naturally corresponds to the classical disjunction of ordinary first-order logic.

Definition 1 (Lax Team Semantics). *Let* $\mathfrak{A}$ *be a* τ-*structure and* X *a team of* $\mathfrak{A}$. *The satisfaction relation* $\models_X$ *for first-order logic is defined as follows:*

$$\mathfrak{A} \models_X x = y \quad \Leftrightarrow \forall s \in X : s(x) = s(y)$$
$$\mathfrak{A} \models_X x \neq y \quad \Leftrightarrow \forall s \in X : s(x) \neq s(y)$$
$$\mathfrak{A} \models_X R(x) \quad \Leftrightarrow \forall s \in X : s(x) \in R^{\mathfrak{A}}$$
$$\mathfrak{A} \models_X \neg R(x) \Leftrightarrow \forall s \in X : s(x) \notin R^{\mathfrak{A}}$$
$$\mathfrak{A} \models_X (\psi \wedge \theta) \Leftrightarrow \mathfrak{A} \models_X \psi \text{ and } \mathfrak{A} \models_X \theta$$
$$\mathfrak{A} \models_X (\psi \vee \theta) \Leftrightarrow \mathfrak{A} \models_Y \psi \text{ and } \mathfrak{A} \models_Z \theta \text{ for some } Y, Z \subseteq X \text{ s.t. } Y \cup Z = X$$
$$\mathfrak{A} \models_X \forall x \psi \quad \Leftrightarrow \mathfrak{A} \models_{X[A/x]} \psi$$
$$\mathfrak{A} \models_X \exists x \psi \quad \Leftrightarrow \mathfrak{A} \models_{X[F/x]} \psi \text{ holds for some } F : X \to \mathcal{P}(A) \setminus \{\emptyset\}.$$

The so-called *strict* team semantics is obtained from the previous definition by adding the following two requirements.

(i) Disjunction: $Y \cap Z = \emptyset$.
(ii) Existential quantification: $F(s)$ is singleton for all $s \in X$.

Proposition 1 *[36]. Let* $\mathfrak{A}$ *be a* τ-*structure,* X *a team of* $\mathfrak{A}$, *and* φ *a formula of* $\mathsf{FO}(\tau)$. *Then*

$$\mathfrak{A} \models_X \varphi \Leftrightarrow \forall s \in X : \mathfrak{A} \models_s \varphi,$$

where $\models_s$ *denotes the ordinary satisfaction relation of first-order logic defined via models and assignments as usual, and* $\models_X$ *denotes the satisfaction relation of either lax or strict team semantics.*

For a model $\mathfrak{A}$ and a *sentence* φ (i.e., a formula with no free variables), the satisfaction relation $\models$ is defined as:

$$\mathfrak{A} \models \varphi \text{ if } \mathfrak{A} \models_{\{\emptyset\}} \varphi,$$

where $\{\emptyset\}$ denotes the singleton team of empty assignment.

Team semantics enables extending first-order logic with various dependency notions. The following dependency atoms were introduced in team semantics setting in [9,13,36].

Definition 2 (Dependency atoms). *Let $\mathfrak{A}$ be a model and X a team of $\mathfrak{A}$. If $\boldsymbol{x}, \boldsymbol{y}$ are variable sequences, then $=(\boldsymbol{x}, \boldsymbol{y})$ is a dependence atom with the satisfaction relation:*

$$\mathfrak{A} \models_X =(\boldsymbol{x}, \boldsymbol{y}) \text{ if for all } s, s' \in X \text{ s.t. } s(\boldsymbol{x}) = s'(\boldsymbol{x}), \text{ it holds that } s(\boldsymbol{y}) = s'(\boldsymbol{y}).$$

If $\boldsymbol{x}, \boldsymbol{y}$ are variable sequences of the same length, then $\boldsymbol{x} \subseteq \boldsymbol{y}$ is an inclusion atom with the satisfaction relation:

$$\mathfrak{A} \models_X \boldsymbol{x} \subseteq \boldsymbol{y} \text{ if for all } s \in X \text{ there exists } s' \in X \text{ such that } s(\boldsymbol{x}) = s'(\boldsymbol{y}).$$

If $\boldsymbol{x}, \boldsymbol{y}, \boldsymbol{z}$ are variable sequences, then $\boldsymbol{y} \perp_{\boldsymbol{x}} \boldsymbol{z}$ is a conditional independence atom with the satisfaction relation:

$$\mathfrak{A} \models_X \boldsymbol{y} \perp_{\boldsymbol{x}} \boldsymbol{z} \text{ if for all } s, s' \in X \text{ such that } s(\boldsymbol{x}) = s'(\boldsymbol{x}) \text{ there exists } s'' \in X$$
$$\text{such that } s''(\boldsymbol{x}) = s(\boldsymbol{x}), \, s''(\boldsymbol{y}) = s(\boldsymbol{y}), \text{ and } s''(\boldsymbol{z}) = s'(\boldsymbol{z}).$$

Note that in the previous definition it is allowed that some or all of the vectors of variables have length 0. For example, $\mathfrak{A} \models_X =(\boldsymbol{x})$ holds iff $\forall s \in X : s(\boldsymbol{x}) = \boldsymbol{c}$ holds for some fixed tuple $\boldsymbol{c}$, and $\mathfrak{A} \models_X \boldsymbol{y} \perp_{\boldsymbol{x}} \boldsymbol{z}$ holds always if either of the vectors $\boldsymbol{y}$ or $\boldsymbol{z}$ is of length 0.

We write FO for first-order logic, and given a set of atoms $\mathcal{C}$, we write $\mathsf{FO}(\mathcal{C})$ (omitting the set parentheses of $\mathcal{C}$) for the logic obtained by adding the atoms of $\mathcal{C}$ to FO. For instance, $\mathsf{FO}(=(\cdot))$ denotes then dependence logic.

Often in literature dependence atoms are defined such that $\boldsymbol{y}$ is a single variable, i.e., the widely used form is $=(\boldsymbol{x}, y)$. The definition above yields the strongest form of *functional dependence*. Moreover the atom $=(\boldsymbol{x}, \boldsymbol{y})$ can be equivalently rewritten as a conjunction of dependence atoms of type $=(\boldsymbol{x}, y)$.

3   Multiteam Semantics

In this section we generalise team semantics with the concept of multisets. Multisets and multiteam semantics can be used, e.g., in applications to database theory to model reasoning with databases with duplicates. In practice, for multitude of reasons, the existence of duplicates in databases is very common. Again as previously noted, we restrict attention to finite sets and finite multisets. In the following definition, occurrences of "zero multiplicities" are allowed for notational convenience.

Definition 3 (Multiset). *A multiset is a pair (A, m) where A is a set and m : $A \to \mathbb{N}$ is a (multiplicity) function. The function m determines the* multiplicities *of the elements in the multiset (A, m). A multiset (X, m) is a* multiteam *if the underlining set X is a team. The* domain *(or the* codomain*) of the multiteam (X, m) is the domain (codomain) of the team X.*

For each multiset (A, m), we define the *canonical set representative* $[(A, m)]_{\text{cset}}$ of (A, m) as follows:

$$[(A, m)]_{\text{cset}} := \{(a, i) \mid a \in A, 0 < i \le m(a)\}.$$

We say that (A, m) is finite whenever $[(A, m)]_{\text{cset}}$ is finite. We say that a multiset (A, m) is a submultiset of a multiset (B, n), $(A, m) \subseteq (B, n)$, if and only if $[(A, m)]_{\text{cset}} \subseteq [(B, n)]_{\text{cset}}$. Furthermore, we define that $(A, m) = (B, n)$ if and only if both $(A, m) \subseteq (B, n)$ and $(B, n) \subseteq (A, m)$ hold.

The *disjoint union* $(A, m) \uplus (B, n)$ of (A, m) and (B, n) is the multiset (C, k), where $C := A \cup B$ and $k : C \to \mathbb{N}$ is the function defined as follows:

$$k(s) := \begin{cases} m(s) + n(s) & \text{if } s \in A \text{ and } s \in B, \\ m(s) & \text{if } s \in A \text{ and } s \notin B, \\ n(s) & \text{if } s \notin A \text{ and } s \in B. \end{cases}$$

We write $|(A, m)|$ to denote the size of the multiset (A, m), i.e., $|(A, m)| := \sum_{a \in A} m(a)$. The set of non-empty submultisets of a multiset (A, m) is the set

$$\mathcal{P}^+((A, m)) := \{(C, l) \mid (C, l) \subseteq (A, m) \text{ s.t. } l(c) \ge 1 \text{ for each } c \in C\} \setminus \{(\emptyset, \emptyset)\}.$$

Let (X, m) be a multiteam, (A, n) a finite multiset, and $F : [(X, m)]_{\text{cset}} \to \mathcal{P}^+((A, n))$ a function. We denote by $(X, m)[(A, n)/x]$ the modified multiteam defined as

$$\biguplus_{s \in X} \biguplus_{a \in A} \{(s(a/x), m(s) \cdot n(a))\}.$$

By $X[F/x]$ we denote the multiteam defined as

$$\biguplus_{s \in X} \biguplus_{1 \le i \le m(s)} \{(s(b/x), l(b)) \mid (B, l) = F((s, i)), b \in B\}.$$

A τ-multistructure is a tuple $\mathfrak{A} = ((A, m), (R_i^{\mathfrak{A}})_{R_i \in \tau})$ where (A, m) is a multiset and, for each $R_i \in \tau$, $R_i^{\mathfrak{A}}$ is an $\text{ar}(R_i)$-ary relation over the set $\{a \in A \mid m(a) \ge 1\}$. A multiteam (X, m) over $\mathfrak{A}$ is a multiteam with codomain A.

Next we define multiteam semantics for first-order logic.

Definition 4 (Multiteam Semantics). *Let $\mathfrak{A}$ be a τ-multistructure, (A, n) the domain of $\mathfrak{A}$, and (X, m) a multiteam over $\mathfrak{A}$. The satisfaction relation $\models_{(X,m)}$ is defined as follows:*

$$\mathfrak{A} \models_{(X,m)} x = y \quad \Leftrightarrow \forall s \in X : \text{ if } m(s) \geq 1 \text{ then } s(x) = s(y)$$
$$\mathfrak{A} \models_{(X,m)} x \neq y \quad \Leftrightarrow \forall s \in X : \text{ if } m(s) \geq 1 \text{ then } s(x) \neq s(y)$$
$$\mathfrak{A} \models_{(X,m)} R(\boldsymbol{x}) \quad \Leftrightarrow \forall s \in X : \text{ if } m(s) \geq 1 \text{ then } s(\boldsymbol{x}) \in R^{\mathfrak{A}}$$
$$\mathfrak{A} \models_{(X,m)} \neg R(\boldsymbol{x}) \Leftrightarrow \forall s \in X : \text{ if } m(s) \geq 1 \text{ then } s(\boldsymbol{x}) \notin R^{\mathfrak{A}}$$
$$\mathfrak{A} \models_{(X,m)} (\psi \wedge \theta) \Leftrightarrow \mathfrak{A} \models_{(X,m)} \psi \text{ and } \mathfrak{A} \models_{(X,m)} \theta$$
$$\mathfrak{A} \models_{(X,m)} (\psi \vee \theta) \Leftrightarrow \mathfrak{A} \models_{(Y,k)} \psi \text{ and } \mathfrak{A} \models_{(Z,l)} \theta \text{ for some multisets}$$
$$(Y,k), (Z,l) \subseteq (X,m) \, s.t. \, (X,m) \subseteq (Y,k) \uplus (Z,l).$$
$$\mathfrak{A} \models_{(X,m)} \forall x \psi \quad \Leftrightarrow \mathfrak{A} \models_{(X,m)[(A,n)/x]} \psi$$
$$\mathfrak{A} \models_{(X,m)} \exists x \psi \quad \Leftrightarrow \mathfrak{A} \models_{(X,m)[F/x]} \psi \text{ holds for some function}$$
$$F \colon [(X,m)]_{\text{cset}} \to \mathcal{P}^+\big((A,n)\big).$$

The so-called *strict multiteam semantics* is obtained from the previous definition by adding the following two requirements.

(i) Disjunction: $(Y,n) \uplus (Z,k) = (X,m)$.
(ii) Existential quantification: for all $s \in X$ and $0 < i \leq m(s)$, $F\big((s,i)\big) = (B,n)$ for some singleton $B = \{b\}$ and $n(b) = 1$.

This alternative semantics is discussed in Sect. 3.3. Otherwise in the paper we restrict attention to the multiteam semantics given in Definition 4, sometimes referred to as *lax multiteam semantics*. The following proposition shows that multiteam semantics and team semantics for first-order logic coincide when the multisets in multistructures are essentially sets. The proof of the proposition is self evident.

Proposition 2 Let $\mathfrak{A}$ be a multistructure with domain (A, n), and (X, m) a multiteam over $\mathfrak{A}$ such that $n(a) = m(s) = 1$ for all $a \in A$ and $s \in X$. Define $\mathfrak{B} := (A, (R^{\mathfrak{A}})_{R \in \tau})$. Then for every $\varphi \in \mathsf{FO}$ it holds that

$$\mathfrak{A} \models_{(X,m)} \varphi \text{ if and only if } \mathfrak{B} \models_X \varphi.$$

Next we generalise inclusion and conditional independence atoms to multiteams by introducing their probabilistic versions. For a multiteam (X, m) of codomain A, a tuple of variables $\boldsymbol{x}$ from $\text{Dom}(X)$, and $\boldsymbol{a} \in A^{|\boldsymbol{x}|}$, we denote by $(X, m)_{\boldsymbol{x}=\boldsymbol{a}}$ the multiteam (X, n) where n agrees with m on all assignments $s \in X$ with $s(\boldsymbol{x}) = \boldsymbol{a}$, and otherwise n maps s to 0.

Definition 5. Let $\mathfrak{A}$ be a multistructure with domain (A, n), and (X, m) a multiteam over $\mathfrak{A}$. If $\boldsymbol{x}, \boldsymbol{y}$ are variable sequences of the same length, then $\boldsymbol{x} \leq \boldsymbol{y}$ is a probabilistic inclusion atom *with the following semantics:*

$$\mathfrak{A} \models_{(X,m)} \boldsymbol{x} \leq \boldsymbol{y} \text{ if } |(X,m)_{\boldsymbol{x}=s(\boldsymbol{x})}| \leq |(X,m)_{\boldsymbol{y}=s(\boldsymbol{x})}| \text{ for all } s \colon \text{Var}(\boldsymbol{x}) \to A.$$

If $\boldsymbol{x}, \boldsymbol{y}, \boldsymbol{z}$ are variable sequences, then $\boldsymbol{y} \perp\!\!\!\perp_{\boldsymbol{x}} \boldsymbol{z}$ is a probabilistic conditional independence atom *with the satisfaction relation defined as*

$$\mathfrak{A} \models_{(X,m)} \boldsymbol{y} \perp\!\!\!\perp_{\boldsymbol{x}} \boldsymbol{z} \tag{1}$$

if for all $s \colon \text{Var}(\boldsymbol{xyz}) \to A$ *it holds that*

$$|(X,m)_{\boldsymbol{xy}=s(\boldsymbol{xy})}| \cdot |(X,m)_{\boldsymbol{xz}=s(\boldsymbol{xz})}| = |(X,m)_{\boldsymbol{xyz}=s(\boldsymbol{xyz})}| \cdot |(X,m)_{\boldsymbol{x}=s(\boldsymbol{x})}|.$$

We call atoms of the form $x \perp_\emptyset y$ *probabilistic marginal independence atoms*, written as the shorthand $x \perp y$. Note that we obtain the following satisfaction relation for $x \perp y$:

$$\mathfrak{A} \models_{(X,m)} x \perp y \text{ if for all } s : \text{Var}(xy) \to A, \tag{2}$$
$$\frac{|(X,m)_{x=s(x)}| \cdot |(X,m)_{y=s(y)}|}{|(X,m)|} = |(X,m)_{xy=s(xy)}|.$$

The study of database dependencies is very interesting also in the practical point of view as many interesting properties of datasets can be revealed. Further the investigation of conditional independence can yield methods to be used to decompose datasets for speeding up different processing tasks on the data.

Multiteams (X, m) induce a natural probability distribution p over the assignments of X. Namely, we define $p : X \to [0, 1]$ such that

$$p(s) = \frac{m(s)}{\sum_{s \in X} m(s)}.$$

The probability that a tuple of (random) variables x takes value a, written $\Pr(x = a)$, is then

$$\sum_{\substack{s \in X, \\ s(x)=a}} p(s).$$

It is now easy to see that $\mathfrak{A} \models_{(X,m)} y \perp_x z$ iff for all abc,

$$\Pr(y = b, z = c | x = a) = \Pr(y = b | x = a) \Pr(z = c | x = a),$$

that is, the probability of $y = b$ is independent of the probability of $z = c$, given $x = a$. Analogously, a probabilistic inclusion atom $x \leq y$ indicates that $\Pr(x = a) \leq \Pr(y = a)$ for all values a, and a probabilistic independence atom of the form $x \perp x$ that $\Pr(x = a) = 1$ for some value a. Note that such atoms have been studied in the literature under the name of *constancy atoms* [9].

One can also study the usual dependency notions of database theory in the multiteam semantics setting.

Definition 6. *Let $\mathfrak{A}$ be a multistructure, (X, m) a multiteam over $\mathfrak{A}$, and φ of the form $=(x, y)$, $x \subseteq y$, or $y \perp_x z$. Then the satisfaction relation $\models_{(X,m)}$ is defined as follows:*

$$\mathfrak{A} \models_{(X,m)} \varphi \text{ iff } \mathfrak{A} \models_{X^+} \varphi,$$

where X^+ is the team $\{s \in X \mid m(s) \geq 1\}$.

First we notice that the known translation of dependence atoms to independence atoms (see Grädel and Väänänen [13]) works also in the probabilistic case.

Proposition 3. *Let $\mathfrak{A}$ be a multistructure, (X, m) a multiteam over $\mathfrak{A}$, and x, y tuples of variables. Then $\mathfrak{A} \models_{(X,m)} y \perp_x y$ iff $\mathfrak{A} \models_{(X,m)} =(x, y)$.*

Proof. From the truth definition we obtain that

$$\mathfrak{A} \models_{(X,m)} y \perp\!\!\!\perp_x y \Leftrightarrow \text{ for all } s : \text{Var}(xy) \to A \text{ with } (X,m)_{xy=s(xy)} \neq \emptyset, \quad (3)$$
$$|(X,m)_{xy=s(xy)}| = |(X,m)_{x=s(x)}|.$$

The result then follows since $\mathfrak{A} \models_{(X,m)} =(x,y)$ iff the right-hand side of Eq. (3) holds. $\qquad\square$

Note that the restriction of Proposition 3 to marginal independence states that

$$\mathfrak{A} \models_{(X,m)} x \perp\!\!\!\perp x \quad \Leftrightarrow \quad \mathfrak{A} \models_{(X,m)} =(x).$$

It is left open whether one can define inclusion or conditional independence atoms in $\mathsf{FO}(\perp\!\!\!\perp_c, \leq)$. However, over constant multiplicity functions conditional independence atoms φ coincide with their probabilistic counterparts whenever $\text{Var}(\varphi) = \text{Dom}(X)$. In the following, we denote by $^x A$ the team of all assignments $\text{Var}(x) \to A$.

Lemma 1. *Let $\mathfrak{A}$ be a multistructure and (X,m) a multiteam over $\mathfrak{A}$. Then*

(i) $\mathfrak{A} \models_{(X,m)} y \perp\!\!\!\perp_x z \quad \Leftrightarrow \quad \mathfrak{A} \models_{(X,m)} (y \setminus x \perp\!\!\!\perp_x z \setminus x),$

(ii) $\mathfrak{A} \models_{(X,m)} y \perp\!\!\!\perp_x z \quad \Leftrightarrow \quad \mathfrak{A} \models_{(X,m)} (y \setminus z \perp\!\!\!\perp_x z \setminus y) \wedge (y \cap z \perp\!\!\!\perp_x y \cap z).$

Proof. **Case (i).** The truth definition in Eq. (1) is symmetric, and hence it suffices to show that $\mathfrak{A} \models_{(X,m)} yx \perp\!\!\!\perp_x z \Leftrightarrow \mathfrak{A} \models_{(X,m)} y \perp\!\!\!\perp_x z$ whenever x is listed in x. This follows since $^{xyxz}A = {}^{xyz}A$, and the Eq. (1) remains the same after removing x.

Case (ii). Let us first show that $\mathfrak{A} \models_{(X,m)} y \perp\!\!\!\perp_x z$ implies $\mathfrak{A} \models_{(X,m)} (y \cap z \perp\!\!\!\perp_x y \cap z)$. For this, it remains to show that $\mathfrak{A} \models_{(X,m)} yu \perp\!\!\!\perp_x z$ implies $\mathfrak{A} \models_{(X,m)} y \perp\!\!\!\perp_x z$, for u not listed in xyz. This follows since for all $s \in {}^{xyz}A$,

$$|(X,m)_{xz=s(xz)}| \cdot |(X,m)_{xy=s(xy)}|$$
$$= |(X,m)_{xz=s(xz)}| \cdot \Sigma_{a \in A}|(X,m)_{xyu=s(xy)a}|$$
$$= \Sigma_{a \in A}(|(X,m)_{xz=s(xz)}| \cdot |(X,m)_{xyu=s(xy)a}|)$$
$$= \Sigma_{a \in A}(|(X,m)_{x=s(x)}| \cdot |(X,m)_{xyzu=s(xyz)a}|)$$
$$= |(X,m)_{x=s(x)}| \cdot \Sigma_{a \in A}|(X,m)_{xyzu=s(xyz)a}|$$
$$= |(X,m)_{x=s(x)}| \cdot |(X,m)_{xyz=s(xyz)}|,$$

where in the third equation we apply the assumption that $\mathfrak{A} \models_{(X,m)} ya \perp\!\!\!\perp_x z$.

For the claim it now suffices to show that $\mathfrak{A} \models_{(X,m)} y \perp\!\!\!\perp_x z \Leftrightarrow \mathfrak{A} \models_{(X,m)} (y \setminus z \perp\!\!\!\perp_x z \setminus y)$ whenever $\mathfrak{A} \models_{(X,m)} (y \cap z \perp\!\!\!\perp_x y \cap z)$. This follows directly from the truth definition since by Eq. (3) for all $s \in {}^{xv}A$ with $(X,m)_{xv=s(xv)} \neq \emptyset$:

$$|(X,m)_{xv=s(xv)}| = |(X,m)_{x=s(x)}|,$$

for $v := x \cap y$. $\qquad\square$

If x, y, z are pairwise disjoint, then $y \perp\!\!\!\perp_x z$ corresponds to the generalised embedded multivalued dependency $x \multimap\!\!\!\rightarrow y \mid z$ that is defined over extended relational data models (i.e., relational data models equipped with a multiplicity function) using semantics that coincide with that of Definition 5 [39,40]. It was shown by Wong [39] that the generalised multivalued dependency $x \multimap\!\!\!\rightarrow y$ holds in an extended relational data model if and only if the underlying relational model satisfies the multivalued dependency $x \twoheadrightarrow y$. This is stated in the following theorem reformulated into the framework of this article.

Theorem 1 *[39].* *Let $\mathfrak{A}$ be a multistructure, X a team over $\mathfrak{A}$, and $y \perp\!\!\!\perp_x z$ a probabilistic conditional independence atom such that $\mathrm{Var}(y \perp\!\!\!\perp_x z) = \mathrm{Dom}(X)$ and x, y, z are pairwise disjoint. Let 1 denote the constant function that maps all assignments of X to 1. Then $\mathfrak{A} \models_{(X,1)} y \perp\!\!\!\perp_x z$ iff $\mathfrak{A} \models_{(X,1)} y \perp_x z$.*

Using Lemma 1, the restriction that x, y, z are disjoint can be now removed.

Proposition 4. *Let $\mathfrak{A}$ be a multistructure, X a team over $\mathfrak{A}$, and $y \perp\!\!\!\perp_x z$ a probabilistic conditional independence atom such that $\mathrm{Var}(y \perp\!\!\!\perp_x z) = \mathrm{Dom}(X)$. Then $\mathfrak{A} \models_{(X,1)} y \perp\!\!\!\perp_x z$ iff $\mathfrak{A} \models_{(X,1)} y \perp_x z$.*

Proof. First note that by Proposition 3 and Lemma 1, $y \perp\!\!\!\perp_x z$ is equivalent in multiteam semantics to $(y \setminus xz \perp\!\!\!\perp_x z \setminus xy) \wedge =(x, y \cap z)$. Moreover, it is known that in team semantics $y \perp_x z$ is equivalent to $y \setminus xz \perp_x z \setminus xy \wedge =(x, y \cap z)$ [13]. Hence, the claim follows by Theorem 1. □

Note that $y \perp\!\!\!\perp_x z$ implies $y \perp_x z$ also over arbitrary multiplicity functions since non-emptiness of $(X, m)_{xy=s(xy)}$ and $(X, m)_{xz=s(xz)}$ implies non-emptiness of $(X, m)_{xyz=s(xyz)}$ by the truth definition in Eq. (1). The converse however does not hold; the multiteam (Y, m) depicted in Fig. 1 satisfies $x \perp y$ but violates $x \perp\!\!\!\perp y$.

A Diversion: Implication Problems. Results similar in spirit to Proposition 4 have been studied in connection to *implication problems* which is a central notion in causal reasoning and database dependency theory. The finite implication problem of independence atoms $y \perp_x z$ is defined as follows. Given a finite collection $\Sigma \cup \{\varphi\}$ of independence atoms, determine whether for all finite $\mathfrak{A}, X$:

$$\mathfrak{A} \models_X \Sigma \Rightarrow \mathfrak{A} \models_X \varphi.$$

If the above holds, we write $\Sigma \models \varphi$. The implication problem of other types of dependencies is defined analogously. Furthermore, the problem for the atoms $y \perp\!\!\!\perp_x z$ can be defined similarly by replacing teams by multiteams. The implication problems of embedded multivalued dependencies (i.e., the atoms $y \perp_x z$) and $y \perp\!\!\!\perp_x z$ have been extensively studied, e.g., for both atoms the problem is not finitely axiomatisable and for the former the problem is known to be undecidable. On the other hand, there are interesting restricted cases where the implication problems are finitely axiomatisable and equivalent, i.e., for all inputs $\Sigma \cup \{\varphi\}$, $\Sigma \models \varphi$ iff $\Sigma^p \models \varphi^p$, where Σ^p and ϕ^p are defined by replacing $y \perp_x z$ by $y \perp\!\!\!\perp_x z$. This holds, for example, for marginal independence atoms and for the so-called *saturated* atoms φ of the form $y \perp\!\!\!\perp_x z$ (or equivalently for $\varphi = y \perp_x z$) which, as in Theorem 1, satisfy $\mathrm{Var}(y \perp\!\!\!\perp_x z) = \mathrm{Dom}(X)$ (see the survey by Wong et al. [40]). Relationships between fragments of conditional independence statements and embedded multivalued dependencies have recently been studied, e.g., in [30–33]. It is also worth noting that the passage from set to multisets has interesting consequences also for the study of implication problems of database dependencies. For example, while key constraints can be expressed by functional dependencies under team semantics, this is no longer true under multiteam semantics [23].

Conditional independence is an important notion for expressing structural aspects of probability distributions. *Context specific independence* is a variant of $y \perp\!\!\!\perp_x z$ expressing independence in a context where the values of some variables of x are restricted to range over a subset of all possible values [2,35]. The next simple example shows how disjunction can be used to express context specific independence statements in $\mathrm{FO}(\perp\!\!\!\perp_c)$. The example shows that combining $y \perp\!\!\!\perp_x z$ with the logical connectives and quantifiers available in $\mathrm{FO}(\perp\!\!\!\perp_c)$ provide us with powerful means to define interesting generalisations of conditional independence. The definability of context specific independence using disjunction has been pointed out in [3].

Example 1. Let $A = \{0,1\}$ and X be a multiteam of A with domain $\mathrm{Dom}(X) = \{x_0, x_1, \ldots, x_n\}$. Now in X the variable x_0 is said to be contextually independent of x_2 given $x_1 = 0$, denoted by

$$x_0 \perp x_2 \mid x_1 = 0, \tag{4}$$

if for all $s : \{x_0, x_1, x_2\} \to A$ such that $s(x_1) = 0$ it holds that

$$|(X,m)_{x_0 x_1 = s(x_0 x_1)}| \cdot |(X,m)_{x_1 x_2 = s(x_1 x_2)}|$$
$$= |(X,m)_{x_0 x_1 x_2 = s(x_0 x_1 x_2)}| \cdot |(X,m)_{x_1 = s(x_1)}|.$$

It is now straightforward to check that Eq. (4) can be equivalently expressed by the $\mathrm{FO}(\perp\!\!\!\perp_c)$-formula $(x_1 \neq c) \vee (x_1 = c \wedge (x_0 \perp\!\!\!\perp_{x_1} x_2))$, where c is a constant symbol interpreted as 0.

3.1 Probabilistic Notions in Multiteam Semantics

In this section we investigate some properties of the probabilistic logics we have defined so far.

The set of *free variables* of a formula $\varphi \in \mathrm{FO}(\mathcal{C})$, denoted by $\mathrm{Fr}(\varphi)$, is defined in the obvious manner as in first-order logic. In particular, we define

$$\mathrm{Fr}(x \subseteq y) := \mathrm{Fr}(x \leq y) := \mathrm{Fr}\big(=(x,y)\big) := \{x,y\}$$
$$\mathrm{Fr}(y \perp\!\!\!\perp_x z) := \mathrm{Fr}(y \perp_x z) := \{x,y,z\}.$$

For $V \subseteq \text{Dom}(X)$, we define $(X, m) \upharpoonright V := (X \upharpoonright V, n)$ where

$$n(s) := \sum_{\substack{s' \in X, \\ s' \upharpoonright V = s}} m(s').$$

The following locality principle holds by easy structural induction.

Proposition 5 (Locality). *Let $\mathfrak{A}$ be a multistructure, (X, m) a multiteam, and V a set of variables such that $\text{Fr}(\varphi) \subseteq V \subseteq \text{Dom}(X)$. Then for all $\varphi \in \text{FO}(\leq, \perp$ $\perp_c, =(\cdot), \subseteq, \perp_c)$ it holds that $\mathfrak{A} \models_{(X,m)} \varphi$ iff $\mathfrak{A} \models_{(X,m) \upharpoonright V} \varphi$.*

The notion of flatness is generalised to the multiteam setting as follows.

Definition 7 (Weak Flatness). *We say that a formula φ is* weakly flat *if for all multistructures $\mathfrak{A}$ and for all multiteams (X, m) it holds that*

$$\mathfrak{A} \models_{(X,m)} \varphi \quad \Leftrightarrow \quad \mathfrak{A} \models_{(X,n)} \varphi,$$

where n agrees with m on all s with $m(s) = 0$, and otherwise maps all s to 1. The multiteam (X, n) is then called the weak flattening *of (X, m). A logic is called* weakly flat *if every formula of this logic is weakly flat.*

Dependence, conditional independence, and inclusion atoms are insensitive to multiplicities, and using structural induction one can prove the following proposition.

Proposition 6. $\text{FO}(=(\cdot), \subseteq, \perp_c)$ *is weakly flat.*

On the other hand, probabilistic dependencies do not satisfy weak flatness as shown in the next example.

Example 2. For instance (Y, m), illustrated in Fig. 1, does not satisfy $\boldsymbol{x} \perp\!\!\!\perp \boldsymbol{y}$ but its weak flattening (Y, n) does.

Analogously, the probabilistic inclusion atom is not weakly flat, and therefore neither of these atoms can be expressed in $\text{FO}(=(\cdot), \subseteq, \perp_c)$.

$$(Y, m) = \quad \begin{array}{c|ccc} & x & y & m(s_i) \\ \hline s_0 & 0 & 0 & 2 \\ s_1 & 0 & 1 & 1 \\ s_2 & 1 & 0 & 1 \\ s_3 & 1 & 1 & 1 \end{array} \qquad (Y, n) = \quad \begin{array}{c|ccc} & x & y & n(s_i) \\ \hline s_0 & 0 & 0 & 1 \\ s_1 & 0 & 1 & 1 \\ s_2 & 1 & 0 & 1 \\ s_3 & 1 & 1 & 1 \end{array}$$

Fig. 1. Assignments for multiteams in Example 2.

A formula φ is called *union closed* (in the multiteam setting) if for all multistructures $\mathfrak{A}$ and all multiteams $(X, m), (Y, n)$: if $\mathfrak{A} \models_{(X,m)} \varphi$ and $\mathfrak{A} \models_{(Y,n)} \varphi$, then $\mathfrak{A} \models_{(Z,h)} \varphi$, where $(Z,h) = (X,m) \uplus (Y,n)$. A logic is called union closed if all its formulae are union closed. It is easy to show by induction on the structure of formulae that probabilistic inclusion logic satisfies union closure.

Proposition 7. FO($\leq, \subseteq$) *is union closed.*

3.2 Probabilistic Notions in Team Semantics

In this section we examine probabilistic independence and inclusion logic in the (set) team semantics setting. Note that all the models considered in this section are usual first-order structures.

Satisfaction of probabilistic atoms in team semantics setting is defined by adding a constant multiplicity function.

Definition 8. *Let $\mathfrak{A}$ be a model, X a team over $\mathfrak{A}$, and φ be a probabilistic atom of the form $y \perp\!\!\!\perp_x z$ or $x \leq y$. Then the satisfaction relation $\models_X$ is defined as follows:*

$$\mathfrak{A} \models_X \varphi \text{ iff } \mathfrak{A} \models_{(X,1)} \varphi,$$

where 1 is the constant function that maps all assignments of X to 1.

The next theorem shows that, since probabilistic inclusion and independence atoms are expressible (in the team semantics setting) in FO($\perp_c$) relative to teams of fixed domain, their addition does not increase the expressive power of FO($\perp_c$).

Theorem 2. *Let $\varphi \in$ FO($\leq, \perp\!\!\!\perp_c, =(\cdot), \subseteq, \perp_c$) be a sentence. Then there exists a sentence $\varphi' \in$ FO($\perp_c$) such that for all models $\mathfrak{A}$ it holds that $\mathfrak{A} \models \varphi$ iff $\mathfrak{A} \models \varphi'$.*

Proof. First note that inclusion and dependence atoms can be expressed in FO($\perp_c$) [9,13]. Also it is easy to see that one can construct existential second-order logic sentences that capture probabilistic inclusion and conditional independence atoms over teams of fixed domain. Namely, for all φ of the form $y \perp\!\!\!\perp_x z$ or $x \leq y$ and all $V \supseteq \text{Fr}(\varphi)$, there exists an ESO sentence $\varphi^*(R)$, where R is a k-ary relation symbol for $k = |\text{Var}(\varphi)|$, such that for all $\mathfrak{A}$ and X with $\text{Dom}(X) = V$,

$$\mathfrak{A} \models_X \varphi \Leftrightarrow (\mathfrak{A}, \text{Rel}(X)) \models \varphi^*(R),$$

where $\text{Rel}(X) = \{(s(x_1), \ldots, s(x_k)) \mid s \in X\}$. All ESO-definable properties of teams translate into FO($\perp_c$) [9], and hence the formula φ' can be constructed from φ by replacing each probabilistic atom with a correct FO($\perp_c$)-translation. $\square$

Note that probabilistic inclusion atoms are not closed under (set) unions in team semantics, and hence they cannot be expressed in FO($\subseteq$) as shown in the following example.

Example 3. Let $\mathfrak{A}$ be a first-order structure with domain $\{0,1,2\}$, and $s := \{(x,0), (y,1), (z,0)\}$, $s' := \{(x,1), (y,0), (z,1)\}$, and $s'' := \{(x,0), (y,1), (z,2)\}$ be assignments. Define $X := \{s,s'\}$ and $Y := \{s',s''\}$. Now $\mathfrak{A} \models_X x \leq y$, $\mathfrak{A} \models_Y x \leq y$, but $\mathfrak{A} \not\models_{X \cup Y} x \leq y$.

3.3   Strict Multiteam Semantics

We briefly consider properties of related logics under strict multiteam semantics.

Proposition 8. *Over strict multiteam semantics* $FO(=(\cdot))$ *is weakly flat.*

The logics $FO(\perp_c)$ and $FO(\subseteq)$ are not weakly flat under strict multiteam semantics as shown in the next example.

Example 4. For instance (X, m), illustrated in Fig. 2, satisfies $(x \subseteq z) \vee (y \subseteq z)$ in strict semantics but its weak flattening (X, n) does not.

	(X, m)					(X, n)			
	x	y	z	$m(s_i)$		x	y	z	$n(s_i)$
s_1	0	0	1	2	s_1	0	0	1	1
s_2	1	2	0	1	s_2	1	2	0	1
s_3	2	1	0	1	s_3	2	1	0	1

Fig. 2. Assignments for teams in Example 4.

Similarly, one can show that $FO(\leq, \subseteq)$ is not union closed under strict multiteam semantics. Moreover one can show that Propositions 2 and 5 hold also under strict multiteam semantics.

4   Approximate Operators

Now we will turn to define an existential and a universal *approximate* operator which allows one to state truth of formulas not with respect to the *full* team but with respect to a *ratio* of the team. The main motivator for this approach is the important application in database theory to be able to model the truth of properties in databases that may contain some faulty data. Moreover, in practice, duplicates occur frequently in databases for a multitude of reasons. Thus the study of database dependencies, such as inclusion dependencies and foreign key constraints, in combination with approximate operators is an important topic as it explains inherent properties of a given dataset. In this section we consider multiteam semantics.

Definition 9. *Let* $\mathfrak{A}$ *be a multistructure, and* (X, m) *a multiteam over* $\mathfrak{A}$, *and* $p \in [0, 1]$ *a rational number.*

$$\mathfrak{A} \models_{(X,m)} \langle p \rangle \varphi \Leftrightarrow \exists (Y, n) \subseteq (X, m), |(Y, n)| \geq p \cdot |(X, m)| : \mathfrak{A} \models_{(Y,n)} \varphi,$$
$$\mathfrak{A} \models_{(X,m)} [p] \varphi \Leftrightarrow \forall (Y, n) \subseteq (X, m), |(Y, n)| \geq p \cdot |(X, m)| : \mathfrak{A} \models_{(Y,n)} \varphi$$

The previous definition generalises the notion of approximate dependence atoms $=_p(\cdot)$, introduced by Väänänen [38], in the following sense: $=_{1-p}(\boldsymbol{x}, y)$ is equivalent to the formula $\langle p \rangle =(\boldsymbol{x}, y)$.

In the following we observe that distributivity does not hold in general with respect to $\langle p \rangle$.

		$(X, 1)$					$(Y, 1)$						$(Z, \cdot)$		
	x	y	z	$1(s_i)$		x	y	z	$1(s_i)$		x	y	$m(s_i)$	$n(s_i)$	$k(s_i)$	$\ell(s_i)$
s_1	0	0	1	1	s_1	0	0	1	1	s_1	0	1	1	0	1	0
s_2	0	1	0	1	s_3	0	1	2	1	s_2	1	0	1	1	1	0
s_3	0	1	2	1						s_3	0	0	1	1	0	1

Fig. 3. Assignments for multiteams in Examples 5 and 6.

Proposition 9. *It is not true that* $\langle p \rangle(\varphi \vee \psi) \equiv \langle p \rangle \varphi \vee \langle p \rangle \psi$.

Proof. Let $\mathfrak{A}$ be the multistructure over the empty vocabulary with domain $(\{0, 1, 2\}, 1)$, where 1 is the constant 1 multiplicity function. Then $\mathfrak{A} \models_{(X,1)} \langle \frac{2}{3} \rangle (x = y \vee x = z)$ but $\mathfrak{A} \not\models_{(X,1)} \langle \frac{2}{3} \rangle (x = y) \vee \langle \frac{2}{3} \rangle (x = z)$, where $(X, 1)$ is the multiteam depicted in the Fig. 3. $\square$

The next simple observation states the distributivity of $[p]$ with respect to conjunction $\wedge$, as well as the merger of two $\langle p \rangle$-operators and two $[q]$-operators, respectively.

Observation 1. *The following equivalences hold:*

1. $[p](\varphi \wedge \psi) \equiv [p]\varphi \wedge [p]\psi$,
2. $\langle p \rangle(\langle q \rangle \varphi) = \langle p \cdot q \rangle \varphi$,
3. $[p]([q]\varphi) = [p \cdot q]\varphi$.

The next two examples show that both downward closure and union closure are violated by the approximate operator.

Example 5. Let $\mathfrak{A}$ be the multistructure over the empty vocabulary with domain $(\{0, 1, 2\}, 1)$, where 1 is the constant 1 multiplicity function. Then $\mathfrak{A} \models_{(X,1)} \langle \frac{1}{3} \rangle (x = y)$ but $\mathfrak{A} \not\models_{(Y,1)} \langle \frac{1}{3} \rangle (x = y)$, where $(Y, 1) \subseteq (X, 1)$ are the multiteams depicted in the Fig. 3.

Example 6. Let $\mathfrak{A}$ be the multistructure over the empty vocabulary with domain $(\{0, 1\}, 1)$, where 1 is the constant 1 multiplicity function. The multiteams $(Z, m), (Z, n), (Z, k), (Z, \ell)$ are depicted in the Fig. 3. Now $\mathfrak{A} \models_{(Z,k)} [\frac{2}{3}](x \le y)$ and $\mathfrak{A} \models_{(Z,\ell)} [\frac{2}{3}](x \le y)$. However $\mathfrak{A} \not\models_{(Z,n)} x \le y$ and thus $\mathfrak{A} \not\models_{(Z,m)} [\frac{2}{3}](x \le y)$ even though $(Z, k) \uplus (Z, l) = (Z, m)$.

Proposition 10. *Let $\mathcal{L}$ be a logic and $\varphi \in \mathcal{L}$ a formula. Then $\langle p \rangle$ preserves union closure (whereas $[p]$ does not), i.e., $\langle p \rangle \varphi$ is union closed whenever φ is.*

Proof. Let $\mathfrak{A}$ be a multistructure and X, Y be multiteams of $\mathfrak{A}$. Assume that $\mathfrak{A} \models_X \langle p \rangle \varphi$ and $\mathfrak{A} \models_Y \langle p \rangle \varphi$. Then there are multiteams $X' \subseteq X$ and $Y' \subseteq Y$ such that $|X'| \geq p|X|$, $|Y'| \geq p|Y|$, and both $\mathfrak{A} \models_{X'} \varphi$ and $\mathfrak{A} \models_{Y'} \varphi$. Hence $|X' \uplus Y'| = |X'| + |Y'| \geq p|X| + p|Y| = p(|X| + |Y|) = p|X \uplus Y|$ and thus $\mathfrak{A} \models_{X \uplus Y} \langle p \rangle \varphi$. □

Yet locality holds for this logic as witnessed by the following proposition. The proof is by induction.

Proposition 11 (Locality). *Let $\mathfrak{A}$ be a multistructure, (X, m) a multiteam, and V be a set of variables such that $\mathrm{Fr}(\varphi) \subseteq V \subseteq \mathrm{Dom}(X)$. Then for all $\varphi \in \mathsf{FO}(\langle p \rangle, [p], \leq, \perp\!\!\!\perp_c, =(\cdot), \subseteq, \perp_c)$, it holds that $\mathfrak{A} \models_{(X,m)} \varphi$ iff $\mathfrak{A} \models_{(X,m) \upharpoonright V} \varphi$.*

5   On the Complexity of Approximate Dependence Logic

In the following we study computational complexity of model checking in dependence logic enriched with the operator $\langle p \rangle$. The results hold under both team and multiteam semantics. To simplify notation, we work with team semantics in this section. Analogously to [5], our results can be seen as a first step towards a systematic classification of the syntactic fragments of approximate dependence logic for which data complexity of model-checking is tractable/intractable.

We first define the model checking problem in the context of team semantics. We consider only Boolean queries, that is we define the model checking problem for a logic $\mathcal{L}$ as follows: given a model $\mathfrak{A}$, a team X of $\mathfrak{A}$, and a formula φ of $\mathcal{L}$, decide whether $\mathfrak{A} \models_X \varphi$ holds. There are three parameters to this problem: the model $\mathfrak{A}$, the team X, and the formula φ. Depending on which of these parameters are fixed, a different variant of the model checking problem arises. Here we consider two of these variants: the variant with a fixed formula (this is called *data complexity*), and a variant in which nothing is fixed (this is called *combined complexity*).

The following two theorems reveal that already very simple formulas of approximate dependence logic witness the NP-completeness of the data complexity of the logic.

Theorem 3 *Model checking for $\langle p \rangle (=(x, y) \wedge =(u, v))$ is NP-complete.*

Proof We sketch the proof here, a full proof can be found in the arXiv version of this article. For the lower bound we give a polynomial many-one reduction from 3SAT inspired by a similar proof of Jarmo Kontinen [25]. Start with a formula $\varphi = \bigwedge_{i=1}^{m} \bigvee_{j=1}^{3} \ell_{i,j}$ where $\ell_{i,j}$ is the jth literal in the ith clause, i.e., either a variable x (said of parity 0) or its negation $\neg x$ (of parity 1). In the following we will construct a tuple (X, ψ) from φ such that $\varphi \in \mathsf{3SAT}$ if and only if $\mathfrak{A} \models_X \psi$. First we define the team X to be the set

$$X = \{(i, j, x, p) \mid \text{in } i\text{th clause the } j\text{th literal is the variable } x \text{ with parity } p\}.$$

Technically the team can be seen as an encoding of the given formula. For instance the formula $\varphi = (x_1 \vee \neg x_2 \vee x_3) \wedge (\neg x_1 \vee \neg x_2 \vee \neg x_3)$ would yield the team $X = \{(1,1,x_1,0),(1,2,x_2,1),(1,3,x_3,0),(2,1,x_1,1),(2,2,x_2,1),(2,3,x_3,1)\}$. The formula ψ is defined as

$$\langle \tfrac{1}{3} \rangle (=(\text{clause}, \text{literal}) \wedge =(\text{variable}, \text{parity})).$$

Then intuitively speaking ψ states that one has to decide for each clause a satisfying literal and do this consistently, i.e., the corresponding assignment has to be consistent. At first one selects exactly one third of the elements in X such that for each clause a literal is chosen (i.e., *clause* will determine the value of *literal*). Then the *parity* of each *variable* is consistently chosen (i.e., *variable* will determine the value of *parity*). It is straightforward to show that $\mathfrak{A} \models_X \psi$ iff φ is satisfiable.

For the NP upper bound, first observe that we can simply guess a subset X' of X such that $|X'| \geq \frac{1}{3}|X|$. Then we just have to check whether $\mathfrak{A} \models_{X'} =(\text{clause}, \text{literal}) \wedge =(\text{variable}, \text{parity})$ holds. This can be clearly done in polynomial time.   □

The next theorem shows that NP-hard properties can be defined using very simple formulas even if the operator $\langle p \rangle$ is restricted to appear only in front of dependence atoms. It is worth noting that the data complexity of formulas addressed in Theorem 4 without the operator $\langle p \rangle$ is in NL by the results of [25].

Theorem 4. *Model checking for* $=(x,y) \vee (\langle p \rangle =(x,y) \wedge =(u,v))$ *is* NP-*complete.*

Proof. The upper bound is due to the same argument as in the proof of Theorem 3: use nondeterminism to tame the $\langle p \rangle$ operator. The rest is just standard technique as for D, see the book of Väänänen [36].

Now we turn to the lower bound, again we will just sketch the proof. Here we will reduce from 3SAT through Max-2SAT, a well-known NP-hard optimisation problem whose decision variant is NP-complete. The problem asks given a 2CNF-formula φ and a number $k \in \mathbb{N}$, if at least k of the clauses of φ can be simultaneously satisfied [12]. Garey et al. describe a reduction f from 3SAT to the decision variant of Max-2SAT such that $\varphi \in$ 3SAT iff at least $\frac{7}{10}$ of the clauses of $f(\varphi)$ can be satisfied.

We will exploit this known reduction in the following way. The team X is constructed in the same way as in the proof of Theorem 3. The formula then is

$$=(\text{clause}, \text{literal}) \vee (=(\text{clause}, \text{literal}) \wedge \langle \tfrac{7}{10} \rangle =(\text{variable}, \text{parity})).   □$$

Currently the $\langle p \rangle$ operator is defined with respect to some value of $p \in [0,1]$. We saw that it depicts the behaviour of a *ratio*. Yet we want to shortly discuss a different approach for this setting. Instead we define $\langle p \rangle$ for values of $p \in \mathbb{N}$ hence p is now a natural number with the following meaning. A team X satisfies a formula $\langle p \rangle \varphi$ if there exists a team $Y \subseteq X$ of size $\geq p$ such that $Y \models \varphi$—similarly for $[p]$ the meaning would be that every team $Y \subseteq X$ of size $\geq p$ satisfies φ.

Sticking to this approach would allow one to state a similar result as for Theorems 3 and 4 but now for *combined complexity* as follows. Here one would just explicitly state the number of rows to be removed from the team, i.e., setting p to m in the constructed formula in the proof of Theorem 3. Regarding Theorem 4 in this setting the formula $f(\varphi)$ increases the number of clauses by factor 10 and therefore requires to set p to $\frac{7}{10} \cdot 10 \cdot m = 7 \cdot m$ where m is the number of clauses of the given 3CNF formula φ.

6 Conclusion

To the best of the authors knowledge this article is the first serious approach in defining team semantics with respect to multisets for first-order dependence logic. We also initiate the study of probabilistic analogues of independence and inclusion logic. Additionally the paper provides a first step into the study of a general approximation operator in the team semantics framework. We show several foundational properties of these newly defined formalisms and present some first computational complexity results for approximate dependence logic (ADL). For ADL we show that the introduction of approximate operators enables us to encode NP-hard properties into the model checking problem (data complexity) of this logic even with only two dependence atoms, a single approximate operator, and a single conjunction. This shows how strong and elegant this kind of approximate notion really is. It is an interesting open question to study the computational properties of the analogously defined approximate inclusion logic.

Heretofore a broad field around intuitionistic logic [4] has developed. Intuitionistic logic can be seen as classical propositional logic without the law of excluded middle. One of the main concepts here is the *intuitionistic implication* $\rightarrow$. In the setting of team semantics it is defined as follows. Let $\mathfrak{A}$ be a structure and X be a team. Then $\mathfrak{A} \models_X \varphi \rightarrow \psi$ is true if and only if for all subsets $X' \subseteq X$ it holds that $\mathfrak{A} \models_{X'} \varphi$ implies $\mathfrak{A} \models_{X'} \psi$. The intuitionistic implication has been studied in the context of dependence logic, see e.g., the work of Yang [41]. An approximate variant of this operator in our setting will yield a nice resemblance to the $[p]$ operator. The slight and quite natural adjustment of intuitionistic implication to our setting is then: $\mathfrak{A} \models_X \varphi \rightarrow_p \psi$ if and only if for all subsets $X' \subseteq X$ with $|X'| \geq p \cdot |X|$ (and $p \in [0,1] \cap \mathbb{Q}$) it holds that $\mathfrak{A} \models_{X'} \varphi$ implies that $\mathfrak{A} \models_{X'} \psi$. The operator $[p]$ can now be expressed with the help of the intuitionistic *approximate* implication. One can easily verify that $[p]\varphi$ is equivalent to $\top \rightarrow_p \varphi$.

In this article we have considered approximation in the context of multi-team semantics when restricted to the finite. However our definitions can be generalised in a straightforward manner to deal with arbitrary cardinalities.

Acknowledgements. The second and the third author were supported by grants 292767, 275241 and 264917 of the Academy of Finland. The fourth author is supported by the DFG grant ME 4279/1-1. The last author was supported by the Foundations' Post Doc Pool via Jenny and Antti Wihuri Foundation. We also thank the anonymous referees for their helpful suggestions.

References

1. Böttcher, S., Link, S., Zhang, L.: Pulling conjunctive query equivalence out of the bag. In: Proceedings of 23rd ACM CIKM, pp. 41–50. ACM (2014)
2. Boutilier, C., Friedman, N., Goldszmidt, M., Koller, D.: Context-specific Independence in Bayesian Networks. In: Proceedings of 12th UAI, pp. 115–123. Morgan Kaufmann Publishers Inc., San Francisco, CA, USA (1996)
3. Corander, J., Kontinen, J., Väänänen, J.: Logical approach to context-specific independence. (in preparation)
4. Dalen, D.: Logic and Structure, 4th edn. Springer, Heidelberg (2004)
5. Durand, A., Kontinen, J., de Rugy-Altherre, N., Väänänen, J.: Tractability frontier of data complexity in team semantics. In: Proceedings of 6th GandALF, EPTCS, vol. 193, pp. 73–85 (2015)
6. Durand, A., Kontinen, J., Vollmer, H.: Expressivity and complexity of dependence logic. In: Dependence Logic: Theory and Applications. Springer (2015, to appear)
7. Ebbing, J., Hella, L., Meier, A., Müller, J.-S., Virtema, J., Vollmer, H.: Extended modal dependence logic. In: Libkin, L., Kohlenbach, U., de Queiroz, R. (eds.) WoLLIC 2013. LNCS, vol. 8071, pp. 126–137. Springer, Heidelberg (2013)
8. Galliani, P.: Probabilistic dependence logic (2008), manuscript
9. Galliani, P.: Inclusion and exclusion dependencies in team semantics - on some logics of imperfect information. Ann. Pure Appl. Logic **163**(1), 68–84 (2012)
10. Galliani, P., Hella, L.: Inclusion logic and fixed point logic. In: Proceedings of CSL, pp. 281–295 (2013)
11. Galliani, P., Mann, A.L.: Lottery semantics: a compositional semantics for probabilistic first-order logic with imperfect information. Studia Logica **101**(2), 293–322 (2013)
12. Garey, M.R., Johnson, D.S., Stockmeyer, L.J.: Some simplified np-complete graph problems. Theor. Comput. Sci. **1**, 237–267 (1976)
13. Grädel, E., Väänänen, J.A.: Dependence and independence. Studia Logica **101**(2), 399–410 (2013)
14. Gyssens, M., Niepert, M., Gucht, D.V.: On the completeness of the semigraphoid axioms for deriving arbitrary from saturated conditional independence statements. Inf. Process. Lett. **114**(11), 628–633 (2014)
15. Hannula, M., Kontinen, J.: A finite axiomatization of conditional independence and inclusion dependencies. In: Beierle, C., Meghini, C. (eds.) FoIKS 2014. LNCS, vol. 8367, pp. 211–229. Springer, Heidelberg (2014)
16. Hannula, M., Kontinen, J., Link, S.: On independence atoms and keys. In: Proceedings of 23rd CIKM, pp. 1229–1238. ACM (2014)
17. Hannula, M., Kontinen, J., Virtema, J., Vollmer, H.: Complexity of propositional independence and inclusion logic. In: Italiano, G.F., Pighizzini, G., Sannella, D.T. (eds.) MFCS 2015. LNCS, vol. 9234, pp. 269–280. Springer, Heidelberg (2015)
18. Hintikka, J., Sandu, G.: Informational independence as a semantical phenomenon. Logic. Methodology and Philosophy of Science, vol. 8, pp. 571–589. Elsevier, Amsterdam (1989)

19. Hodges, W.: Compositional semantics for a language of imperfect information. Logic J. IGpPL **5**(4), 539–563 (1997). (electronic)

20. Hyttinen, T., Paolini., G., Väänänen, J.: A logic for arguing about probabilities in measure teams. arXiv e-prints: 1509.01812 (2015)

21. Hyttinen, T., Paolini, G., Väänänen, J.: Quantum team logic and Bell's inequalities. Rev. Symbolic Logic FirstView, 1–21 (2015). http://dx.doi.org/10.1017/S1755020315000192

22. Kivinen, J., Mannila, H.: Approximate inference of functional dependencies from relations. Theor. Comput. Sci. **149**(1), 129–149 (1995)

23. Köhler, H., Link, S.: Armstrong axioms and Boyce-Codd-Heath normal form under bag semantics. Inf. Process. Lett. **110**(16), 717–724 (2010)

24. Kolaitis, P.G.: The query containment problem: set semantics vs. bag semantics. In: Proceedings of 7th AMW, CEUR Workshop Proceedings, vol. 1087 (2013)

25. Kontinen, J.: Coherence and computational complexity of quantifier-free dependence logic formulas. Studia Logica **101**(2), 267–291 (2013)

26. Kontinen, J., Link, S., Väänänen, J.: Independence in database relations. In: Libkin, L., Kohlenbach, U., de Queiroz, R. (eds.) WoLLIC 2013. LNCS, vol. 8071, pp. 179–193. Springer, Heidelberg (2013)

27. Kontinen, J., Müller, J.S., Schnoor, H., Vollmer, H.: A Van Benthem theorem for modal team semantics. In: Proceedings of 24th CSL, LIPIcs, vol. 41, pp. 277–291. Schloss Dagstuhl-Leibniz-Zentrum Fuer Informatik, Dagstuhl, Germany (2015)

28. Krebs, A., Meier, A., Virtema, J.: A team based variant of CTL. In: Proceedings of TIME 2015 (2015)

29. Lamperti, G., Melchiori, M., Zanella, M.: On multisets in database systems. In: Proceedings of WMP, pp. 147–216. Springer, London, UK (2001)

30. Link, S.: Reasoning about saturated conditional independence under uncertainty: axioms, algorithms, and Levesque's situations to the rescue. In: Proceedings of AAAI, AAAI Press (2013)

31. Link, S.: Sound approximate reasoning about saturated conditional probabilistic independence under controlled uncertainty. J. Appl. Logic **11**(3), 309–327 (2013)

32. Link, S.: Frontiers for propositional reasoning about fragments of probabilistic conditional independence and hierarchical database decompositions. Theor. Comput. Sci. **603**, 111–131 (2015)

33. Niepert, M., Gyssens, M., Sayrafi, B., Gucht, D.V.: On the conditional independence implication problem: a lattice-theoretic approach. Artif. Intell. **202**, 29–51 (2013)

34. Papadimitriou, C.H.: Computational Complexity. Addison-Wesley, Boston (1994)

35. Pensar, J., Nyman, H.J., Koski, T., Corander, J.: Labeled directed acyclic graphs: a generalization of context-specific independence in directed graphical models. Data Min. Knowl. Discov. **29**(2), 503–533 (2015)

36. Väänänen, J.: Dependence Logic - A New Approach to Independence Friendly Logic, London Mathematical Society Student Texts, vol. 70. Cambridge University Press, Cambridge (2007)

37. Väänänen, J.: Modal dependence logic. In: Apt, K., van Rooij, R. (eds.) New Perspectives on Games and Interaction. Texts in Logic and Games, vol. 5, pp. 237–254. Amsterdam University Press, Amsterdam (2008)

38. Väänänen, J.: The logic of approximate dependence. arXiv:1408.4437 (2014)

39. Wong, S.K.M.: An extended relational data model for probabilistic reasoning. J. Intell. Inf. Syst. **9**(2), 181–202 (1997)

40. Wong, S.K.M., Butz, C.J., Wu, D.: On the implication problem for probabilistic conditional independency. IEEE Trans. Syst. Man Cybern. Part A: Syst. Hum. **30**(6), 785–805 (2000)
41. Yang, F.: On extensions and variants of dependence logic - a study of intuitionistic connectives in the team semantics setting. Ph.D thesis, Department of Mathematics and Statistics, University of Helsinki (2014)
42. Yang, F., Väänänen, J.: Propositional logics of dependence and independence, Part I. CoRR abs/1412.7998 (2014)

The Complexity of Non-Iterated Probabilistic Justification Logic

Ioannis Kokkinis[✉]

Institute of Computer Science, University of Bern, Bern, Switzerland
kokkinis@inf.unibe.ch

Abstract. The logic PJ is a probabilistic logic defined by adding (non-iterated) probability operators to the basic justification logic J. In this paper we establish upper and lower bounds for the complexity of the derivability problem in the logic PJ. The main result of the paper is that the complexity of the derivability problem in PJ remains the same as the complexity of the derivability problem in the underlying logic J, which is Π_2^p-complete. This implies hat the probability operators do not increase the complexity of the logic, although they arguably enrich the expressiveness of the language.

Keywords: Justification logic · Probabilistic logic · Complexity · Derivability · Satisfiability

1 Introduction

Traditional modal epistemic logic uses formulas of the form $\Box\alpha$ to express that an agent believes α. The language of justification logic [5] 'unfolds' the $\Box$-modality into a family of so-called *justification terms*, which are used to represent evidence for the agent's belief. Hence, instead of $\Box\alpha$, justification logic includes formulas of the form $t : \alpha$ meaning

the agent believes α for reason t.

Artemov [2,3] developed the first justification logic, the Logic of Proofs, to provide intuitionistic logic with a classical provability semantics. There, justification terms represent formal proofs in Peano Arithmetic. However, terms may also represent informal justifications. For instance, our belief in α may be justified by direct observation of α or by learning that a friend heard about α. This general reading of justification led to a big variety of epistemic justification logics for many different applications [6,7,19]. In [15,16] we extended justification logic with probability operators in order to accommodate the idea that

different kinds of evidence for α lead to different degrees of belief in α.

For example it could be the case that the agent learns α from some unreliable source (e.g. from some friend of his) or that the agent reads about α in some

© Springer International Publishing Switzerland 2016
M. Gyssens and G. Simari (Eds.): FoIKS 2016, LNCS 9616, pp. 292–310, 2016.
DOI: 10.1007/978-3-319-30024-5_16

reliable newspaper. In both cases the agent has a justification for α: in the first case he has the statement of his friend and in the second case the text of the newspaper. However, it is natural that the agent does not want to put the same credence in both sources of information. This differentiation in credulity cannot be expressed in classical justification logic. So, the main contribution of justification logics with probability operators (probabilistic justification logics [15,16]) is the ability to compare different sources of information. Uncertain reasoning in justification logic has also been studied in [11,12,21]. See [15,16] for an extended comparison between our approach and the ones from [11,12,21].

Probabilistic logics are logics than can be used to model uncertain reasoning. Although the idea of probabilistic logic was first proposed by Leibnitz, the modern development of this topic started only in the 1970s and 1980s in the papers of Keisler [13] and Nilsson [22]. Following Nilsson's research, Fagin et al. [10] introduced a logic with arithmetical operations built into the syntax so that Boolean combinations of linear inequalities of probabilities of formulas can be expressed. The probabilistic logic of [10] can be considered as a probabilistic logic with classical base. The derivability problem in this logic is proved to be $coNP$-complete, the same as that of classical propositional logic. Following the lines of [10], Ognjanović et al. [23] defined the logic $\mathsf{LPP_2}$, which is also a probabilistic logic with classical base. The logic $\mathsf{LPP_2}$ makes use of an infinitary rule which makes the proof of strong completeness possible (as opposed to the finitary system of [10] which is only simply complete). The $\mathsf{LPP_2}$-derivability problem is again $coNP$-complete.

Following the lines of [23] the logic PJ was defined in [15]. PJ is a probabilistic logic defined over the basic justification logic J.[1] The language of PJ contains formulas of the form $P_{\geq s}\alpha$ meaning

the probability of truthfulness of the justification formula α is at least s.

So, in the logic PJ, statements like "evidence t serves as a justification for α with probability at least 30 %" can be expressed. PJ does not allow iterations of the probability operator. In [16] we study an extension of PJ, the logic PPJ,[2] where iterations of the probability operator as well as justification operators over probability operators are allowed.

The results of [1,8,17,20] showed that, under some reasonable assumptions, the derivability problem for the justification logic J is Π_2^p-complete, i.e. it is complete in the second level of the polynomial hierarchy. In this paper we show that under the same assumptions the derivability problem for the probabilistic justification logic PJ remains in the class Π_2^p-complete. We achieve this, by showing that the satisfiability problem for the logic PJ, which is dual to the derivability problem, belongs to the class Σ_2^p-complete. The methods we use are adaptations from [10] and [17]. As it is the case in [23] and [10] we also make use of some well known results from the theory of linear programming. The main result of the paper is that the probability operators do not increase the

---

[1] J stands for justification, whereas PJ stands for probabilistic justification.
[2] The two P's stand for iterations of the probability operator.

complexity of the logic, although they arguably enrich the expressiveness of the logical framework.

The rest of the paper is organized as follows. In Sect. 2 we briefly recall the justification logic J and the probabilistic justification logic PJ. In Sect. 3 we establish a small model theorem for PJ. In Sect. 4 we present an algorithm that decides the satisfiability problem for the logic PJ and evaluate its complexity. We close the paper in Sect. 5 with some final observations.

An earlier version of the present paper is available in arXiv [14].

2   The Logics J and PJ

In this section we briefly recall the basic justification logic J [5] and the probabilistic justification logic PJ [15].

Justification terms are built according to the following grammar:

$$t ::= c \mid x \mid (t \cdot t) \mid (t + t) \mid \, !t$$

where c is a constant and x is a variable. Tm denotes the set of all terms. For any term t and any non-negative integer n we define:

$$!^0 t := t \qquad \text{and} \qquad !^{n+1} t := \, ! \, (!^n t)$$

Terms are used to provide justifications for formulas. Constants are used as justifications for axioms, whereas variables are used as justifications for arbitrary formulas. The operator $\cdot$ can be used by the agents to apply modus ponens (see axiom (J) in Fig. 1), the operator $+$ is used for concatenation of proofs (see axiom (+) in Fig. 1) and the operator ! is used for stating positive introspection (see rule (AN!) in Fig. 2). That is, if the agent has a justification c for α then he has a justification $!c$ for the fact that c is a justification for α and so on.

Let Prop denote a countable set of atomic propositions. Formulas of the language $\mathcal{L}_J$ (justification formulas) are built according to the following grammar:

$$\alpha ::= p \mid \neg \alpha \mid \alpha \wedge \alpha \mid t : \alpha$$

where $t \in$ Tm and $p \in$ Prop. Any formula of the form $t : \alpha$ for $t \in$ Tm and $\alpha \in \mathcal{L}_J$ will be called a *justification assertion*. We will use the letter p possibly primed or with subscripts to represent an element of Prop and lower-case Greek letters like $\alpha, \beta, \gamma, \ldots$ for $\mathcal{L}_J$-formulas. In Fig. 1 we present the axioms schemes of the logic J.

In order to build justifications for arbitrary formulas in the logic J we need to start by some justifications for the axioms. That is why we need the notion of a constant specification. A *constant specification* is any set CS that satisfies the following condition:

$$\text{CS} \subseteq \{(c, \alpha) \mid c \text{ is a constant and } \alpha \text{ is an instance}$$
$$\text{of some axiom scheme of the logic J}\}$$

A constant specification CS will be called:

$$\boxed{\begin{array}{l} \text{(P) finitely many axiom schemes for classical} \\ \quad \text{propositional logic in the language of } \mathcal{L_J} \\ \text{(J) } \vdash u : (\alpha \to \beta) \to (v : \alpha \to u \cdot v : \beta) \\ \text{(+) } \vdash \big(u : \alpha \lor v : \alpha\big) \to u + v : \alpha \end{array}}$$

Fig. 1. Axioms Schemes of J

Axiomatically Appropriate: if for every instance of a J-axiom, α, there exists some constant c such that $(c, \alpha) \in$ CS, i.e. every instance of a J-axiom scheme is justified by at least one constant.

Schematic: if for every constant c the set

$$\{\alpha \mid (c, \alpha) \in \text{CS}\}$$

consists of all instances of several (possibly zero) axiom schemes, i.e. if every constant specifies certain axiom schemes and only them.

Decidable: if the set CS is decidable. In this paper when we refer to a decidable CS, we will always imply that CS is decidable in *polynomial time*.

Finite: if CS is a finite set.

Almost Schematic: if $\text{CS} = \text{CS}_1 \cup \text{CS}_2$ where $\text{CS}_1 \cap \text{CS}_2 = \emptyset$, CS_1 is a schematic constant specification and CS_2 is a finite constant specification.

Total: if for every constant c and every instance α of a J-axiom scheme, $(c, \alpha) \in$ CS.

Let CS be any constant specification. The deductive system J_{CS} is presented in Fig. 2.

$$\boxed{\begin{array}{l} \quad \text{axioms schemes of J} \\ \qquad + \\ \text{(AN!) } \vdash \ !^{n+1}c : !^n c : \cdots : !c : c : \alpha, \text{ where } (c, \alpha) \in \text{CS and } n \in \mathbb{N} \\ \text{(MP) if } T \vdash \alpha \text{ and } T \vdash \alpha \to \beta \text{ then } T \vdash \beta \end{array}}$$

Fig. 2. System J_{CS}

As usual $T \vdash_\mathsf{L} \alpha$ means that the formula α is provable from the set of formulas T using the rules and axioms of the logic L. When L is clear from the context it will be omitted.

Now we present the semantics for the logic J. The models for a J_{CS} are the so called J_{CS}-evaluations (see Definition 1). We use T to represent the truth value "true" and F to represent the truth value "false". Let $\mathcal{P}(W)$ denote the powerset of the set W.

Definition 1 (J$_{CS}$-Evaluation). *Let* CS *be any constant specification. A* J$_{CS}$-*evaluation is a function* $*$ *such that* $* : \mathsf{Prop} \to \{\mathsf{T}, \mathsf{F}\}$ *and* $* : \mathsf{Tm} \to \mathcal{P}(\mathcal{L}_J)$ *and for* $u, v \in \mathsf{Tm}$, *for a constant* c *and* $\alpha, \beta \in \mathcal{L}_J$ *we have:*

(1) $\left(\alpha \to \beta \in u^* \text{ and } \alpha \in v^*\right) \Longrightarrow \beta \in (u \cdot v)^*$
(2) $u^* \cup v^* \subseteq (u + v)^*$
(3) if $(c, \alpha) \in \mathsf{CS}$ *then for all* $n \in \mathbb{N}$ *we have[3]:*

$$!^{n-1}c : !^{n-2}c : \cdots : !c : c : \alpha \in (!^n c)^*$$

We will usually write t^* *and* p^* *instead of* $*(t)$ *and* $*(p)$ *respectively.*

Now we will define the binary relation $\Vdash$.

Definition 2 (Truth under a J$_{CS}$-Evaluation). *We define what it means for an* $\mathcal{L}_J$-*formula to hold under a* J$_{CS}$-*evaluation* $*$ *inductively as follows:*

$$* \Vdash p \iff p^* = \mathsf{T}$$
$$* \Vdash \neg\alpha \iff * \nVdash \alpha$$
$$* \Vdash \alpha \wedge \beta \iff \left(* \Vdash \alpha \text{ and } * \Vdash \beta\right)$$
$$* \Vdash t : \alpha \iff \alpha \in t^*$$

We have the following theorem.

Theorem 1 (Completeness of J[4,19]). *Let* CS *be any constant specification. Let* $\alpha \in \mathcal{L}_J$. *Then we have:*

$$\vdash_{\mathsf{J}_{CS}} \alpha \iff \Vdash_{CS} \alpha.$$

where $\Vdash_{CS} \alpha$ *means that* α *holds under any* J$_{CS}$-*evaluation.*

Let S be the set of all rational numbers from the interval $[0, 1]$. The formulas of the language $\mathcal{L}_P$ (the so called probabilistic formulas) are built according to the following grammar:

$$A ::= P_{\geq s}\alpha \mid \neg A \mid A \wedge A$$

where $s \in \mathsf{S}$, and $\alpha \in \mathcal{L}_J$. We use capital Latin letters like $A, B, C, \ldots$ for $\mathcal{L}_P$-formulas. We employ the standard abbreviations for classical connectives. Additionally, we set:

$$P_{<s}\alpha \equiv \neg P_{\geq s}\alpha \qquad\qquad P_{\leq s}\alpha \equiv P_{\geq 1-s}\neg\alpha$$
$$P_{>s}\alpha \equiv \neg P_{\leq s}\alpha \qquad\qquad P_{=s}\alpha \equiv P_{\geq s}\alpha \wedge P_{\leq s}\alpha$$

The axioms schemes of PJ are presented in Fig. 3. For any constant specification CS the deductive system PJ$_{CS}$ is presented in Fig. 4. Definitions 3–5 describe the semantics for the logic PJ.

---

[3] We agree to the convention that the formula $!^{n-1}c : !^{n-2}c : \cdots : !c : c : \alpha$ represents the formula α for $n = 0$.

> (P)   finitely many axiom schemes for classical
>   propositional logic in the language of $\mathcal{L}_P$
> (PI) $\vdash P_{\geq 0}\alpha$
> (WE) $\vdash P_{\leq r}\alpha \rightarrow P_{<s}\alpha$, where $s > r$
> (LE) $\vdash P_{<s}\alpha \rightarrow P_{\leq s}\alpha$
> (DIS) $\vdash P_{\geq r}\alpha \wedge P_{\geq s}\beta \wedge P_{\geq 1}\neg(\alpha \wedge \beta) \rightarrow P_{\geq \min(1,r+s)}(\alpha \vee \beta)$
> (UN) $\vdash P_{\leq r}\alpha \wedge P_{<s}\beta \rightarrow P_{<r+s}(\alpha \vee \beta)$, where $r + s \leq 1$

Fig. 3. Axioms Schemes of PJ

> axiom schemes of PJ
>
> $+$
>
> (MP) if $T \vdash A$ and $T \vdash A \rightarrow B$ then $T \vdash B$
> (CE) if $\vdash_{\mathsf{JCS}} \alpha$ then $\vdash_{\mathsf{PJ_{CS}}} P_{\geq 1}\alpha$
> (ST) if $T \vdash A \rightarrow P_{\geq s-\frac{1}{k}}\alpha$ for every integer $k \geq \frac{1}{s}$ and $s > 0$
>   then $T \vdash A \rightarrow P_{\geq s}\alpha$

Fig. 4. System PJ$_{\mathsf{CS}}$

Definition 3 (Algebra Over a Set). *Let W be a non-empty set and let H be a non-empty subset of $\mathcal{P}(W)$. H will be called an* algebra over W *iff the following hold:*

- $W \in H$
- $U, V \in H \Longrightarrow U \cup V \in H$
- $U \in H \Longrightarrow W \setminus U \in H$

Definition 4 (Finitely Additive Measure). *Let H be an algebra over W and $\mu : H \rightarrow [0,1]$. We call μ a* finitely additive measure *iff the following hold:*

(1) $\mu(W) = 1$
(2) for all $U, V \in H$:

$$U \cap V = \emptyset \Longrightarrow \mu(U \cup V) = \mu(U) + \mu(V)$$

Definition 5 (PJ$_{\mathsf{CS}}$-Model). *Let* CS *be any constant specification. A* PJ$_{\mathsf{CS}}$ - *model, or simply a model, is a structure $M = \langle W, H, \mu, * \rangle$ where:*

- *W is a non-empty set of objects called worlds.*
- *H is an algebra over W.*
- *$\mu : H \rightarrow [0,1]$ is a finitely additive measure.*
- *$*$ is a function from W to the set of all J$_{\mathsf{CS}}$-evaluations, i.e. $*(w)$ is a J$_{\mathsf{CS}}$- evaluation for each world $w \in W$. We will usually write $*_w$ instead of $*(w)$.*

Definition 6 (Measurable Model). *Let* $M = \langle W, H, \mu, * \rangle$ *be a model and* $\alpha \in \mathcal{L}_J$. *We define the following set:*

$$[\alpha]_M = \{w \in W | *_w \Vdash \alpha\}$$

We will omit the subscript M, i.e. we will simply write $[\alpha]$, if M is clear from the context. A $\mathsf{PJ_{CS}}$-model $M = \langle W, H, \mu, * \rangle$ is measurable iff $[\alpha]_M \in H$ for every $\alpha \in \mathcal{L}_J$. The class of measurable $\mathsf{PJ_{CS}}$-models will be denoted by $\mathsf{PJ_{CS,Meas}}$.

Definition 7 (Truth in a $\mathsf{PJ_{CS,Meas}}$-model). *Let CS be any constant specification. Let $M = \langle W, H, \mu, * \rangle$ be a $\mathsf{PJ_{CS,Meas}}$-model. We define what it means for an $\mathcal{L}_P$-formula to hold in M inductively as follows[4]:*

$$M \models P_{\geq s}\alpha \Longleftrightarrow \mu([\alpha]_M) \geq s$$
$$M \models \neg A \Longleftrightarrow M \not\models A$$
$$M \models A \wedge B \Longleftrightarrow (M \models A \text{ and } M \models B)$$

In the sequel we may refer to $\mathsf{PJ_{CS,Meas}}$-models simply as models if there is no danger for confusion. We have the following theorem.

Theorem 2 (Strong Completeness for PJ[15]). *Any $\mathsf{PJ_{CS}}$ is sound and strongly complete with respect to $\mathsf{PJ_{CS,Meas}}$-models, i.e. for any $T \subseteq \mathcal{L}_P$ and any $A \in \mathcal{L}_P$:*

$$T \vdash_{\mathsf{PJ_{CS}}} A \Longleftrightarrow T \models_{\mathsf{PJ_{CS}}} A$$

Let CS be any constant specification. A formula $A \in \mathcal{L}_P$ is satisfied in $M \in \mathsf{PJ_{CS,Meas}}$ iff $M \models A$. A will be called $\mathsf{PJ_{CS,Meas}}$-satisfiable or simply satisfiable if there is a $\mathsf{PJ_{CS,Meas}}$-model that satisfies A. We define the $\mathsf{PJ_{CS,Meas}}$-satisfiability problem to be the decision problem defined as follows:

"For a given $A \in \mathcal{L}_P$ and a given CS is A $\mathsf{PJ_{CS,Meas}}$-satisfiable?"

A formula $\alpha \in \mathcal{L}_J$ is satisfied in a $\mathsf{J_{CS}}$-evaluation $*$ iff $* \Vdash \alpha$. α will be called $\mathsf{J_{CS}}$-satisfiable or simply satisfiable if there is some $\mathsf{J_{CS}}$-evaluation $*$ that satisfies α. We define the $\mathsf{J_{CS}}$-satisfiability problem to be the decision problem defined as follows:

"For a given $\alpha \in \mathcal{L}_J$ and a given CS is α $\mathsf{J_{CS}}$-satisfiable?"

3   Small Model Property

The goal of this section is to prove a small model property for the logic PJ. The small model property will be the most important tool for establishing the upper bound for the complexity of PJ.

---

[4] Observe that the satisfiability relation of a $\mathsf{J_{CS}}$-evaluation is represented with $\Vdash$ whereas the satisfiability relation of a model is represented with $\models$.

Definition 8 (Subformulas). *The set* $\mathsf{subf}(\cdot)$ *is defined recursively as follows:*
For $\mathcal{L}_\mathsf{J}$ *-formulas:*

- $\mathsf{subf}(p) := \{p\}$
- $\mathsf{subf}(t : \alpha) := \{t : \alpha\} \cup \mathsf{subf}(\alpha)$
- $\mathsf{subf}(\neg\alpha) := \{\neg\alpha\} \cup \mathsf{subf}(\alpha)$
- $\mathsf{subf}(\alpha \wedge \beta) := \{\alpha \wedge \beta\} \cup \mathsf{subf}(\alpha) \cup \mathsf{subf}(\beta)$

For $\mathcal{L}_\mathsf{P}$ *-formulas:*

- $\mathsf{subf}(P_{\geq s}\alpha) := \{P_{\geq s}\alpha\} \cup \mathsf{subf}(\alpha)$
- $\mathsf{subf}(\neg A) := \{\neg A\} \cup \mathsf{subf}(A)$
- $\mathsf{subf}(A \wedge B) := \{A \wedge B\} \cup \mathsf{subf}(A) \cup \mathsf{subf}(B)$

Observe that for $A \in \mathcal{L}_\mathsf{P}$ *we have that* $\mathsf{subf}(A) \subseteq \mathcal{L}_\mathsf{P} \cup \mathcal{L}_\mathsf{J}$.

Definition 9 (Atoms). *Let* A *be an* $\mathcal{L}_\mathsf{P}$- *or an* $\mathcal{L}_\mathsf{J}$-*formula. Let* X *be the set that contains all the atomic propositions and the justification assertions from the set* $\mathsf{subf}(A)$. *An atom of* A *is any formula of the following form:*

$$\bigwedge_{B \in X} \pm B \tag{1}$$

where $\pm B$ *denotes either* B *or* $\neg B$. *We will use the lowercase Latin letter* a *for atoms, possibly with subscripts.*

Let A be an $\mathcal{L}_\mathsf{P}$- or an $\mathcal{L}_\mathsf{J}$-formula. Assume that A is either of the form $\bigwedge_i B_i$ or of the form $\bigvee_i B_i$. Then $C \in A$ means that for some i, $B_i = C$.

Definition 10 (Sizes). *The size function* $|\cdot|$ *is defined as follows:*
For $\mathcal{L}_\mathsf{P}$ *-formulas: (recursively)*

- $|P_{\geq s}\alpha| := 2$
- $|\neg A| := 1 + |A|$
- $|A \wedge B| := |A| + 1 + |B|$

For Sets
Let W be a set. $|W|$ is the cardinal number of W.
For Non-negative Integers
Let r be an non-negative integer. We define the size of r to be equal to the length of r written in binary, i.e.:

$$|r| := \begin{cases} 1 & ,r = 0 \\ \lfloor \log_2(r) + 1 \rfloor & ,r \geq 1 \end{cases}$$

where $\lfloor \cdot \rfloor$ *is the function that returns the greatest integer that is less than or equal to its argument.*
For Non-negative Rational Numbers
Let $r = \frac{s_1}{s_2}$, where s_1 and s_2 are relatively prime non-negative integers with $s_2 \neq 0$, be a non-negative rational number. We define:

$$|r| := |s_1| + |s_2|.$$

Let $A \in \mathcal{L}_P$ we define:

$$||A|| := \max \left\{ |s| \mid P_{\geq s}\alpha \in \mathsf{subf}(A) \right\}$$

Lemma 1 was originally proved in [23] for the logic LPP_2. The proof for the logic PJ is given in [15].

Lemma 1. *For any constant specification* CS, *we have:*

$$\vdash_{\mathsf{Jcs}} \alpha \leftrightarrow \beta \Longleftrightarrow \vdash_{\mathsf{PJcs}} P_{\geq s}\alpha \leftrightarrow P_{\geq s}\beta$$

A proof for Theorem 3 can be found in [9, p.145].

Theorem 3. *Let S be a system of r linear equalities. Assume that the vector[5] x is a solution of S such that all of x's entries are non-negative. Then there is a vector x^* such that:*

(1) x^* is a solution of S.
(2) all the entries of x^* are non-negative.
(3) at most r entries of x^* are positive.

Theorem 4 establishes some properties for the solutions of a linear system.

Theorem 4. *Let S be a linear system of n variables and of r linear equalities and/or inequalities with integer coefficients each of size at most l. Assume that the vector $x = x_1, \ldots, x_n$ is a solution of S such that for all $i \in \{1, \ldots, n\}$, $x_i \geq 0$. Then there is a vector $x^* = x_1^*, \ldots, x_n^*$ with the following properties:*

(1) x^* is a solution of S.
(2) for all $i \in \{1, \ldots, n\}$, $x_i^* \geq 0$.
(3) at most r entries of x^* are positive.
(4) for all $i \in \{1, \ldots, n\}$, if $x_i^* > 0$ then $x_i > 0$.
(5) for all i, x_i^* is a non-negative rational number with size bounded by

$$2 \cdot \left(r \cdot l + r \cdot \log_2(r) + 1 \right) .$$

Proof. In S we replace the variables that correspond to the entries of x that are equal to zero (if any) with zeros. This way we obtain a new linear system S_0, with r linear equalities and/or inequalities and $m \leq n$ variables. x is a solution[6] of S_0. It also holds that any solution of S_0 is a solution[7] of S.

---

[5] We will always use bold font for vectors.

[6] In the proof of Theorem 4 all vectors have n entries. The entries of the vectors are assumed to be in one to one correspondence with the variables that appear in the original system S.

Let y be a solution of a linear system T. If y has more entries than the variables of T we imply that entries of y that correspond to variables that appear in T compose a solution of T.

[7] Assume that system T has less variables than system T'. When we say that any solution of T is a solution of T' we imply that the missing variables are set to 0.

Assume that the system $\mathcal{S}_0$ contains an inequality of the form

$$b_1 \cdot y_1 + \ldots + b_m y_m \lozenge c \tag{2}$$

for $\lozenge \in \{<, \leq, \geq, >\}$ where $y_1, \ldots, y_m$ are variables of $\mathcal{S}$ and $b_1, \ldots, b_m, c$ are constants that appear in $\mathcal{S}$. $\boldsymbol{x}$ is a solution of Eq. (2). We replace the inequality Eq. (2) in $\mathcal{S}_0$ with the following equality:

$$b_1 \cdot y_1 + \ldots + b_m y_m = b_1 \cdot x_1 + \ldots + b_l \cdot x_m$$

We repeat this procedure for every inequality of $\mathcal{S}_0$. This way we obtain a system of linear equalities which we call $\mathcal{S}_1$. It is easy to see that $\boldsymbol{x}$ is a solution of $\mathcal{S}_1$ and that any solution of $\mathcal{S}_1$ is also a solution of $\mathcal{S}_0$ and thus of $\mathcal{S}$.

Now we will transform $\mathcal{S}_1$ to another linear system by applying the following algorithm.

Algorithm

We set $i = 1$, $e_i = r$, $v_i = m$, $\boldsymbol{x}^i = \boldsymbol{x}$ and we execute the following steps:

(i) If $e_i = v_i$ then go to step (ii). Otherwise go to step (iii).
(ii) If the determinant of $\mathcal{S}_i$ is non-zero then stop. Otherwise go to step (v).
(iii) If $e_i < v_i$ then go to step (iv), else go to step (v).
(iv) We know that the vector $\boldsymbol{x}^i$ is a non-negative solution for the system $\mathcal{S}_i$. From Theorem 3 we obtain a solution $\boldsymbol{x}^{i+1}$ for the system $\mathcal{S}_i$ which has at most e_i entries positive. In $\mathcal{S}_i$ we replace the variables that correspond to zero entries of the solution $\boldsymbol{x}^{i+1}$ with zeros. We obtain a new system which we call $\mathcal{S}_{i+1}$ with e_{i+1} equalities and v_{i+1} variables. $\boldsymbol{x}^{i+1}$ is a solution of $\mathcal{S}_{i+1}$ and any solution of $\mathcal{S}_{i+1}$ is a solution of $\mathcal{S}_i$. We set $i := i + 1$ and we go to step (i).
(v) From any set of equalities that are linearly dependent we keep only one equation. We obtain a new system which we call $\mathcal{S}_{i+1}$ with e_{i+1} equalities and $v_{i+1} := v_i$ variables. We set $i := i + 1$ and $\boldsymbol{x}^{i+1} := \boldsymbol{x}^i$. We go to step (i).

Let I be the final value of i after the execution of the algorithm. Since the only way for our algorithm to terminate is through step (ii) it holds that system $\mathcal{S}_I$ is an $e_I \times e_I$ system of linear equalities with non-zero determinant (for $e_I \leq r$). System $\mathcal{S}_I$ is obtained from system $\mathcal{S}_1$ by replacing some variables that correspond to zero entries of the solution with zeros. So any solution of $\mathcal{S}_I$ is also a solution of system $\mathcal{S}_1$ and thus a solution of $\mathcal{S}$. From the algorithm we have that $\boldsymbol{x}^I$ is a solution of $\mathcal{S}_I$. Since $\mathcal{S}_I$ has a non-zero determinant Cramer's rule can be applied. Hence the vector $\boldsymbol{x}^I$ is the unique solution of system $\mathcal{S}_I$. Let x_i^I be an entry of $\boldsymbol{x}^I$. x_i^I will be equal to the following rational number

$$\frac{\begin{vmatrix} a_{11} & \cdots & a_{1e_I} \\ \vdots & \ddots & \vdots \\ a_{e_I 1} & \cdots & a_{e_I e_I} \end{vmatrix}}{\begin{vmatrix} b_{11} & \cdots & b_{1e_I} \\ \vdots & \ddots & \vdots \\ b_{e_I 1} & \cdots & b_{e_I e_I} \end{vmatrix}}$$

where all the a_{ij} and b_{ij} are integers that appear in the original system $\mathcal{S}$. By properties of the determinant we know that the numerator and the denominator of the above rational number will each be at most equal to $r! \cdot (2^l - 1)^r$. So we have that:

$$|x_i^I| \leq 2 \cdot \left(\log_2(r! \cdot (2^l - 1)^r) + 1\right) \qquad \Longrightarrow$$
$$|x_i^I| \leq 2 \cdot \left(\log_2(r^r \cdot 2^{l \cdot r}) + 1\right) \qquad \Longrightarrow$$
$$|x_i^I| \leq 2 \cdot (r \cdot \log_2(r) + l \cdot r + 1)$$

As we already mentioned the final vector $\boldsymbol{x^I}$ is a solution of the original linear system $\mathcal{S}$. We also have that all the entries of $\boldsymbol{x^I}$ are non-negative, at most r of its entries are positive and the size of each entry of $\boldsymbol{x^I}$ is bounded by $2 \cdot (r \cdot \log_2 r + r \cdot l + 1)$. Furthermore, since the variables that correspond to zero entries of the original vector $\boldsymbol{x}$ were replaced by zeros, we have that for every i, if the i-th entry of $\boldsymbol{x^I}$ is positive then the i-th entry of $\boldsymbol{x}$ is positive too. So $\boldsymbol{x^I}$ is the requested vector $\boldsymbol{x^*}$.

The following theorem is an adaptation of the small model theorem from [10]. Similar techniques have also been used in [23] to obtain decidability for the logic LPP$_2$.

Theorem 5 (Small Model Property). *Let* CS *be any constant specification and let* $A \in \mathcal{L}_{\mathsf{P}}$. *If* A *is* PJ$_{\mathsf{CS,Meas}}$*-satisfiable then it is satisfiable in a* PJ$_{\mathsf{CS,Meas}}$*-model* $M = \langle W, H, \mu, * \rangle$ *such that:*

(1) $|W| \leq |A|$
(2) $H = \mathcal{P}(W)$
(3) *For every* $w \in W$, $\mu(\{w\})$ *is a rational number with size at most*

$$2 \cdot \left(|A| \cdot ||A|| + |A| \cdot \log_2(|A|) + 1\right)$$

(4) *For every* $V \in H$

$$\mu(V) = \sum_{w \in V} \mu(\{w\})$$

(5) *For every atom of* A, a, *there exists at most one* $w \in W$ *such that* $*_w \Vdash a$.

Proof. Let CS be any constant specification and let $A \in \mathcal{L}_{\mathsf{P}}$. Let $a_1, \ldots, a_n$ be all the atoms of A. By propositional reasoning (in the logic PJ$_{\mathsf{CS}}$) we can prove that:

$$\mathsf{PJ_{CS}} \vdash A \leftrightarrow \bigvee_{i=1}^{K} \bigwedge_{j=1}^{l_i} P_{\Diamond_{ij} s_{ij}}(\beta^{ij})$$

where all the $P_{\Diamond_{ij} s_{ij}}(\beta^{ij})$ appear in A and $\Diamond_{ij} \in \{\geq, <\}$.

By using propositional reasoning again (but this time in the logic J$_{\mathsf{CS}}$) we can prove that each β^{ij} is equivalent to a disjunction of some atoms of A. So, by using Lemma 1 we have that:

$$\mathsf{PJ_{CS}} \vdash A \leftrightarrow \bigvee_{i=1}^{K} \bigwedge_{j=1}^{l_i} P_{\Diamond_{ij} s_{ij}}(a^{ij})$$

where each α^{ij} is a disjunction of some atoms of A. By Theorem 2 we have that for any $M \in \mathsf{PJ_{CS,Meas}}$:

$$M \models A \Longleftrightarrow M \models \bigvee_{i=1}^{K} \bigwedge_{j=1}^{l_i} P_{\Diamond_{ij}s_{ij}}(\alpha^{ij}) \tag{3}$$

Assume that A is satisfiable. By Eq. (3) there must exist some i such that

$$\bigwedge_{j=1}^{l_i} P_{\Diamond_{ij}s_{ij}}(\alpha^{ij})$$

is satisfiable. Let $M' = \langle W', H', \mu', *' \rangle$ be a model such that:

$$M' \models \bigwedge_{j=1}^{l_i} P_{\Diamond_{ij}s_{ij}}(\alpha^{ij}) \tag{4}$$

For every $k \in \{1, \dots, n\}$ we define:

$$x_k = \mu'([a_k]_{M'}) \tag{5}$$

In every world of M' some atom of A must hold. Thus, we have:

$$W' = \bigcup_{k=1}^{n} [a_k]_{M'}$$

And since $\mu'(W') = 1$ we get:

$$\mu'\left(\bigcup_{k=1}^{n} [a_k]_{M'}\right) = 1 \tag{6}$$

The a_k's are atoms of the same formula, so we have:

$$k \neq k' \implies [a_k]_{M'} \cap [a_{k'}]_{M'} = \emptyset \tag{7}$$

By Eqs. (6) and (7) and the fact that μ' is a finitely additive measure we get:

$$\sum_{k=1}^{n} \mu'([a_k]_{M'}) = 1$$

and by Eq. (5):

$$\sum_{k=1}^{n} x_k = 1 \tag{8}$$

Let $j \in \{1, \dots, l_i\}$. From Eq. (4) we get:

$$M' \models P_{\Diamond_{ij}s_{ij}}(\alpha^{ij}).$$

This implies that $\mu'([\alpha^{ij}]_{M'}) \Diamond_{ij} s_{ij}$, i.e.

$$\mu'\left(\left[\bigvee_{a_k \in \alpha^{ij}} a_k\right]_{M'}\right) \Diamond_{ij} s_{ij}$$

which implies that

$$\mu'\left(\bigcup_{a_k \in \alpha^{ij}} [a_k]_{M'}\right) \Diamond_{ij} s_{ij}$$

By Eq. (7) and the additivity of μ' we have that:

$$\sum_{a_k \in \alpha^{ij}} \mu'([a_k]_{M'}) \Diamond_{ij} s_{ij}$$

and by Eq. (5):

$$\sum_{a_k \in \alpha^{ij}} x_k \Diamond_{ij} s_{ij}.$$

So we have that

$$\text{for every } j \in \{1, \ldots, l_i\}, \sum_{a_k \in \alpha^{ij}} x_k \Diamond_{ij} s_{ij} \tag{9}$$

Let $\mathcal{S}$ be the following linear system:

$$\sum_{k=1}^{n} z_k = 1$$

$$\sum_{a_k \in \alpha^{i1}} z_k \Diamond_{i1} s_{i1}$$

$$\vdots$$

$$\sum_{a_k \in \alpha^{il_i}} z_k \Diamond_{il_i} s_{il_i}$$

where the variables of the system are $z_1, \ldots, z_n$. We have the following:

(i) By Eqs. (8) and (9) the vector $\boldsymbol{x} = x_1, \ldots, x_n$ is a solution of $\mathcal{S}$.
(ii) From Eq. (5) every x_k is non-negative.
(iii) Every s_{ij} is a rational number with size at most $\|A\|$.
(iv) System $\mathcal{S}$ has at most $|A|$ equalities and inequalities.

From (i)–(iv) and Theorem 4 we have that there exists a vector $\boldsymbol{y} = y_1, \ldots, y_n$ such that:

(I) $\boldsymbol{y}$ is a solution of $\mathcal{S}$.
(II) every y_i is a non-negative rational number with size at most

$$2 \cdot (|A| \cdot \|A\| + |A| \cdot \log_2(|A|) + 1).$$

(III) at most $|A|$ entries of $\boldsymbol{y}$ are positive.

(IV) for all i, if $y_i > 0$ then $x_i > 0$.

Assume that $y_1, \ldots, y_N$ are the positive entries of $\boldsymbol{y}$ where

$$N \leq |A| \tag{10}$$

We define the quadruple $M = \langle W, H, \mu, * \rangle$ as follows:

(a) $W = \{w_1, \ldots, w_N\}$, for some $w_1, \ldots, w_N$.

(b) $H = \mathcal{P}(W)$.

(c) for all $V \in H$:

$$\mu(V) = \sum_{w_k \in V} y_k.$$

(d) Let $i \in \{1, \ldots, N\}$. We define $*_{w_i}$ to be some $\mathsf{J_{CS}}$-evaluation that satisfies the atom a_i. Since y_i is positive, by (IV), x_i is positive too, i.e. $\mu'([a_i]_{M'}) > 0$, which means that $[a_i]_{M'} \neq \emptyset$, i.e. that the atom a_i is $\mathsf{J_{CS}}$-satisfiable.

It holds:

$$\mu(W) = \sum_{w_k \in W} y_k$$

$$= \sum_{k=1}^{n} y_k$$

$$\overset{(I)}{=} 1$$

Let $U, V \in H$ such that $U \cap V = \emptyset$. It holds:

$$\mu(U \cup V) = \sum_{w_k \in U \cup V} y_k$$

$$= \sum_{w_k \in U} y_k + \sum_{w_k \in V} y_k$$

$$= \mu(U) + \mu(V)$$

Thus μ is a finitely additive measure. By Definitions 5 and 6 we have that $M \in \mathsf{PJ_{CS,Meas}}$.

We will now prove the following statement:

$$(\forall 1 \leq k \leq n)\left[w_k \in [\alpha^{ij}]_M \Longleftrightarrow a_k \in \alpha^{ij}\right] \tag{11}$$

Let $k \in \{1, \ldots, n\}$. We prove the two directions of Eq. (11) separately.

($\Longrightarrow$:) Assume that $w_k \in [\alpha^{ij}]$. This means that $*_{w_k} \Vdash \alpha^{ij}$. Assume that $a_k \notin \alpha^{ij}$. Then, since α^{ij} is a disjunction of atoms of A, there must exist some $a_{k'} \in \alpha^{ij}$, with $k \neq k'$, such that $*_{w_k} \Vdash a_{k'}$. However, by definition we have that $*_{w_k} \Vdash a_k$. But this is a contradiction, since a_k and $a_{k'}$ are different atoms of the same formula, which means that they cannot be satisfied by the same J_{CS}-evaluation. Hence, $a_k \in \alpha^{ij}$.

($\Longleftarrow$:) Assume that $a_k \in \alpha^{ij}$. We know that $*_{w_k} \Vdash a_k$, which implies that $*_{w_k} \Vdash \alpha^{ij}$, i.e. $w_k \in [\alpha^{ij}]_M$.

Hence, Eq. (11) holds. Now, we will prove the following statement:

$$(\forall 1 \le j \le l_i)\left[M \models P_{\Diamond_{ij}s_{ij}}\alpha^{ij}\right] \tag{12}$$

Let $j \in \{1, \dots, l_i\}$. It holds

$$M \models P_{\Diamond_{ij}s_{ij}}(\alpha^{ij}) \qquad\qquad \Longleftrightarrow$$
$$\mu([\alpha^{ij}]_M)\ \Diamond_{ij}\ s_{ij} \qquad\qquad \Longleftrightarrow$$
$$\sum_{w_k \in [\alpha^{ij}]_M} y_k \Diamond_{ij}\ s_{ij} \qquad\qquad \overset{(11)}{\Longleftrightarrow}$$
$$\sum_{a_k \in \alpha^{ij}} y_k\ \Diamond_{ij}\ s_{ij}$$

The last statement holds because of (I). Thus, Eq. (12) holds.

By Eq. (12) we have that $M \models \bigwedge_{j=1}^{l_i} P_{\Diamond_{ij}s_{ij}}(\alpha^{ij})$, which implies that

$$M \models \bigvee_{i=1}^{K} \bigwedge_{j=1}^{l_i} P_{\Diamond_{ij}s_{ij}}(\alpha^{ij}),$$

which, by Eq. (3), implies that $M \models A$.

Let $w_k \in W$. It holds:

$$\mu(\{w_k\}) = \sum_{w_i \in \{w_k\}} y_i = y_k \tag{13}$$

Now we will show that conditions (1)–(5) in the theorem's statement hold.

- Condition (1) holds because of (a) and Eq. (10).
- Condition (2) holds because of (b).
- Condition (3) holds because of Eq. (13) and (II).
- For every $V \in H$, because of Eq. (13), we have:

$$\mu(V) = \sum_{w_k \in V} y_k = \sum_{w_k \in V} \mu(\{w_k\}) \tag{14}$$

Hence condition (4) holds.
- By (d) every world of M satisfies a unique atom of α. Thus condition (5) holds.

So M is the model in question.

4   Complexity

Lemmas 2 and 3 can be proved by straightforward induction on the complexity of the formula. Lemma 2 tells us that if two J_{CS}-evaluations agree on some atom of a justification formula then they agree on the formula itself.

Lemma 2. *Let* CS *be any constant specification. Let* $\alpha \in \mathcal{L}_J$ *and let* a *be an atom of* α. *Let* $*_1, *_2$ *be two* J_{CS}-*evaluations and assume that*

$$*_1 \Vdash a \iff *_2 \Vdash a.$$

Then we have:

$$*_1 \Vdash \alpha \iff *_2 \Vdash \alpha.$$

Lemma 3. *Let* $\alpha \in \mathcal{L}_J$ *and let* a *be an atom of* α. *Let* $*$ *be a* J_{CS}-*evaluation and assume that* $* \Vdash a$. *The decision problem*

$$does \ * \ satisfy \ \alpha?$$

belongs to the complexity class P.

Kuznets [17] presented an algorithm for the J_{CS}-satisfiability problem for a total constant specification CS. Kuznets' algorithm is divided in two parts: the saturation algorithm and the completion algorithm. Let $\alpha \in \mathcal{L}_J$ be the formula that is tested for satisfiability.

- The saturation algorithm produces a set of requirements that should be satisfied by any J_{CS}-evaluation that satisfies α. The saturation algorithm operates in NP-time[8].
- The completion algorithm determines whether a J_{CS}-evaluation that satisfies α exists or not. The completion algorithm operates in $coNP$-time.

If the saturation and the completion algorithm are taken together, then we obtain a Σ_2^p-algorithm for the J_{CS}-satisfiability problem (for a total CS). The completion algorithm (adjusted to our notation) is stated in Theorem 6.

Theorem 6. *Let* CS *be a total constant specification. Let* a *be an atom of some* $\mathcal{L}_J$-*formula. The decision problem*

$$is \ a J_{CS}\text{-}satisfiable?$$

belongs to the complexity class $coNP$.

Now we are ready to prove the upper bound for the complexity of the $PJ_{CS,Meas}$-satisfiability problem.

Theorem 7. *Let* CS *be a total constant specification. The* $PJ_{CS,Meas}$-*satisfiability problem belongs to the complexity class* Σ_2^p.

---

[8] A reader unfamiliar with notions of computational complexity theory may consult a textbook on the field, like [24].

Proof. First we will describe an algorithm that decides the problem in question and we will explain its correctness. Then we will evaluate the complexity of the algorithm.

Algorithm

Let $A \in \mathcal{L}_\mathsf{P}$. It suffices to guess a small model $M = \langle W, H, \mu, * \rangle$ that satisfies A and also satisfies the conditions (1)–(5) that appear in the statement of Theorem 5. We guess M as follows: we guess n atoms of A, call them $a_1, \ldots, a_n$, and we also choose n worlds, $w_1, \ldots, w_n$, for $n \leq |A|$. Using Theorem 6 we verify that for each $i \in \{1, \ldots, n\}$ there exists a $\mathsf{J_{CS}}$-evaluation $*_i$ such that $*_i \Vdash a_i$. We define $W = \{w_1, \ldots, w_n\}$. For every $i \in \{1, \ldots, n\}$ we set $*_{w_i} = *_i$. Since we are only interested in the satisfiability of justification formulas that appear in A, by Lemma 2, the choice of the $*_{w_i}$ is not important (as long as $*_{w_i}$ satisfies a_i).

We assign to every $\mu(\{w_i\})$ a rational number with size at most:

$$2 \cdot \left(|A| \cdot ||A|| + |A| \cdot \log_2(|A|) + 1 \right).$$

We set $H = \mathcal{P}(W)$. For every $V \in H$ we set:

$$\mu(V) = \sum_{w_i \in V} \mu(\{w_i\}).$$

It is then straightforward to see that the conditions (1)–(5) that appear in the statement of Theorem 5 hold.

Now we have to verify that our guess is correct, i.e. that $M \models A$. Assume that $P_{\geq s}\alpha$ appears in A. In order to see whether $P_{\geq s}\alpha$ holds we need to calculate the measure of the set $[\alpha]_M$ in the model M. The set $[\alpha]_M$ will contain every $w_i \in W$ such that $*_{w_i} \Vdash \alpha$. Since $*_{w_i}$ satisfies an atom of A it also satisfies an atom of α. So, by Lemma 3, we can check whether $*_{w_i}$ satisfies α in polynomial time. If $\sum_{w_i \in [\alpha]_M} \mu(\{w_i\}) \geq s$ then we replace $P_{\geq s}\alpha$ in A with the truth value T, otherwise with the truth value F. We repeat the above procedure for every formula of the form $P_{\geq s}\alpha$ that appears in A. At the end we have a formula that is constructed only from the connectives $\neg$, $\wedge$ and the truth constants T and F. Using a truth table we can verify in polynomial time that the formula is true. This, of course implies that $M \models A$.

Complexity Evaluation

All the objects that are guessed in our algorithm have size that is polynomial on A. Also the verification phase of our algorithm can be made in polynomial time. Furthermore the application of Theorem 6 is possible with an NP-oracle (an NP-oracle can obviously decide $coNP$ problems too). Thus our algorithm is an NP^{NP} algorithm and since $\Sigma_2^p = NP^{NP}$ the claim of the Theorem follows.

5   Final Remarks and Conclusion

As a continuation of [15] and [16] we showed that results for justification logic and probabilistic logic can be nicely combined. Recall that the probabilistic justification logic PJ is obtained by adding probability operators to the justification logic J. In [17] it was proved that under some assumptions on the constant specification the complexity of the satisfiability problem for the logic J belongs to the class Σ_2^p. By Theorem 7 we have that, under the same assumptions on the constant specification, the complexity of the satisfiability problem for the logic PJ remains in the same complexity class. Hence, the probabilistic operators do not increase the complexity of the satisfiability problem, although they increase the expressiveness of the language.

As it is pointed out in [18], Theorem 6 holds for a decidable almost schematic constant specification. Theorem 7 uses Theorem 6 as an oracle. So, obviously Theorem 7 holds for a decidable almost schematic constant specification too.

The upper complexity bound we established is tight. By a result from [20] which was later strengthened in [8] and [1] we have that for a decidable, schematic and axiomatically appropriate constant specification CS the $\mathsf{J_{CS}}$-satisfiability problem is Σ_2^p-hard. For any $\alpha \in \mathcal{L}_\mathsf{J}$ it is not difficult to prove that:

$$\alpha \text{ is } \mathsf{J_{CS}}\text{-satisfiable} \Longleftrightarrow P_{\geq 1}\alpha \text{ is } \mathsf{PJ_{CS,Meas}}\text{-satisfiable} \tag{15}$$

Hence, the $\mathsf{J_{CS}}$-satisfiability problem can be reduced to the $\mathsf{PJ_{CS,Meas}}$-satisfiability problem, which implies that the $\mathsf{PJ_{CS,Meas}}$-satisfiability problem is Σ_2^p-hard too. Thus the $\mathsf{J_{CS}}$-satisfiabilty problem as well as the $\mathsf{PJ_{CS,Meas}}$-satisfiability problem are Σ_2^p-complete.

Observe that by Theorem 2 and our previous remarks we have that, for a decidable schematic and axiomatically appropriate constant specification, the derivability problem for the logic $\mathsf{PJ_{CS}}$ is Π_2^p-complete.

In [16] the probabilistic justification logic PPJ is defined. PPJ is a natural extension of PJ that supports iterations of the probability operator as well as justifications over probabilities. An interesting open problem related to the present work is to determine complexity bounds for PPJ.

Funding

The author is supported by the SNSF project 153169, *Structural Proof Theory and the Logic of Proofs*.

Acknowledgements. The author is grateful to Antonis Achilleos, Thomas Studer and the anonymous referees for valuable comments and remarks that helped him improve the quality of the paper substantially.

References

1. Achilleos, A.: Nexp-completeness and universal hardness results for justification logic, cSR 2015, pp. 27–52 (2015)
2. Artemov, S.N.: Operational modal logic. Technical report, MSI 95–29, Cornell University, December 1995
3. Artemov, S.N.: Explicit provability and constructive semantics. Bull. Symbolic Logic **7**(1), 1–36 (2001)
4. Artemov, S.N.: The ontology of justifications in the logical setting. Studia Logica **100**(1–2), 17–30 (2012). Published online, February 2012

5. Artemov, S.N., Fitting, M.: Justification logic. In: Zalta, E.N. (ed.) The Stanford Encyclopedia of Philosophy. Fall 2012 edn. (2012). http://plato.stanford.edu/archives/fall2012/entries/logic-justification/
6. Bucheli, S., Kuznets, R., Studer, T.: Justifications for common knowledge. J. Appl. Non-Classical Logics **21**(1), 35–60 (2011)
7. Bucheli, S., Kuznets, R., Studer, T.: Partial realization in dynamic justification logic. In: Beklemishev, L.D., de Queiroz, R. (eds.) WoLLIC 2011. LNCS, vol. 6642, pp. 35–51. Springer, Heidelberg (2011)
8. Buss, S.R., Kuznets, R.: Lower complexity bounds in justification logic. Ann. Pure Appl. Logic **163**(7), 888–905 (2012)
9. Chvátal, V.: Linear programming. W. H. Freeman and Company, New York (1983)
10. Fagin, R., Halpern, J., Megiddo, N.: A logic for reasoning about probabilities. Inf. Comput. **87**, 78–128 (1990)
11. Fan, T., Liau, C.: A logic for reasoning about justified uncertain beliefs. In: Yang, Q., Wooldridge, M. (eds.) Proceedings of IJCAI 2015, pp. 2948–2954. AAAI Press (2015)
12. Ghari, M.: Justification logics in a fuzzy setting. ArXiv e-prints, July 2014
13. Keisler, J.: Hyperfinite model theory. In: Gandy, R.O., Hyland, J.M.E. (eds.) Logic Colloquim 1976, p. 510. North-Holland (1977)
14. Kokkinis, I.: On the complexity of probabilistic justification logic. ArXiv e-prints (2015)
15. Kokkinis, I., Maksimović, P., Ognjanović, Z., Studer, T.: First steps towards probabilistic justification logic. Logic J. IGPL **23**(4), 662–687 (2015)
16. Kokkinis, I., Ognjanović, Z., Studer, T.: Probabilistic justification logic. In: Artemov, S., Nerode, A. (eds.) Symposium on Logical Foundations in Computer Science 2016 (2016, to appear)
17. Kuznets, R.: On the complexity of explicit modal logics. In: Clote, P.G., Schwichtenberg, H. (eds.) CSL 2000. LNCS, vol. 1862, pp. 371–383. Springer, Heidelberg (2000)
18. Kuznets, R.: Complexity Issues in Justification Logic. Ph.D. thesis, City University of New York, May 2008. http://gradworks.umi.com/33/10/3310747.html
19. Kuznets, R., Studer, T.: Justifications, ontology, and conservativity. In: Bolander, T., Braüner, T., Ghilardi, S., Moss, L. (eds.) Advances in Modal Logic, vol. 9, pp. 437–458. College Publications, London (2012)
20. Milnikel, R.S.: Derivability in certain subsystems of the logic of proofs is Π_2^p-complete. Ann. Pure Appl. Logic **145**(3), 223–239 (2007)
21. Milnikel, R.S.: The logic of uncertain justifications. Ann. Pure Appl. Logic **165**(1), 305–315 (2014)
22. Nilsson, N.: Probabilistic logic. Artif. Intell. **28**, 7187 (1986)
23. Ognjanović, Z., Rašković, M., Marković, Z.: Probability logics. Logic Comput. Sci. **12**(20), 35–111 (2009). Zbornik radova subseries
24. Papadimitriou, C.H.: Computational Complexity. Addison-Wesley, Reading (1994)

Relational Complexity and Higher Order Logics

José Maria Turull-Torres[1,2](✉)

[1] Depto. de Ingeniería e Investigaciones Tecnológicas,
Universidad Nacional de La Matanza, Buenos Aires, Argentina
J.M.Turull@massey.ac.nz
[2] Massey University, Palmerston North, New Zealand

Abstract. Relational machines (RM) were introduced as abstract machines that compute queries to relational database instances (dbi's), that are *generic* (i.e., that preserve isomorphisms). As RM's cannot discern between tuples that are equivalent in first order logic with k variables, *Relational Complexity* was introduced as a complexity theory where the input dbi to a query is measured as its $size_k$, i.e., as the number of classes in the equivalence relation of equality of FO^k types of k-tuples in the dbi. We describe the basic notions of Relational Complexity, and survey known characterizations of some of its main classes through different fixed point logics and through fragments of second and third order logics.

1 Introduction

Relational machines (RM) were introduced in [3] (there called *loosely coupled generic machines*) as abstract machines that compute queries to (finite) relational structures, or relational database instances (dbi's) as functions from such structures to relations, that are *generic* (i.e., that preserve isomorphisms), and hence are more appropriate than Turing machines (TM) for query computation. RM's are TM's endowed with a relational store that holds the input structure, as well as *work* relations, and that can be accessed through first order logic (FO) queries (sentences) and updates (formulas with free variables). As the set of those FO formulas for a given machine is fixed, an RM can only distinguish between tuples (i.e., sequences of elements in the domain of the dbi) when the differences between them can be expressed with FO formulas with k variables, where k is the maximum number of variables in any formula in the finite control of the given RM. Note that the same is true for FO queries (i.e., relational calculus), or equivalently relational algebra queries.

On the other hand, it has been proved that RM's have the same computation, or expressive power, as the (effective fragment of the) well known infinitary logic with finitely many variables $L^\omega_{\infty\omega}$ [1], in the context of Finite Model Theory, i.e., with sentences interpreted by finite relational structures or database instances—dbi's. This logic extends FO with conjunctions and disjunctions of sets of formulas of arbitrary (infinite) cardinality, while restricting the number of variables in each (infinitary) formula to be finite. This is a very important logic

© Springer International Publishing Switzerland 2016
M. Gyssens and G. Simari (Eds.): FoIKS 2016, LNCS 9616, pp. 311–333, 2016.
DOI: 10.1007/978-3-319-30024-5_17

in *Descriptive Complexity theory*, in which among other properties, equivalence is characterized by pebble (Ehrenfeucht-Fraïssé) games, and on ordered dbi's it can express all computable queries (see [20], among others). Hence, a nice characterization of the *discerning* power of RM's is also given by those games.

Consequently, k-*ary* RM's are incapable of computing the size of the input structure though, however, they can compute its $size_k$. A k-ary RM, for a positive integer k, is an RM in which the FO formulas in its finite control have at most k different variables, and the $size_k$ of a structure (or dbi) is the number of equivalence classes in the relation $\equiv^k$ of equality of FO^k types in the set of k-tuples of the structure, for $1 \leq k$.

Then, it was a natural consequence to define a new notion of complexity suitable for RM's. *Relational Complexity* was introduced in the original work in 1991 as a complexity theory where the (finite relational) input structure $\mathcal{A}$ to a query is measured as its $size_k$, for some $k \geq 1$, instead of the size of its encoding, as in Computational Complexity. Roughly, two k-tuples in $\mathcal{A}$ have the same FO^k types if they both satisfy in $\mathcal{A}$ exactly the same FO formulas with up to k variables, r of them being free, for all $0 \leq r \leq k$. That is, if the two tuples have the same properties in the structure $\mathcal{A}$, considering only the properties that can be expressed in FO^k. In that way, relational complexity classes mirroring computational complexity classes like P, NP, $PSPACE$, and $EXPTIME$, etc., were defined in [2,3], and denoted as P_r, NP_r, $PSPACE_r$, and $EXPTIME_r$, respectively.

Beyond the study of RM's as a model of computation for queries to relational databases, Relational Complexity turned out to be a theoretical framework in which we can characterize *exactly* the expressive power of the well known *fixed point quantifiers* (FP) of a wide range of types. Those quantifiers have been typically added to first order logic, thus forming the so called *fixed point logics*, where the different types of fixed point quantifiers add to FO different kinds of iterations of first-order operators [2,20].

In [2], Abiteboul, Vardi, and Vianu introduced new fixed point quantifiers, and organized a wide range of them as either deterministic (det), non-deterministic (ndet), or alternating (alt), and either inflationary (inf) or non-inflationary (ninf), according to the type of iteration implied by the semantics of each such quantifier. In the same article, they proved the following equivalences: det-inf-FP = P_r, ndet-inf-FP = NP_r, alt-inf-FP = det-ninf-FP = ndet-ninf-FP = $PSPACE_r$, and alt-ninf-FP = $EXPTIME_r$.

Those characterizations of *relational* complexity classes are actually very interesting and meaningful, given that it was already known that if we restrict the input to only *ordered* structures, the following equivalences with *computational* complexity classes hold: det-inf-FP = P, ndet-inf-FP = NP, det-ninf-FP = ndet-ninf-FP = alt-inf-FP = $PSPACE$, and alt-ninf-FP = $EXPTIME$ [2,20].

Regarding the characterization of relational complexity classes with other logics, Dawar introduced in [6] the logic SO^ω as a semantic restriction of second order logic (SO) where the valuating relations for the quantified second order variables are "unions" of complete FO^k types for r-tuples for some constants

$k \geq r \geq 1$, that depend on the quantifiers[1]. That is, the relations are *closed* under the relation $\equiv^k$ of equality of FO^k types in the set of r-tuples of the structure.

In [6], it was also proved that the existential fragment of SO^ω, $\Sigma_1^{1,\omega}$, characterizes exactly the non-deterministic fixed point logic $(FO + NFP)$, and, hence, by the equivalences mentioned above, it turned out that $\Sigma_1^{1,\omega}$ captures NP_r, analogously to the well-known relationship $\Sigma_1^1 = NP$ [10]. Continuing the analogy, the characterization of the relational polynomial time hierarchy PH_r with *full* SO^ω was stated without proof in [6], and later proved by us jointly with Ferrarotti in [13].

In [5], aiming to characterize higher relational complexity classes, and as a natural continuation of the study of the logic SO^ω, in a joint work with Arroyuelo, we defined a variation of third order logic (TO) denoted as TO^ω, under finite interpretations. We defined it as a semantic restriction of TO where the (second order) relations which form the tuples in the third order relations that valuate the quantified third order variables are *closed* under the relation $\equiv^k$ as above. In [5], we also introduced a variation of the non deterministic relational machine, which we denoted 3-NRM (for third order NRM), where we allow TO relations in the relational store of the machine. We defined the class $NEXPTIME_{3,r}$ as the class of 3-NRM's that work in time exponential in the $size_k$ (see above) of the input dbi. We then proved that the existential fragment of TO^ω, denoted $\Sigma_1^{2,\omega}$, captures $NEXPTIME_{3,r}$.

Finally, in [22], we proved a stronger result: we showed that the existential fragment of TO^ω *also captures* the relational complexity class $NEXPTIME_r$.

As it turned out that $NEXPTIME_r = NEXPTIME_{3,r}$, an interesting consequence of our result is that RM's in their original formulation are strong enough as to *simulate* the existence of TO relations in their relational store and, hence, to also *simulate* the existence of TO^ω formulas in their finite control (without TO^ω or SO^ω quantifiers, as in 3-NRM's in [5], see below). That is, for every 3-NRM that works in time $NEXPTIME_{3,r}$, i.e., relational *third order* exponential time, in the $size_k$ of their input, there is an NRM that computes the same query, and that works in time $NEXPTIME_r$, i.e., relational exponential time in the $size_k$ of their input. Nevertheless, we think that we still need 3-NRM's and $NEXPTIME_{3,r}$ to work with *oracle* NRM's with third order relations.

Beyond the natural theoretical relevance in creating and studying new logics as computation models, and thus getting information on new aspects of the problems that can be expressed in them, an important application of the creation of new logics to Complexity Theory is the search for lower bounds of problems w.r.t. those logics, aiming to separate computational complexity classes.

This article is organized as follows. In Sect. 2, we give the basic definitions for the setting of Finite Model Theory, and for Second Order Logic. In Sect. 3, we describe various fixed point quantifiers and their expressibility on ordered structures. In Sect. 4, we describe relational machines, the infinitary logic $L_{\infty\omega}^\omega$, and the notion of type of a tuple; we also introduce relational complexity and give

---

[1] In the sense of [12] these relations are *redundant* relations.

the relationship between some of the main classes and the different fixed point logics. In Sect. 5, we describe the logic SO^ω and its expressibility w.r.t. relational complexity. In Sect. 6, we describe the logic TO^ω, define third order NRM's, and give the expressive power of $\Sigma_1^{2,\omega}$ w.r.t. relational complexity. Finally, in Sect. 7, we mention a few examples of known lower bounds for problems w.r.t. some of those logics.

2 Preliminaries

Basic Definitions

We assume a basic knowledge of Logic and Finite Model Theory (refer to [20]). By a logic, we mean, informally, the usual notion in FMT. We only consider vocabularies of the form $\sigma = \langle R_1, \ldots, R_s \rangle$ (i.e., *purely relational*), where the arities of the relation symbols are $r_1, \ldots, r_s \geq 1$, respectively. We can also have constant symbols in σ, and in some definitions we will consider them, but to make the presentation simpler, in the main results and definitions we will avoid them here. We assume that the vocabularies always contain a symbol for *equality* ($=$). We consider only *finite* σ structures, or relational database instances (dbi's) denoted as $\mathcal{A} = \langle A, R_1^{\mathcal{A}}, \ldots, R_s^{\mathcal{A}} \rangle$, where A is the domain, also denoted $dom(\mathcal{A})$, and $R_1^{\mathcal{A}}, \ldots, R_s^{\mathcal{A}}$ are (second order) relations in $A^{r_1}, \ldots, A^{r_s}$, respectively. We denote as $Str[\sigma]$ the class of *finite* σ structures. By a *set* (or *class*) of structures we mean a set of structures closed under isomorphisms. We will use lower case Roman letters like x and y for individual (i.e., FO) variables, upper case Roman letters like X and Y for second order relation variables, and calligraphic upper case letters like $\mathcal{X}$ and $\mathcal{Y}$ for third order relation variables. By $\varphi(x_1, \ldots, x_r)$ we denote a formula of a logic whose free variables are *exactly* $\{x_1, \ldots, x_r\}$. Technically, to avoid the consideration of the free variables in a formula as a set, we always assume that the set of variables in a logic is totally ordered. Then, whenever we write $\varphi(x_1, \ldots, x_r)$ we assume that the sequence $\langle x_1, \ldots, x_r \rangle$ follows *that order*. Note that, however, in the formula we may use the variables in any arbitrary order.

Queries or Global Relations

Let σ be a vocabulary. Two σ structures $\mathcal{A}$ and $\mathcal{B}$ are *isomorphic*, written $\mathcal{A} \cong \mathcal{B}$, if there is an *isomorphism* from $\mathcal{A}$ to $\mathcal{B}$, i.e., a bijection $h : A \to B$ that preserves relations and constants in σ, that is, (i) for r-ary $R \in \sigma$, and $a_1, \ldots, a_r \in A$, $(a_1, \ldots, a_r) \in R^{\mathcal{A}}$ if and only if $(h(a_1), \ldots, h(a_r)) \in R^{\mathcal{B}}$, and (ii) for every constant $c \in \sigma$, $h(c^{\mathcal{A}}) = c^{\mathcal{B}}$. Let $r \geq 1$, and let R be a relation symbol of arity r. A *query* or *global relation of arity* r and vocabulary σ is a function $q : Str[\sigma] \to Str[\langle R \rangle]$ such that: (i) q preserves isomorphisms, i.e., for every pair of σ structures $\mathcal{A}$ and $\mathcal{B}$, and for every isomorphism $h : dom(\mathcal{A}) \longrightarrow dom(\mathcal{B})$, $q(\mathcal{B}) = h(q(\mathcal{A}))$, (ii) for every σ structure $\mathcal{A}$, $dom(q(\mathcal{A})) \subseteq dom(\mathcal{A})$ (that is, all the elements which form the output to the query q when evaluated on a given structure $\mathcal{A}$ must belong to the domain of that structure). A *Boolean query* is a function $q : Str[\sigma] \rightarrow \{\text{TRUE}, \text{FALSE}\}$ that preserves isomorphisms, i.e., for every pair of σ structures $\mathcal{A}$ and $\mathcal{B}$, if they are isomorphic, then $q(\mathcal{A}) = q(\mathcal{B})$. A *computable* (or *recursive*) query is a query that is recursive in some encoding of $\mathcal{A}$. If $\varphi(x_1, \ldots, x_r)$ is a σ formula as above, of some logic $\mathcal{L}$, $\mathcal{A}$ is a σ structure, and $a_1, \ldots, a_r$ are elements of the domain of $\mathcal{A}$, with $\mathcal{A} \models \varphi(x_1, \ldots, x_r)[a_1, \ldots, a_r]$ we denote that φ is TRUE, when interpreted by $\mathcal{A}$, under a valuation v where for $1 \leq i \leq r$ it is $v(x_i) = a_i$. Now we consider the set of all such valuations as follows: $\varphi^{\mathcal{A}} = \{(a_1, \ldots, a_r) : a_1, \ldots a_r \in dom(\mathcal{A}) \land \mathcal{A} \models \varphi(x_1, \ldots, x_r)[a_1, \ldots, a_r]\}$ That is, $\varphi^{\mathcal{A}}$ is the *relation defined by* φ *in the structure* $\mathcal{A}$, and its arity is given by the number of free variables in φ. Formally, a σ formula $\varphi(x_1, \ldots, x_r)$, *expresses a* σ query q if, for every σ structure $\mathcal{A}$, $q(\mathcal{A}) = \varphi^{\mathcal{A}}$. Similarly, a sentence φ expresses a Boolean query q if, for every σ structure $\mathcal{A}$, $q(\mathcal{A}) = \text{TRUE}$ if and only if $\mathcal{A} \models \varphi$. In both cases, we say that q is *definable*, or *expressible*, in $\mathcal{L}$ (by φ). A logic $\mathcal{L}$ *captures* a complexity class $\mathcal{C}$ if and only if every class of relational structures definable in $\mathcal{L}$ is in $\mathcal{C}$ and vice versa. We denote as $\mathcal{L}_1 \subseteq \mathcal{L}_2$ the fact that every query expressible in $\mathcal{L}_1$ is also expressible in $\mathcal{L}_2$; correspondingly, we also use the relations $\subset, \supseteq, \supset$, and $=$ with the obvious meaning.

Second Order Logic

Second order logic (SO) is an extension of first order logic (FO) which allows to quantify over relations. In addition to the symbols of FO, its alphabet contains, for each $n \geq 1$, countably many n-ary *relation variables*. As usual, we will use upper case letters to denote SO relation variables. We define the set of SO formulas of vocabulary σ to be the set generated by the rules for FO formulas *extended* by: (i) if X is a relation variable of arity n and $t_1, \ldots, t_n$ are terms, i.e., individual variables or constants, then $X(t_1, \ldots, t_n)$ is a formula; (ii) if φ is a formula and X is a relation variable, then $\exists X(\varphi)$ and $\forall X(\varphi)$ are formulas. The free occurrence of a relation variable in an SO formula is defined in the obvious way and the notion of satisfaction is extended canonically. The informal semantics of $\exists X(\varphi)$ and $\forall X(\varphi)$ over a relational structure $\mathcal{A}$, where X is a relation variable of arity r, is "there is at least one relation $R \subseteq A^r$ such that φ is true when X is interpreted by R", and "for every relation $R \subseteq A^r$, φ is true when X is interpreted by R", respectively. Given a (finite) σ structure $\mathcal{A}$, a formula $\varphi \equiv \varphi(x_1, \ldots, x_n, X_1, \ldots, X_k)$ with free individual variables $x_1, \ldots, x_n$ and free relation variables $X_1, \ldots, X_k$, elements $a_1, \ldots, a_n \in A$, and relations $R_1, \ldots, R_k$, over $\mathcal{A}$ of arities corresponding to $X_1, \ldots, X_k$, respectively, we say that $\mathcal{A} \models \varphi[a_1, \ldots, a_n, R_1, \ldots, R_k]$ if the elements $a_1, \ldots, a_n$ together with $R_1, \ldots, R_k$ satisfy φ in $\mathcal{A}$. It is well known that every SO formula is logically equivalent to one in *prenex normal form* in which each SO quantifier precedes all FO quantifiers. Let $m \geq 1$, then such a formula is called Σ^1_m, if the string of SO quantifiers consists of m consecutive blocks, where in each block all quantifiers are of the same type (i.e., all universal or all existential), adjacent blocks contain quantifiers of different type, and the first block is *existential*. Π^1_m is defined in the same way, except that the first block consists of *universal* quantifiers. The following well known results give the expressive power of SO and the fragments defined above:

Theorem 1 [10]. *(i)* Σ_1^1 *captures NP; (ii)* Π_1^1 *captures co-NP.*

Theorem 2 [21]. *Let* $m \geq 1$. *(i)* Σ_m^1 *captures* Σ_m^P; *(ii)* Π_m^1 *captures* Π_m^P; *(iii)* SO *captures PH.*

3   Fixed Point Quantifiers

We will mainly follow the definitions and notation as in [2,20]. Let $\varphi(x_1, \ldots, x_k, X)$ be a σ formula, where X is a k-ary relation symbol. Together with a dbi $\mathcal{A} \in Str[\sigma]$, φ gives rise to an operation: F^φ : $\mathcal{P}(dom(\mathcal{A})^k) \rightarrow \mathcal{P}(dom(\mathcal{A})^k)$ defined by $F^\varphi(R) = \{(a_1, \ldots, a_k) : \mathcal{A} \models \varphi(x_1, \ldots, x_k, X)[a_1, \ldots, a_k, R]\}$, where R is a k-ary relation in $dom(\mathcal{A})$. Let S be a k-ary relation in $dom(\mathcal{A})$. Then S is a *fixed point* of the operator F^φ if $F^\varphi(S) = S$. We denote the sequence $\phi, F^\varphi(\phi), F^\varphi(F^\varphi(\phi)), \ldots$ by $F_0^\varphi, F_1^\varphi, F_2^\varphi, \ldots$ and $F_{n+1}^\varphi = F^\varphi(F_n^\varphi)$. If there is an n_0 such that $F_{n_0}^\varphi = F_{n_0+1}^\varphi$, we denote $F_{n_0}^\varphi$ by F_∞^φ. Note that, for all $n \geq n_0$, $F_n^\varphi = F_\infty^\varphi$. However, such n_0 might not exist, i.e., the sequence $F_0^\varphi, F_1^\varphi, F_2^\varphi, \ldots$ might have *no fixed point*. We define next three important properties of the operator F^φ: (i) F^φ is *inductive* if $F_0^\varphi \subseteq F_1^\varphi \subseteq F_2^\varphi \subseteq \ldots$; (ii) F^φ is *inflationary* if for all $R \subseteq dom(\mathcal{A})^k : R \subseteq F^\varphi(R)$; (iii) F^φ is *monotone* if for all $R, S \subseteq dom(\mathcal{A})^k$: $R \subseteq S \Rightarrow F^\varphi(R) \subseteq F^\varphi(S)$. If F^φ is *inflationary* or *monotone*, then F^φ is *inductive*. If F^φ is *inductive*, then $F_\infty^\varphi = \bigcup_{i=0}^{n_0} F_i^\varphi$, is the fixed point of the sequence $F_0^\varphi, F_1^\varphi, F_2^\varphi, \ldots$. If F^φ is *monotone*, then the operator F_∞^φ has a *least fixed point*, which is defined as $\bigcap \{Y : F^\varphi(Y) = Y\}$. Furthermore, the least fixed point of the operator F^φ is $F_\infty^\varphi = \bigcup_{i=0}^{n_0} F_i^\varphi$, as above. That is, F_∞^φ is a fixed point of F^φ, and it is included in all fixed points of F^φ. F_∞^φ is *the* fixed point of the sequence $\phi, F^\varphi(\phi), F^\varphi(F^\varphi(\phi)), \ldots$, and it is included in all the fixed points of the operator F^φ. Though it is undecidable whether, for a given $\varphi \in$ FO, F^φ is monotone (see [20]), it has been proved that if φ *is positive on* X, then F^φ is *monotone* (see [20]). Then, we define the *least fixed point* (**LFP**) of F^φ restricting φ to be *positive* on X. That is, the least fixed point of F^φ, where φ is positive on X, is F_∞^φ, as defined above. *Least (deterministic) fixed point logic* (LFP) is the closure of FO under the operation of taking deterministic least fixed points (i.e., where the deterministic fixed point is applied to formulas that are *positive* on the relation variable binded by the quantifier). We denote this logic as (FO + *LFP*).

We define the *non-inflationary (deterministic) fixed point* of F^φ as $F_\infty^\varphi = \bigcup_{i=0}^{n_0} F_i^\varphi$, as above, if the fixed point of the sequence $F_0^\varphi, F_1^\varphi, F_2^\varphi, \ldots$ exists. If it does not exist, then we define $F_\infty^\varphi = \phi$. *Non-inflationary (deterministic) fixed point logic* (det-ninf-FP) is the closure of FO under the operation of taking deterministic non-inflationary fixed points (i.e., where the deterministic fixed point is applied to *arbitrary formulas*). We denote this logic as (FO + det-ninf-FP).

In particular, φ is *inflationary in* X if it is of the form $X(\bar{x}) \vee \psi$, where ψ is an arbitrary formula (we will call formulas of such form *inflationary formulas*).

Hence it is also *inductive*, and the dbi I being finite, the sequence has always a fixed point, which is $F_\infty^\varphi = \bigcup_{i=0}^{n_0} F_i^\varphi$, as above. *Inflationary (deterministic) fixed point logic* (det-inf-FP) is the closure of FO under the operation of taking deterministic inflationary fixed points (i.e., where the deterministic fixed point is applied to *inflationary formulas*). We denote this logic as (FO + det-inf-FP). Note that det-ninf-FP fixpoint logic is an extension of det-inf-FP. In [17], it was proved that on finite structures it is *IFP = LFP*. For that reason, we will not consider *LFP* in this context.

Note that the three fixed point quantifiers defined above are obtained by *deterministically* iterating FO operators. In [2], new fixed point quantifiers were introduced, namely, the *non-deterministic* and the *alternating* fixed point quantifiers, including also their inflationary and non inflationary versions. Then, considering both the new fixed point quantifiers, as well as those described above, they parameterized all of them as either *deterministic* (det), *non-deterministic* (ndet), or *alternating* (alt), according to the power of their iteration construct, and either *inflationary* (inf) or *non-inflationary* (ninf), according to the power of their FO operators.

Non-deterministic Fixed Point

Given $k \geq 1$ and two FO formulas $\varphi_0(x_1, \ldots, x_k, X)$ and $\varphi_1(x_1, \ldots, x_k, X)$ of a same vocabulary σ, we define a sequence of stages $F_b^{(\varphi_0,\varphi_1)}$ indexed by binary strings $b \in \{0,1\}^*$, as follows: (i) $F_\lambda^{(\varphi_0,\varphi_1)} = \emptyset$ for the empty string λ; (ii) $F_{b\cdot 0}^{(\varphi_0,\varphi_1)} = F_b^{(\varphi_0,\varphi_1)} \cup F^{\varphi_0}(F_b^{(\varphi_0,\varphi_1)})$; (iii) $F_{b\cdot 1}^{(\varphi_0,\varphi_1)} = F_b^{(\varphi_0,\varphi_1)} \cup F^{\varphi_1}(F_b^{(\varphi_0,\varphi_1)})$. The *non-deterministic fixed point* of the sequence is $\bigcup_{b \in \{0,1\}^*} F_b^{(\varphi_0,\varphi_1)}$. The *non-deterministic inflationary fixed point logic* (ndet-inf-FP) is the closure of FO under the operation of taking non-deterministic inflationary fixed points (i.e., where both, φ_0 and φ_1, are *inflationary formulas*). The *non-deterministic non-inflationary fixed point logic*(ndet-ninf-FP) is the closure of FO under the operation of taking non-deterministic non-inflationary fixed points (i.e., where both, φ_0 and φ_1, are *arbitrary formulas*). If no non-deterministic fixed point exists, then we define the non-deterministic fixed point to be the *empty set*. In the two cases there is the restriction that *negation cannot be applied* to the fixed point operator. We denote these two logics as (FO + ndet-inf-FP) and (FO + ndet-ninf-FP), respectively.

Alternating Fixed Point

Let $k \geq 1$ and let $\varphi_0(x_1, \ldots, x_k, X)$ and $\varphi_1(x_1, \ldots, x_k, X)$ be two FO formulas of a same vocabulary σ. This pair of operators generates *convergent trees* of stages that are obtained by successively applying, until convergence is reached, either one of φ_0 and φ_1, or both of φ_0 and φ_1. Formally, a convergent tree is a labeled binary tree such that (i) the root is labeled by the empty relation; (ii) if a node x with label S_x is at an odd level of the tree, then x has one child x', labeled by $F^{\varphi_0}(S_x)$ or $F^{\varphi_1}(S_x)$; (iii) if a node x with label S_x is at an even level of the tree, then x has two children x_1 and x_2 labeled by $F^{\varphi_0}(S_x)$ and $F^{\varphi_1}(S_x)$, respectively; (iv) if x is a leaf with label S_x, then $F^{\varphi_0}(S_x) = F^{\varphi_1}(S_x) = S_x$. We take the *intersection* of the labels of the leaves of a convergent stage tree to be a *local alternating fixed point* of the pair φ_0, φ_1. Note that the pair φ_0, φ_1 can have more than one local alternating fixed point or none. We define the *alternating fixed point* of the pair φ_0, φ_1 as the *union* of all local alternating fixed points of the pair φ_0, φ_1. If no local alternating fixed point exists, then we define the alternating fixed point to be the *empty set*. The *number* of stages of an alternating fixed point is taken to be the *maximum* over all convergent trees. The *alternating inflationary fixed point logic* (alt-inf-FP) is the closure of FO under the operation of taking alternating inflationary fixed points (i.e., where both, φ_0 and φ_1, are *inflationary formulas*). The *alternating non-inflationary fixed point logic* (alt-ninf-FP) is the closure of FO under the operation of taking alternating non-inflationary fixed points (i.e., where both, φ_0 and φ_1, are *arbitrary formulas*). We denote these two logics as (FO+alt-inf-FP) and (FO+alt-ninf-FP), respectively.

Expressibility of the Fixed Point Logics

In the context of Descriptive Complexity Theory, the following characterizations with *computational complexity classes* have been proved, when we restrict the input structures (or dbi's) to those with a *total order* relation in its signature (relational schema) [2, 20]. We denote, as usual, the restriction to ordered structures by adding $\leq$ to the identification of the logic.

(i) $P = (\text{FO} + \leq + \text{det-inf-FP})$;
(ii) $NP = (\text{FO} + \leq + \text{ndet-inf-FP})$;
(iii) $PSPACE = (\text{FO} + \leq + \text{det-ninf-FP})$
            $= (\text{FO} + \leq + \text{ndet-ninf-FP})$
            $= (\text{FO} + \leq + \text{alt-inf-FP})$;
(iv) $EXPTIME = (\text{FO} + \leq + \text{alt-ninf-FP})$.

4 Relational Machines and the Infinitary Logic $L^{\omega}_{\infty\omega}$

A *deterministic relational machine* is a Turing machine augmented with a *finite* set of *fixed arity* relations forming a *relational store* (rs). Designated relations contain initially the input structure, and one specific relation holds the output at the end of the computation. A relational machine uses a *finite* set of first-order formulas to interact with the rs. An RM is an eleven-tuple $\langle Q, \Sigma, \delta, q_0, \flat, F, \tau, \sigma, T, \Omega, \Phi \rangle$, where: Q is the finite set of states; Σ is the tape alphabet; $\flat \in \Sigma$ is the blank symbol; $q_0 \in Q$ is the initial state; $F \subseteq Q$ is the set of accepting final states; τ is the vocabulary of the rs; $\sigma \subset \tau$ is the vocabulary of the input structure; $T \in \tau - \sigma$ is the output relation; Ω is a finite set of first-order sentences of vocabulary τ; Φ is a finite set of first-order formulas of vocabulary τ; $\delta : Q \times \Sigma \times \Omega \to \Sigma \times Q \times \{R, L\} \times \Phi \times \tau$ is a partial function called the transition function. Note that, as equality is always included in σ, we can access the domain of the input structure with the formula $x = x$. *Transitions* are based on the current state; the contents of the current tape cell; and the answer to a Boolean first-order query evaluated on the τ-structure held in the *rs*. If the transition function is undefined, the computation *stops*. The *arity* of an RM M, denoted as $arity(M)$, is $\max(\{|var(\varphi)| : \varphi \in \Omega \cup \Phi\})$. We will always assume that the arity of M is *greater than or equal to* the arities of all the relation symbols in its *rs*. A *relational language* is a class of structures of a relational vocabulary that is closed under isomorphisms. The relational language *accepted* by M, denoted $L(M)$ is the set of input structures accepted by M.

A *non-deterministic relational machine* is an eleven-tuple, $\langle Q, \Sigma, \delta, q_0, \flat, F, \sigma, \tau, T, \Omega, \Phi \rangle$, where each component is as in the deterministic case, with the exception that the transition function is defined by $\delta : Q \times \Sigma \times \Omega \rightarrow \mathcal{P}(\Sigma \times Q \times \{R, L\} \times \Phi \times \tau)$, where, for any set A, $\mathcal{P}(A)$ denotes the powerset of A.

A *relational oracle machine* is a relational machine with a distinguished set of relations in its *rs*, called *oracle relations*, and three distinguished states $q_?$, the *query* state, and q_{YES}, q_{NO}, the *answer* states. Similarly to the case of oracle Turing machines, the computation of an oracle relational machine requires that a *relational oracle language* be fixed previously to the computation. Let $\mathcal{C}$ be a relational language. The computation of a relational oracle machine M with oracle $\mathcal{C}$ and distinguished set of oracle relation symbols σ^o, proceeds like in an ordinary relational machine, except for transitions from the query state. From the query state M transfers into the state q_{YES} if the relational structure of vocabulary σ^o formed by the *domain* of the input structure and the distinguished set of oracle relations currently held in the *rs*, belongs to $\mathcal{C}$; otherwise, M transfers into the state q_{NO}.

The Infinitary Logic $L_{\infty\omega}^{\omega}$

Let $s \geq 1$ and $m \geq 0$. We denote by $\mathrm{FO}^{s;m}$ the fragment of FO containing only formulas whose free and bound variables are among $\{v_1, \ldots, v_s\}$, and of quantifier rank $\leq m$. We further denote by FO^s the fragment of FO given by $\bigcup_{m \geq 0} \mathrm{FO}^{s;m}$. The infinitary logic $L_{\infty\omega}^s$ extends FO by allowing arbitrary (infinite) disjunctions and conjunctions, while restricting the free and bound variables in the formulas to be among $\{v_1, \ldots, v_s\}$. We define $L_{\infty\omega}^{\omega} = \bigcup_{s \geq 1} L_{\infty\omega}^s$. As non-recursive queries can be expressed in these logics, we will denote with $L_{\infty\omega}^s|rec$ and $L_{\infty\omega}^{\omega}|rec$ the fragments of $L_{\infty\omega}^s$ and $L_{\infty\omega}^{\omega}$, respectively, that express only *recursive queries*. The semantics is a direct extension of the semantics of FO, with $\bigvee \Psi$ and $\bigwedge \Psi$ being interpreted as the disjunction and conjunction, respectively, over all formulas in Ψ. If all formulas in Ψ are sentences, then $\mathcal{A} \models \bigvee \Psi$ if and only if for some $\psi \in \Psi$, $\mathcal{A} \models \psi$. and similarly for $\bigwedge \Psi$. Note that the q.r. of a formula in these logics can be *infinite*. In our setting, we only consider $L_{\infty\omega}^{\omega}$ formulas over finite signatures.

Proposition 1 [20]. *Let σ be a relational vocabulary such that it contains a binary relation symbol $\leq$, that is interpreted as a* total order *in any given σ*

structure, and such that the arity of the relation symbol of maximum arity *in* σ *is* k, *for* $k \geq 2$. *Then, over* finite σ *structures, (i) every Boolean query is expressible in* $L_{\infty\omega}^k$; *(ii) every* k-ary *query is expressible in* $L_{\infty\omega}^k$.

Proposition 2 [1,4]. *(i)* $RM = L_{\infty\omega}^\omega|_{rec}$, *i.e., the relational machines can compute exactly the same queries that can be expressed in the effective fragment of the logic* $L_{\infty\omega}^\omega$; *(ii) relational machines are complete on* ordered *input structures; (iii) relational machines* collapse to FO *on vocabularies with only* unary *relation symbols (and equality).*

Note that since $L_{\infty\omega}^\omega$ has the so called *0-1 Law* (i.e., all Boolean queries expressible in the logic are either asymptotically true, or asymptotically false, see [20] among other sources), the equivalence (i) means that the relational machines also have the *0-1 Law*. This implies, for instance, that a query as simple as the parity query cannot be computed by an RM, while on the other hand from the perspective of Computational Complexity it is a regular language, and hence is in $DSPACE(O(1))$.

Types, or Properties of Tuples

For any l-tuple $\bar{a} = (a_1, \ldots, a_l)$ of elements in A, with $1 \leq l \leq k$, we define the FO^k type of $\bar{a}$, denoted $tp_{\mathcal{A}}^{FO^k}(\bar{a})$, to be the set of FO^k formulas $\varphi \in FO^k$ with free variables among $x_1, \ldots, x_l$, such that $\mathcal{A} \models \varphi[a_1, \ldots, a_l]$. If τ is an FO^k type, tuple $\bar{a}$ *realizes* τ in $\mathcal{A}$ if and only if $\tau = tp_{\mathcal{A}}^{FO^k}(\bar{a})$. Let $\mathcal{A}$ and $\mathcal{B}$ be σ structures and let $\bar{a}$ and $\bar{b}$ be two l-tuples on $\mathcal{A}$ and $\mathcal{B}$ respectively, we write $(\mathcal{A}, \bar{a}) \equiv^k (\mathcal{B}, \bar{b})$, to denote that $tp_{\mathcal{A}}^{FO^k}(\bar{a}) = tp_{\mathcal{B}}^{FO^k}(\bar{b})$. If $\mathcal{A} = \mathcal{B}$, we also write $\bar{a} \equiv^k \bar{b}$. We denote as $size_k(\mathcal{A})$ the number of equivalence classes in $\equiv^k$ in $\mathcal{A}$.

Characterization of Equality of FO^k Types

Equality of FO^k-types can be characterized by pebble games and by k-Back and Forth Systems of partial isomorphisms. We will first give a brief explanation of the games, together with the basic notions that we need to define them, and then the characterization theorem. Let σ be a vocabulary, and $\mathcal{A}$, $\mathcal{B}$ σ structures. Let f be a *partial function* with $do(f) \subseteq A$ and $rg(f) \subseteq B$, where $do(f)$ and $rg(f)$ denote the domain and the range of f, respectively. Then f is said to be a *partial isomorphism* (or *p.i.*) from $\mathcal{A}$ to $\mathcal{B}$ if (i) f is injective; (ii) for every constant symbol c in σ: $c^{\mathcal{A}} \in do(f)$ and $f(c^{\mathcal{A}}) = c^{\mathcal{B}}$; (iii) for every relation symbol R of arity r in σ, and for all $a_1, \ldots, a_r \in do(f)$: $(a_1, \ldots, a_r) \in R^{\mathcal{A}} \Leftrightarrow (f(a_1), \ldots, f(a_r)) \in R^{\mathcal{B}}$. We denote by $Part(\mathcal{A}, \mathcal{B})$ the set of all partial isomorphisms $f : \mathcal{A} \to \mathcal{B}$, and by $Part^k(\mathcal{A}, \mathcal{B})$ the set of all partial isomorphisms $f : \mathcal{A} \to \mathcal{B}$ with $|f| \leq k$. That is, a partial isomorphism is an isomorphism between two sub-structures. Let $\mathcal{A}$, $\mathcal{B}$ be two structures of a relational signature τ. Let A, B be their respective domains, and let $*$ not belong to neither domain. For $\bar{a} \in (A \cup \{*\})^s$ with $\bar{a} = (a_1, \ldots, a_s)$, let $supp(\bar{a}) = \{i : a_i \in A\}$ be the support of $\bar{a}$, and if $a \in A$, let $\bar{a}\frac{a}{i}$ denote $(a_1, \ldots, a_{i-1}, a, a_{i+1}, \ldots, a_s)$. For

$\bar{a} \in (A \cup \{*\})^s$ and $\bar{b} \in (B \cup \{*\})^s$ we say that $\bar{a} \mapsto \bar{b}$ is an *s-partial isomorphism* from $\mathcal{A}$ to $\mathcal{B}$, if $supp(\bar{a}) = supp(\bar{b})$ and $\bar{a}' \mapsto \bar{b}'$ is a partial isomorphism from $\mathcal{A}$ to $\mathcal{B}$, where $\bar{a}'$ and $\bar{b}'$ are the subsequences of $\bar{a}$ and $\bar{b}$ with indices in the support. Now we define the *s-pebble game with m rounds* $P_m^s(\mathcal{A}, \bar{a}, \mathcal{B}, \bar{b})$. There are s pairs of *pebbles* (α_i, β_i), with $1 \leq i \leq s$. Initially, α_i is placed on a_i if $a_i \in A$, and off the board if $a_i = *$, and similarly for β_i with b_i and B. Each *play* consists of m rounds. In each round the Spoiler selects a structure, $\mathcal{A}$ or $\mathcal{B}$, and a pebble for this structure (being off the board or already placed on an element). If he selects $\mathcal{A}$ and α_i, he places α_i on some element of $\mathcal{A}$, and then the Duplicator places β_i on some element of $\mathcal{B}$. If the Spoiler selects $\mathcal{B}$ and β_i, he places β_i on some element of $\mathcal{B}$, and the Duplicator places α_i on some element of $\mathcal{A}$. Note that there may be several pebbles on the same element. The *Duplicator wins* the game if for each $j \leq m$ we have that $\bar{e} \mapsto \bar{f}$ is an *s*-partial isomorphism, where $\bar{e} = (e_1, \ldots, e_s)$ are the elements marked by $\alpha_1, \ldots, \alpha_s$ after the j-th round ($e_i = *$ in case α_i is off the board) and where $\bar{f} = (f_1, \ldots, f_s)$ are the corresponding elements given by $\beta_1, \ldots, \beta_s$. For $j = 0$ this means that $\bar{a} \mapsto \bar{b}$ is an s-partial isomorphism. The *Spoiler wins* the game if the Duplicator does not win the game. The *s-pebble game with infinitely many rounds* $P_\infty^s(\mathcal{A}, \bar{a}, \mathcal{B}, \bar{b})$ is defined similarly. From now on, when writing $\mathcal{A} \models \varphi[\bar{a}]$ for $\bar{a} \in (A \cup \{*\})^s$ we tacitly assume that the free variables of φ have indices in $supp(\bar{a})$. We give now the characterization theorem (see [9], among other sources).

Theorem 3 (e.g., [9]). Let $k \geq 1$. Let $\mathcal{A}$ and $\mathcal{B}$ be two σ structures, and $\bar{a} \in (A \cup \{*\})^k$ and $\bar{b} \in (B \cup \{*\})^k$ with $supp(\bar{a}) = supp(\bar{b})$. The following are equivalent:

(i) The Duplicator wins $P_\infty^k(\mathcal{A}, \bar{a}, \mathcal{B}, \bar{b})$.

(ii) $\bar{a}$ satisfies in $\mathcal{A}$ the same FO^k formulas as $\bar{b}$ in $\mathcal{B}$: $tp_\mathcal{A}^{FO^k}(\bar{a}) = tp_\mathcal{B}^{FO^k}(\bar{b})$.

(iii) $\bar{a}$ satisfies in $\mathcal{A}$ the same $L_{\infty\omega}^k$ formulas as $\bar{b}$ in $\mathcal{B}$: $tp_\mathcal{A}^{L_{\infty\omega}^k}(\bar{a}) = tp_\mathcal{B}^{L_{\infty\omega}^k}(\bar{b})$.

Examples of FO^k Types

1. Let $\mathcal{T}$ be an out tree, where all leaves are at the same depth, and all internal nodes have output degrees ≥ 2. Let a, b be two nodes of the same depth in $\mathcal{T}$, with *output degrees 2 and 3*, respectively. Then, we have the following:

    (a) $tp_\mathcal{T}^{FO^2}(a) = tp_\mathcal{T}^{FO^2}(b)$: Consider the pebble game $P_\infty^2(\mathcal{T}, a, \mathcal{T}, b)$. Note that, never minding which copy of $\mathcal{T}$ the Spoiler chooses, and where he places the two pebbles, the Duplicator can answer by playing in the other copy of $\mathcal{T}$, placing the corresponding two pebbles in such a way that the sub-graphs induced in the two copies of $\mathcal{T}$ by the pebbled nodes are isomorphic, does winning the game.

    (b) $tp_\mathcal{T}^{FO^3}(a) = tp_\mathcal{T}^{FO^3}(b)$: Consider the pebble game $P_\infty^3(\mathcal{T}, a, \mathcal{T}, b)$. If the Spoiler chooses the second copy of $\mathcal{T}$, and puts pebbles β_1, β_2 and β_3 in the three children of b, the Duplicator wins the game by playing in the first copy of $\mathcal{T}$, putting pebbles α_1, α_2 and α_3 in 3 arbitrary nodes with no edges among them. For instance, putting the 3 pebbles in the three children of b in the first copy of $\mathcal{T}$.

(c) $tp_{\mathcal{T}}^{FO^4}(a) \neq tp_{\mathcal{T}}^{FO^4}(b)$: Consider the pebble game $P_\infty^4(\mathcal{T}, a, \mathcal{T}, b)$. Initially, in the first copy of $\mathcal{T}$, pebble α_1 is placed on a, and pebbles α_2, α_3 and α_4 are off the board. Correspondingly, in the second copy of $\mathcal{T}$, pebble β_1 is placed on b, and pebbles β_2, β_3 and β_4 are off the board. After the first three rounds the *Spoiler* wins the game by choosing the second copy of $\mathcal{T}$, and placing pebbles β_2, β_3 and β_4 on the three children of b.

2. For each $n \geq 1$ there is an FO^3 formula $\varphi_n(x, y)$ that in digraphs expresses that there is *a path* of *length at most* n, with possible repetitions of nodes, from x to y [20]: Let $\varphi_1(x, y) \equiv E(x, y)$, and let $\varphi_{n+1}(x, y) \equiv \exists z(E(x, z) \wedge \exists x((x = z) \wedge \varphi_n(x, y)))$. Let $\mathcal{T}$ be tree as defined above. For each $n \geq 1$ there is an FO^2 formula $\psi_n(x)$ that in $\mathcal{T}$ expresses that there is *a path* of *length* n from x to a leaf: Let $\psi_1(x) \equiv \exists y(E(x, y) \wedge \neg \exists x(E(y, x)))$, and let $\psi_{n+1}(x) \equiv \exists y(E(x, y) \wedge \exists x((x = y) \wedge \psi_n(x)))$. Now, let c, d be two nodes of *different depth* in $\mathcal{T}$. Then, even if their output degrees in $\mathcal{T}$ are equal, we have that

(a) $tp_{\mathcal{T}}^{FO^2}(c) \neq tp_{\mathcal{T}}^{FO^2}(d)$: Consider the pebble game $P_\infty^2(\mathcal{T}, c, \mathcal{T}, d)$. Suppose that node c is at a shorter distance to a leaf than d is. Then, the *Spoiler* wins the game by choosing the second copy of $\mathcal{T}$, and *walking* with the pebbles β_1 and β_2 along the path from d to a leaf. Note that the *Duplicator* will have to stop walking with pebbles α_1 and α_2 in the first copy of $\mathcal{T}$ before the *Spoiler* does, since the path from c to a leaf is shorter.

Equality and Ordering of FO^k Types in det-inf-FP

We can always build an (FO + det-inf-FP) formula that expresses that two l-tuples of a given structure have the *same* FO^k types, for any $1 \leq l \leq k$.

Proposition 3 [8]. *Let σ be a relational vocabulary. For every $k \geq 1$ and every $1 \leq l \leq k$, there is a det-inf-FP formula $\eta(\bar{x}, \bar{y})$ with $2l$ free variables such that, for every structure $\mathcal{A}$ in $Str[\sigma]$ and for all $\bar{a}, \bar{b} \in A^l$, $\mathcal{A} \models \eta(\bar{x}, \bar{y})[\bar{a}, \bar{b}]$ if and only if $tp_{\mathcal{A}}^{FO^k}(\bar{a}) = tp_{\mathcal{A}}^{FO^k}(\bar{b})$.*

Suppose we have a set X partitioned into subsets $X_1, \ldots, X_m$. Now consider a binary relation $\prec$ on X given by $x \prec y \Leftrightarrow x \in X_i, y \in X_j$, and $i < j$. Relations obtained in such way are called *strict preorders*. With each strict preorder $\prec$ we associate an equivalence relation whose equivalence classes are precisely $X_1, \ldots, X_m$. It can be defined by the formula $\neg(x \prec y) \wedge \neg(y \prec x)$. Similarly, we can also define the relation $\preceq$, getting a preorder that is not strict. The corresponding equivalence relation to $\preceq$ is defined by the formula $(x \preceq y) \wedge (y \preceq x)$. Note that a preorder induces a *total order* in the set of subsets $X_1, \ldots, X_m$, and similarly, a strict preorder induces a *strict total order* in that set. We can also build an (FO + det-inf-FP) formula that expresses that defines a strict preorder $\prec^{k,l}$, whose equivalence relation is the relation $\equiv^{k,l}$ of equality of FO^k types for l-tuples in a given structures, for any $1 \leq l \leq k$.

Theorem 4 [8]. *Let σ be a relational vocabulary. For every $k \geq l \geq 1$ there is a det-inf-FP formula $\chi(\bar{x}, \bar{y})$ with $|\bar{x}| = |\bar{y}| = l$, such that on every structure $\mathcal{A}$ in*

$Str[\sigma]$, the formula χ defines a strict preorder $\prec^k$, whose equivalence relation is the equality of FO^k types for l-tuples, denoted by $\equiv^k$.

Relational Machines and FO^k Types

Let $k \geq 1$, let σ be a relational vocabulary, and let $\mathcal{A}$ be a σ structure. We say that a relation R of arity $r \leq k$ is *closed under* $\equiv^k$ if and only if, for every pair of r-tuples $\bar{a}$ and $\bar{b}$ over $\mathcal{A}$, if $\bar{a} \in R^{\mathcal{A}}$ and $\bar{a} \equiv^k \bar{b}$, then $\bar{b} \in R^{\mathcal{A}}$. Note that all the relations that form a σ structure are *closed* under $\equiv_k$, since k is $\geq$ than all the arities in σ (see Fact 9 in [13]).

Proposition 4 [2]. *Let $1 \leq r \leq k$. For every pair of r-tuples $\bar{a}$ and $\bar{b}$ over a relational structure $\mathcal{A}$, $\bar{a} \equiv^k \bar{b}$ if and only if no k-ary relational machine can distinguish between $\bar{a}$ and $\bar{b}$ over $\mathcal{A}$. That is, for every k-ary RM M, and for every s-ary relation symbol $R \in \tau$, with $1 \leq s \leq k$, in every configuration $(q, w, \mathcal{A})$ in the computation of M on $\mathcal{A}$: $\mathcal{A} \models R(x_{i_1}, \ldots, x_{i_s})[\bar{a}]$ if and only if $\mathcal{A} \models R(x_{i_1}, \ldots, x_{i_s})[\bar{b}]$, with $1 \leq i_1, \ldots, i_s \leq r$.*

Since k-ary relational machines cannot distinguish between tuples which are $\equiv^k$-equivalent, they *cannot* compute the size of their input structures. However, they *can* compute the number of $\equiv^k$-classes.

Proposition 5 [2]. *For each $k \geq 1$ and relational vocabulary σ, there is a deterministic relational machine M_σ of arity $2k$ that outputs on its Turing machine tape, for an input structure $\mathcal{A}$ of vocabulary σ, a string of length $size_k(\mathcal{A})$ in time polynomial in $size_k(\mathcal{A})$.*

Lemma 1 [2]. *For every relational vocabulary σ and every $k \geq 1$, there is a deterministic relational machine $M_{\prec^k}$ of arity $k' \geq 2k$, such that on any input structure $\mathcal{A}$ of vocabulary σ, $M_{\prec^k}$ computes the preorder $\preceq^k$ of T.11.20, working in time bounded by a polynomial in $size_{k'}(\mathcal{A})$.*

Relational Complexity

Let M be a relational machine. If M is *deterministic*, then the *computation time* of M on an input structure $\mathcal{A}$ is the number of transitions that M makes before accepting or rejecting $\mathcal{A}$, and the *computation space* is the number of tape cells scanned. Note that the *rs is not considered* for the computation space. If M is *non-deterministic*, then we only consider accepting computations. In that case, the *computation time* of M on an input structure $\mathcal{A}$ is the number of transitions in the *shortest accepting* computation of M on $\mathcal{A}$, and the *computation space* is the *minimum* number of tape cells that are scanned in any accepting computation of M on $\mathcal{A}$. If M *rejects* $\mathcal{A}$, then both the *computation time* of M on $\mathcal{A}$, and the *computation space* of M on $\mathcal{A}$, are 1. Let $L(M)$ be the relational language accepted by a relational machine M of arity k. Let t and s be functions on the natural numbers such that $t(n) \geq n + 1$ and $s(n) \geq 1$. $L(M) \in DTIME_r(t(n))$ if M is deterministic and its computation time on any input structure $\mathcal{A}$ is bounded above by $t(size_k(\mathcal{A}))$; $L(M) \in NTIME_r(t(n))$ if M is non-deterministic and its computation time on any input structure $\mathcal{A}$ is bounded above by $t(size_k(\mathcal{A}))$; $L(M) \in DSPACE_r(s(n))$ if M is deterministic and its computation space on any input structure $\mathcal{A}$ is bounded above by $s(size_k(\mathcal{A}))$; $L(M) \in NSPACE_r(s(n))$ if M is non-deterministic and its computation space on any input structure $\mathcal{A}$ is bounded above by $s(size_k(\mathcal{A}))$. We define the following classes: (i) P_r of the relational languages decidable by relational machines working in polynomial time in the k-size of their input structures: $P_r = \bigcup_{c \in \mathbb{N}} DTIME_r(n^c)$; (ii) NP_r of the relational languages decidable by non-deterministic relational machines working in polynomial time in the k-size of their input structures: $NP_r = \bigcup_{c \in \mathbb{N}} NTIME_r(n^c)$; (iii) $PSPACE_r$ of the relational languages decidable by relational machines working in polynomial space in the k-size of their input structures: $PSPACE_r = \bigcup_{c \in \mathbb{N}} DSPACE_r(n^c)$. The time complexity of oracle relational machines is defined precisely in the same way as with ordinary relational machines. Each query step counts as *one* ordinary step. Thus if $\mathcal{C}$ is any deterministic or non-deterministic relational time complexity class and $\mathcal{A}$ is a relational language, we can define $\mathcal{C}^{\mathcal{A}}$ to be the class of all relational languages accepted by relational machines of the same time bound as in $\mathcal{C}$, only that the machines have now an oracle $\mathcal{A}$. The levels of the *relational polynomial time hierarchy* are defined as follows: $\Delta_0^{P_r} = \Sigma_0^{P_r} = \Pi_0^{P_r} = P_r$, and, for $m > 0$, $\Delta_{m+1}^{P_r} = P_r^{\Sigma_m^{P_r}}$, $\Sigma_{m+1}^{P_r} = NP_r^{\Sigma_m^{P_r}}$, $\Pi_{m+1}^{P_r} = coNP_r^{\Sigma_m^{P_r}}$. Then, $PH_r = \bigcup_{m \in \mathbb{N}} \Sigma_m^{P_r}$.

Relational Complexity Classes and Fixed Point Quantifiers

In [2], the following characterizations of relational complexity classes with fixed point logics were proved (on *arbitrary* finite structures):

Proposition 6 [2].

(i) $P_r = (\mathrm{FO} + det - \inf - \mathrm{FP})$;
(ii) $NP_r = (\mathrm{FO} + ndet - \inf - \mathrm{FP})$;
(iii) $PSPACE_r = (\mathrm{FO} + det - ninf - \mathrm{FP})$
$\qquad = (\mathrm{FO} + ndet - ninf - \mathrm{FP})$
$\qquad = (\mathrm{FO} + alt - \inf - \mathrm{FP})$;
(iv) $EXPTIME_r = (\mathrm{FO} + alt - ninf - \mathrm{FP})$.

Comparing the expressive power among these fixed point logics, and with $L_{\infty\omega}^{\omega}$, we have the following picture:

Proposition 7 [2,6,19].

$(\mathrm{FO} + LFP) = (\mathrm{FO} + det - \inf - \mathrm{FP})$
$\subseteq (\mathrm{FO} + ndet - \inf - \mathrm{FP})$
$\subseteq (\mathrm{FO} + det - ninf - \mathrm{FP}) = (\mathrm{FO} + ndet - ninf - \mathrm{FP}) = (\mathrm{FO} + alt - \inf - \mathrm{FP})$
$\subseteq (\mathrm{FO} + alt - ninf - \mathrm{FP})$
$\subset L_{\infty\omega}^{\omega}$.

Note that the different fixed point quantifiers, added to FO, character-ize *exactly* the corresponding computational complexity classes, when the input structures are required to be *ordered* (see above). That is, for $\mathcal{C} \in \{P, NP, PSPACE, EXPTIME\}$, $\mathcal{C}_r \in \{P_r, NP_r, PSPACE_r, EXPTIME_r\}$, $\alpha \in \{det, ndet, alt\}$, and $\beta \in \{inf, ninf\}$, the following holds: $\mathcal{C} = (\text{FO} + \leq + \alpha\text{-}\beta\text{-FP})$ if and only if $\mathcal{C}_r = (\text{FO} + \alpha\text{-}\beta\text{-FP})$.

5   The Restricted Second Order Logic SO^ω

We denote by $\Sigma_m^{1,\omega}[\sigma]$ the class of formulas of the form $\exists^{k_{11}} Y_{11}^{r_{11},k_{11}} \ldots$
$\exists^{k_{1l_1}} Y_{1l_1}^{r_{1l_1},k_{1l_1}} \forall^{k_{21}} Y_{21}^{r_{21},k_{21}} \ldots \forall^{k_{2l_2}} Y_{2l_2}^{r_{2l_2},k_{2l_2}} \ldots Q^{k_{t1}} Y_{t1}^{r_{t1},k_{t1}} \ldots Q^{k_{tl_t}} Y_{tl_t}^{r_{tl_t},k_{tl_t}}(\phi)$,
where the quantifiers $Q^{k_{t1}}, \ldots, Q^{k_{tl_t}}$ are $\forall^{k_{t1}}, \ldots, \forall^{k_{tl_t}}$, if t is even, or $\exists^{k_{t1}}, \ldots, \exists^{k_{tl_t}}$, if t is odd, ϕ is an FO formula in the vocabulary $\sigma \cup \{Y_{11}^{r_{11},k_{11}}, \ldots, Y_{tl_t}^{r_{tl_t},k_{tl_t}}\}$, with $r_{11} \leq k_{11}, \ldots, r_{tl_t} \leq k_{tl_t}$, respectively. We use upper case Roman letters $X_i^{r,k}$ for SO^ω variables, where $r \geq 1$ is their arity, and $k \geq r$ (see below). In this article we will often drop the superindex k, when it is clear from the context. We define $SO^\omega = \bigcup_{m \geq 1} \Sigma_m^{1,\omega}$. The second order quanti-fier $\exists^k$ has the following semantics: let $\mathcal{A}$ be a σ structure; then $\mathcal{A} \models \exists^k Y^{r,k} \varphi$ if there is an r-ary (second order) relation $R^{r,k}$ on A that is closed under the relation $\equiv^k$ in $\mathcal{A}$, such that $(\mathcal{A}, R) \models \varphi$. Note that a *valuation* in this setting also assigns to each SO^ω variable $X^{r,k}$ a (second order) relation on A of arity r that is closed under $\equiv^k$ in $\mathcal{A}$.

Remark 1. SO^ω is defined in this way because *it seems* that equivalence between arbitrary formulas with SO^ω quantifiers, and prenex formulas *does not hold* in this logic (as opposite to SO). Let $\psi \equiv \exists x \forall^k R(\varphi(x, R))$, the usual trans-lation strategy in SO to a prenex formula yields the following formula $\hat{\psi} \equiv \exists^{k'} X \, \forall^k R(\forall x(X(x) \Rightarrow \varphi(x, R)))$. However, in SO^ω the two formulas are not equivalent, since for a given structure we might not get a valuation which assigns to X sets with only elements of a single FO type. That is, it might be the case that every set assigned to X has elements of more than one FO type.

As to expressibility of the logics, clearly, $SO^\omega \subseteq SO$, and for every $m \geq 1$, $\Sigma_m^{1,\omega} \subseteq \Sigma_m^1$ and $\Pi_m^{1,\omega} \subseteq \Pi_m^1$. Equality is obtained if the input structures are (totally) ordered:

Theorem 5 [6]. *On* ordered *structures, for every* $m \geq 1$, $\Sigma_m^{1,\omega} = \Sigma_m^1$ *and* $\Pi_m^{1,\omega} = \Pi_m^1$.

The following result establishes the relationship between SO^ω and the infini-tary logic $L_{\infty\omega}^\omega$. Note that as $L_{\infty\omega}^\omega$ has the *0-1 Law* (see above) this means that SO^ω also has the *0-1 Law*.

Corollary 1 [6]. $SO^\omega \subset L_{\infty\omega}^\omega$.

Example 1. The following $\Sigma_1^{1,\omega}$ formula expresses *connectivity on undirected graphs*: $\exists^3 R^2 \big["E \subseteq R" \wedge \forall x(R(x,x)) \wedge "R \bowtie E \subseteq R" \wedge \forall xy\big(R(x,y) \Rightarrow [E(x,y) \vee x = y \vee \exists z(R(x,z) \wedge E(z,y))]\big) \wedge \forall xy(R(x,y))\big]$, where "$E \subseteq R$" is short for $\forall xy(R(x,y) \Rightarrow E(x,y))$ and" $R \bowtie E \subseteq R$" is short for $\forall xyz((R(x,z) \wedge E(z,y)) \Rightarrow R(x,y))$. Note that, with the intended semantics, R is always closed under $\equiv^3$, since whenever two tuples $(x_1,y_1),(x_2,y_2)$ are $\equiv^3$, and $(x_1,y_1) \in R$, then for every $1 \le i \le |V|$ there is a path of length i between x_1 and y_1 if and only if there is a path of length i between x_2 and y_2, since for all i that property is expressible in FO^3 (see Examples of FO^k Types above). Hence also $(x_2,y_2) \in R$.

SO^ω and Relational Complexity

The following characterizations of relational complexity classes with SO^ω or its fragments have been proved.

Theorem 6 [6]. $\mathrm{SO}^\omega \subseteq (\mathrm{FO} + \det - \mathrm{ninf} - \mathrm{FP})$.

Theorem 7 [6]. $\Sigma_1^{1,\omega} = (\mathrm{FO} + \mathrm{ndet} - \inf - \mathrm{FP})$.

Corollary 2 [6]. $\Sigma_1^{1,\omega}$ *captures* NP_r.

In [13], a direct proof of the relationship $\Sigma_1^{1,\omega} = NP_r$ was given, continuing with the characterization of the relational polynomial time hierarchy:

Theorem 8 [13]. *(i) For* $m \ge 1$: $\Sigma_m^{P_r} = \Sigma_m^{1,\omega}$; *(ii)* $PH_r = \mathrm{SO}^\omega$.

6   The Restricted Third Order Logic TO^ω

The logic TO^ω was introduced in [5] as a fragment of Third Order Logic (TO). A *third order relation type* is a w-tuple $\tau = (r_1, \ldots, r_w)$ where $w, r_1, \ldots, r_w \ge 1$. In addition to the symbols of SO^ω, the alphabet of TO^ω contains, for every $k \ge 1$, a *third-order quantifier* $\exists^k$, and, for every relation type τ such that $r_1, \ldots, r_w \le k$, a countably infinite set of third order variables, denoted as $\mathcal{X}_1^{\tau,k}, \mathcal{X}_2^{\tau,k}, \ldots$, and called TO^ω *variables*. In this article, we will often drop either one or the two superindices, when they are clear from the context. Let σ be a relational vocabulary. A TO^ω *atomic formula* of vocabulary σ, on the TO^ω variable $\mathcal{X}^{\tau,k}$, is a formula of the form $\mathcal{X}^{\tau,k}(V_1, \ldots, V_w)$, where $V_1, \ldots, V_w$ are either second order variables of the form $X_i^{r_i,k}$, or relation symbols in σ, and whose arities are respectively $r_1, \ldots, r_w \le k$. Note that all the relations that form a σ structure are closed under $\equiv^k$, since k is $\ge$ than all the arities in σ (see above, and Fact 9 in [13]). Let $m \ge 1$. We denote by $\Sigma_m^{2,\omega}[\sigma]$ the class of formulas of the form $\exists^{k_{3,11}} \mathcal{X}_{11}^{\tau_{11},k_{3,11}} \ldots \exists^{k_{3,1s_1}} \mathcal{X}_{1s_1}^{\tau_{1s_1},k_{3,1s_1}} \forall^{k_{3,21}} \mathcal{X}_{21}^{\tau_{21},k_{3,21}} \ldots$ $\forall^{k_{3,2s_2}} \mathcal{X}_{2s_2}^{\tau_{2s_2},k_{3,2s_2}} \ldots Q^{k_{3,m1}} \mathcal{X}_{m1}^{\tau_{m1},k_{3,m1}} \ldots Q^{k_{3,ms_m}} \mathcal{X}_{ms_m}^{\tau_{ms_m},k_{3,ms_m}} (\psi)$, where, for $i,j \ge 1$, $\tau_{ij} = (r_{ij,1}, \ldots, r_{ij,w_{ij}})$, and $r_{ij,1}, \ldots, r_{ij,w_{ij}} \le k_{3,ij}$, Q is either $\exists^k$ or $\forall^k$, for some k, depending on whether m is odd or even, respectively, and ψ is an SO^ω

formula with the addition of TO^ω atomic formulas. As usual, $\forall^k \mathcal{X}^{\tau,k}(\psi)$ abbreviates $\neg \exists^k \mathcal{X}^{\tau,k}(\neg \psi)$. We define $\mathrm{TO}^\omega = \bigcup_{m \geq 1} \Sigma_m^{2,\omega}$. A TO^ω *relation* $\mathcal{R}^{\tau,k}$ *of type* τ *and closed under* $\equiv^k$ on a σ structure $\mathcal{A}$ is a set of w tuples $(R_1^{r_1,k}, \ldots, R_w^{r_w,k})$ of (second order) relations on $\mathcal{A}$ with respective arities $r_1, \ldots, r_w \leq k$, closed under $\equiv^k$. The third order quantifier $\exists^k$ has the following semantics: let $\mathcal{A}$ be a σ structure; then $\mathcal{A} \models \exists^k \mathcal{X}^{\tau,k} \varphi$ if there is a TO^ω relation $\mathcal{R}^{\tau,k}$ of type τ on I closed under the relation $\equiv^k$ in $\mathcal{A}$, such that $(\mathcal{A}, \mathcal{R}) \models \varphi$. Here $(\mathcal{A}, \mathcal{R})$ is the *third order* $(\sigma \cup \{\mathcal{X}^{\tau,k}\})$ structure expanding $\mathcal{A}$, in which $\mathcal{X}$ is interpreted as $\mathcal{R}$. Note that a *valuation* in this setting also assigns to each TO^ω variable $\mathcal{X}^{\tau,k}$ a third order relation $\mathcal{R}^{\tau,k}$ on A of type τ, closed under $\equiv^k$ in $\mathcal{A}$. We do not allow free second or third order variables in the logics SO^ω and TO^ω. Note that allowing elements (from the domain of the structure) in a third order relation type would change the semantics of TO^ω, since we could use a third order relation of such type to simulate a second order relation *not closed* under $\equiv^k$.

An Example in $\Sigma_1^{2,\omega}$

We give next a sketch of an example of a non-trivial query in $\Sigma_1^{2,\omega}$. Consider the query *the graph G is undirected, connected, with $|V| \geq 2$, and its diameter is even*. We quantify two third order relations, $\mathcal{X}^{(2,2)}$ and $\preceq^{(2,2,2,2)}$, that form a totally ordered set of pairs of (second order) relations, where, for $0 \leq i \leq m$, in the first component of the i-th pair (R_1, R_2), we have all the pairs of nodes (x, y) such that the minimum distance between them is i. Note that all these relations are closed under $\equiv_3$ since with 3 variables we can say in FO that there is a path of length d between two nodes, for every $d \geq 0$ (see [20]). Then, if two pairs of nodes are $\equiv_3$, either they are both in the relation or none of them are, which is correct since the set of distances of all the paths between the two pairs of nodes is the same. We use the second relation in each pair (S_2 and R_2) as Boolean flags, where $\emptyset$ means *off* and $V \times V$ means *on*. Then, along the sequence of pairs of relations, we switch the flags *on* and *off*, starting in *on*. Note that if the position of the last pair of relations in the sequence is i (i.e., m), it means that the diameter of G is i, and, if the flag is *on*, then the diameter is *even*: $\varphi \equiv \exists^3 \mathcal{X}^{(2,2)} \preceq^{(2,2,2,2)} \forall^3 R_1^2 R_2^2 S_1^2 S_2^2 \left[(\text{``} \preceq \text{ is a total order in } \mathcal{X}\text{''}) \wedge \right.$ $(\text{``the succesor of } (S_1, S_2) \text{ is } (R_1, R_2)\text{''}] \Rightarrow [\text{``}R_2 \text{ is the complement of } S_2\text{''} \wedge \text{``the pairs in } R_1 \text{ are formed by extending the pairs in } S_1 \text{ with the edges in } E\text{''}]) \wedge$ $([\text{``}(S_1, S_2) \prec (R_1, R_2)\text{''}] \Rightarrow [\text{``no pair in } R_1 \text{ is in } S_1, \text{ i.e., the distances at every stage are minimal''}]) \wedge ([\text{``}(R_1, R_2) \text{ is the first pair in } \preceq\text{''}] \Rightarrow [\text{``the flag is } on\text{''} \wedge \text{``}R_1 \text{ is `='''}]) \wedge ([\text{``}(R_1, R_2) \text{ is the last pair in } \preceq\text{''}] \Rightarrow [\text{``the flag is } on\text{''} \wedge \text{``the pairs in } R_1 \text{ cannot be extended with edges in } E, \text{ i.e., there are no minimum distances bigger than the ones in } R_1\text{''}]) \wedge [\text{``}G \text{ is connected''}] \wedge [\text{``}G \text{ is undirected''}] \wedge [\text{``}|V| \geq 2\text{''}]$. Finally, note that the $\Sigma_1^{1,\omega}$ formula for "G is connected" is given in Example 1. Recall that usually in FMT undirected graphs are represented as symmetric directed graphs.

Remark 2. Note that this query *does not* actually *need* the expressive power of $\Sigma_1^{2,w}$. It is not difficult to describe a (deterministic) RM working in relational polynomial time (P_r) that computes this query. Then, by the characterization of NP_r (which includes P_r) with the logic $\Sigma_1^{1,w}$ (Corollary 2), we know there is also a $\Sigma_1^{1,w}$ formula expressing the same query, though most likely more complicated than the above $\Sigma_1^{2,w}$ formula.

3-NRM's

The third order non-deterministic relational machine (3-NRM) was introduced in [5] as a variation of the non-deterministic relational machine NRM, where third order relations are allowed in the relational store. The relational complexity class $NEXPTIME_{3,r}$ was also introduced there, to represent the 3-NRM's that work in non-deterministic relational time. A *third order non-deterministic relational machine*, denoted as 3-NRM, of *arity* k, for $k \geq 1$, is an 11-tuple $\langle Q, \Sigma, \delta, q_0, \mathfrak{b}, F, \sigma, \tau, T, \Omega, \Phi \rangle$ where: Q is the finite set of internal states; $q_0 \in Q$ is the initial state; Σ is the finite tape alphabet; $\mathfrak{b} \in \Sigma$ is the symbol denoting blank; $F \subseteq Q$ is the set of accepting states; τ is the finite vocabulary of the *rs* (its *relational store*), with finitely many TO^ω relation symbols $\mathcal{R}_i^{\tau_i, k'}$ of any arbitrary type $\tau_i = (r_{i1}, \ldots, r_{iw})$, with $1 \leq r_{i1}, \ldots, r_{iw} \leq k' = k$, and finitely many SO^ω relation symbols $R_i^{r_i, k''}$ of arities $r_i \leq k'' = k$; $T \in \tau$ is the output relation; σ is the vocabulary of the input structure; Ω is a finite set of TO^ω formulas with up to k FO variables, with *no SO^ω or TO^ω quantifiers*, and with no free variables of any order (i.e., all the SO^ω and TO^ω relation symbols are in τ); Φ is a finite set of TO^ω formulas with up to k *FO* variables that are not sentences, with *no SO^ω or TO^ω quantifiers*, and where the free variables are either *all* FO variables, or *all* SO^ω variables; $\delta : Q \times \Sigma \times \Omega \to \mathcal{P}(\Sigma \times Q \times \{R, L\} \times \Phi \times \tau)$ is the transition function. In any pair in δ, if φ, S occur in the 5-tuple of its second component, for Φ and τ, then either S is a TO^ω relation symbol $\mathcal{R}_i^{\tau_i, k'}$ in *rs* and φ has $|\tau_i|$ SO^ω free variables $X_1^{r_1, k''}, \ldots, X_{|\tau_i|}^{r_{|\tau_i|}, k''}$ with arities according to τ_i, and $1 \leq r_1, \ldots, r_{|\tau_i|} \leq k'' = k' = k$, or S is an SO^ω relation symbol $R_i^{r_i, k''}$ in *rs* and φ has $1 \leq r_i \leq k'' = k$ FO free variables. At any stage of the computation of a 3-NRM on an input σ structure $\mathcal{A}$, there is one relation in its *rs* of the corresponding relation type (or arity) in $\mathcal{A}$ for each relation symbol in τ, so that in each transition there is a (finite) τ-structure $\mathcal{A}$ in the *rs*, which we can *query* and/or *update* through the formulas in Ω and Φ, respectively, and a finite Σ string in its tape, which we can access as in Turing machines. The concept of *computation* is analogous to that in the Turing machine. We define the complexity class $NEXPTIME_{3,r}$ as the class of the *relational languages* or *Boolean queries* (i.e., sets of finite structures of a given relational vocabulary, closed under isomorphisms) that are decidable by 3-NRM machines of *some* arity k', that work in non-deterministic exponential time in the number of equivalence classes in $\equiv^{k'}$ of the input structure. In symbols: $NEXPTIME_{3,r} = \bigcup_{c \in N} NTIME_{3,r}(2^{c \cdot (size_k)})$ (as usual, this notation *does not mean* that the arity of the 3-NRM must be k).

$\Sigma_1^{2,\omega}$ Captures $NEXPTIME_{3,r}$

The following results were proved in [5]:

Theorem 9 [5]. $NEXPTIME_{3,r} \subseteq \Sigma_1^{2,\omega}$. *That is, given a 3-NRM M in $NTIME_{3,r}(2^{c \cdot (size_k)})$, for some positive integer c and with input vocabulary σ, that computes a Boolean query q, we can build a formula $\varphi_M \in \Sigma_1^{2,\omega}$ such that, for every σ-structure I, M accepts I if and only if $I \models \varphi_M$.*

Theorem 10 [5]. $\Sigma_1^{2,\omega} \subseteq NEXPTIME_{3,r}$. *That is, every class of relational structures definable in $\Sigma_1^{2,\omega}$ is in $NTIME_{3,r}(2^{c \cdot (size_k)})$.*

$\Sigma_1^{2,\omega}$ also Captures $NEXPTIME_r$

Then, in [22], we proved the following results. The next corollary is a consequence of the first theorem above by the following two immediate facts: (i) an NRM is a special case of a 3-NRM, with no third order relations in its rs, and (ii) an NRM M is in $NEXPTIME_r$ if and only if M, as a 3-NRM, is in $NEXPTIME_{3,r}$.

Corollary 3 [22]. $NEXPTIME_r \subseteq \Sigma_1^{2,\omega}$. *That is, given an NRM M that works in $NTIME_r(2^{c \cdot (size_k)})$, for some positive integer c, and with input vocabulary σ that computes a Boolean query q we can build a formula $\varphi_M \in \Sigma_1^{2,\omega}$ such that, for every σ-structure I, M accepts I if and only if $I \models \varphi_M$.*

The general idea for the proof of the theorem below, is similar to that of the last theorem above. The most important difference is that as we cannot hold third order relations in the rs of the NRM M_φ, we use the bit strings $b_{\bar{R}^{r_3,i}}^3$, $b_{\bar{R}^{r_3,i}}^2$ and $b_{\mathcal{X}_i^{r_i,k_3,i}}^1$ (that represent at different levels a third order relation $\mathcal{X}_i^{r_i,k_3,i}$, and which we use in [5] to guess the relations) instead (see the proofs in [5,22]). Then, to evaluate the FO formula ϕ (in the vocabulary $\sigma \cup \{Y_{11}^{r_2,11}, \ldots, Y_{tl_t}^{r_2,tl_t}\}$, and with atomic TO^ω formulas), we cannot do it in just one step as in [5]. Instead, we use the syntax tree of ϕ, and evaluate one node of it at a time in the finite control of M_φ, in a bottom up direction.

Theorem 11 [22]. $\Sigma_1^{2,\omega} \subseteq EXPTIME_r$. *That is, every class of relational structures definable in $\Sigma_1^{2,\omega}$ is in $\bigcup_{c \in N} NTIME_r(2^{c \cdot (size_k)})$.*

RM's Can Simulate the Existence of TO Relations in Their rs

From the results above, we have the following:

Corollary 4 [22]. *Let M_3 be a 3-NRM that works in $NTIME_{3,r}(2^{c \cdot (size_k)})$, for some positive integer c, that computes a Boolean query q. Then, there is a NRM M_2 that works in $NTIME_r(2^{d \cdot (size_k)})$, for some positive integer d, that also computes q.*

This is very interesting, since in the general case it is *much easier* to define an NRM using TO relations in its rs, and TO formulas to access it, than restricting the machine to SO relations in its rs, and SO formulas. Then, to prove that a given query is computable by an NRM it suffices to show that it can be computed by a 3-NRM. Note however, that we think that we still *need* 3-NRM's as well as the third order relational complexity class $NEXPTIME_{3,r}$, if we need to work with *oracle* NRM's with third order relations, since as the oracle cannot access the tape of the base machine (see [13]), there seems to be no way to pass the bit strings that represent TO relations from the base machine to the oracle. Recall that it has been proved that RM's have the same computation, or expressive power, as the (effective fragment of the) well known infinitary logic with finitely many variables $\mathcal{L}^\omega_{\infty\omega}$ (see above). On the other hand, analogously to the well known result that states that the computation power of deterministic and non-deterministic Turing machines is the same, it is straightforward to see that any NRM M_n can be simulated by a (deterministic) RM M_d working in relational time exponentially higher, just by checking in M_d all possible transitions instead of guessing one in each non-deterministic step of the transition relation of M_n. Then, the following is immediate:

Corollary 5 [22]. $\Sigma_1^{2,\omega} \subseteq L^\omega_{\infty\omega}|_{rec}$.

7   Some Inexpressibility Results

An important application of the creation of new logics to Complexity Theory is the search for lower bounds of problems w.r.t. those logics, aiming to separate computational complexity classes. We will next give a few examples of problems for which lower bounds w.r.t. the logics $\mathcal{L}^\omega_{\infty\omega}$, $\Sigma_1^{1,\omega}$, $\Sigma_1^{1,F}$, SO^ω, SO^F (see below), and some fragments of them have been proved.

The Fragment SO^F of SO

In [15], the logic SO^F was introduced and defined as a semantic restriction of SO where, for $r \geq 1$, the valuating r-ary relations for the quantified SO variables of arity r are *closed* under the relation $\equiv_F$ of equality of FO types in the set of r-tuples of the structure. It was shown there that $SO^\omega \subset SO^F \subset SO$ (see below), and that its existential fragment $\Sigma_1^{1,F}$ is *not* included in $\mathcal{L}^\omega_{\infty\omega}$, as opposite to $\Sigma_1^{1,\omega}$ which is (see above). Then, we have the following result:

Corollary 6. $\Sigma_1^{1,F} \subsetneq \Sigma_1^{2,\omega}$.

A structure is *rigid* if its only automorphism is the identity function. In a finite rigid structure each element realizes a different FO type and thus, for $k \geq 1$, every k-tuple also realizes a different FO type for k-tuples. A rigid structure is FO^k *rigid* if every element is definable in FO with up to k different variables.

A class $\mathcal{C}$ of rigid structures is FO^k *rigid* if every structure in $\mathcal{C}$ is FO^k *rigid*. Then the following is immediate, since in rigid structures, for $k \geq 1$, the relation $\equiv_F$ is the identity relation in the set of k-tuples of the structure, and hence the SO^F quantifiers $\exists^F$ and $\forall^F$ have the same semantics as the SO quantifiers $\exists$ and $\forall$, respectively. Compare this result with Theorem 5 on ordered structures for SO^ω: note that the class of rigid structures strictly includes the class of ordered structures.

Corollary 7. *On classes of rigid structures, for every $m \geq 1$, $\Sigma_m^{1,F} = \Sigma_m^1$ and $\Pi_m^{1,F} = \Pi_m^1$.*

- As an example, let us consider the class of *odd-multipedes* from [18]. Recall that all the structures in the class are rigid, but there is *no* k such that all of them are FO^k rigid. Then the following is immediate:

  Fact 12. *Let $\mathcal{O}$ be the class of odd-multipedes. Then, on $\mathcal{O}$, $SO^F = SO$, but $SO^\omega \subset SO$.*

- In [7], rigidity was proved to be *not expressible* in $L_{\infty\omega}^\omega$, and hence, by the results above, *not expressible* in SO^ω either. As it is easily expressible in $\Sigma_1^{1,F}$ [15], rigidity is a query which *separates* SO^F from SO^ω, then $\Sigma_1^{1,F} \not\subseteq SO^\omega$, and $SO^\omega \subset SO^F$.
- In [6], the following *NP-complete* problems were proved to be *expressible* in $\Sigma_1^{1,\omega}$: non-deterministic finite automata inequivalence, restricted to finite languages, or to unary alphabets. Considering the results above, those two *NP-complete* problems are also in $\Sigma_1^{1,F}$.
- In [6], the *NP-complete* problem *3-colorability* was proved to be *not expressible* in $L_{\infty\omega}^\omega$, and, hence, by the results above, *not expressible* in $\Sigma_1^{1,\omega}$ either.
- In [11], it was proved that the problem *3-colorability* on the class of *bunch graphs* is still *NP-complete*, and that it is *not expressible* in $L_{\infty\omega}^\omega$, and hence, by the results above, is *not expressible* in SO^ω either, but it *is expressible* in $\Sigma_1^{1,F}$. Hence, the following also holds: $\Sigma_1^{1,\omega} \subset \Sigma_1^{1,F}$.
- The following *NP-complete* problems have been proved to be *not expressible* in $L_{\infty\omega}^\omega$, and hence, by the results above, *not expressible* in $\Sigma_1^{1,\omega}$ either (see in [6] the references: Lovasz and Gacs [21], Immerman [15], and Dahlhaus [7]): *Satisfability, Hamiltonicity* and *Clique*.
- Let $\mathcal{C}$ be a class of structures of an unary signature with equality. Then SO^F on $\mathcal{C}$ is equivalent to FO [16]. Hence, among all the queries not expressible in FO on such vocabularies, parity is *not expressible* in SO^F either. Then, $SO^F \subset SO$.
- The query 2-colorability *is expressible* in monadic existential SO, but is *not expressible* in monadic SO^F [14].
- The property of having exactly one FO type for elements *is expressible* in monadic SO^F, but is *not expressible* in monadic existential SO [14].

References

1. Abiteboul, S., Vardi, M.Y., Vianu, V.: Computing with infinitary logic. Theor. Comput. Sci. **149**(1), 101–128 (1995)
2. Abiteboul, S., Vardi, M.Y., Vianu, V.: Fixpoint logics, relational machines, and computational complexity. J. ACM **44**(1), 30–56 (1997)
3. Abiteboul, S., Vianu, V.: Generic computation and its complexity. In: Proceedings of the 23rd Annual ACM Symposium on Theory of Computing, New Orleans, LA, 5–8 May 1991, pp. 209–219 (1991)
4. Abiteboul, S., Vianu, V.: Computing with first-order logic. J. Comput. Syst. Sci. **50**(2), 309–335 (1995)
5. Arroyuelo, J., Turull-Torres, J.M.: The existential fragment of third order logic and third order relational machines. In: Proceedings of the XX Argentine Conference on Computer Science, CACIC 2014, Buenos Aires, Argentina, 20–24 October 2014, pp. 324–333 (2014)
6. Dawar, A.: A restricted second order logic for finite structures. Inf. Comput. **143**(2), 154–174 (1998)
7. Dawar, A., Grädel, E.: Properties of almost all graphs and generalized quantifiers. Fundam. Inform. **98**(4), 351–372 (2010)
8. Dawar, A., Lindell, S., Weinstein, S.: Infinitary logic and inductive definability over finite structures. Inf. Comput. **119**(2), 160–175 (1995)
9. Ebbinghaus, H., Flum, J.: Finite Model Theory. Perspectives in Mathematical Logic. Springer, Heidelberg (1995)
10. Fagin, R.: Generalized first-order spectra and polynomial-time recognizable sets. In: Karp, R.M. (ed.) Complexity of Computation, New York City, 18–19 April 1973. SIAM-AMS Proceedings, vol. 7, pp. 21–74 (1974)
11. Ferrarotti, F.A., Grosso, A.L., Turull-Torres, J.M.: Semantic restrictions over second-order logic. In: Schewe, K.-D., Thalheim, B. (eds.) SDKB 2013. LNCS, vol. 7693, pp. 174–197. Springer, Heidelberg (2013)
12. Ferrarotti, F.A., Paoletti, A.L., Torres, J.M.T.: Redundant relations in relational databases: a model theoretic perspective. J. UCS **16**(20), 2934–2955 (2010)
13. Ferrarotti, F.A., Turull Torres, J.M.: The relational polynomial-time hierarchy and second-order logic. In: Schewe, K.-D., Thalheim, B. (eds.) SDKB 2008. LNCS, vol. 4925, pp. 48–76. Springer, Heidelberg (2008)
14. Grosso, A.L.: SOF: A Logic Where Relation Variables Are Interpreted with Unions of FO Types. Doctoral thesis (in Spanish). Universidad Nacional De San Luis, Argentina (2013)
15. Grosso, A.L., Turull-Torres, J.M.: A second-order logic in which variables range over relations with complete first-order types. In: Ochoa, S.F., Meza, F., Mery, D., Cubillos, C. (eds.) SCCC 2010, Proceedings of the XXIX International Conference of the Chilean Computer Science Society, Antofagasta, Chile, 15–19 November 2010, pp. 270–279. IEEE Computer Society (2010)
16. Grosso, A.L., Turull-Torres, J.M.: Expressibility of the logic SOF on classes of structures of bounded FO types. In: Proceedings of the XVIII Argentine Conference on Computer Science, CACIC 2012, Bahía Blanca, Argentina, 8–12 October 2012, pp. 1389–1398 (2012)
17. Gurevich, Y., Shelah, S.: Fixed-point extensions of first-order logic. Ann. Pure Appl. Logic **32**, 265–280 (1986)
18. Gurevich, Y., Shelah, S.: On finite rigid structures. J. Symb. Log. **61**(2), 549–562 (1996)

19. Kolaitis, P.G., Vardi, M.Y.: Infinitary logics and 0–1 laws. Inf. Comput. **98**(2), 258–294 (1992)
20. Libkin, L.: Elements of Finite Model Theory. Texts in Theoretical Computer Science. An EATCS Series. Springer, Heidelberg (2004)
21. Stockmeyer, L.J.: The polynomial-time hierarchy. Theor. Comput. Sci. **3**(1), 1–22 (1976)
22. Turull-Torres, J.M.: Capturing relational NEXPTIME with a fragment of existential third order logic. J. Comput. Sci. Technol. (2015, to appear). Selected article from the XXI Argentine Conference on Computer Science, CACIC 2015, Junín, Argentina, 5–9 October 2015 (2015)

A Logic for Non-deterministic Parallel Abstract State Machines

Flavio Ferrarotti[1](✉), Klaus-Dieter Schewe[1], Loredana Tec[1], and Qing Wang[2]

[1] Software Competence Center Hagenberg, 4232 Hagenberg, Austria
{flavio.ferrarotti,klaus-dieter.schewe,loredana.tec}@scch.at
[2] Research School of Computer Science, The Australian National University,
Canberra, ACT 0200, Australia
qing.wang@anu.edu.au

Abstract. We develop a logic which enables reasoning about single steps of non-deterministic parallel Abstract State Machines (ASMs). Our logic builds upon the unifying logic introduced by Nanchen and Stärk for reasoning about hierarchical (parallel) ASMs. Our main contribution to this regard is the handling of non-determinism (both bounded and unbounded) within the logical formalism. Moreover, we do this without sacrificing the completeness of the logic for statements about single steps of non-deterministic parallel ASMs, such as invariants of rules, consistency conditions for rules, or step-by-step equivalence of rules.

1 Introduction

Gurevich's Abstract State Machines (ASMs) provide not only a formal theory of algorithms, but also are the basis for a general software engineering method based in the specification of higher-level ground models and step-by-step refinement. Chapter 9 in the book [5] gives a summary of many application projects that have developed complex systems solutions on the grounds of ASMs. A major advantage of the ASM method and a key for its success resides in the fact that it provides, not only a simple and precise framework to communicate and document design ideas, but also an accurate and checkable overall understanding of complex systems. In this context, formal verification of dynamic properties for given ASMs is a fundamentally important task, in particular in the case of modelling safety critical systems, where there is a need to ensure the integrity and reliability of the system. Clearly, a logical calculi appropriate for the formalisation and reasoning about dynamic properties of ASMs is an essential and valuable tool for this endeavour.

Numerous logics have been developed to deal with specific features of ASM verification such as correctness and deadlock-freeness (see Sect. 9.4.3 in the book [5]) for detailed references), but a complete logic for ASMs was only developed

Work supported by the **Austrian Science Fund (FWF: [P26452-N15])**. Project: *Behavioural Theory and Logics for Distributed Adaptive Systems.*

M. Gyssens and G. Simari (Eds.): FoIKS 2016, LNCS 9616, pp. 334–354, 2016.
DOI: 10.1007/978-3-319-30024-5_18

in [13] by Nanchen and Stärk. The logic formalizes properties of a single step of an ASM, which permits to define Hilbert-style proof theory and to show its completeness. In this work the treatment of non-determinism was deliberately left out. Same as parallelism, which is on the other hand captured by the logic for ASMs of Nanchen and Stärk, non-determinism is also a prevalent concept in the design and implementation of software systems, and consequently a constitutive part of the ASM method for systems development [5]. Indeed, nondeterminism arises in the specification of many well known algorithms and software applications. Examples range from graph algorithms, such as minimum spanning tree and shortest path, to search techniques whose objective is to arrive at some admissible goal state (as in the n-queens and combinatorial-assignment problems [7]), and learning strategies such as converging on some classifier that labels all data instances correctly [14]. Non-deterministic behavior is also common in cutting edge fields of software systems. Distributed systems frequently need to address non-deterministic behaviour such as changing role (if possible) as strategic response to observed problems concerning load, input, throughput, etc. Also, many cyber-physical systems and hybrid systems such as railway transportation control systems [2] and systems used in high-confidence medical healthcare devices exhibit highly non-deterministic behaviour.

Notice that although we could say that there is a kind of latent parallelism in non-determinism, they represent completely different behaviours and thus both are needed to faithfully model the behaviour of complex systems, more so in the case of the ASM method where the ability to model systems at every level of abstraction is one of its main defining features. For instance, while a nondeterministic action can evaluate to multiple behaviors, only if at least one of these behaviors does not conflict with concurrent tasks, then there is an admissible execution of the action in parallel with these tasks.

The ASM method allows for two different, but complementary, approaches to non-determinism. The first approach assumes that choices are made by the environment via monitored functions that can be viewed as external oracles. In this case, non-deterministic ASMs are just interactive ASMs. The second approach assumes the ASMs themselves rather than the environment, to have the power of making non-deterministic choices. In this case the one-step transition function of the ASMs is no longer a function but a binary relation. This is also the approach followed by non-deterministic Turing machines. However, in the case of non-deterministic Turing machines the choice is always bounded by the transition relation. For ASMs the non-determinism can also be unbounded, i.e., we can choose among an infinite number of possibilities. Clearly, unbounded nondeterminism should also be allowed if we want our ASMs to be able to faithfully model algorithms at any level of abstraction.

In this work we develop a logic which enables reasoning about single steps of non-deterministic parallel ASMs, i.e., ASMs which include the well known **choose** and **forall** rules [5]. This builds upon the complete logic introduced in the work of Nanchen and Stärk [13] for reasoning about single steps of hierarchical ASMs. Hierarchical ASMs capture the class of synchronous and deterministic parallel algorithms in the precise sense of the ASM thesis of Blass and Gurevich [3,4] (see also [6]). Our main contribution to this regard is the handling of non-determinism (both bounded and unbounded) within the logical formalism. More importantly, this is done without sacrificing the completeness of the logic. As highlighted by Nanchen and Stärk [13], non-deterministic transitions manifest themselves as a difficult task in the logical formalisation for ASMs.

The paper is organized as follows. The next section introduces the required background from ASMs. Section 3 formalises the model of non-deterministic parallel ASM used through this work. In Sect. 4 we introduce the syntax and semantics of the proposed logic for non-deterministic parallel ASMs. Section 5 presents a detailed discussion regarding consistency and update sets, and the formalisation of a proof system. In Sect. 6 we use the proof system to derive some interesting properties of our logic, including known properties of the ASM logic in [13]. In Sect. 7 we present our main result, namely that the proposed logic is complete for statements about single steps of non-deterministic parallel ASMs, such as invariants of rules, consistency conditions for rules, or step-by-step equivalence of rules. We conclude our work in Sect. 8.

2   Preliminaries

The concept of Abstract State Machines (ASMs) is well known [5]. In its simplest form an ASM is a finite set of so-called *transition rules* of the form **if** *Condition* **then** *Updates* **endif** which transforms abstract states. The condition or guard under which a rule is applied is an arbitrary first-order logic sentence. *Updates* is a finite set of assignments of the form $f(t_1, \ldots, t_n) := t_0$ which are executed in parallel. The execution of $f(t_1, \ldots, t_n) := t_0$ in a given state proceeds as follows: first all parameters $t_0, t_1, \ldots t_n$ are evaluated to their values, say $a_0, a_1, \ldots, a_n$, then the value of $f(a_1, \ldots, a_n)$ is updated to a_0, which represents the value of $f(a_1, \ldots, a_n)$ in the next state. Such pairs of a function name f, which is fixed by the signature, and optional argument $(a_1, \ldots, a_n)$ of dynamic parameters values a_i, are called *locations*. They represent the abstract ASM concept of memory units which abstracts from particular memory addressing. Location value pairs (ℓ, a), where ℓ is a location and a a value, are called *updates* and represent the basic units of state change.

The notion of ASM *state* is the classical notion of *first-order structure* in mathematical logic. For the evaluation of first-order terms and formulae in an ASM state, the standard interpretation of function symbols by the corresponding functions in that state is used. As usually in this setting and w.l.o.g., we treat predicates as characteristic functions and constants as 0-ary functions.

The notion of the ASM *run* is an instance of the classical notion of the computation of transition systems. An ASM computation step in a given state consists in executing *simultaneously* all updates of all transition rules whose guard is true in the state, if these updates are consistent, in which case the result of their execution yields a next state. In the case of inconsistency, the computation does not yield a next state. A set of updates is *consistent* if it contains no pairs (ℓ, a), (ℓ, b) of updates to a same location ℓ with $a \neq b$.

Simultaneous execution, as obtained in one step through the execution of a set of updates, provides a useful instrument for high-level design to locally describe a global state change. This synchronous parallelism is further enhanced by the transition rule **forall** x **with** φ **do** r **enddo** which expresses the simultaneous execution of a rule r for each x satisfying a given condition φ.

Similarly, non-determinism as a convenient way of abstracting from details of scheduling of rule executions can be expressed by the rule **choose** x **with** φ **do** r **enddo**, which means that r should be executed with an arbitrary x chosen among those satisfying the property φ.

The following example borrowed from [5] clearly illustrates the power of the **choose** and **forall** rules.

Example 1. The following ASM generates all and only the pairs $vw \in A^*$ of different words v, w of same length (i.e., $v \neq w$ and $|v| = |w|$).

```
choose n, i with i < n do
    choose a, b with a ∈ A ∧ b ∈ A ∧ a ≠ b do
        v(i) := a
        w(i) := b
        forall j with j < n ∧ j ≠ i do
            choose a, b with a ∈ A ∧ b ∈ A do
                v(j) := a
                w(j) := b
            enddo
        enddo
    enddo
enddo
```

When all possible choices are realized, the set of reachable states of this ASM is the set of all "vw" states with $v \neq w$ and $|v| = |w|$.

3 Non-deterministic Parallel ASMs

It is key for the completeness of our logic to make sure that the ASMs do not produce infinite update sets. For that we formally define ASM states as simple metafinite structures [8] instead of classical first-order structures, and restrict the variables in the **forall** rules to range over the finite part of such metafinite states. Nevertheless, the class of algorithms that are captured by these ASM machines coincides with the class of parallel algorithms that satisfy the postulates of the parallel ASM thesis of Blass and Gurevich [3,4] (see [6] for details).

A *metafinite structure* S consists of: a finite first-order structure S_1 –the *primary part* of S; a possibly infinite first-order structure S_2 –the *secondary part* of S; and a finite set of functions which map elements of S_1 to elements of S_2 –the *bridge functions*. A signature Υ of metafinite structures comprises a sub-signature Υ_1 for the primary part, a sub-signature Υ_2 for the secondary part and a finite set $\mathcal{F}_b$ of bridge function names. The *base set* of a state S is a nonempty set of values $B = B_1 \cup B_2$, where B_1 is the finite domain of S_1, and B_2 is the possibly infinite domain of S_2. Function symbols f in Υ_1 and Υ_2 are interpreted as functions f^S over B_1 and B_2, respectively. The interpretation of a n-ary function symbol $f \in \mathcal{F}_b$ defines a function f^S from B_1^n to B_2. As usual, we distinguish between *updatable* dynamic functions and static functions.

Let $\Upsilon = \Upsilon_1 \cup \Upsilon_2 \cup \mathcal{F}_b$ be a signature of metafinite states. Fix a countable set $\mathcal{X} = \mathcal{X}_1 \cup \mathcal{X}_2$ of first-order variables. Variables in $\mathcal{X}_1$, denoted with standard lowercase letters $x, y, z, \ldots$, range over the primary part of a meta-finite state (i.e., the finite set B_1), whereas variables in $\mathcal{X}_2$, denoted with typewriter-style lowercase letters $\mathtt{x}, \mathtt{y}, \mathtt{z}, \ldots$, range over B_2. The set of first-order terms $\mathcal{T}_{\Upsilon,\mathcal{X}}$ of vocabulary Υ is defined in a similar way than in meta-finite model theory [8]. That is, $\mathcal{T}_{\Upsilon,\mathcal{X}}$ is constituted by the set $\mathcal{T}_p$ of *point terms* and the set $\mathcal{T}_a$ of *algorithmic terms*. The set of point terms $\mathcal{T}_p$ is the closure of the set $\mathcal{X}_1$ of variables under the application of function symbols in Υ_1. The set of algorithmic terms $\mathcal{T}_a$ is defined inductively: Every variable in $\mathcal{X}_2$ is an algorithmic term in $\mathcal{T}_a$; If $t_1, \ldots, t_n$ are point terms in $\mathcal{T}_p$ and f is an n-ary bridge function symbol in $\mathcal{F}_b$, then $f(t_1, \ldots, t_n)$ is an algorithmic term in $\mathcal{T}_a$; if $t_1, \ldots, t_n$ are algorithmic terms in $\mathcal{T}_a$ and f is an n-ary function symbol in Υ_2, then $f(t_1, \ldots, t_n)$ is an algorithmic term in $\mathcal{T}_a$; nothing else is an algorithmic term in $\mathcal{T}_b$.

Let S be a meta finite state of signature Υ. A *valuation* or *variable assignment* ζ is a function that assigns to every variable in $\mathcal{X}_1$ a value in the base set B_1 of the primary part of S and to every variable in $\mathcal{X}_2$ a value in the base set B_2 of the secondary part of S. The value $val_{S,\zeta}(t)$ of a term $t \in \mathcal{T}_{\Upsilon,\mathcal{X}}$ in the state S under the valuation ζ is defined as usual in first-order logic. The *first-order logic of metafinite structures* (states) is defined as the first-order logic with equality which is built up from equations between terms in $\mathcal{T}_{\Upsilon,\mathcal{X}}$ by using the standard connectives and first-order quantifiers. Its semantics is defined in the standard way. The truth value of a first-order formula of meta finite structures φ in S under the valuation ζ is denoted as $[\![\varphi]\!]_{S,\zeta}$.

In our definition of ASM rule, we use the fact that function arguments can be read as tuples. Thus, if f is an n-ary function and $t_1, \ldots, t_n$ are arguments for f, we write $f(t)$ where t is a term which evaluates to the tuple $(t_1, \ldots, t_n)$, instead of $f(t_1, \ldots, t_n)$. This is not strictly necessary, but it greatly simplifies the presentation of the technical details in this paper. Let t and s denote terms in $\mathcal{T}_p$, let $\mathtt{t}$ and $\mathtt{s}$ denote terms in $\mathcal{T}_a$ and let φ denote a first-order formula of metafinite structures of vocabulary Υ. The set of *ASM rules* over Υ is inductively defined as follows:

- *update rule 1:* $f(t) := s$ (where $f \in \Upsilon_1$);
- *update rule 2:* $f(\mathtt{t}) := \mathtt{s}$ (where $f \in \Upsilon_2$);
- *update rule 3:* $f(t) := \mathtt{s}$ (where $f \in \mathcal{F}_b$);
- *conditional rule:* **if** φ **then** r **endif**
- *forall rule:* **forall** x **with** φ **do** r **enddo**
- *bounded choice rule:* **choose** x **with** φ **do** r **enddo**
- *unbounded choice rule:* **choose** $\mathtt{x}$ **with** φ **do** r **enddo**

- *parallel rule*: **par** r_1 r_2 **endpar** (execute the rules r_1 and r_2 in parallel);
- *sequence rule*: **seq** r_1 r_2 **endseq** (first execute rule r_1 and then rule r_2).

If r is an ASM rule of signature Υ and S is a state of Υ, then we associate to them a set $\Delta(r, S, \zeta)$ of update sets which depends on the variable assignment ζ. Let $\zeta[x \mapsto a]$ denote the variable assignment which coincides with ζ except that it assigns the value a to x. We formally define in Fig. 1 the sets of update sets yielded by the ASM rules. Items 1–3 in Fig. 1 correspond to the update rules 1–3, respectively. Each update rules yields a set which contains a single update set, which in turns contains a single update to a function of S. Depending on whether the function name f belongs to Υ_1, Υ_2 or $\mathcal{F}_b$, the produced update corresponds to a function in the primary or secondary part of S or to a bridge function, respectively. The choice rules introduce non-determinism. The bounded choice rule yields a finite set of update sets, since x range over the (finite) primary part of S (see item 6 in Fig. 1). The unbounded choice rule yields a possibly infinite set of update sets (see item 7 in Fig. 1). In this latter case, $\mathbf{x}$ range over the (possible infinite) secondary part of S and it might happen that there are infinite valuations for $\mathbf{x}$ that satisfy the condition φ, each resulting in a different update set. All other rules only rearrange updates into different update sets. Update sets are explained in more detail in Sect. 5.2.

Remark 1. For every state S, ASM rule r and variable assignment ζ, we have that every $\Delta \in \Delta(r, S, \zeta)$ is a finite set of updates. This is a straightforward consequence of the fact that the variable x in the definition of the **forall** rule ranges over the (finite) primary part of S, and it is also the case in the ASM thesis for parallel algorithms of Blass and Gurevich [3,4] where it is implicitly assumed that the **forall** rule in the parallel ASMs range over finite hereditary multisets. See our work in [6] for a detailed explanation. Regarding the set $\Delta(r, S, \zeta)$ of update sets, we note that it might be infinite since the unbounded choice rule can potentially produce infinitely many update sets. In fact, this is the case if we consider the first unbounded choice rule in Example 1.

Formally, a *non-deterministic parallel* ASM M over a signature Υ of metafinite states consists of: (a) a set $\mathcal{S}$ of metafinite states over Υ, (b) non-empty subsets $\mathcal{S}_I \subseteq \mathcal{S}$ of *initial states* and $\mathcal{S}_F \subseteq \mathcal{S}$ of *final states*, and (c) a *closed* ASM rule r over Υ, i.e., a rule r in which all free variables in the first-order formulae of the rule are bounded by **forall** or **choose** constructs.

Every non-deterministic parallel ASM M defines a corresponding *successor relation* δ over $\mathcal{S}$ which is determined by the main rule r of M. A pair of states (S_1, S_2) belongs to δ iff there is a consistent update set $\Delta \in \Delta(r, S)$ (the valuation ζ is omitted from $\Delta(r, S, \zeta)$ since r is closed) such that S_2 is the unique state resulting from updating S_1 with Δ. A *run* of an ASM M is a finite sequence $S_0, \ldots, S_n$ of states with $S_0 \in \mathcal{S}_I$, $S_n \in \mathcal{S}_F$, $S_i \notin \mathcal{S}_F$ for $0 < i < n$, and $(S_i, S_{i+1}) \in \delta$ for all $i = 0, \ldots, n-1$.

The following example, adapted from [10], illustrates a parallel ASMs with bounded non-determinism.

1. $\Delta(f(t) := s, S, \zeta) = \{\{(f, (a), b)\}\}$ for $a = vals_{S,\zeta}(t) \in B_1$ and $b = vals_{S,\zeta}(s) \in B_1$

2. $\Delta(f(\mathbf{t}) := \mathbf{s}, S, \zeta) = \{\{(f, (a), b)\}\}$ for $a = vals_{S,\zeta}(\mathbf{t}) \in B_2$ and $b = vals_{S,\zeta}(\mathbf{s}) \in B_2$

3. $\Delta(f(t) := \mathbf{s}, S, \zeta) = \{\{(f, (a), b)\}\}$ for $a = vals_{S,\zeta}(t) \in B_1$ and $b = vals_{S,\zeta}(\mathbf{s}) \in B_2$

4. $\Delta(\text{if } \varphi \text{ then } r \text{ endif}, S, \zeta) = \begin{cases} \Delta(r, S, \zeta) & \text{if } [\![\varphi]\!]_{S,\zeta} = \text{true} \\ \{\emptyset\} & \text{otherwise} \end{cases}$

5. $\Delta(\text{forall } x \text{ with } \varphi \text{ do } r \text{ enddo}, S, \zeta) =$
   $$\{\Delta_1 \cup \cdots \cup \Delta_n \mid \Delta_i \in \Delta(r, S, \zeta[x \mapsto a_i])\},$$
   where $\{a_1, \ldots, a_n\} = \{a_i \in B_1 \mid [\![\varphi]\!]_{S,\zeta[x \mapsto a_i]} = true\}$

6. $\Delta(\text{choose } x \text{ with } \varphi \text{ do } r \text{ enddo}, S, \zeta) =$
   $$\bigcup_{a_i \in B_1} \{\Delta(r, S, \zeta[x \mapsto a_i]) \mid [\![\varphi]\!]_{S,\zeta[x \mapsto a_i]} = \text{true}\}$$

7. $\Delta(\text{choose } \mathbf{x} \text{ with } \varphi \text{ do } r \text{ enddo}, S, \zeta) =$
   $$\bigcup_{a_i \in B_2} \{\Delta(r, S, \zeta[x \mapsto a_i]) \mid [\![\varphi]\!]_{S,\zeta[x \mapsto a_i]} = \text{true}\}$$

8. $\Delta(\text{par } r_1 \ r_2 \text{ endpar}, S, \zeta) =$
   $$\{\Delta_1 \cup \Delta_2 \mid \Delta_1 \in \Delta(r_1, S, \zeta) \text{ and } \Delta_2 \in \Delta(r_2, S, \zeta)\}$$

9. $\Delta(\text{seq } r_1 \ r_2 \text{ endseq}, S, \zeta) =$
   $$\{\Delta_1 \oslash \Delta_2 \mid \Delta_1 \in \Delta(r_1, S, \zeta) \text{ is consistent and } \Delta_2 \in \Delta(r_2, S + \Delta_1, \zeta)\} \cup$$
   $$\{\Delta_1 \in \Delta(r_1, S, \zeta) \mid \Delta_1 \text{ is inconsistent}\},$$
   where $\Delta_1 \oslash \Delta_2 = \Delta_2 \cup \{(\ell, a) \in \Delta_1 \mid \ell \neq \ell' \text{ for all } (\ell', a') \in \Delta_2\}$

Fig. 1. Sets of Update Sets of Non-deterministic Parallel ASMs

Example 2. We consider metafinite states with: (a) a primary part formed by a connected weighted graph $G = (V, E)$, (b) a secondary part formed by the set of natural numbers $\mathbb{N}$, and (c) a bridge function *weight* from the set of edges in E to $\mathbb{N}$. Apart from the static (Boolean) function symbols V and E, the vocabulary of the primary part of the states also includes dynamic function symbols *label* and T, and static function symbols *first* and *second*, the last two for extracting the first and second element of an ordered pair, respectively. Since G is an undirected graph, we have that $(x, y) \in E$ iff $(y, x) \in E$.

The non-deterministic parallel ASM in this example, which we denote as M, formally expresses Kruskal's algorithm [12] for computing the *minimum spanning tree* in a connected, weighted graph. Recall that a spanning tree T of a graph G is a tree such that every pair of nodes in G are connected via edges in T. We say that T is minimum if the sum of the weights of all its edges is the least among all spanning trees of G. We assume that in every initial state of M, $label(x) = x$ for every $x \in V$ and that $T((x, y)) = false$ for every $(x, y) \in E$.

The condition in the first **choose** rule is simply ensuring that the chosen edge x is eligible, i.e., that the nodes $first(x)$ and $second(x)$ that make up the endpoints of the edge x have different labels, and that x has minimal weight among the set of eligible edges. The following two update rules simply add the edge x to the tree T. The second **choose** rule reflects the fact that from the point of view of the correctness of the algorithm, it does not matter which endpoint y of the edge x we choose at this stage. Finally, the **forall** rule simply relabels (as expected) every node with the same label than the endpoint y of x (including the node y itself) with the label of the opposite endpoint of x.

```
choose x with E(x) ∧ label(first(x)) ≠ label(second(x))∧
        ∀y (E(y) ∧ label(first(y)) ≠ label(second(y))) → weight(y) ≥ weight(x)) do
    T(x) := true
    T((second(x), first(x))) := true
    choose y with y = first(x) ∨ y = second(x) do
        forall z with label(z) = label(y) do
            if label(y) = label(first(x)) then label(z) := label(second(x)) endif
            if label(y) = label(second(x)) then label(z) := label(first(x)) endif
        enddo
    enddo
enddo
```

4 A Logic for Non-deterministic Parallel ASMs

The logic for non-deterministic parallel ASMs (denoted $\mathcal{L}$) is a dynamic first-order logic extended with membership predicates over finite sets, an update set predicate and a multi-modal operator. $\mathcal{L}$ is defined over many sorted first-order structures which have:

- a *finite individual sort* with variables $x_1, x_2, \ldots$ which range over a finite domain D_1,
- an *individual sort* with variables $x_1, x_2, \ldots$, which range over a (possibly infinite) domain D_2, and
- a *predicate sort* with variables $x_1^1, x_2^1, \ldots$, which range over the domain P_1 formed by all finite subsets (relations) on $\mathcal{F}_{dyn} \times (D_1 \cup D_2) \times (D_1 \cup D_2)$.
- a *predicate sort* with variables $x_1^2, x_2^2, \ldots$, which range over the domain P_2 formed by all finite subsets (relations) on $\mathcal{F}_{dyn} \times (D_1 \cup D_2) \times (D_1 \cup D_2) \times D_1$.

A signature Σ of the logic $\mathcal{L}$ comprises a finite set F_1 of names for functions on D_1, a finite set F_2 of names for functions on D_2, and a finite set F_b of names for functions which take arguments from D_1 and return values on D_2.

We define terms of $\mathcal{L}$ by induction. Variables $x_1, x_2, \ldots$ and $x_1, x_2, \ldots$ are terms of the first and second individual sort, respectively. Variables $x_1^1, x_2^1, \ldots$ and $x_1^2, x_2^2, \ldots$ are terms of the first and second predicate sort, respectively. If f is an n-ary function name in F_1 and $t_1, \ldots, t_n$ are terms of the first individual sort, then $f(t_1, \ldots, t_n)$ is a term of the first individual sort. If f is an n-ary function name in F_2 and $t_1, \ldots, t_n$ are terms of the second individual sort, then

$f(t_1, \ldots, t_n)$ is a term of the second individual sort. If f is an n-ary function name in F_b and $t_1, \ldots, t_n$ are terms of the first individual sort, then $f(t_1, \ldots, t_n)$ is a term of the second individual sort.

The formulae of $\mathcal{L}$ are those generated by the following grammar:

$$\varphi, \psi ::= s = t \mid s_a = t_a \mid \neg\varphi \mid \varphi \wedge \psi \mid \forall x(\varphi) \mid \forall \mathbf{x}(\varphi) \mid \forall x^1(\varphi) \mid \forall x^2(\varphi) \mid$$
$$\in^1(x^1, f, t_0, s_0) \mid \in^2(x^2, f, t_0, s_0, s) \mid \mathrm{upd}(r, x^1) \mid [x^1]\varphi$$

where s and t denote terms of the first individual sort, s_a and t_a denote terms of the second individual sort, f is a dynamic function symbol, r is an ASM rule and, t_0 and s_0 denote terms of either the first or the second individual sort.

The interpretation of terms and the semantics of the first-order formulae is defined in the standard way. This includes equality which is used under a fixed interpretation and only between terms of a same individual sort.

The update set predicate $\mathrm{upd}(r, x^1)$ states that the *finite* update set represented by x^1 is generated by the rule r. Let S be a state of some signature Σ of the logic $\mathcal{L}$. Let ζ be a variable assignment over S which maps each variable of the first and second individual sort to a value in D_1 and D_2, respectively, and maps each variable of the first and second predicate sort to a value in P_1 and P_2, respectively. The truth value of $\mathrm{upd}(r, x^1)$ is defined by $[\![\mathrm{upd}(r, x^1)]\!]_{S,\zeta} = true$ iff $val_{S,\zeta}(x^1) \in \Delta(r, S, \zeta)$.

The set membership predicate $\in^1(x^1, f, t_0, s_0)$ indicates that (f, t_0, s_0) is an update in the update set represented by x^1 while the auxiliary set membership predicate $\in^2(x^2, f, t_0, s_0, s)$ is used to keep track of which parallel branch produced each update in x^2. Their truth values are formally defined as follows: $[\![\in^1 (x^1, f, t_0, s_0)]\!]_{S,\zeta} = true$ iff $(f, val_{S,\zeta}(t_0), val_{S,\zeta}(s_0)) \in val_{S,\zeta}(x^1)$ $[\![\in^2 (x^2, f, t_0, s_0, s)]\!]_{S,\zeta} = true$ iff $(f, val_{S,\zeta}(t_0), val_{S,\zeta}(s_0), val_{S,\zeta}(s)) \in val_{S,\zeta}(x^2)$

Finally, we use $[x^1]\varphi$ to express the evaluation of φ over the successor state obtained by applying the updates in x^1 to the current state. Its truth value is defined by: $[\![[x^1]\varphi]\!]_{S,\zeta} = true$ iff $\Delta = \zeta(x^1)$ is inconsistent or $[\![\varphi]\!]_{S+\Delta,\zeta} = true$ for $\zeta(x^1) = \Delta \in \Delta(r, S, \zeta)$. That is, when $\Delta = \zeta(x^1)$ is inconsistent, successor states for the current state S do not exist and thus $S + \Delta$ is undefined. In this case, $[x^1]\varphi$ is interpreted as *true*. With the use of the modal operator $[\,]$ for an update set $\Delta = \zeta(x^1)$ (i.e., $[x^1]$), $\mathcal{L}$ is empowered to be a multi-modal logic.

We say that a formula φ of $\mathcal{L}$ is *static* if all the function symbols which appear in φ are static and say that it is *pure* if it is generated by the following grammar: $\varphi, \psi ::= s = t \mid s_a = t_a \mid \neg\varphi \mid \varphi \wedge \psi \mid \forall x(\varphi) \mid \forall \mathbf{x}(\varphi)$.

Since metafinite states are just a special kind of two sorted first-order structures in which one of the sorts is finite, we can identify every metafinite state S of $\mathcal{L}$ with a corresponding many sorted first-order structure S' of the class used in definition of $\mathcal{L}$. This can be done by taking the domains D_1 and D_2 of the individual sorts of S' to be the base sets B_1 and B_2 of S, respectively, the sets F_1, F_2 and F_b of function names of the signature Σ of S' to be the sets Υ_1, Υ_2 and $\mathcal{F}_b$ of the signature Υ of S, respectively, and the interpretation in S' of the function names in Σ to coincide with the interpretation in S of the corresponding function symbols in Υ. Following this transformation we have that for every state S, every corresponding pair of many sorted first-order structure S' and S'' are isomorphic by an isomorphism which is the identity among elements of the individual sorts. Thus, we can talk of *the* many sorted structure S corresponding to a state S and, when it is clear from the context, we can even talk of the state S meaning the many sorted structure S.

In what follows, we use the somehow clearer and more usual syntax of second-order logic to denote the set membership predicates and the quantification over the predicate sorts. Thus we use upper case letters $X, Y, \ldots$ and $\mathcal{X}, \mathcal{Y}, \ldots$ to denote variables $x_1^1, x_2^1, \ldots$ and $x_1^2, x_2^2, \ldots$ of the first and second predicate sorts, respectively, and we write $\forall X(\varphi)$, $\forall \mathcal{X}(\varphi)$, $[X]\varphi$, $X(f, t_0, s_0)$, $\mathcal{X}(f, t_0, s_0, s)$ and $\mathrm{upd}(r, X)$ instead of $\forall x^1 (\varphi)$, $\forall x^2 (\varphi)$, $[x^1]\varphi$, $\in^1 (x^1, f, t_0, s_0)$, $\in^1 (x^1, f, t_0, s_0, s)$ and $\mathrm{upd}(r, x^1)$, respectively. Furthermore, in our formulae we use disjunction $\vee$, implication $\to$, double implication $\leftrightarrow$ and existential quantification $\exists$. All of them are defined as abbreviations in the usual way.

Example 3. $\mathcal{L}$ can express properties of the ASM in Example 2 such as:

- If r yields in the current state S an update set Δ with an update $(T, x, true)$, then in the successor state $S + \Delta$ the vertices of x have a same label.

$\forall X(\mathrm{upd}(r, X) \quad \to \quad \forall x(X(T, x, true) \quad \to \quad [X](label(first(x)) \quad = label(second(x)))))$

- Each update set yielded by r updates T in no more than one location.

$\forall X(\mathrm{upd}(r, X) \to \neg(\exists xy(X(T, x, true) \wedge X(T, y, true) \wedge x \neq y)))$

- If an edge x meets in a state S the criteria of the first **choose** rule in r, then there is an update set $\Delta \in \Delta(r, S)$ such that $T(x) = true$ holds in $S + \Delta$.

$\forall x(E(x) \wedge label(first(x)) \neq label(second(x)) \wedge$
$\quad \forall y(E(y) \wedge label(first(y)) \neq label(second(y)) \to weight(y) \geq weight(x))$
$\quad \to \exists X(\mathrm{upd}(r, X) \wedge [X](T(x) = true)))$

5 A Proof System

In this section we develop a proof system for the logic $\mathcal{L}$ for non-deterministic parallel ASMs.

Definition 1. *We say that a state S is a* model *of a formula φ (denoted as $S \models \varphi$) iff $\llbracket \varphi \rrbracket_{S,\zeta} = true$ holds for every variable assignment ζ. If Ψ is a set of formulae, we say that S models Ψ (denoted as $S \models \Psi$) iff $S \models \varphi$ for each $\varphi \in \Psi$. A formula φ is said to be a* logical consequence *of a set Ψ of formulae (denoted as $\Psi \models \varphi$) if for every state S, if $S \models \Psi$, then $S \models \varphi$. A formula φ is said to be* valid *(denoted as $\models \varphi$) if $\llbracket \varphi \rrbracket_{S,\zeta} = true$ in every state S for every variable assignment ζ. A formula φ is said to be* derivable *from a set Ψ of formulae (denoted as $\Psi \vdash_{\mathfrak{R}} \varphi$) if there is a deduction from formulae in Ψ to φ by using a set $\mathfrak{R}$ of axioms and inference rules.*

We will define such a set $\mathfrak{R}$ of axioms and rules in Subsect. 5.3. Then we simply write $\vdash$ instead of $\vdash_{\mathfrak{R}}$. We also define equivalence between two ASM rules. Two equivalent rules r_1 and r_2 are either both defined or both undefined.

Definition 2. *Let r_1 and r_2 be two ASM rules. Then r_1 and r_2 are equivalent (denoted as $r_1 \equiv r_2$) if for every state S it holds that $S \models \forall X(\mathrm{upd}(r_1, X) \leftrightarrow \mathrm{upd}(r_2, X))$.*

5.1   Consistency

In [13] Nanchen and Stärk use a predicate $\mathrm{Con}(r)$ as an abbreviation for the statement that the rule r is consistent. As every rule r in their work is deterministic, there is no ambiguity with the reference to the update set associated with r, i.e., each deterministic rule r generates exactly one (possibly empty) update set. Thus a deterministic rule r is consistent iff the update set generated by r is consistent. However, in our logic $\mathcal{L}$, the presence of non-determinism makes the situation less straightforward.

Let r be an ASM rule and Δ be an update set. Then the consistency of an update set Δ, denoted by the formula $\mathrm{conUSet}(X)$ (where X represents Δ), can be expressed as:

$$\mathrm{conUSet}(X) \equiv \bigwedge_{f \in \mathcal{F}_{dyn}} \forall xyz((X(f,x,y) \wedge X(f,x,z)) \to y = z) \qquad (1)$$

Then $\mathrm{con}(r, X)$ is an abbreviation of the following formula which expresses that an update set Δ (represented by the variable X) generated by the rule r is consistent.

$$\mathrm{con}(r, X) \equiv \mathrm{upd}(r, X) \wedge \mathrm{conUSet}(X) \qquad (2)$$

As the rule r may be non-deterministic, it is possible that r yields several update sets. Thus, we develop the consistency of ASM rules in two versions:

– A rule r is *weakly consistent* (denoted as $\mathrm{wcon}(r)$) if at least one update set generated by r is consistent. This can be expressed as follows:

$$\mathrm{wcon}(r) \equiv \exists X(\mathrm{con}(r, X)) \qquad (3)$$

– A rule r is *strongly consistent* (denoted as $\mathrm{scon}(r)$) if every update set generated by r is consistent. This can be expressed as follows:

$$\mathrm{scon}(r) \equiv \forall X(\mathrm{upd}(r, X) \Rightarrow \mathrm{con}(r, X)) \qquad (4)$$

In the case that a rule r is deterministic, the weak notion of consistency coincides with the strong notion of consistency, i.e., $\mathrm{wcon}(r) \leftrightarrow \mathrm{scon}(r)$.

U1. $\mathrm{upd}(f(t) := s, X) \leftrightarrow X(f, t, s) \wedge \forall xy(X(f, x, y) \rightarrow x = t \wedge y = s) \wedge$
$$\bigwedge_{f \neq f' \in \mathcal{F}_{dyn},} \forall xy(\neg X(f', x, y))$$

U2. $\mathrm{upd}(\mathbf{if}\,\varphi\,\mathbf{then}\,r\,\mathbf{endif}, X) \leftrightarrow (\varphi \wedge \mathrm{upd}(r, X)) \vee (\neg\varphi \wedge \bigwedge_{f \in \mathcal{F}_{dyn}} \forall xy(\neg X(f, x, y))$

U3. $\mathrm{upd}(\mathbf{forall}\,x\,\mathbf{with}\,\varphi\,\mathbf{do}\,r\,\mathbf{enddo}, X) \leftrightarrow$
$$\exists \mathcal{X}\Big(\forall x\big((\varphi \rightarrow \exists Y(\mathrm{upd}(r, Y) \wedge \bigwedge_{f \in \mathcal{F}_{dyn}} \forall y_1 y_2(Y(f, y_1, y_2) \leftrightarrow \mathcal{X}(f, y_1, y_2, x))))\wedge$$
$$(\neg\varphi \rightarrow \bigwedge_{f \in \mathcal{F}_{dyn}} \forall y_1 y_2(\neg\mathcal{X}(f, y_1, y_2, x))))\wedge$$
$$\bigwedge_{f \in \mathcal{F}_{dyn}} \forall x_1 x_2(X(f, x_1, x_2) \leftrightarrow \exists x_3(\mathcal{X}(f, x_1, x_2, x_3))))\Big)$$

U4. $\mathrm{upd}(\mathbf{par}\,r_1\,r_2\,\mathbf{endpar}, X) \leftrightarrow \exists Y_1 Y_2(\mathrm{upd}(r_1, Y_1) \wedge \mathrm{upd}(r_2, Y_2)\wedge$
$$\bigwedge_{f \in \mathcal{F}_{dyn}} \forall xy(X(f, x, y) \leftrightarrow (Y_1(f, x, y) \vee Y_2(f, x, y)))$$

U5. $\mathrm{upd}(\mathbf{choose}\,x\,\mathbf{with}\,\varphi\,\mathbf{do}\,r\,\mathbf{enddo}, X) \leftrightarrow \exists x(\varphi \wedge \mathrm{upd}(r, X))$

U6. $\mathrm{upd}(\mathbf{choose}\,\mathbf{x}\,\mathbf{with}\,\varphi\,\mathbf{do}\,r\,\mathbf{enddo}, X) \leftrightarrow \exists\mathbf{x}(\varphi \wedge \mathrm{upd}(r, X))$

U7. $\mathrm{upd}(\mathbf{seq}\,r_1\,r_2\,\mathbf{endseq}, X) \leftrightarrow \big(\mathrm{upd}(r_1, X) \wedge \neg\mathrm{con}(X)\big)\vee$
$$(\exists Y_1 Y_2(\mathrm{upd}(r_1, Y_1) \wedge \mathrm{con}(Y_1) \wedge [Y_1]\mathrm{upd}(r_2, Y_2)\wedge$$
$$\bigwedge_{f \in \mathcal{F}_{dyn}} \forall xy(X(f, x, y) \leftrightarrow ((Y_1(f, x, y) \wedge \forall z(\neg Y_2(f, x, z))) \vee Y_2(f, x, y)))))$$

Fig. 2. Axioms for Predicate $\mathrm{upd}(r, X)$

5.2 Update Sets

We present the axioms for the predicate $upd(r, X)$ in Fig. 2. To simplify the presentation, we give the formulae only for the case in which all the function symbols in $\mathcal{F}_{dyn}$ correspond to functions on the primary part (finite individual sort) of the state. To deal with dynamic function symbols corresponding to function of the secondary part and to bridge functions, we only need to slightly change the formulae by replacing some of the first-order variables in $\mathcal{X}_1$ by first-order variables in $\mathcal{X}_2$. For instance, if f is a bridge function symbol, we should write $\forall xy(X(f, x, \mathrm{y}) \rightarrow x = t \wedge \mathrm{y} = \mathbf{s})$ instead of $\forall xy(X(f, x, y) \rightarrow x = t \wedge y = s)$.

In the following we explain Axioms **U1-U7** in turn. We assume a state S of some signature Υ and base set $B = B_1 \cup B_2$, where B_1 is the base set of the *finite* primary part of S. We also assume a variable assignment ζ.

As in our case an ASM rule may be non-deterministic, a straightforward extension from the formalisation of the **forall** and **par** rules used in the logic for

ASMs in [13] would not work for Axioms **U3** and **U4**. The axioms correspond to the definition of update sets in Fig. 1.

- Axiom **U1** says that X is an update yielded by the assignment rule $f(t) := s$ iff it contains exactly one update which is (f, t, s).
- Axiom **U2** asserts that, if the formula φ evaluates to *true*, then X is an update set yielded by the conditional rule **if** φ **then** r **endif** iff X is an update set yielded by the rule r. Otherwise, the conditional rule yields only an empty update set.
- Axiom **U3** states that X is an update set yielded by the rule **forall** x **with** φ **do** r **enddo** iff X coincides with $\Delta_{a_1} \cup \cdots \cup \Delta_{a_n}$, where $\{a_1, \ldots, a_n\} = \{a_i \in B_1 \mid val_{S, \zeta[x \mapsto a_i]}(\varphi) = true\}$ and Δ_{a_i} (for $1 \leq i \leq n$) is an update set yielded by the rule r under the variable assignment $\zeta[x \mapsto a_i]$. Note that the update sets $\Delta_{a_1}, \ldots, \Delta_{a_n}$ are encoded into $\mathcal{X}$.
- Axiom **U4** states that X is an update set yielded by the parallel rule **par** r_1 r_2 **endpar** iff it corresponds to the union of an update set yielded by r_1 and an update set yielded by r_2.
- Axioms **U5** asserts that X is an update set yielded by the rule **choose** x **with** φ **do** r **enddo** iff it is an update set yielded by the rule r under a variable assignment $\zeta[x \mapsto a]$ which satisfies φ.
- Axiom **U6** is similar to Axiom **U5**, but for the case of the **choose x with** φ **do** r **enddo** rule.
- Axiom **U7** asserts that X is an update set yielded by a sequence rule **seq** r_1 r_2 **endseq** iff it corresponds to either an inconsistent update set yielded by rule r_1, or to an update set formed by the updates in an update set Y_2 yielded by rule r_2 in a successor state $S + Y_1$, where Y_1 encodes a consistent set of updates produced by rule r_1, plus the updates in Y_1 that correspond to locations other than the locations updated by Y_2.

The following lemma is an easy consequence of the axioms in Fig. 2.

Lemma 1. *Every formula in the logic $\mathcal{L}$ can be replaced by an equivalent formula not containing any subformulae of the form* $upd(r, X)$.

Remark 2. The inclusion of the parameter X in the predicate $upd(r, X)$ is important because a rule r in a non-deterministic parallel ASM rule may be associated with multiple update sets, and thus we need a way to specify which update set yielded by rule r is meant.

5.3   Axioms and Inference Rules

Now we can present a set of axioms and inference rules which constitute a proof system for the logic $\mathcal{L}$. To avoid unnecessary repetitions of almost identical axioms and rules, we describe them only considering variables of the first individual sort, but the exact same axioms and inference rules are implicitly assumed for the case of variables of the second individual sort as well as for variables of the predicate sorts. In the definition of the set of axioms and rules, we sometimes use $\varphi[t/x]$ to denote the substitution of a term t for a variable x in a formula φ. That is, $\varphi[t/x]$ is the result of replacing all free instances of x by t in φ provided that no free variable of t becomes bound after substitution.

Formally, the set $\mathfrak{R}$ of axioms and inference rules is formed by:

- The axioms **U1-U7** in Fig. 2 which assert the properties of $\mathrm{upd}(r, X)$.
- Axiom **M1** and Rules **M2-M3** from the axiom system K of modal logic, which is the weakest normal modal logic system [11]. Axiom **M1** is called *Distribution Axiom* of K, Rule **M2** is called *Necessitation Rule* of K and Rule **M3** is the inference rule called *Modus Ponens* in the classical logic. By using these axiom and rules together, we are able to derive all modal properties that are valid in Kripke frames.

  M1 $[X](\varphi \rightarrow \psi) \rightarrow ([X]\varphi \rightarrow [X]\psi)$

  M2 $\varphi \vdash [X]\varphi$ **M3** $\varphi, \varphi \rightarrow \psi \vdash \psi$

- Axiom **M4** asserts that, if an update set Δ is not consistent, then there is no successor state obtained after applying Δ over the current state and thus $[X]\varphi$ (for X interpreted by Δ) is interpreted as true for any formula φ. As applying a consistent update set Δ over the current state is deterministic, Axiom **M5** describes the deterministic accessibility relation in terms of $[X]$.

  M4 $\neg\mathrm{conUSet}(X) \rightarrow [X]\varphi$ $quad$**M5** $\neg[X]\varphi \rightarrow [X]\neg\varphi$

- Axiom **M6** is called *Barcan Axiom*. It originates from the fact that all states in a run of a non-deterministic parallel ASM have the same base set, and thus the quantifiers in all states always range over the same set of elements.

  M6 $\forall x([X]\varphi) \rightarrow [X]\forall x(\varphi)$

- Axioms **M7** and **M8** assert that the interpretation of static or pure formulae is the same in all states of non-deterministic parallel ASMs, since they are not affected by the execution of any ASM rule r.

  M7 $\mathrm{con}(r, X) \wedge \varphi \rightarrow [X]\varphi$ for static or pure φ

  M8 $\mathrm{con}(r, X) \wedge [X]\varphi \rightarrow \varphi$ for static or pure φ

- Axiom **A1** asserts that, if a consistent update set Δ (represented by X) does not contain any update to the location (f, x), then the content of (f, x) in a successor state obtained after applying Δ is the same as its content in the current state. Axiom **A2** asserts that, if a consistent update set Δ does contain an update which changes the content of the location (f, x) to y, then the content of (f, x) in the successor state obtained after applying Δ is y.

  A1 $\mathrm{conUSet}(X) \wedge \forall z(\neg X(f, x, z)) \wedge f(x) = y \rightarrow [X]f(x) = y$

  A2 $\mathrm{conUSet}(X) \wedge X(f, x, y) \rightarrow [X]f(x) = y$

- The following are axiom schemes from classical logic.

  P1 $\varphi \rightarrow (\psi \rightarrow \varphi)$

  P2 $(\varphi \rightarrow (\psi \rightarrow \chi)) \rightarrow ((\varphi \rightarrow \psi) \rightarrow (\varphi \rightarrow \chi))$

  P3 $(\neg\varphi \rightarrow \neg\psi) \rightarrow (\psi \rightarrow \varphi)$

- The following four inference rules describe when the universal and existential quantifiers can be added to or deleted from a statement. Rules **UI**, **EG**, **UG** and **EI** are usually known as *Universal Instantiation*, *Existential Generalisation*, *Universal Generalisation* and *Existential Instantiation*, respectively.

UI $\forall x(\varphi) \vdash \varphi[t/x]$ if φ is pure or t is static.

EG $\varphi[t/x] \vdash \exists x(\varphi)$ if φ is pure or t is static.

UG $\varphi[t_a/x] \vdash \forall x(\varphi)$ if $\varphi[t_a/x]$ holds for every element a in the domain of x and corresponding term t_a representing a, and further φ is pure or every t_a is static.

EI $\exists x(\varphi) \vdash \varphi[t/x]$ if t represents a valuation for x which satisfies φ, and further φ is pure or t is static.

- The following are the equality axioms from first-order logic with equality. Axiom **EQ1** asserts the reflexivity property while Axiom **EQ2** asserts the substitutions for functions.

EQ1 $t = t$ for static term t

EQ2 $t_1 = t_{n+1} \wedge \cdots \wedge t_n = t_{2n} \rightarrow f(t_1, \ldots, t_n) = f(t_{n+1}, \ldots, t_{2n})$ for any function f and static terms t_i $(i = 1, \ldots, 2n)$.

- The following axiom is taken from dynamic logic, asserting that executing a **seq** rule equals to executing rules sequentially.

DY1 $\exists X(\text{upd}(\textbf{seq } r_1 \ r_2 \ \textbf{endseq}, X) \wedge [X]\varphi) \leftrightarrow$
$$\exists X_1(\text{upd}(r_1, X_1) \wedge [X_1]\exists X_2(\text{upd}(r_2, X_2) \wedge [X_2]\varphi))$$

- Axiom **E** is the extensionality axiom.

E $r_1 \equiv r_2 \rightarrow \exists X_1 X_2((\text{upd}(r_1, X_1) \wedge [X_1]\varphi) \leftrightarrow (\text{upd}(r_2, X_2) \wedge [X_2]\varphi))$

The following soundness theorem for the proof system is relatively straightforward, since the non-standard axioms and rules are just a formalisation of the definitions of the semantics of rules, update sets and update multisets.

Theorem 1. *Let φ be a formula from $\mathcal{L}$ and let Φ be a set of formulae also from $\mathcal{L}$ (all of them of the same vocabulary as φ). If $\Phi \vdash \varphi$, then $\Phi \models \varphi$.*

6 Derivation

In this section we present some properties of the logic for non-deterministic parallel ASMs which are implied by the axioms and rules from the previous section. This includes properties known for the logic for ASMs [13]. In particular, the logic for ASMs uses the modal expressions $[r]\varphi$ and $\langle r \rangle \varphi$ with the following semantics:

- $[\![r]\varphi]\!]_{S,\varsigma} = true$ iff $[\![\varphi]\!]_{S+\Delta,\varsigma} = true$ for all consistent $\Delta \in \Delta(r, S, \varsigma)$.
- $[\![\langle r \rangle \varphi]\!]_{S,\varsigma} = true$ iff $[\![\varphi]\!]_{S+\Delta,\varsigma} = true$ for at least one consistent $\Delta \in \Delta(r, S, \varsigma)$.

Instead of introducing modal operators $[]$ and $\langle\rangle$ for a non-deterministic parallel ASM rule r, we use the modal expression $[X]\varphi$ for an update set yielded by a possibly non-deterministic rule. The modal expressions $[r]\varphi$ and $\langle r \rangle \varphi$ in the logic for ASMs can be treated as the shortcuts for the following formulae in our logic:

$$[r]\varphi \equiv \forall X(\text{upd}(r, X) \rightarrow [X]\varphi). \tag{5}$$

$$\langle r \rangle \varphi \equiv \exists X(\text{upd}(r, X) \wedge [X]\varphi). \tag{6}$$

Lemma 2. *The following axioms and rules used in the logic for ASMs are derivable in $\mathcal{L}$, where the rule r in Axioms (c) and (d) is assumed to be defined and deterministic: (a) $([r](\varphi \to \psi) \to [r]\varphi) \to [r]\psi$; (b) $\varphi \to [r]\varphi$; (c) $\neg wcon(r) \to [r]\varphi$; (d) $[r]\varphi \leftrightarrow \neg[r]\neg\varphi$.*

Proof. We prove each property in the following.

- (a): By Eq. 5, we have that $[r](\varphi \to \psi) \wedge [r]\varphi \equiv \forall X(\text{upd}(r, X) \to [X](\varphi \to \psi)) \wedge \forall X(\text{upd}(r, X) \to [X]\varphi)$. By the axioms from classical logic, this is in turn equivalent to $\forall X(\text{upd}(r, X) \to ([X](\varphi \to \psi) \wedge [X]\varphi))$. Then by Axiom **M1** and axioms from the classical logic, we get $\forall X(\text{upd}(r, X) \to ([X](\varphi \to \psi) \wedge [X]\varphi)) \to \forall X(\text{upd}(r, X) \to [X]\psi)$. Therefore, $([r](\varphi \to \psi) \to [r]\varphi) \to [r]\psi$ is derivable.
- (b): By Rule **M2**, we have that $\varphi \to [X_i]\varphi$. Since X is free in $\varphi \to [X]\varphi$, this holds for every possible valuation of X. Thus using Rule **UG** (applied to the variable X of the first predicate sort) and the axioms from classical logic, we can clearly derive $\varphi \to \forall X(\text{upd}(r, X) \to [X]\varphi)$.
- (c): By Eq. 3, we have $\neg wcon(r) \leftrightarrow \neg\exists X(\text{con}(r, X))$. In turn, by Eq. 2, we get $\neg wcon(r) \leftrightarrow \neg\exists X(\text{upd}(r, X) \wedge \text{conUSet}(X))$. Since a rule r in the logic for ASMs is deterministic, we get $\neg wcon(r) \leftrightarrow \neg\text{conUSet}(X)$. By Axiom **M4**, we get $\neg wcon(r) \to [r]\varphi$.
- (d): By Eq. 5, we have $\neg[r]\neg\varphi \equiv \exists X(\text{upd}(r, X) \wedge \neg[X]\neg\varphi)$. By applying Axiom **M5** to $\neg[X]\neg\varphi$, we get $\neg[r]\neg\varphi \equiv \exists X(\text{upd}(r, X) \wedge [X]\varphi)$. When the rule r is deterministic, the interpretation of $\forall X(\text{upd}(r, X) \to [X]\varphi)$ coincides with he interpretation of $\exists X(\text{upd}(r, X) \wedge [X]\varphi)$ and therefore $[r]\varphi \leftrightarrow \neg[r]\neg\varphi$.

Note that the formula $\text{Con}(R)$ in **Axiom 5** in [13] (i.e., in $\neg\text{Con}(R) \to [R]\varphi$) corresponds to the weak version of consistency (i.e., $wcon(r)$) in the theory of $\mathcal{L}$.

Lemma 3. *The following properties are derivable in $\mathcal{L}$: (e) $\text{con}(r, X) \wedge [X]f(x) = y \to X(f, x, y) \vee (\forall z(\neg X(f, x, z)) \wedge f(x) = y)$; (f) $\text{con}(r, X) \wedge [X]\varphi \to \neg[X]\neg\varphi$; (g) $[X]\exists x(\varphi) \to \exists x([X]\varphi)$; (h) $[X]\varphi_1 \wedge [X]\varphi_2 \to [X](\varphi_1 \wedge \varphi_2)$.*

Proof. (e) is derivable by applying Axioms **A1** and **A2**. (f) is a straightforward result of Axiom **M5**. (g) can be derived by applying Axioms **M5** and **M6**. Regarding (h), it is derivable by using Axioms **M1-M3**.

Lemma 4. *For terms and variables of the appropriate types, the following properties in [9] are derivable in $\mathcal{L}$.*

- $x = t \to (y = s \leftrightarrow [f(t) := s]f(x) = y)$
- $x \neq t \to (y = f(x) \leftrightarrow [f(t) := s]f(x) = y)$

Following the approach of defining the predicate joinable in [13], we define the predicate joinable over two non-deterministic parallel ASMs rules. As we consider non-deterministic parallel ASMs rules, the predicate joinable(r_1, r_2) means that there exists a pair of update sets without conflicting updates, which are yielded by rules r_1 and r_2, respectively. Then, based on the use of predicate joinable, the properties in Lemma 5 are all derivable.

$$\text{joinable}(r_1, r_2) \equiv \exists X_1 X_2 (\text{upd}(r_1, X_1) \wedge \text{upd}(r_2, X_2) \wedge$$
$$\bigwedge_{f \in \mathcal{F}_{dyn}} \forall xyz (X_1(f, x, y) \wedge X_2(f, x, z) \rightarrow y = z)) \qquad (7)$$

Lemma 5. *The following properties for weak consistency are derivable in* $\mathcal{L}$.

(i) $\text{wcon}(f(t) := s)$ *(j)* $\text{wcon}(f(t) := \mathbf{s})$ *(k)* $\text{wcon}(f(\mathbf{t}) := \mathbf{s})$

(j) $\text{wcon}(\boldsymbol{if}\,\varphi\,\boldsymbol{then}\,r\,\boldsymbol{endif}) \leftrightarrow \neg\varphi \vee (\varphi \wedge \text{wcon}(r))$

(l) $\text{wcon}(\boldsymbol{forall}\,x\,\boldsymbol{with}\,\varphi\,\boldsymbol{do}\,r\,\boldsymbol{enddo}) \leftrightarrow$
$$\forall x(\varphi \rightarrow \text{wcon}(r) \wedge \forall y(\varphi[y/x] \rightarrow \textit{joinable}(r, r[y/x])))$$

(m) $\text{wcon}(\boldsymbol{par}\,r_1\,r_2\,\boldsymbol{endpar}) \leftrightarrow \text{wcon}(r_1) \wedge \text{wcon}(r_2) \wedge \textit{joinable}(r_1, r_2)$

(n) $\text{wcon}(\boldsymbol{choose}\,x\,\boldsymbol{with}\,\varphi\,\boldsymbol{do}\,r\,\boldsymbol{enddo}) \leftrightarrow \exists x(\varphi \wedge \text{wcon}(r))$

(o) $\text{wcon}(\boldsymbol{choose}\,\mathbf{x}\,\boldsymbol{with}\,\varphi\,\boldsymbol{do}\,r\,\boldsymbol{enddo}) \leftrightarrow \exists \mathbf{x}(\varphi \wedge \text{wcon}(r))$

(p) $\text{wcon}(\boldsymbol{seq}\,r_1\,r_2\,\boldsymbol{endseq}) \leftrightarrow \exists X(\text{con}(r_1, X) \wedge [X]\text{wcon}(r_2))$

We omit the proof of the previous lemma as well as the proof of the remaining lemmas in this section, since they are lengthy but relatively easy exercises.

Lemma 6. *The following properties for the formula* $[r]\varphi$ *are derivable in* $\mathcal{L}$.

(q) $[\boldsymbol{if},\varphi,\boldsymbol{then},r,\boldsymbol{endif}]\psi \leftrightarrow (\varphi \wedge [r]\psi) \vee (\neg\varphi \wedge \psi)$

(r) $[\boldsymbol{choose}\,x\,\boldsymbol{with}\,\varphi\,\boldsymbol{do}\,r\,\boldsymbol{enddo}]\psi \leftrightarrow \forall x(\varphi \rightarrow [r]\psi)$

(s) $[\boldsymbol{choose}\,\mathbf{x}\,\boldsymbol{with}\,\varphi\,\boldsymbol{do}\,r\,\boldsymbol{enddo}]\psi \leftrightarrow \forall \mathbf{x}(\varphi \rightarrow [r]\psi)$

Lemma 7 states that a parallel composition is commutative and associative while a sequential composition is associative.

Lemma 7. *The following properties are derivable in* $\mathcal{L}$.

(t) $\boldsymbol{par}\,r_1\,r_2\,\boldsymbol{endpar} \equiv \boldsymbol{par}\,r_2\,r_1\,\boldsymbol{endpar}$

(u) $\boldsymbol{par}\,(\boldsymbol{par}\,r_1\,r_2\,\boldsymbol{endpar})\,r_3\,\boldsymbol{endpar} \equiv \boldsymbol{par}\,r_1\,(\boldsymbol{par}\,r_2\,r_3\,\boldsymbol{endpar})\,\boldsymbol{endpar}$

(v) $\boldsymbol{seq}\,(\boldsymbol{seq}\,r_1\,r_2\,\boldsymbol{endseq})\,r_3\,\boldsymbol{endseq} \equiv \boldsymbol{seq}\,r_1\,(\boldsymbol{seq}\,r_2\,r_3\,\boldsymbol{endseq})\,\boldsymbol{endseq}$

Lemma 8. *The extensionality axiom for transition rules in the logic for ASMs is derivable in* $\mathcal{L}$: $r_1 \equiv r_2 \rightarrow ([r_1]\varphi \leftrightarrow [r_2]\varphi)$.

7 Completeness

We can prove the completeness of $\mathcal{L}$ by using a similar strategy to that used in [13]. That is, we can show that $\mathcal{L}$ is a definitional extension of a complete logic. However, the logic for hierarchical ASMs in [13] is a definitional extension of first-order logic. In the case of the logic $\mathcal{L}$, the proof is more complicated since we have to deal with set membership predicates and corresponding predicate sorts. The key idea is to show instead that $\mathcal{L}$ is a *definitional extension* of first-order logic extended with two membership predicates with respect to finite sets, which in turns constitutes itself a complete logic.

In the remaining of this section, we will use $\mathcal{L}^{\in}$ to denote the logic obtained by restricting the formulae of $\mathcal{L}$ to those produced by the following grammar:

$$\varphi, \psi ::= s = t \mid s_a = t_a \mid \neg\varphi \mid \varphi \wedge \psi \mid \forall x(\varphi) \mid \forall \mathbf{x}(\varphi) \mid \forall x^1(\varphi) \mid \forall x^2(\varphi) \mid$$
$$\in^1(x^1, f, t_0, s_0) \mid \in^2(x^2, f, t_0, s_0, s).$$

Let us define the theory of $\mathcal{L}^{\in}$ as the theory obtained by taking the union of a sound and complete axiomatisation of first-order logic and the sound and complete axiomatisation of the properties of finite sets introduced in [1]. Clearly, such theory of $\mathcal{L}^{\in}$ is a conservative extension of the first-order theory, in the sense that if Φ is a set of pure first-order formulae and φ is a pure first-order formula (not containing subformulae of the form $\in^n(x^n, t_1, \ldots, t_n)$) and $\Phi \vdash \varphi$ holds in the theory of $\mathcal{L}^{\in}$, then there already exists a derivation using the axiomatisation for first-order logic. Indeed, due to the soundness of the axioms and rules in the theory of $\mathcal{L}^{\in}$, we obtain $\Phi \models \varphi$, which is a pure statement about models for first-order logic. Thus the known completeness for first-order logic gives $\Phi \vdash \varphi$ in an axiomatisation for first-order logic, hence the claimed conservativism of the extension. Since then the theory of $\mathcal{L}^{\in}$ proves no new theorems about first-order logic, all the new theorems belong to the theory of properties of finite sets and thus can be derived by using the axiomatisation in [1] (which also form part of the axiomatisation of $\mathcal{L}^{\in}$), we get the following key result.

Theorem 2. *Let φ be a formula and Φ be a set of formulae in the language of $\mathcal{L}^{\in}$ (all of the same vocabulary). If $\Phi \models \varphi$, then $\Phi \vdash \varphi$.*

Finally, we need to show that all the formulae in $\mathcal{L}$ which are not formulae of $\mathcal{L}^{\in}$ can be translated into formulae of $\mathcal{L}^{\in}$ based on derivable equivalences in the theory of $\mathcal{L}$. First, we reduce the general atomic formulae in $\mathcal{L}$ to atomic formulae of the form $x = y$, $\mathbf{x} = \mathbf{y}$, $f(x) = y$, $f(x) = \mathbf{y}$, $f(\mathbf{x}) = \mathbf{y}$, $\in^1(x^1, f, x, y)$, $\in^1(x^1, f, x, \mathbf{y})$, $\in^1(x^1, f, \mathbf{x}, \mathbf{y})$, $\in^2(x^2, f, x, y, z)$, $\in^2(x^2, f, x, \mathbf{y}, z)$ and $\in^2(x^2, f, \mathbf{x}, \mathbf{y}, z)$. Let t, s and s' denote point terms and let t_a and s_a denote algorithmic terms. This can be done by using the following equivalences.

$$s = t \leftrightarrow \exists x(s = x \wedge x = t)$$
$$s_a = t_a \leftrightarrow \exists \mathbf{x}(s_a = \mathbf{x} \wedge \mathbf{x} = t_a)$$
$$f(s) = y \leftrightarrow \exists x(s = x \wedge f(x) = y)$$
$$f(s) = \mathbf{y} \leftrightarrow \exists x(s = x \wedge f(x) = \mathbf{y})$$
$$f(s_a) = \mathbf{y} \leftrightarrow \exists \mathbf{x}(s_a = \mathbf{x} \wedge f(\mathbf{x}) = \mathbf{y})$$
$$\in^1(x^1, f, t, s) \leftrightarrow \exists xy(t = x \wedge s = y \wedge \in^1(x^1, f, x, y))$$
$$\in^1(x^1, f, t, s_a) \leftrightarrow \exists xy(t = x \wedge s_a = \mathbf{y} \wedge \in^1(x^1, f, x, \mathbf{y}))$$
$$\in^1(x^1, f, t_a, s_a) \leftrightarrow \exists \mathbf{x}y(t_a = \mathbf{x} \wedge s_a = \mathbf{y} \wedge \in^1(x^1, f, \mathbf{x}, \mathbf{y}))$$
$$\in^2(x^2, f, t, s, s') \leftrightarrow \exists xyz(t = x \wedge s = y \wedge s' = z \wedge \in^2(x^2, f, x, y, z))$$
$$\in^2(x^2, f, t, s_a, s') \leftrightarrow \exists xyz(t = x \wedge s_a = \mathbf{y} \wedge s' = z \wedge \in^2(x^2, f, x, \mathbf{y}, z))$$
$$\in^2(x^2, f, t_a, s_a, s') \leftrightarrow \exists \mathbf{x}yz(t_a = \mathbf{x} \wedge s_a = \mathbf{y} \wedge s' = z \wedge \in^2(x^2, f, \mathbf{x}, \mathbf{y}, z))$$

The translation of modal formulae into $\mathcal{L}^\in$ distributes over negation, Boolean connectives and quantifiers. We eliminate atomic formulae of the form $\mathrm{upd}(r, x^1)$ using Axioms **U1-U7**, and the modal operator in formulae of the form $[x^1]\varphi$, where φ is already translated to $\mathcal{L}^\in$, using the following derivable equivalences.

$$[x^1]x = y \leftrightarrow (\mathrm{conUSet}(x^1) \to x = y); \qquad [x^1]\mathbf{x} = \mathbf{y} \leftrightarrow (\mathrm{conUSet}(x^1) \to \mathbf{x} = \mathbf{y});$$

$$[x^1]f(x) = y \leftrightarrow (\mathrm{conUSet}(x^1) \to \in^1(x^1, f, x, y) \vee (\forall z(\neg \in^1(x^1, f, x, z)) \wedge f(x) = y))$$

$$[x^1]f(x) = \mathbf{y} \leftrightarrow (\mathrm{conUSet}(x^1) \to \in^1(x^1, f, x, \mathbf{y}) \vee (\forall \mathbf{z}(\neg \in^1(x^1, f, x, \mathbf{z})) \wedge f(x) = \mathbf{y}))$$

$$[x^1]f(\mathbf{x}) = \mathbf{y} \leftrightarrow (\mathrm{conUSet}(x^1) \to \in^1(x^1, f, \mathbf{x}, \mathbf{y}) \vee (\forall \mathbf{z}(\neg \in^1(x^1, f, \mathbf{x}, \mathbf{z})) \wedge f(\mathbf{x}) = \mathbf{y}));$$

$$[x^1]\in^1(x^1, f, x, y) \leftrightarrow (\mathrm{conUSet}(x^1) \to \in^1(x^1, f, x, y));$$

$$[x^1]\in^1(x^1, f, x, \mathbf{y}) \leftrightarrow (\mathrm{conUSet}(x^1) \to \in^1(x^1, f, x, \mathbf{y}));$$

$$[x^1]\in^1(x^1, f, \mathbf{x}, \mathbf{y}) \leftrightarrow (\mathrm{conUSet}(x^1) \to \in^1(x^1, f, \mathbf{x}, \mathbf{y}));$$

$$[x^1]\in^2(x^2, f, x, y, z) \leftrightarrow (\mathrm{conUSet}(x^1) \to \in^2(x^2, f, x, y, z));$$

$$[x^1]\in^2(x^2, f, \mathbf{x}, \mathbf{y}, z) \leftrightarrow (\mathrm{conUSet}(x^1) \to \in^2(x^2, f, x, \mathbf{y}, z));$$

$$[x^1]\in^2(x^2, f, \mathbf{x}, \mathbf{y}, z) \leftrightarrow (\mathrm{conUSet}(x^1) \to \in^2(x^2, f, \mathbf{x}, \mathbf{y}, z));$$

$$[x^1]\neg\varphi \leftrightarrow (\mathrm{conUSet}(x^1) \to \neg[x^1]\varphi); \qquad [x^1](\varphi \wedge \psi) \leftrightarrow ([x^1]\varphi \wedge [x^1]\psi);$$

$$[x^1]\forall x(\varphi) \leftrightarrow \forall x([x^1]\varphi); \qquad\qquad\qquad [x^1]\forall \mathbf{x}(\varphi) \leftrightarrow \forall \mathbf{x}([x^1]\varphi);$$

$$[x^1]\forall y^1(\varphi) \leftrightarrow \forall y^1([x^1]\varphi); \qquad\qquad\quad [x^1]\forall x^2(\varphi) \leftrightarrow \forall x^2([x^1]\varphi).$$

Our main technical result then follows from Theorem 2 and the fact that the described translation from formulae φ of $\mathcal{L}$ to formulae φ' of $\mathcal{L}^\in$ satisfies the properties required for $\mathcal{L}$ to be a definitional extension of $\mathcal{L}^\in$, i.e., (a) $\varphi \leftrightarrow \varphi'$ is derivable in $\mathcal{L}$ and (b) φ' is derivable in $\mathcal{L}^\in$ whenever φ is derivable $\mathcal{L}$.

Theorem 3. *Let φ be a formula and Φ a set of formulae in the language of $\mathcal{L}$ (all of the same vocabulary). If $\Phi \models \varphi$, then $\Phi \vdash \varphi$.*

8   Conclusion

Non-deterministic transitions manifest themselves as a difficult task in the logical formalisation for ASMs. Indeed, Nanchen and Stärk analysed potential problems to several approaches they tried by taking non-determinism into consideration and concluded [13]:

> Unfortunately, the formalisation of consistency cannot be applied directly to non-deterministic ASMs. The formula $\mathrm{Con}(r)$ (as defined in Sect. 8.1.2 of [5]) expresses the property that the *union of all possible* update sets of (an ASM rule) r in a given state is consistent. This is clearly not what is meant by consistency. Therefore, in a logic for ASMs with **choose** one had to add $\mathrm{Con}(r)$ as an atomic formula to the logic.

However, we observe that this conclusion is not necessarily true, as finite update sets can be made explicit in the formulae of a logic to capture non-deterministic transitions. In doing so, the formalisation of consistency defined in [13] can still be applied to such an explicitly specified update set Δ yielded by a rule r in the form of the formula $\mathrm{con}(r, \Delta)$ as discussed in Subsect. 5.1. We thus solve this problem by the addition of the modal operator $[\Delta]$ for an update set generated by a non-deterministic parallel ASM rule. The approach works well, because in the parallel ASMs the number of possible parallel branches, although unbounded, is still finite. Therefore the update sets produced by these machines are restricted to be finite as well. This is implicitly assumed in the parallel ASM thesis of Blass and Gurevich [3,4] and it is made explicit in the new parallel ASM thesis that we propose in [6].

The proof systems that we develop in this work for the proposed logic for non-deterministic parallel ASMs, extends the proof system developed in [13] in two different ways. First, an ASM rule may be associated with a set of different update sets. Applying different update sets may lead to a set of different successor states to the current state. As the logic for non-deterministic parallel ASMs includes formulae denoting explicit update sets and variables that are bounded to update sets, our proof system allows us to reason about the interpretation of a formula over all successor states or over some successor state after applying an ASM rule over the current state. Secondly, in addition to capturing the consistency of an update set yielded by an ASM rule, our proof system also develops two notions of consistency (weak and strong consistency) w.r.t. a given rule. When the rule is deterministic, these two notions coincide.

We plan as future work to embed our one-step logic into a complex dynamic logic and demonstrate how desirable properties of ASM runs can be formalised in such a logic. Of course, there is no chance of obtaining a complete proof theory for full ASM runs, but there is clearly many potential practical benefits from the perspective of the ASM method for systems development [5].

References

1. Ågotnes, T., Walicki, M.: Complete axiomatisations of properties of finite sets. Logic J. IGPL **16**(3), 293–313 (2008)
2. Alur, R.: Principles of Cyber-Physical Systems. MIT Press, Cambridge (2015)
3. Blass, A., Gurevich, Y.: Abstract state machines capture parallel algorithms. ACM Trans. Comp. Logic **4**(4), 578–651 (2003)
4. Blass, A., Gurevich, Y.: Abstract state machines capture parallel algorithms: correction and extension. ACM Trans. Comp. Logic **9**(3), 1–32 (2008)
5. Börger, E., Stärk, R.F.: Abstract State Machines: a Method for High-Level System Design and Analysis. Springer, New York (2003)
6. Ferrarotti, F., Schewe, K., Tec, L., Wang, Q.: A new thesis concerning synchronised parallel computing - simplified parallel ASM thesis. CoRR abs/1504.06203 (2015). http://arxiv.org/abs/1504.06203
7. Floyd, R.W.: Nondeterministic algorithms. J. ACM **14**(4), 636–644 (1967). http://doi.acm.org/10.1145/321420.321422
8. Grädel, E., Gurevich, Y.: Metafinite model theory. Inf. Comput. **140**(1), 26–81 (1998)
9. Groenboom, R., Renardel de Lavalette, G.: A formalization of evolving algebras. In: Proceedings of Accolade95. Dutch Research School in Logic (1995)
10. Huggins, J.K., Wallace, C.: An abstract state machine primer. Technical report 02–04, Computer Science Department, Michigan Technological University (2002)

11. Hughes, G., Cresswell, M.: A New Introduction to Modal Logic. Burns & Oates, London (1996)
12. Kruskal, J.B.: On the shortest spanning subtree of a graph and the travelling salesman problem. Proc. Amer. Math. Soc. **2**, 48–50 (1956)
13. Stärk, R., Nanchen, S.: A logic for abstract state machines. J. Univ. Comput. Sci. **7**(11) (2001)
14. Vapnik, V.N.: The Nature of Statistical Learning Theory. Springer, New York (1995)

Author Index

Ahmetaj, Shqiponja   169

Bauters, Kim   24
Beierle, Christoph   65, 83
Biskup, Joachim   211
Bliem, Bernhard   95

Cruz-Filipe, Luís   235

Durand, Arnaud   271

Eichhorn, Christian   65

Ferrarotti, Flavio   334
Fischl, Wolfgang   169

Godo, Lluís   24
Goethals, Bart   131

Hannula, Miika   271
Huynh, Van-Nam   115

Kern-Isberner, Gabriele   65
Kobti, Ziad   255
Kokkinis, Ioannis   292
Kontinen, Juha   271
Kröll, Markus   169

Liu, Weiru   24

Meier, Arne   271
Moguillansky, Martín O.   3
Müller, Emmanuel   131

Nguyen, Thu-Hien Thi   115
Nunes, Isabel   235

Pichler, Reinhard   169
Pivert, Olivier   42
Prade, Henri   42

Rácz, Gábor   149
Ravve, Elena V.   191

Sali, Attila   149
Schewe, Klaus-Dieter   149, 334
Schneider-Kamp, Peter   235
Šimkus, Mantas   169
Skritek, Sebastian   169

Tec, Loredana   334
Turull-Torres, José Maria   311

Van Brussel, Thomas   131
Virtema, Jonni   271

Wang, Qing   334
Woltran, Stefan   95

Zadeh, Pooya Moradian   255

Printed in the United States
By Bookmasters